MATHEMATICAL FOUNDATIONS FOR MANAGEMENT SCIENCE AND SYSTEMS ANALYSIS

OPERATIONS RESEARCH
AND INDUSTRIAL ENGINEERING

Consulting Editor: J. William Schmidt
Virginia Polytechnic Institute and State University, Blacksburg, Virginia

---

Applied Statistical Methods
I. W. Burr

Mathematical Foundations of Management Science
and Systems Analysis
J. William Schmidt

In preparation:

Urban Systems Models
Walter Helly

Mathematical Foundations for Management Science and Systems Analysis

J. WILLIAM SCHMIDT
VIRGINIA POLYTECHNIC INSTITUTE AND STATE UNIVERSITY

ACADEMIC PRESS   New York and London
A Subsidiary of Harcourt Brace Jovanovich, Publishers

COPYRIGHT © 1974, BY ACADEMIC PRESS, INC.
ALL RIGHTS RESERVED.
NO PART OF THIS PUBLICATION MAY BE REPRODUCED OR
TRANSMITTED IN ANY FORM OR BY ANY MEANS, ELECTRONIC
OR MECHANICAL, INCLUDING PHOTOCOPY, RECORDING, OR ANY
INFORMATION STORAGE AND RETRIEVAL SYSTEM, WITHOUT
PERMISSION IN WRITING FROM THE PUBLISHER.

ACADEMIC PRESS, INC.
111 Fifth Avenue, New York, New York 10003

United Kingdom Edition published by
ACADEMIC PRESS, INC. (LONDON) LTD.
24/28 Oval Road, London NW1

Library of Congress Cataloging in Publication Data

Schmidt, Joseph William.
    Mathematical foundations for management science and
systems analysis.

    (Operations research and industrial engineering)
    Includes bibliographies.
    1.   Operations research.   2.   System analysis.
I.   Title.
T57.6.S32   658.4'03   73-2078
ISBN 0–12–627050–3

PRINTED IN THE UNITED STATES OF AMERICA

TO JOAN

Without her continued patience and encouragement, this project would never have come to fruition.

CONTENTS

PREFACE xi
ACKNOWLEDGMENTS xiii

Chapter 1 Probability and Random Variables

1.1 Introduction 1
1.2 Fundamental Concepts 2
    1.2.1 Random Variables 2
    1.2.2 Basic Probability 6
1.3 Probability Distributions 18
    1.3.1 Discrete Random Variables 29
    1.3.2 Continuous Random Variables 34
1.4 Joint Distributions 40
1.5 Derived Distributions 48
    1.5.1 Functions of a Single Random Variable 49
    1.5.2 Functions of Several Random Variables 56
1.6 Expectation 74
    1.6.1 Moment Generating Functions 81
1.7 Applications 90
    1.7.1 Statistical Methods 90
    1.7.2 Mathematical Modeling 98
    1.7.3 Engineering Applications 101
    Problems 109
    References 112

Chapter 2 Matrix Algebra

2.1 Introduction 114
2.2 Definitions and Basic Operations 114
    2.2.1 Matrix Multiplication 116
2.3 The Transpose of a Matrix 121
    2.3.1 Symmetric, Skew-Symmetric, Scalar, Identity, and Periodic Matrices 123

2.4	The Determinant of a Matrix	126
	2.4.1 Cofactor Method	132
2.5	Rank of a Matrix	136
2.6	Inverse of a Square Matrix	137
2.7	Vectors	142
2.8	Vector Spaces	147
2.9	Convex Sets	155
2.10	Characteristic Values and Quadratic Forms	161
2.11	Applications	179
	2.11.1 Finite Markov Chains	179
	2.11.2 Experimental Design	186
	2.11.3 Production Systems	190
	2.11.4 Linear Programming	193
	2.11.5 Engineering Applications	200
	Problems	203
	References	206

Chapter 3 Real Analysis

3.1	Introduction	207
3.2	Sets	207
3.3	Real Numerical Sets	216
3.4	Functions, Limits, Continuity, and the Derivative	220
	3.4.1 Partial Differentiation	252
3.5	Integration	257
	Problems	277
	References	280

Chapter 4 Classical Optimization Theory

4.1	Introduction	281
4.2	Functions of a Single Variable	287
4.3	Functions of Several Variables	302
4.4	Optimization Subject to Constraints	315
	4.4.1 Equality Constraints	316
	4.4.2 Lagrange Multipliers	324
	4.4.3 Inequality Constraints	336
4.5	Applications	341
	4.5.1 Hypothesis Testing	341
	4.5.2 Production and Inventory Systems	344
	4.5.3 The Gradient Method	346
	4.5.4 Engineering Applications	351
	Problems	356
	References	361

CONTENTS

Chapter 5 Calculus of Finite Differences

5.1	Introduction	363
5.2	The Divided Difference	364
5.3	Optimization of Discrete Functions	381
5.4	The Antidifference and Summation of Series	392
5.5	Higher-Order Differences	400
5.6	Difference Equations	405
	5.6.1 Homogeneous Linear Difference Equations with Constant Coefficients	411
	5.6.2 Nonhomogeneous Linear Difference Equations with Constant Coefficients	421
5.7	Applications	425
	5.7.1 Numerical Methods: Interpolation and Approximation	425
	5.7.2 Numerical Methods: Integration	429
	5.7.3 Numerical Methods: Differentiation	433
	5.7.4 Economic Systems: Inventory Models	434
	5.7.5 Economic Systems: Quality Control Models	437
	5.7.6 Queueing Models: Single Channel	441
	5.7.7 Queueing Models: Single Channel with Impatient Arrivals	444
	5.7.8 Engineering Applications	448
	Problems	452
	References	456

Chapter 6 Complex Variables and Transform Methods

6.1	Introduction	457
6.2	Complex Variables	457
6.3	Functions of a Complex Variable	461
6.4	Complex Integration	467
	6.4.1 The Method of Residues	489
6.5	The Fourier Transforms	512
6.6	The Laplace Transform	522
6.7	Applications	531
	6.7.1 The Characteristic Function	531
	6.7.2 The Z Transform	540
	6.7.3 Transient State Analysis of Queueing Systems	546
	6.7.4 Replacement Analysis and Renewal Processes	548
	6.7.5 Engineering Applications	553
	Problems	561
	References	566

Appendix

Table 1	Cumulative Distribution Function of the Standard Normal Random Variable	567
Table 2	Antidifference Table	573
Table 3	Table of Laplace Transforms	575

INDEX 577

PREFACE

This text is written for students in operations research, management science, or systems analysis. It is particularly slanted toward the student with a limited background in mathematics. The topical coverage of the text represents an attempt to present, in a single volume, the fundamental mathematical background necessary for advanced study in areas dealing with the quantitative analysis of management systems. While emphasis is placed upon an understanding of the mathematical concepts presented and the computational methods resulting, application of these methods to practical problems are also given through examples.

Chapter 1 deals with the essentials of random variables, probability, distribution theory, and mathematical expectation. A knowledge of the material presented in Chapter 1 is essential to the analysis of problems in quality control, reliability, maintenance, queueing, inventory, traffic flow, and is required for an understanding of basic statistical methods and experimental analysis and design. Chapter 2 introduces the student to the fundamentals of matrix algebra, vectors and vector spaces, convex sets, and quadratic forms. The material in this chapter has important applications in classical optimization theory, mathematical programming, and experimental design. Chapter 3, Real Analysis, covers limits, continuity, differentiation, and integration. The material in Chapter 3 is presented to facilitate an understanding of the topics covered in Chapters 4, Classical Optimization Theory, and 6, Complex Variables and Transform Methods. The foundations of optimization theory are presented in Chapter 4. In this chapter optimization of functions of one or more continuous variables is treated where the space of feasible solutions is unconstrained or is restricted by equality or inequality constraints. Since the analysis of many management systems has as its objective the optimum design and control of the system considered, classical optimization theory is of paramount importance in operations research and systems analysis.

Chapter 6 is devoted to the calculus of finite differences, introducing the student to the divided difference, the antidifference and summation of series, optimization of functions of a discrete variable, and difference equations and their solutions. Finite calculus may be viewed as the discrete analog of infinitesimal calculus. As such, its importance in dealing with systems involving discrete variables parallels that of infinitesimal calculus in the treatment of systems consisting of continuous variables. In addition, the methods of numerical interpolation, numerical differentiation, and numerical integration are founded in the calculus of finite differences and are introduced at the end of Chapter 5. An elementary discussion of functions of a complex variable is presented in the first half of Chapter 6. The last half of Chapter 6 is devoted to Fourier and Laplace transforms, characteristic functions, and the Z transform. The initial discussion of complex variable theory is included so that the student will have the background necessary for proper understanding of the transforms presented and the methods discussed for their inversion which, in many cases, require integration in the complex plane. The application of transform techniques are particularly important in treating discrete and continuous control systems and in modeling stochastic systems, in general, such as those encountered in queueing, reliability, quality control, and inventory control.

The topical coverage and the manner of presentation are such that the text is best suited for students with an introductory knowledge of calculus. The text may be used in a one-quarter or one-semester course or in a two-quarter or two-semester sequence of courses, depending upon the depth of understanding to be achieved. The text has been used in a single course where the methods of analysis and the application of the methods presented are stressed while de-emphasizing a thorough understanding of the theory underlying the methodology. An understanding of the methods presented may be achieved by treating the theorems included in the chapters as results; thus excluding most if not all proofs. In such a course, the material in Chapter 3 would typically be excluded from the course. If the text is to be used for a sequence of two courses, theory, methodology, and applications may be successfully presented.

Regardless of whether the text is used in a single course or in a sequence of two courses, the author has found that the student achieves a maximum of understanding if the examples given in the text are emphasized in lecture presentations. Learning is further reinforced by the assignment of homework from the problems given at the end of each chapter. Where theory and methodology are to be covered, the author has given homework assignments covering both theory and methodology (including application), where two to three homework problems are devoted to methodology for each assignment filling in or extending the theory presented.

ACKNOWLEDGMENTS

I would like to express my indebtedness to the faculty and graduate students of the Department of Industrial Engineering and Operations Research at Virginia Polytechnic Institute and State University for their many helpful comments regarding the content and organization of the text. In particular, I am especially grateful to Dr. G. K. Bennett and Dr. P. M. Ghare for their editorial commentary after using the original manuscript in their course. I would like to thank Dr. A. A. B. Pritsker for his observations after reviewing the manuscript, particularly with regard to the examples included in the text. I would like to thank Dr. M. H. Agee for his continuous encouragement.

I would like to thank Miss D. F. Nolen, Mrs. C. G. Strickler, Mrs. F. D. Ebert, Mrs. H. L. Hall, and Mrs. B. B. Belcher for their considerable typing assistance in preparing the manuscript for this text. With the possible exception of the author, no one was happier to see this text go to press than Dot Nolen and Margie Strickler.

Finally, I would like to thank my wife, Joan, and our children Suzanne, Billy, Patty, and the twins, Kurt and Cathy, for their encouragement and patience throughout the preparation of this text.

CHAPTER 1

PROBABILITY AND RANDOM VARIABLES

1.1 Introduction

In attempting to analyze even the simplest of systems, one must frequently deal with random variables. For example, to model the cost of operating a given inventory system it is often necessary to account for the random nature of product demand and order lead time (Hadley and Whitin, 1963). To analyze the performance of service systems, we are normally forced to treat the time between successive customer arrivals and the time required to provide service to each as random variables (Cooper, 1972; Morse, 1962; Saaty, 1961). To determine the effectiveness of a quality control system, we must model the random variation present in the characteristic or characteristics to be controlled. Product design usually requires consideration of the reliability of the product where the life of the product is a random variable (Shooman, 1968). The analysis of vehicular traffic flow usually leads to a model that includes such random variables as velocity, distance between successive vehicles, turning movements, and gap acceptance in merging situations. The emission of particles from a radioactive source is considered to be a random variable as is the spatial distribution of particles in a gas. The list of systems of which random variation is an integral part could be continued without end. In fact, it is difficult to imagine a physical system in which random variation does not play at least a minor role.

Since random variables play such an important part in systems analysis, it is important that the prospective analyst understands the basic concepts of probability theory. The purpose of this chapter is to present these basic concepts and their application to the analysis of practical problems. However, a complete treatment of probability theory cannot be accomplished in a single chapter and the reader should see Brownlee (1960), Freeman (1963), Parzen (1960), or Wilks (1962) for a complete treatment of the subject.

1.2 Fundamental Concepts

1.2.1 Random Variables

To develop the fundamental concepts of probability theory we must first establish a working definition of a *random variable*. Consider repeated trials of an experiment such as flipping a coin or tossing a set of dice. The *outcome* of any trial of the first experiment is either a head or a tail while the outcome of a toss of dice is a number between two and twelve inclusive. Since the outcome of such experiments cannot be predicted with certainty, these experiments are usually referred to as *random experiments*. The outcome of the dice tossing experiment is numerical valued while the outcome of the coin flipping experiment is not. However, if we arbitrarily assign a zero to a tail and a one to a head, then the outcome of the coin flipping experiment is also numerical valued. To generalize, suppose that we specify a rule that assigns a distinct real numerical value to each possible outcome of a random experiment. Then the variable that assumes these numerical values at trials of the random experiment is called a random variable. We will use capital letters, such as X, Y, or Z, to denote a random variable, and lower case letters, x, y, or z, to denote values that the corresponding random variable may assume. The totality of all possible outcomes is called the *sample space*. As indicated in the preceding definition, there are two concepts which are important in defining a random variable. First, a random variable must assume real numerical values; second, the numerical value resulting from any trial of the random experiment must be unpredictable.

As in the dice tossing experiment, there is often a natural association between the observed outcome and the numerical value assigned. This natural association results whenever the outcome of a random experiment is measurable. For example, the time required to serve a customer, the monthly demand for a given product, and the weight of an item are all measurable, and the resulting measurement would normally be treated as the numerical value assigned to the outcome observed. When there is no natural association between the real numbers and the outcome of a random experiment, then the assignent of numerical values to outcomes may be arbitrary. As an illustration, suppose that a unit of product is selected from a manufacturing line and inspected. Let us assume that the unit is classified as good or bad as a result of the inspection. The outcome of the experiment is then "good" or "bad." We might assign 0 if the unit is bad and 1 if the unit is good. However, we might assign -10 to the outcome "good" and $+150$ to the outcome "bad." Either definition of the random variable is valid, although the 0–1 convention is normally adopted when the outcome of a random experiment is binary in nature.

Example 1.1 Define the sample space and random variables corresponding to the possible outcomes of the following random experiments.

1.2 FUNDAMENTAL CONCEPTS

a. The color arising from a given play of roulette; red, black, or green.
b. The number arising from a given play of roulette; 00, 0, 1, 2, ..., 36.
c. The evaluation received by a randomly selected student; poor, below average, average, above average, outstanding.
d. The time spent by a customer waiting for service.

The sample space for the random experiment in part a are the colors red, black, and green. The numerical values assigned to the colors red, black, and green can be arbitrarily chosen. Therefore, let X be the random variable with values as follows

$$x = \begin{cases} 1, & \text{red} \\ 2, & \text{black} \\ 3, & \text{green} \end{cases}$$

The sample space for the experiment in part b is the 38 possible outcomes 00, 0, 1, ..., 36. In part b, a natural numbering system could be used if it were not for the fact that the real numbers 0 and 00 are indistinguishable. Therefore, if the outcome of a play of roulette is 00 we will assign a value of -1 to the outcome. Otherwise, the value assigned to the outcome is the number resulting from the play. Therefore the random variable x may assume integer values between -1 and 36 inclusive.

In part c, as in part a, there is no logical association between the evaluation of a given student and the real number system. Therefore, if X is the random variable representing the outcome of an evaluation let

$$x = \begin{cases} -2, & \text{poor} \\ -1, & \text{below average} \\ 0, & \text{average} \\ 1, & \text{above average} \\ 2, & \text{outstanding} \end{cases}$$

The sample space consists of the five possible evaluations.

In part d there is a logical association between the time spent waiting for service and the real number system, although the analyst must define the units used to measure time. If the time spent waiting is measured in minutes then the random variable representing waiting time, X, is simply the amount of time spent waiting measured in minutes and may assume any value greater than or equal to zero. In this case the sample space is all real numbers greater than or equal to zero.

Many situations arise in which the analyst is not concerned merely with the outcomes of a random experiment, but rather with sets or groupings of outcomes. For example, suppose that items of product are selected at random from a manufacturing process and inspected. The purpose of the inspection is to determine whether the selected items are defective or not. Let us assume that any item may contain defects of three types; A, B, and C. Let A, B, and C represent the presence of each defect respectively and let $\bar{A}$, $\bar{B}$, and $\bar{C}$ indicate the absence of the respective defects. The possible outcomes which may result from the inspection of a randomly selected item are then given by

$$\begin{array}{ccc} \bar{A} & \bar{B} & \bar{C} \\ \bar{A} & \bar{B} & C \\ \bar{A} & B & \bar{C} \\ A & \bar{B} & \bar{C} \\ \bar{A} & B & C \\ A & \bar{B} & C \\ A & B & \bar{C} \\ A & B & C \end{array}$$

However, as stated, the analyst is interested only in whether or not the item is good or bad. An item is considered good if it contains none of the three defects and is bad if it contains one or more of the defects. Therefore the outcomes $\bar{A}\,\bar{B}\,\bar{C}$, $\bar{A}\,\bar{B}\,C$, $\bar{A}\,B\,\bar{C}$, $A\,\bar{B}\,\bar{C}$, $\bar{A}\,B\,C$, $A\,\bar{B}\,C$, and $A\,B\,\bar{C}$ characterize a bad item and $A\,B\,C$ characterizes a good item. We have thus grouped the outcomes of the experiment into *events*. Therefore, *an event is a group or set of outcomes of a random experiment*. As indicated in the prior illustration, the definition of an event or events may be based on the subjective judgement of the analyst.

Example 1.2 An experiment consists of tossing a set of dice. An outcome is defined as the combination of points appearing on the dice. Three events may result after each toss of the dice; win, lose, or tie. A win occurs if the sum of the points on the dice is 7 or 11. A tie occurs if the points showing on on the dice are the same. Any other outcome results in a loss. Define the outcomes of the experiment and the events in terms of the outcomes.

There are 36 possible outcomes of the random experiment—all possible paired combinations of the numbers 1 through 6 inclusive. The events, win, lose, and tie, are defined in the Table 1.1.

TABLE 1.1

Event	Outcomes, die #1–die #2
Win	1-6, 2-5, 3-4, 4-3, 5-2, 6-1, 5-6, 6-5
Tie	1-1, 2-2, 3-3, 4-4, 5-5, 6-6
Lose	1-2, 1-3, 1-4, 1-5, 2-1, 2-3, 2-4, 2-6, 3-1, 3-2, 3-5, 3-6, 4-1, 4-2, 4-5, 4-6, 5-1, 5-3, 5-4, 6-2, 6-3, 6-4

1.2 FUNDAMENTAL CONCEPTS

Now suppose that we redefine the events pertaining to the random experiment described in Example 1.2. Let the events be the occurrence of a 1, 2, 3, 4, 5, or a 6 at any trial of the experiment. Let us denote these events $E_1, E_2, \ldots, E_6$. If the toss of the dice yields a 4 and a 4 then event E_4 occurs. If a 4 and a 6 result the events E_4 and E_6 result. Thus, in this case, two events may occur at any trial of the experiment. One can easily imagine experiments where three or more events might occur at any given trial.

Definition 1.1 Let A and B be two events such that the occurrence of one eliminates the possibility of the occurrence of the other at any given trial of a random experiment. Then, A and B are said to be *mutually exclusive events*.

Definition 1.2 Let A and B be events such that the occurrence of one has no effect on the occurrence of the other at any given trial of a random experiment. Then A and B are said to be *independent events*.

Definition 1.3 Events which are not independent are said to be *dependent*.

The reader will note that mutually exclusive events are also dependent since the occurrence of one affects (eliminates) the possibility that the other will also occur. An example of mutually exclusive events is given in Example 1.2. That is, it is not possible to win and lose, win and tie, lose and tie, or win, lose, and tie at a given trial of the experiment.

Example 1.3 Classify the following events as dependent or independent for the random experiments described. If the events are dependent, determine whether or not they are mutually exclusive.

 a. The random experiment consists of flipping a coin and rolling a die simultaneously. Event A is a head on the coin and event B is a 4 on the die.
 b. The random experiment consists of counting the total number of customers entering a supermarket during an eight-hour day. Event A is the arrival of 100 customers during the day and event B is the arrival of 200 customers during the day.
 c. The random experiment consists of randomly drawing two cards from a standard deck of playing cards one after the other. The first card is not replaced before the second is drawn. Event A is an ace on the first draw and event B is an ace on the second draw.
 d. The random experiment and the events are the same as those described in part c except that the first card is replaced before the second is drawn.

In part a, the occurrence of a head on the coin should have no effect on the occurrence of a 4 on the die. Therefore these events would be considered to be independent.

In part b, if the total number of customers arriving is 100 then this total could not be 200. Thus the arrival of either 100 or 200 customers eliminates the possibility of occurrence of the other. Hence, events A and B are dependent and mutually exclusive.

In part c, the chance of randomly selecting an ace on the first draw is 4 in 52. If an ace is in fact selected on the first draw then there are 3 aces left for possible selection on the second draw since the first ace is not replaced in the deck. Thus the chance of selecting an ace on the second draw is 3 in 51 if the first draw was an ace. On the other hand, if an ace was not selected on the first draw, then four aces remain in the deck for the second draw and the chance of selecting an ace on the second draw is 4 in 51. Hence, the chance of selecting an ace on the second draw is dependent upon whether an ace resulted from the first draw and events A and B are dependent. However, since the selection of an ace on the first draw does not eliminate the selection of an ace on the second, the events are not mutually exclusive.

In part d, the random experiment is the same as that described in part c except that the first card is replaced in the deck before the second is drawn. Therefore the chance of selecting an ace on the second draw is 4 in 52 regardless of what was selected on the first draw. Thus events A and B are independent.

1.2.2 Basic Probability

In our discussion of random variables we emphasized the fact that random implies that the value which the random variable assumes cannot be predicted with certainty. However, we are frequently able to assess the relative likelihood that a given value will arise. For example, suppose that we had collected data on the number of automobiles sold by a particular dealer. The collected data is summarized in Table 1.2 where f_i represents the number of days on which i cars were sold and the data collection period is 200 days.

Table 1.2
SUMMARY OF AUTOMOBILE SALES PER DAY

Sales per Day, i	Sales Frequency, f_i
0	34
1	90
2	40
3	20
4	10
5	5
6	1

Now, suppose that we wished to predict the sales on a randomly selected day given only the information in the preceding table. We could estimate that sales are

1.2 FUNDAMENTAL CONCEPTS

likely to lie between 0 and 6 since the only information available from the past indicates sales probably will not exceed 6. In addition, we might estimate the likelihood or chance of i sales on the randomly selected day to be $f_i/200$. If we let f_i be the frequency of occurrence of the ith value of the random variable based upon F trials of a random experiment, then f_i/F is called the estimated *probability* of occurrence of the ith value, $\hat{P}(i)$. We refer to $\hat{P}(i)$ as an estimate of the likelihood or probability of occurrence of i since $\hat{P}(i)$ might change if we increase F. If $P(i)$ is the true probability of occurrence of i, then, intuitively, we would expect $\hat{P}(i)$ to approach $P(i)$ as F is increased. These concepts lead us to the definition of the probability of occurrence of a given value of a random variable.

Definition 1.4 Let f_i be the frequency of occurrence of the value i of a random variable based upon F trials of a random experiment. Then the probability of occurrence of i, $P(i)$ is defined as

$$P(i) = \lim_{F \to \infty} \frac{f_i}{F} \tag{1.1}$$

where $\lim_{F \to \infty} f_i/F$ is the value of f_i/F as the number of trials of the random experiment approaches infinity.

The concept of probability has been defined in terms of the value of a random variable. The definition was given in these terms since we frequently describe the outcome of a random experiment in numerical terms, that is, as a random variable. However, we can define probability in terms of the outcome of a random experiment without ever defining a random variable corresponding to the various outcomes.

Definition 1.5 Let f_i be the frequency of occurrence of the ith outcome, 0_i, of a given random experiment based upon F trials of the experiment. Let $P(0_i)$ be the probability of occurrence of the outcome 0_i. Then

$$P(0_i) = \lim_{F \to \infty} \frac{f_i}{F} \tag{1.2}$$

Definitions 1.4 and 1.5 have essentially the same meaning. The only difference is that in Definition 1.4 we have assigned a numerical value to each outcome while in Definition 1.5 we have not. As an illustration, let a random experiment consist of flipping a fair coin. The outcome of the experiment is a head or a tail. Since the coin is assumed to be fair, the probability of a head is $\frac{1}{2}$ as is the probability of a tail. Now let us define a random variable X associated with the outcome of a flip of the coin such that $X = 0$ if a tail results and $X = 1$ if a head results. Then

$$P(0) = P(\text{tail}) = \tfrac{1}{2} \quad \text{and} \quad P(1) = P(\text{head}) = \tfrac{1}{2}$$

Thus it is not necessary to define a random variable associated with a random experiment to describe the probabilities of occurrence of the various outcomes comprising the sample space. However, as we shall see, there are certain properties of random phenomena which can be evaluated only when these phenomena are characterized by random variables.

Let us now consider some of the properties of the probability of the occurrence of the outcomes of random experiments.

Theorem 1.1 Let 0_i be the ith outcome out of $N > 0$ possible outcomes for a given random experiment, $i = 1, 2, \ldots, N$. Then

$$0 \leq P(0_i) \leq 1 \tag{1.3}$$

and

$$\sum_{i=1}^{N} P(0_i) = 1 \tag{1.4}$$

Proof We will assume that the result of a trial of the random experiment is a single outcome which can be distinguished from all other possible outcomes. From Eq. (1.2)

$$P(0_i) = \lim_{F \to \infty} \frac{f_i}{F}$$

since $0 \leq f_i \leq F$, $0 \leq f_i/F \leq 1$ for all F. Thus

$$0 \leq P(0_i) \leq 1$$

If the number of possible outcomes of the random experiment is N, then

$$F = \sum_{i=1}^{N} f_i$$

Hence

$$\sum_{i=1}^{N} P(0_i) = \sum_{i=1}^{N} \lim_{F \to \infty} \frac{f_i}{F} = \lim_{F \to \infty} \sum_{i=1}^{N} \frac{f_i}{F} = \lim_{F \to \infty} (1) = 1 \qquad \blacksquare$$

For a more complete discussion of limiting operations the reader should see Chapter 3.

Now suppose that event E occurs whenever the outcome of a random experiment is $0_1, 0_2, \ldots, 0_n$, while E does not occur if any other outcome results. We assume that the set of all possible outcomes are mutually exclusive since only one outcome may result at any trial of the experiment. The probability of occurrence of E may then be described by

$$P(E) = \sum_{i=1}^{n} P(0_i) \tag{1.5}$$

We may demonstrate the validity of Eq. (1.5) as follows. The probability of E may be defined as

$$P(E) = \lim_{F \to \infty} \frac{M}{F} \tag{1.6}$$

1.2 FUNDAMENTAL CONCEPTS

where M is the number of occurrences of event E in F trials of the experiment. Therefore

$$M = \sum_{i=1}^{n} f_i \qquad (1.7)$$

since E occurs whenever $O_1, O_2, \ldots, O_n$ occur. Hence

$$P(E) = \lim_{F \to \infty} \frac{1}{F} \sum_{i=1}^{n} f_i = \sum_{i=1}^{n} \lim_{F \to \infty} \frac{f_i}{F} = \sum_{i=1}^{n} P(O_i)$$

From Theorem 1.1 one can deduce that $0 \leq P(E) \leq 1$. If $\bar{E}$ represents the set of all events excluding E then

$$P(E) + P(\bar{E}) = 1 \qquad (1.8)$$

When event E may not occur as a result of a trial of a random experiment, then $P(E) = 0$. In addition, if event E will always occur at any trial then $P(E) = 1$.

Let us now consider the calculation of the probabilities of combinations of events. If E_1 and E_2 are two events then

$$P(E_1 + E_2) = P(E_1 \text{ or } E_2 \text{ occurs}) \qquad (1.9)$$

$$P(E_1 E_2) = P(E_1 \text{ and } E_2 \text{ occur}) \qquad (1.10)$$

$$P(E_1 | E_2) = P(E_1 \text{ occurs given that } E_2 \text{ occurs}) \qquad (1.11)$$

$P(E_1 + E_2)$ is often called the *total probability* of E_1 and E_2. $P(E_1 E_2)$ is the *joint probability* of E_1 and E_2 and $P(E_1 | E_2)$ is called the *conditional probability* of E_1 given E_2. The three probabilities defined in Eqs. (1-9)–(1.11) are interrelated as shown in the following two theorems.

Theorem 1.2 Let E_1 and E_2 be any two events. Then

$$P(E_1 + E_2) = P(E_1) + P(E_2) - P(E_1 E_2) \qquad (1.12)$$

Proof Let E_1 occur whenever outcomes $A_1, A_2, \ldots$, or A_n occur and E_2 occur whenever outcomes $B_1, B_2, \ldots$, or B_m occur. If none of the outcomes $A_1, A_2, \ldots, A_n$ are included in the set of outcomes $B_1, B_2, \ldots, B_m$, that is, the two sets of outcomes are disjoint, then

$$P(E_1 + E_2) = \sum_{i=1}^{n} P(A_i) + \sum_{i=1}^{m} P(B_i) = P(E_1) + P(E_2) \qquad (1.13)$$

since $P(E_1 + E_2)$ is the sum of the probabilities of the outcomes resulting in E_1 or E_2. In this case E_1 and E_2 are mutually exclusive and $P(E_1 E_2) = 0$. Therefore Eq. (1.13) is equivalent to Eq. (1.12).

Now suppose that k of the outcomes in the two sets are identical. That is, there are k outcomes for which both E_1 and E_2 occur. Define $A_1, A_2, \ldots, A_n$ and $B_1, B_2, \ldots, B_m$ such that $A_i = B_i$, $i = 1, 2, \ldots, k$, where $k \leq n$ and $k \leq m$. Then

$$P(E_1 + E_2) = \sum_{i=1}^{n} P(A_i) + \sum_{i=k+1}^{m} P(B_i)$$

$$= \sum_{i=1}^{n} P(A_i) + \sum_{i=1}^{m} P(B_i) - \sum_{i=1}^{k} P(A_i) \quad (1.14)$$

Since

$$P(E_1 E_2) = \sum_{i=1}^{k} P(A_i) = \sum_{i=1}^{k} P(B_i) \quad (1.15)$$

we have

$$P(E_1 + E_2) = P(E_1) + P(E_2) - P(E_1 E_2) \quad \blacksquare \quad (1.16)$$

Theorem 1.3

$$P(E_1 \mid E_2) = \frac{P(E_1 E_2)}{P(E_2)} \quad (1.17)$$

if $P(E_2) \neq 0$.

Proof Let f_2 be the frequency of occurrence of event E_2 in F trials of a random experiment and let f_{12} be the frequency of occurrence of both E_1 and E_2 in the same F trials, where $f_{12} \leq f_2$. Then

$$P(E_2) = \lim_{F \to \infty} \frac{f_2}{F} \quad (1.18)$$

and

$$P(E_1 E_2) = \lim_{F \to \infty} \frac{f_{12}}{F} \quad (1.19)$$

Now $P(E_1 \mid E_2)$ is the probability that E_1 occurs given that E_2 occurs or has occurred. Therefore $P(E_1 \mid E_2)$ may be estimated as the number of occurrences of both E_1 and E_2 out of the set of occurrences resulting in the course of F trials of a random experiment. Hence, if $\hat{P}(E_1 \mid E_2)$ is the estimated probability of E_1 given E_2 based upon F trials of the random experiment, then

$$\hat{P}(E_1 \mid E_2) = \frac{f_{12}}{f_2} \quad (1.20)$$

and

$$P(E_1 \mid E_2) = \lim_{F \to \infty} \frac{f_{12}}{f_2} = \lim_{F \to \infty} \frac{f_{12}/F}{f_2/F} = \frac{P(E_1 E_2)}{P(E_2)} \quad \blacksquare \quad (1.21)$$

Theorem 1.4 If the events E_1 and E_2 are independent, then

$$P(E_1 E_2) = P(E_1) P(E_2) \quad (1.22)$$

1.2 FUNDAMENTAL CONCEPTS

Proof Since E_1 and E_2 are independent, the occurrence of either does not affect the probability of occurrence of the other. Thus

$$P(E_1 \mid E_2) = P(E_1) \tag{1.23}$$

From Theorem 1.3,

$$P(E_1 \mid E_2) = \frac{P(E_1 E_2)}{P(E_2)} \tag{1.24}$$

Substituting Eq. (1.23) into Eq. (1.24) we have

$$P(E_1)P(E_2) = P(E_1 E_2)$$

Example 1.4 A retailer purchases a particular product from four different vendors, V_1, V_2, V_3, and V_4. A total 1000 shipments have been received over the past two years. Of these, 100 were from V_1, 300 from V_2, 400 from V_3, and 200 from V_4. Occasionally late shipments occur and damaged shipments are sometimes found. The record of damaged and late shipments by the vendor is shown in Table 1.3, where D implies damaged and $\bar{D}$ denotes undamaged. Based upon the data in this table, suppose that a shipment is selected at random.

TABLE 1.3

Delivery Time	Vendor V_1		Vendor V_2		Vendor V_3		Vendor V_4	
	D	$\bar{D}$	D	$\bar{D}$	D	$\bar{D}$	D	$\bar{D}$
On Time	17	78	70	200	50	330	70	110
Late	3	2	20	10	10	10	5	15

a. What is the probability that it was late and damaged?
b. What is the probability that it was late and from vendor V_3?
c. What is the probability that it was delivered on time but was damaged?

Let V_i represent the event that the shipment was from vendor i and $\bar{V}_i$ the event that the shipment was not from vendor i. Let D represent the event that a shipment was damaged, $\bar{D}$ the event that the shipment was undamaged, L the event that the shipment was late, and $\bar{L}$ the event that the shipment was delivered on time.

In part a we must find $P(LD)$. From the data given, of the 245 damaged shipments, 38 were also late. Therefore

$$P(LD) = \frac{38}{1000} = 0.038$$

In part b we are to find $P(LV_3)$. From the data summary, 20 late shipments came from vendor 3. Hence

$$P(LV_3) = \frac{20}{1000} = 0.020$$

In part c we are to find $P(\bar{L}D)$. Of the 1000 shipments received, 207 were delivered on time but were damaged. Thus

$$P(\bar{L}D) = \frac{207}{1000} = 0.207$$

Example 1.5 Find the probability of randomly selecting two aces, one at a time, from a standard deck of playing cards if

    a. sampling is without replacement (the first card is not replaced before drawing the second),

    b. sampling is with replacement (the first card is replaced before drawing the second).

Let A_1 and A_2 be the events representing the selection of aces on the first and second draws respectively. From Eq. (1.17)

$$P(A_1 A_2) = P(A_1)P(A_2 | A_1)$$

When sampling is without replacement, part a, we have

$$P(A_1) = \frac{1}{13} \quad \text{and} \quad P(A_2 | A_1) = \frac{3}{51}$$

Hence

$$P(A_1 A_2) = \left(\frac{1}{13}\right)\left(\frac{3}{51}\right) = \frac{1}{221}$$

In part b sampling is with replacement. Therefore, $P(A_2)$ is unaffected by the occurrence of A_1. That is, A_1 and A_2 are independent.

$$P(A_1 A_2) = P(A_1)P(A_2) = \left(\frac{1}{13}\right)\left(\frac{1}{13}\right) = \frac{1}{169}$$

Example 1.6 Units of a particular product are manufactured in lots of size 1000. In a particular lot there are 100 units having defect A and 200 units having defect B. Defects A and B occur independently. A unit is drawn at random from the lot. Find

    a. $P(AB)$ b. $P(A|B)$ c. $P(B|A)$

From Theorem 1.4

$$P(A\,B) = P(A)P(B) \tag{1.25}$$

since A and B are independent events. Therefore

$$P(A\,B) = \left(\frac{100}{1000}\right)\left(\frac{200}{1000}\right) = \frac{1}{50}$$

From Theorem 1.3,

$$P(A|B) = \frac{P(AB)}{P(B)} \quad \text{and} \quad P(B|A) = \frac{P(AB)}{P(A)}$$

But from Eq. (1.25)

$$P(A\,|\,B) = P(A) = \frac{1}{10} \quad \text{and} \quad P(B\,|\,A) = P(B) = \frac{1}{5}$$

Theorems 1.2–1.4 can be extended to include an arbitrary number of events.

Theorem 1.5 Let $E_1, E_2, \ldots, E_n$ be n events. Then

$$P(E_1 + E_2 + \cdots + E_n) = \sum_{i=1}^{n} P(E_i) - \sum_{\substack{i=1 \\ i \neq j}}^{n} \sum_{j=1}^{n} P(E_i E_j) + \sum_{\substack{i=1 \\ i \neq j}}^{n} \sum_{\substack{j=1 \\ i \neq k}}^{n} \sum_{\substack{k=1 \\ j \neq k}}^{n} P(E_i E_j E_k)$$
$$- \cdots - (-1)^n P(E_1 E_2 \cdots E_n) \quad (1.26)$$

Proof For $n = 2$, we have

$$P(E_1 + E_2) = P(E_1) + P(E_2) - P(E_1 E_2)$$

from Theorem 1.2. For $n = 3$, let $A = E_1 + E_2$. Then, again from Theorem 1.2

$$\begin{aligned} P(E_1 + E_2 + E_3) &= P(A + E_3) \\ &= P(A) + P(E_3) - P(AE_3) \\ &= P(E_1) + P(E_2) + P(E_3) - P(E_1 E_2) - P(AE_3) \end{aligned}$$

But

$$\begin{aligned} P(AE_3) &= P[(E_1 + E_2)E_3] = P[(E_1 \text{ or } E_2) \text{ and } E_3] \\ &= P[E_1 \text{ and } E_3 \text{ or } E_2 \text{ and } E_3] = P[E_1 E_3 + E_2 E_3] \end{aligned} \quad (1.27)$$

Treating $E_1 E_3$ and $E_2 E_3$ as events, we have from Theorem 1.2,

$$P(E_1 E_3 + E_2 E_3) = P(E_1 E_3) + P(E_2 E_3) - P[(E_1 E_3)(E_2 E_3)]$$

But

$$P[(E_1 E_3)(E_2 E_3)] = P(E_1 \text{ and } E_2 \text{ and } E_3) = P(E_1 E_2 E_3)$$

Therefore

$$P(AE_3) = P(E_1 E_3) + P(E_2 E_3) - P(E_1 E_2 E_3) \quad (1.28)$$

and

$$\begin{aligned} P(E_1 + E_2 + E_3) = {} & P(E_1) + P(E_2) + P(E_3) - P(E_1 E_2) \\ & - P(E_1 E_3) - P(E_2 E_3) + P(E_1 E_2 E_3) \end{aligned}$$

Thus Eq. (1.26) holds for $n = 2$ and $n = 3$. We will assume that Eq. (1.26) holds for $n = m$ and show by induction that it also holds for $n = m + 1$ and therefore for general n. If Eq. (1.26) holds for $n = m$, then

$$P(E_1 + E_2 + \cdots + E_m) = \sum_{i=1}^{m} P(E_i) - \sum_{\substack{i=1 \\ i \neq j}}^{m} \sum_{j=1}^{m} P(E_i E_j) + \sum_{\substack{i=1 \\ i \neq j}}^{m} \sum_{\substack{j=1 \\ i \neq k}}^{m} \sum_{\substack{k=1 \\ j \neq k}}^{m} P(E_i E_j E_k)$$
$$- \cdots - (-1)^m P(E_1 E_2 \cdots E_m) \quad (1.29)$$

Let
$$A = \sum_{i=1}^{m} E_i$$
Then
$$P(A + E_{m+1}) = P(A) + P(E_{m+1}) - P(AE_{m+1})$$
But
$$P(AE_{m+1}) = P[(E_1 + E_2 + \cdots + E_m)E_{m+1}]$$
$$= P(E_1 E_{m+1} + E_2 E_{m+1} + \cdots + E_m E_{m+1})$$
from Eq. (1.27). Let
$$B_i = E_i E_{m+1}$$
Then
$$P(B_1 + B_2 + \cdots + B_n)$$
$$= \sum_{i=1}^{m} P(B_i) - \sum_{\substack{i=1 \\ i \neq j}}^{m} \sum_{j=1}^{m} P(B_i B_j) + \cdots - (-1)^m P(B_1 B_2 \cdots B_m)$$
$$- \sum_{i=1}^{m} P(E_i E_{m+1}) + \sum_{\substack{i=1 \\ i \neq j}}^{m} \sum_{j=1}^{m} P(E_i E_j E_{m+1})$$
$$- \cdots + (-1)^m P(E_1 E_2 \cdots E_m E_{m+1})$$
$$= \sum_{i=1}^{m+1} P(E_i) - \sum_{\substack{i=1 \\ i \neq j}}^{m+1} \sum_{j=1}^{m+1} P(E_i E_j) + \sum_{\substack{i=1 \\ i \neq j}}^{m+1} \sum_{\substack{j=1 \\ i \neq k}}^{m+1} \sum_{\substack{k=1 \\ j \neq k}}^{m+1} P(E_i E_j E_k)$$
$$- (-1)^{m+1} P(E_1 E_2 \cdots E_{m+1}) \qquad (1.30)$$

Since Eq. (1.26) holds for general m and $m + 1$, it must hold for all n and the proof is complete. ◨

Theorem 1.6 Let $E_1, E_2, \ldots, E_n$ be n events. Then

$$P(E_1 E_2 \cdots E_n) = P(E_1)P(E_2 | E_1)P(E_3 | E_1 E_2) \cdots P(E_n | E_1 E_2 \cdots E_{n-1}) \qquad (1.31)$$

Proof As in Theorem 1.5 we will use a proof by induction to establish the validity of Eq. (1.31). From Theorem 1.3 we have

$$P(E_1 E_2) = P(E_1)P(E_2 | E_1)$$

Let $A = E_1 E_3$. Then

$$P(AE_3) = P(A)P(E_3 | A) = P(E_1)P(E_2 | E_1)P(E_3 | E_1 E_2)$$

1.2 FUNDAMENTAL CONCEPTS

Assume Eq. (1.31) holds for $n = m$ and let $A = E_1 E_2 \cdots E_m$. Then

$$P(AE_{m+1}) = P(A)P(E_{m+1}|A)$$
$$= P(E_1)P(E_2|E_1)P(E_3|E_1 E_2) \cdots P(E_m|E_1 E_2 \cdots E_{m-1})P(E_{m+1}|A)$$
$$= P(E_1)P(E_2|E_1)P(E_3|E_1 E_2) \cdots P(E_{m+1}|E_1 E_2 \cdots E_m)$$

Since Eq. (1.31) holds for general m and $m + 1$, it holds for all n. ∎

Theorem 1.7 If the events $E_1, E_2, \ldots, E_n$ are independent then

$$P(E_1 E_2 \cdots E_n) = P(E_1)P(E_2) \cdots P(E_n) \qquad (1.32)$$

The condition given in Eq. (1.32) has important implications in probability theory. As we shall see later in this chapter, Eq. (1.32) and extensions thereof can be used to test the independence of random variables. In particular, we will show that we can operate on Eq. (1.32) to produce similar relationships for joint probability density functions, joint probability mass functions, joint distribution functions, and joint expectations.

Example 1.7 Three cards are drawn at random one after another from a standard deck without replacement. Find

   a. the probability of drawing three aces
   b. the probability of drawing at least two aces
   c. the probability of drawing an ace or a king.

Let A_1, A_2, and A_3 represent the selection of an ace on the first, second, and third draws respectively and let K_1, K_2, and K_3 represent the selection of a king on the first, second, and third draws respectively.

In part a we are to calculate $P(A_1 A_2 A_3)$. From Theorem 1.6 we have

$$P(A_1 A_2 A_3) = P(A_1)P(A_2|A_1)P(A_3|A_1, A_2)$$

If the first two cards selected were aces, then the probability of drawing a third ace is $\frac{2}{50}$ and

$$P(A_3|A_1 A_2) = \frac{1}{25}$$

Similarly

$$P(A_2 A_1) = \frac{3}{51} \quad \text{and} \quad P(A_1) = \frac{1}{13}$$

Hence

$$P(A_1 A_2 A_3) = \left(\frac{1}{13}\right)\left(\frac{3}{51}\right)\left(\frac{1}{25}\right) = \frac{1}{5525}$$

In part b the events leading to at least two aces are $A_1 A_2 \bar{A}_3$, $A_1 \bar{A}_2 A_3$, $\bar{A}_1 A_2 A_3$, and $A_1 A_2 A_3$, where $\bar{A}_i$ implies no ace on the ith draw. Therefore

$$P(\text{at least 2 aces}) = P(A_1 A_2 \bar{A}_3 + A_1 \bar{A}_2 A_3 + \bar{A}_1 A_2 A_3 + A_1 A_2 A_3)$$

Now $A_1 A_2 \bar{A}_3$, $A_1 \bar{A}_2 A_3$, $\bar{A}_1 A_2 A_3$, and $A_1 A_2 A_3$ are mutually exclusive sets of events. Thus

$$P(\text{at least 2 aces}) = P(A_1 A_2 \bar{A}_3) + P(A_1 \bar{A}_2 A_3) + P(\bar{A}_1 A_2 A_3) + P(A_1 A_2 A_3)$$

$$= \left(\frac{4}{52}\right)\left(\frac{3}{51}\right)\left(\frac{48}{50}\right) + \left(\frac{4}{52}\right)\left(\frac{48}{51}\right)\left(\frac{3}{50}\right)$$

$$+ \left(\frac{48}{52}\right)\left(\frac{4}{51}\right)\left(\frac{3}{50}\right) + \left(\frac{4}{52}\right)\left(\frac{3}{51}\right)\left(\frac{2}{50}\right)$$

$$= \left(\frac{4}{52}\right)\left(\frac{3}{51}\right)\left(\frac{146}{50}\right)$$

$$= \frac{73}{5525}$$

The probability of drawing an ace or a king can be calculated directly. However, there is a simpler approach to this problem. The total probability of all possible selections of three cards is unity. Therefore

$$P(\text{ace or king}) = 1 - P(\text{no ace and no king})$$

Now

$$P(\text{no ace and no king}) = P[(\bar{A}_1 \bar{K}_1)(\bar{A}_2 \bar{K}_2)(\bar{A}_3 \bar{K}_3)]$$
$$= P(\bar{A}_1 \bar{K}_1) P(\bar{A}_2 \bar{K}_2 | \bar{A}_1 \bar{K}_1) P[\bar{A}_3 \bar{K}_3 | (\bar{A}_1 \bar{K}_1)(\bar{A}_2 \bar{K}_2)]$$

from Theorem 1.6 and

$$P(\bar{A}_1 \bar{K}_1) = \frac{44}{52}, \quad P(\bar{A}_2 \bar{K}_2 | \bar{A}_1 \bar{K}_1) = \frac{43}{51}, \quad P[\bar{A}_3 \bar{K}_3 | (\bar{A}_1 \bar{K}_1)(\bar{A}_2 \bar{K}_2)] = \frac{42}{50}$$

Hence

$$P(\text{no ace and no king}) = \left(\frac{44}{52}\right)\left(\frac{43}{51}\right)\left(\frac{42}{50}\right) = \frac{3311}{5525}$$

and

$$P(\text{ace or king}) = \frac{2214}{5525}$$

Example 1.8 Three ground to air missiles are fired simultaneously at an enemy aircraft. The probability that any one missile destroys the aircraft is $\frac{1}{2}$, where all missiles act independently. What is the probability that the aircraft is destroyed?

Two approaches to the solution of this problem will be presented. Let D_i be the event that the ith missile strikes the aircraft and $\bar{D}_i$ the event that the ith missile fails to strike the aircraft. Therefore, the aircraft is destroyed if $D_1 \bar{D}_2 \bar{D}_3$, $\bar{D}_1 D_2 \bar{D}_3$, $\bar{D}_1 \bar{D}_2 D_3$, $D_1 D_2 \bar{D}_3$, $D_1 \bar{D}_2 D_3$, $\bar{D}_1 D_2 D_3$, and $D_1 D_2 D_3$ occur. Since the missiles act independently the probability of any one of these combinations of events is $(\frac{1}{2})^3$. Since the above sets of three events are mutually exclusive

P(aircraft destroyed)
$$= P(D_1 \bar{D}_2 \bar{D}_3) + P(\bar{D}_1 D_2 \bar{D}_3) + P(\bar{D}_1 \bar{D}_2 D_3)$$
$$+ P(D_1 D_2 \bar{D}_3) + P(D_1 \bar{D}_2 D_3) + P(\bar{D}_1 D_2 D_3) + P(D_1 D_2 D_3)$$
$$= 7(\tfrac{1}{2})^3 = \tfrac{7}{8}$$

The second approach makes use of the fact that the aircraft is destroyed unless $\bar{D}_1 \bar{D}_2 \bar{D}_3$ occurs. Hence

$$P(\text{aircraft destroyed}) = 1 - P(\bar{D}_1 \bar{D}_2 \bar{D}_3) = 1 - (\tfrac{1}{2})^3 = \tfrac{7}{8}$$

Example 1.9 (*Bayes' Theorem*) Event A occurs only when event $E_1, E_2, \ldots,$ or E_n occur. However, $E_1, E_2, \ldots, E_n$ are mutually exclusive events. Find $P(E_i | A)$ if $P(E_i)$ and $P(A | E_i)$, $i = 1, 2, \ldots, n$, are known.

Since A occurs only in conjunction with $E_1, E_2, \ldots,$ or E_n, we have

$$P(A) = \sum_{i=1}^{n} P(AE_i) \tag{1.33}$$

Now

$$P(E_i | A) = \frac{P(AE_i)}{P(A)} \tag{1.34}$$

and

$$P(A | E_i) = \frac{P(AE_i)}{P(E_i)} \tag{1.35}$$

from Theorem 1.3. Equations (1.34) and (1.35) lead to

$$P(E_i)P(A | E_i) = P(A)P(E_i | A) \tag{1.36}$$

Therefore

$$P(E_i | A) = \frac{P(E_i)P(A | E_i)}{P(A)} \quad \text{or} \quad P(E_i | A) = \frac{P(E_i)P(A | E_i)}{\sum_{i=1}^{n} P(AE_i)}$$

But

$$P(AE_i) = P(E_i)P(A | E_i)$$

Hence

$$P(E_i | A) = \frac{P(E_i)P(A | E_i)}{\sum_{i=1}^{n} P(E_i)P(A | E_i)} \tag{1.37}$$

Example 1.10 A manufacturer produces one of its products at four different plants. Let E_i be the event that a unit was produced at plant i, where

$$P(E_1) = 0.10, \qquad P(E_2) = 0.40, \qquad P(E_3) = 0.20, \qquad P(E_4) = 0.30$$

Each plant ships the units to retailers located throughout the USA. The manufacturer pays a penalty whenever a unit is delivered after the agreed delivery date. Let A be the event that a unit is delivered late. The probability of a late delivery given that a unit was made at the ith plant, $P(A | E_i)$, is given by

$$P(A | E_1) = 0.20, \qquad P(A | E_2) = 0.40, \qquad P(A | E_3) = 0.40, \qquad P(A | E_4) = 0.10$$

The manufacturer is informed that a unit of product was delivered late. What is the probability that the unit was manufactured at each of the four plants?
From Eq. (1.37)

$$P(E_i | A) = \frac{P(E_i) P(A | E_i)}{\sum_{i=1}^{n} P(E_i) P(A | E_i)}$$

Now

$$\sum_{i=1}^{4} P(E_i) P(A | E_i) = (0.10)(0.20) + (0.40)(0.40) + (0.20)(0.40) + (0.30)(0.10)$$

$$= 0.29$$

Therefore

$$P(E_1 | A) = \frac{0.02}{0.29} = 0.069, \qquad P(E_2 | A) = \frac{0.16}{0.29} = 0.552$$

$$P(E_3 | A) = \frac{0.08}{0.29} = 0.276, \qquad P(E_4 | A) = \frac{0.03}{0.29} = 0.103$$

1.3 Probability Distributions

When outcomes or events are characterized as random variables we can develop measures of the probability of occurrence of values or sets of values of the random variable through mathematical relationships. One such measure is the *cumulative distribution function* or simply the *distribution function* which is defined as follows.

Definition 1.6 Let $F(x)$ be the cumulative distribution function (CDF) of the random variable X. Then

$$F(x) = P(X \leq x) \tag{1.38}$$

where $P(X \leq x)$ is the probability that the random variable X assumes a value less than or equal to x.

1.3 PROBABILITY DISTRIBUTIONS

Consider the random variable X that may assume integer values, x, between 0 and n inclusive, and let k be an integer. Then

$$F(k) = P(X \le k) = P(x = 0, \text{ or } x = 1, \ldots, \text{ or } x = k) \quad (1.39)$$

If we assume that only one value of the random variable may occur at a given trial of the experiment, then the values which the random variable may assume are mutually exclusive and

$$F(k) = P(x = 0) + P(x = 1) + \cdots + P(x = k) = \sum_{x=0}^{k} P(X = x) \quad (1.40)$$

As implied by Eqs. (1.38) and (1.40), the distribution function may be thought of as an *accumulator* of probability. Since the probability of a given value or set of values of a random variable may not be negative, $F(x)$ is a *nondecreasing function*. Further, since probabilities must lie between 0 and 1 inclusive, $0 \le F(x) \le 1$. Finally, based upon the above results one can show that $F(-\infty) = 0$ and $F(\infty) = 1$. To summarize, the properties of the cumulative distribution function are

a. $0 \le F(x) \le 1$ \hfill (1.41)

b. $F(x)$ is a nondecreasing function of x

c. $F(-\infty) = 0$ \hfill (1.42)

d. $F(\infty) = 1$ \hfill (1.43)

Example 1.11 The number of aircraft landing at a small airfield during 15-minute intervals has been observed to lie between 0 and 5. Let X be the random variable representing the number of landings per 15-minute interval and let $P(X = x)$ be the probability of x landings where

$$P(X = x) = \begin{cases} 0.00, & x < 0 \\ 0.30, & x = 0 \\ 0.30, & x = 1 \\ 0.20, & x = 2 \\ 0.10, & x = 3 \\ 0.06, & x = 4 \\ 0.04, & x = 5 \\ 0.00, & x > 5 \end{cases}$$

Define $F(x)$ for this random variable.

To define $F(x)$ we may use Eq. (1.40). Since it is impossible to have less than zero landings in any 15-minute interval, we have

$$P(X = x) = 0, \quad x < 0 \quad \text{and} \quad F(x) = \sum_{y=-\infty}^{x} P(X = y) = 0, \quad x < 0$$

For $x = 0, 1, \ldots, 5$ we have

$$F(x) = \sum_{y=-\infty}^{x} P(X = y) = \sum_{y=-\infty}^{-1} P(X = y) + \sum_{y=0}^{x} P(X = y)$$

$$= \sum_{y=0}^{x} P(X = y), \quad x = 0, 1, \ldots, 5$$

Therefore

$$F(x) = \begin{cases} 0.30, & x = 0 \\ 0.60, & x = 1 \\ 0.80, & x = 2 \\ 0.90, & x = 3 \\ 0.96, & x = 4 \\ 1.00, & x = 5 \end{cases}$$

For $x > 5$, $F(x)$ is given by

$$F(x) = \sum_{y=-\infty}^{x} P(X=y) = \sum_{y=-\infty}^{-1} P(X=y) + \sum_{y=0}^{5} P(X=y) + \sum_{y=6}^{x} P(X=y)$$

$$= 1 + \sum_{y=6}^{x} P(X=x)$$

But since $P(X = x) = 0$, $x \geq 6$, we have $F(x) = 1$, $x \geq 6$.

We will be primarily concerned with two distinct types of distribution functions. The first case arises when the distribution function increases at and only at discrete values of the random variable on the interval $-\infty$ to $+\infty$. The graph of such distribution functions consists of a series of horizontal straight lines with jumps at discrete values of the random variable as shown in Fig. 1.1. Distribution

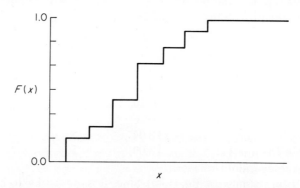

Figure 1.1 Distribution Function for a Discrete Random Variable

functions of this type arise when the random variable of concern may assume discrete values only.

When the distribution function is continuous on the interval $-\infty$ to $+\infty$, then the random variable is said to be continuous. The graph of the distribution function for a continuous random variable may take forms of the type shown in Fig. 1.2.

1.3 PROBABILITY DISTRIBUTIONS

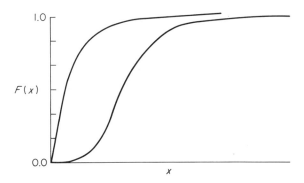

Figure 1.2 Distribution Functions for Continuous Random Variables

Although they are not frequently encountered in practice, distribution functions may be continuous on $(-\infty, \infty)$, except at certain values of the random variables where discontinuities occur. In these cases the random variable is both continuous and discrete and the distribution function is referred to as mixed. A graphical example of a mixed distribution function is shown in Fig. 1.3.

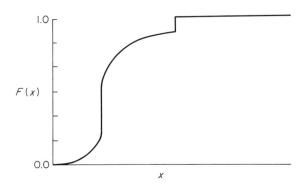

Figure 1.3 Mixed Distribution Function

To summarize, when the distribution function of a random variable is continuous throughout the interval $-\infty$ to $+\infty$, the random variable is referred to as continuous. If the distribution function is continuous and horizontal at all but a discrete set of values of the random variable on the interval $-\infty$ to $+\infty$, then the random variable is said to be discrete. Finally, if the distribution function is continuous at all but a discrete set of values of the random variable on the interval $-\infty$ to $+\infty$, and if the distribution function is an increasing function of the random variable on some interval a to b $(-\infty < a < b < \infty)$, then the distribution function and the random variable are called mixed.

While a random variable may be uniquely characterized by its distribution function, the *probability density function* or simply *density function* of a continuous random variable and the *probability mass function* of a discrete random variable are more frequently used as a basis for describing the characteristics of a random variable. The probability mass function of a discrete random variable and the probability density function of a continuous random variable are defined in terms of the distribution function as follows.

Definition 1.7 Let $F(x)$ and $p(x)$ be the distribution function and the probability mass function of the discrete random variable X. Further, assume that $F(x)$ is discontinuous at $x = x_1, x_2, \ldots, x_n$. Then

$$p(x_i) = F(x_i) - F(x_i - 1) \tag{1.44}$$

and

$$p(x) = 0, \quad x \neq x_i, \quad i = 1, 2, \ldots, n \tag{1.45}$$

Definition 1.8 Let $F(x)$ and $f(x)$ be the distribution function and probability density function of the continuous random variable X. Then

$$f(x) = \frac{d}{dx} F(x), \quad -\infty < x < \infty \tag{1.46}$$

In Definitions 1.7 and 1.8 we define the probability mass function, $p(x)$, and probability density function, $f(x)$, in terms of their corresponding distribution functions. These definitions lead us to corresponding definitions of $F(x)$ in terms of $p(x)$ and $f(x)$. That is, from Eqs. (1.44) and (1.45) we are led to

$$F(x_i) = \sum_{j=1}^{i} p(x_j) \tag{1.47}$$

and

$$F(x) = \sum_{j=1}^{k} p(x_j), \quad x_k < x < x_{k+1} \tag{1.48}$$

In a similar manner, from Eq. (1.46) we have

$$F(x) = \int_{-\infty}^{x} f(y) \, dy \tag{1.49}$$

Example 1.12 Let

$$F(x) = \begin{cases} 0, & x < 1 \\ 1 - p^x, & x = 1, 2, \ldots \\ 1 - p^k, & k < x < k + 1 \end{cases}$$

where $0 < p < 1$ and k is a positive integer. Find the probability mass function of X.

The definition of $F(x)$ implies that $F(x)$ is discontinuous at all integer values

1.3 PROBABILITY DISTRIBUTIONS

of x greater than 0. Further, $F(x)$ is horizontal for all noninteger values of x and for all $x \leq 0$. Therefore

$$p(x) = F(x) - F(x-1)$$

for integer $x > 0$ and hence

$$p(x) = (1-p^x) - (1-p^{x-1}) = p^{x-1}(1-p), \qquad x = 1, 2, \ldots$$

and

$$p(x) = 0, \qquad\qquad\qquad\qquad x \neq 1, 2, \ldots$$

Example 1.13 Let

$$F(x) = \begin{cases} 0, & x < 0 \\ 1 - e^{-x}, & 0 \leq x < \infty \end{cases}$$

Find the density function of X.

From Eq. (1.46) we have

$$f(x) = \frac{d}{dx} F(x)$$

Since

$$\frac{d}{dx} F(x) = 0, \quad x < 0 \qquad \text{and} \qquad \frac{d}{dx} F(x) = e^{-x}, \quad 0 \leq x < \infty$$

we are led to

$$f(x) = \begin{cases} 0, & x < 0 \\ e^{-x}, & 0 \leq x < \infty \end{cases}$$

Example 1.14 The density function of the continuous random variable X is defined by

$$f(x) = \begin{cases} 0, & x < a \\ 1/(b-a), & a \leq x \leq b \\ 0, & x > b \end{cases}$$

Find the distribution function of X.

From Eq. (1.49)

$$F(x) = \int_{-\infty}^{x} f(y)\, dy$$

For $x < a$, $F(x) = 0$. For $a \leq x \leq b$,

$$F(x) = \int_{-\infty}^{x} f(y)\, dy = \int_{-\infty}^{a} f(y)\, dy + \int_{a}^{x} f(y)\, dy = 0 + \int_{a}^{x} \frac{dy}{b-a} = \frac{x-a}{b-a}$$

Finally, for $x > b$,

$$F(x) = \int_{-\infty}^{x} f(y)\, dy = \int_{-\infty}^{a} f(y)\, dy + \int_{a}^{b} f(y)\, dy + \int_{0}^{x} f(y)\, dy = 0 + 1 + 0 = 1$$

Hence

$$F(x) = \begin{cases} 0, & x < a \\ (x-a)/(b-a), & a \le x \le b \\ 1, & x > b \end{cases}$$

In addition to defining probabilities of the type $P(X \le x)$, the distribution function may be used to determine the probability that the value of a random variable will fall within a specified interval. There are four such probabilities which are of interest, where $a < b$.

1. $P(a \le x \le b)$
2. $P(a < x \le b)$
3. $P(a \le x < b)$
4. $P(a < x < b)$

To illustrate the use of the distribution function to determine these probabilities, let us define $P(a < x \le b)$ in terms of $F(x)$ when x is a discrete and then a continuous random variable. Assume that X is a discrete random variable such that $p(x) \ne 0$ for integer x on the interval 0 to n inclusive. For integer a and b and $a \ge 0, b \le n$

$$P(a < x \le b) = P(X = a+1) + P(X = a+2) + \cdots + P(X = b)$$

$$= \sum_{x=a+1}^{b} p(x) = \sum_{x=0}^{b} p(x) - \sum_{x=0}^{a} p(x) = F(b) - F(a) \quad (1.50)$$

For $a < 0$ and $b \le n$, $F(a) = 0$ and

$$P(a < x \le b) = F(b) \quad (1.51)$$

For $a \ge 0$ and $b > n$, $F(b) = 1$ and

$$P(a < x \le b) = 1 - F(a) \quad (1.52)$$

Finally, for $a < 0$ and $b > n$,

$$P(a < x \le b) = 1 \quad (1.53)$$

Hence Eq. (1.50) holds for all integers a and b on the interval $-\infty$ to ∞.

Continuing with this example, if $k < x < k+1$ where k is an integer, then

$$F(x) = \sum_{y=0}^{k} p(y) \quad (1.54)$$

Hence, for noninteger a and integer b,

$$P(a < x \le b) = F(b) - F(k), \quad k < a < k+1$$
$$= F(b) - F(a) \quad (1.55)$$

For integer a and noninteger b,

$$P(a < x \le b) = F(k) - F(a), \quad k < b < k+1$$
$$= F(b) - F(a) \quad (1.56)$$

1.3 PROBABILITY DISTRIBUTIONS

Finally, for noninteger a and b and integer j and k,

$$P(a < x \le b) = F(k) - F(j), \quad k < b < k+1, \quad j < a < j+1$$
$$= F(b) - F(a) \tag{1.57}$$

Therefore, Eq. 1.50 holds for all a and b, both integer and noninteger.

By arguments similar to those just given, one may show that Eq. (1.50) is valid for all discrete random variables, X, whether or not X is integer valued. That is, X may be a discrete random variable which may assume only the values $\frac{1}{8}, \frac{1}{4}, 3, 4\frac{1}{3}, 8\frac{1}{4}$, and $21\frac{7}{8}$. In this case X is discrete but is not restricted to integer values. Although discrete random variables of this type sometimes arise in practice, they will not be of significant concern here.

If X is a discrete random variable such that

$$p(b - \Delta_1) \ne 0 \tag{1.58}$$
$$p(a - \Delta_2) \ne 0 \tag{1.59}$$

where $a < b$, and $\Delta_i > 0$, $i = 1, 2$, and $p(x) = 0$ for $b - \Delta_1 < x < b$ and $a - \Delta_2 < x < a$, then it can be shown that

$$P(a \le x \le b) = F(b) - F(a - \Delta_2) = P(x \le b) - P(x < a) \tag{1.60}$$
$$P(a < x \le b) = F(b) - F(a) = P(x \le b) - P(x \le a) \tag{1.61}$$
$$P(a \le x < b) = F(b - \Delta_1) - F(a - \Delta_2) = P(x < b) - P(x < a) \tag{1.62}$$
$$P(a < x < b) = F(b - \Delta_1) - F(a) = P(x < b) - P(x \le a) \tag{1.63}$$

where the random variable x is integer valued.

When X is a continuous random variable, $F(x)$ is defined by Eq. (1.49). For $a < b$ we have

$$F(a) = \int_{-\infty}^{a} f(x)\, dx, \quad F(b) = \int_{-\infty}^{b} f(x)\, dx$$

But

$$F(b) = \int_{-\infty}^{a} f(x)\, dx + \int_{a}^{b} f(x)\, dx \tag{1.64}$$

Thus

$$\int_{a}^{b} f(x)\, dx = F(b) - F(a) = P(x \le b) - P(x \le a) = P(a < x \le b) \tag{1.65}$$

Since X is a continuous random variable,

$$P(x \le k) = \int_{-\infty}^{k} f(x)\, dx, \quad P(x < k) = \int_{-\infty}^{k} f(x)\, dx$$

and
$$P(x \le k) = P(x < k) \tag{1.66}$$
Therefore, for continuous random variables,
$$P(a \le x \le b) = P(a < x \le b) = P(a \le x < b)$$
$$= P(a < x < b) = \int_a^b f(x)\,dx \tag{1.67}$$

Example 1.15 The number of automobile tires purchased by a customer, X, has the distribution function given by

$$F(x) = \begin{cases} 0, & x < 1 \\ \dfrac{1 - (\frac{1}{2})^x}{1 - (\frac{1}{2})^4}, & x = 1, 2, 3, 4 \\ \dfrac{1 - (\frac{1}{2})^k}{1 - (\frac{1}{2})^4}, & k < x < k+1 \\ 1, & x > 4 \end{cases}$$

where k is a positive integer. Find

a. the probability that a customer purchases 2 or less tires.
b. the probability that a customer purchases at least 2 tires.
c. the probability that a customer purchases between 1 and 3 tires inclusive.
d. the probability that a customer purchases 2 or 3 tires.

The probability that a customer purchases 2 or less tires is given by $F(2)$. Hence

$$P(2 \text{ or less tires purchased}) = F(2) = \frac{1 - (\frac{1}{2})^2}{1 - (\frac{1}{2})^4} = \frac{3/4}{15/16} = \frac{4}{5}.$$

For part b, the probability that a customer purchases at least 2 tires is given by
$$P(\text{at least 2 tires purchased}) = P(2 \le x < \infty)$$
From Eq. (1.62)
$$P(2 \le x < \infty) = F(\infty) - F(2 - \Delta_2)$$
In this case, the value of Δ_2 such that $p(2 - \Delta_2) \neq 0$ and $p(x) = 0$ for $2 - \Delta_2 < x < 2$ is unity. Therefore
$$P(2 \le x < \infty) = F(\infty) - F(1) = 1 - \frac{1 - (\frac{1}{2})}{1 - (\frac{1}{2})^4} = 1 - \frac{8}{15} = \frac{7}{15}$$

In part c we have
$$P(1 \le x \le 3) = F(3) - F(0)$$

1.3 PROBABILITY DISTRIBUTIONS

since $\Delta_2 = 1$. Hence

$$P(1 \leq x \leq 3) = F(3) - 0 = \frac{1 - (\frac{1}{2})^3}{1 - (\frac{1}{2})^4} = \frac{14}{15}$$

For part d

$$P(2 \text{ or } 3 \text{ tires purchased}) = P(2 \leq x \leq 3) = F(3) - F(1)$$

$$= \frac{1 - (\frac{1}{2})^3}{1 - (\frac{1}{2})^4} - \frac{1 - (\frac{1}{2})}{1 - (\frac{1}{2})^4} = \frac{4}{5} - \frac{8}{15} = \frac{4}{15}$$

Example 1.16 The probability density function of time between successive customer arrivals to a service station, X, is given by

$$f(x) = \begin{cases} 0, & x < 0 \\ 10 \exp[-10x], & 0 \leq x < \infty \end{cases}$$

where x is measured in hours. Find

a. $P(0.1 < x < 0.5)$
b. $P(x < 1)$
c. $P(0.2 < x < 0.3 \text{ or } 0.5 < x < 0.7)$

From Eq. (1.67) we have for part a,

$$P(0.1 < x < 0.5) = \int_{0.1}^{0.5} 10 \exp[-10x] \, dx = -\exp[-10x] \Big|_{0.1}^{0.5}$$

$$= \exp[-1] - \exp[-5]$$

or

$$P(0.1 < x < 0.5) = 0.905 - 0.607 = 0.298$$

For part b,

$$P(x < 1) = F(1) = \int_0^1 10 \exp[-10x] \, dx = -\exp[-10x] \Big|_0^1$$

$$= 1 - \exp[-10]$$

$$= 0.632$$

In part c we wish to find the probability that x lies between 0.2 and 0.3 or between 0.5 and 0.7. Since x cannot fall in each of these intervals simultaneously, the events that x falls in the intervals 0.2 to 0.3 and 0.5 to 0.7 are mutually exclusive. Hence

$$P(0.2 < x < 0.3 \text{ or } 0.5 < x < 0.7)$$

$$= P(0.2 < x < 0.3) + P(0.5 < x < 0.7)$$

$$= F(0.3) - F(0.2) + F(0.7) - F(0.5)$$

$$= \int_{0.2}^{0.3} 10 \exp[-10x] \, dx + \int_{0.5}^{0.7} 10 \exp[-10x] \, dx = 0.188$$

Definition 1.7 and Eq. (1.60) give us methods for determining the probability that a given value of a discrete random variable will occur. From Eq. (1.60),

$$P(X = y) = P(y \leq x \leq y) = F(y) - F(y - \Delta_2) = p(y) \quad (1.68)$$

from Definition 1.7. Now suppose that we wish to know $P(X = y)$, where X is a continuous random variable. From Eq. 1.67, if $f(x)$ is the density function of X, then

$$P(X = y) = \int_{y}^{y} f(x)\, dx = 0 \quad (1.69)$$

Thus, the probability of observing a specific value of a continuous random variable is zero. Hence, when we wish to determine probabilities associated with continuous random variables we must consider the probability that the value of the random variable will fall in some interval rather than at a point.

In our definitions of the distribution function, $F(x)$, the probability density function, $f(x)$, and the probability mass function, $p(x)$, each function was defined over the range $-\infty < x < \infty$. In Example 1.16, $f(x)$ was defined as

$$f(x) = \begin{cases} 0, & x < 0 \\ 10 \exp[-10x], & 0 \leq x < \infty \end{cases}$$

Henceforth we will define these functions only for those values of x for which they are nonzero. Thus the above density function would be given by

$$f(x) = 10 \exp[-10x], \quad 0 \leq x < \infty$$

and is assumed to be zero otherwise.

There are several discrete and continuous random variables that are of particular importance in probability theory and its applications. Among the more important discrete random variables are the Bernoulli, binomial, geometric, negative binomial, hypergeometric, Poisson, and rectangular. The Bernoulli, binomial, hypergeometric, and Poisson random variables have important applications in quality control, particularly in the area of acceptance sampling (Duncan, 1965). The Poisson random variable is frequently used to describe such phenomena as the number of customer arrivals and product demands in a given period of time (Hadley and Whitin, 1963; Morse, 1962; Saaty, 1961). In addition, the Poisson random variable plays an important role in stochastic processes (Karlin, 1966; Parzen, 1962). Under appropriate conditions, the Poisson probability mass function may be used to approximate the binomial and hypergeometric probability mass functions. The geometric random variable is sometimes used to characterize such phenomena as the number units of product requested given that a demand for product has occurred (Hadley and Whitin, 1963) and the number of arrivals at a service system when bulk arrivals occur. The Bernoulli random variable is an important building block for other random variables since the binomial, geometric, and negative binomial random variables may be expressed in terms of the Bernoulli. In addition, the Bernoulli random variable is often used in simulating sampling experiments (Schmidt and Taylor, 1970). Finally, the rectangular random variable is both important and fundamental since it may be used to describe the occurrence of equally likely events and because it has wide application in system simulation (Schmidt and Taylor, 1970). The characteristics of several important discrete random variables are summarized in Table 1.4.

Continuous random variables play an important role in modeling physical systems and in statistical hypothesis testing. Probably the most well-known continuous random variable is the normal. While the normal random variable can be used to characterize many physical phenomena, its most important applications lie in statistical estimation and hypothesis testing where it provides the foundation for many if not most parametric statistical tests (Ostle, 1963). In addition, as a result of the central limit theorem (Parzen, 1960), it can be shown that, under certain conditions, the distribution of the sum of random variables approaches the normal distribution as the number of random variables included in the sum approaches infinity. Other random variables which are of importance in hypothesis testing are the t, χ^2, and F (Bowker and Lieberman, 1959; Brownlee, 1960; Graybill, 1961; Ostle, 1963).

In the areas of reliability (Shooman, 1968), maintenance (Morse, 1962), inventory control (Hadley and Whitin, 1963), queueing theory (Morse, 1962; Saaty, 1961), and stochastic processes (Karlin, 1966; Parzen, 1962) in general, the exponential random variable has significant application. Its importance arises, at least in part, from the fact that it alone, among continuous random variables, is without memory, as we shall show later in this chapter. Another important continuous random variable, which has wide application in the areas already mentioned in conjunction with the exponential, is the gamma random variable. Under certain constraints, the gamma random variable can be represented as the sum of exponential random variables. In addition, the family of χ^2 random variables falls in the more general family of gamma random variables. The uniform random variable is the continuous analogue of the discrete rectangular random variable, and forms an important basis for Monte Carlo simulation (Schmidt and Taylor, 1970). Other continuous random variables of somewhat lesser but still significant importance in the applications of probability theory include the beta and Weibull. The beta random variable is often used to describe the proportion of defective units in manufactured product lots while the Weibull random variable has significant applications in reliability engineering (Shooman, 1968). The characteristics of several important continuous random variables are summarized in Table 1.5.

1.3.1 Discrete Random Variables

(*a*) *Bernoulli* Consider a random experiment in which two outcomes may occur. To the first outcome we assign a 0 and to the second we assign a 1. Let

$$P(X = 1) = p, \quad P(X = 0) = 1 - p$$

TABLE 1.4
SOME IMPORTANT DISCRETE RANDOM VARIABLES AND THEIR CHARACTERISTICS

Random Variable, X	Probability Mass Function, $p(x)$	Moment Generating Function, $M(t, X)$	Mean, m	Variance, σ^2
Bernoulli	$p^x(1-p)^{1-x}, \quad x = 0, 1$	$[p\exp[t] + (1-p)]$	p	$p(1-p)$
Binomial	$\binom{n}{x}p^x(1-p)^{n-x}, \quad x = 0, 1, \ldots, n$	$[p\exp[t] + (1-p)]^n$	np	$np(1-p)$
Geometric	$p(1-p)^{x-1}, \quad x = 1, 2, \ldots$	$\dfrac{p\exp[t]}{1 - (1-p)\exp[t]}$	$\dfrac{1}{p}$	$\dfrac{1-p}{p^2}$
Negative Binomial	$\binom{x-1}{r-1}p^r(1-p)^{x-r}, \quad x = r, r+1, \ldots$	$\left[\dfrac{p\exp[t]}{1 - (1-p)\exp[t]}\right]^r$	$\dfrac{r}{p}$	$\dfrac{r(1-p)}{p^2}$
Hypergeometric	$\dfrac{\binom{M}{x}\binom{N-M}{n-x}}{\binom{N}{n}}, \quad x = 0, 1, \ldots, n$		$\dfrac{Mn}{N}$	$\dfrac{(N-n)(Mn)(N-M)}{N^2(N-1)}$
Poisson	$\dfrac{\lambda^x}{x!}\exp[-\lambda], \quad x = 0, 1, \ldots$	$\exp[-\lambda(1 - \exp[t])]$	λ	λ
Rectangular	$\dfrac{1}{b-a+1}, \quad x = a, a+1, \ldots, b$	$\dfrac{\exp(b+1)t - \exp[at]}{(b-a+1)(\exp[t] - 1)}$	$\dfrac{a+b}{2}$	$\dfrac{(b+a)^2 + 2(b-a)}{12}$

1.3 PROBABILITY DISTRIBUTIONS

TABLE 1.5
Some Important Continuous Random Variables and Their Characteristics

Random Variable, X	Probability Density Function, $f(x)$	Moment Generating Function, $M(t, X)$	Mean, m	Variance, σ^2
Normal	$\dfrac{1}{\sigma\sqrt{2\pi}}\exp\left[-\dfrac{(x-m)^2}{2\sigma^2}\right]$, $-\infty < x < \infty$	$\exp[mt] + \dfrac{\sigma^2 t^2}{2}$	m	σ^2
χ^2	$\dfrac{1}{2^{n/2}\Gamma(n/2)} x^{(n/2)-1} \exp[-x/2]$, $0 < x < \infty$	$(1-2t)^{-n/2}$	n	$2n$
t	$\dfrac{1}{\sqrt{n\pi}}\dfrac{\Gamma[[n+1]/2]}{\Gamma(n/2)}\left(1+\dfrac{x^2}{n}\right)^{-(n+1)/2}$, $-\infty < x < \infty$		0	$\dfrac{n}{n-2}, n > 2$
Exponential	$\lambda\exp[-\lambda x]$, $0 < x < \infty$	$\left(1-\dfrac{t}{\lambda}\right)^{-1}$	$\dfrac{1}{\lambda}$	$\dfrac{1}{\lambda^2}$
Gamma	$\dfrac{\lambda^n}{\Gamma(n)} x^{n-1} \exp[-\lambda x]$, $0 < x < \infty$	$\left(1-\dfrac{t}{\lambda}\right)^{-n}$	$\dfrac{n}{\lambda}$	$\dfrac{n}{\lambda^2}$
Beta	$\dfrac{\Gamma(a+b)}{\Gamma(a)\Gamma(b)} x^{a-1}(1-x)^{b-1}$, $0 < x < 1$		$\dfrac{a}{a+b}$	$\dfrac{ab}{(a+b)^2(a+b+1)}$
Weibull	$\dfrac{a}{b-c}\left(\dfrac{x-c}{b-c}\right)^{a-1}\exp\left[-\left(\dfrac{x-c}{b-c}\right)^a\right]$, $c < x < \infty$		$c + (b-c)\Gamma\left(\dfrac{a+1}{a}\right)$	$(b-c)^2\left\{\Gamma\left(\dfrac{a+2}{a}\right) - \left[\Gamma\left(\dfrac{a+1}{a}\right)\right]^2\right\}$
Uniform	$\dfrac{1}{b-a}$, $a < x < b$	$\dfrac{\exp[bt]-\exp[at]}{t(b-a)}$	$\dfrac{a+b}{2}$	$\dfrac{(b-a)^2}{12}$

where $0 < p < 1$. Then X is called a Bernoulli random variable with probability mass function given by

$$p(x) = p^x(1-p)^{1-x}, \quad x = 0, 1 \tag{1.70}$$

(b) *Binomial* Let $Y_1, Y_2, \ldots, Y_n$ be independent Bernoulli random variables each with parameter p. Now define the random variable X as

$$X = \sum_{i=1}^{n} Y_i \tag{1.71}$$

The random variable X is called a binomial random variable with probability mass function given by

$$p(x) = \binom{n}{x} p^x(1-p)^{n-x}, \quad x = 0, 1, \ldots, n \tag{1.72}$$

where $0 < p < 1$ and where

$$\binom{n}{x} = \frac{n!}{x!(n-x)!} \tag{1.73}$$

(c) *Geometric* As in the case of the binomial random variable, the geometric random variable is related to the Bernoulli random variable. Let $Y_i, i = 1, 2, \ldots$, be a sequence of independent Bernoulli random variables each with parameter p. Now let X be the random variable representing the number of Bernoulli trials of the experiment necessary to obtain the first 1. That is, the value of the random variable X is x if

$$y_i = 0, \quad i = 1, 2, \ldots, x-1 \quad \text{and} \quad y_x = 1$$

X is called a geometric random variable and has probability mass function given by

$$p(x) = p(1-p)^{x-1}, \quad x = 1, 2, \ldots \tag{1.74}$$

where $0 < p < 1$.

(d) *Negative Binomial* Again we encounter a random variable which is related to the Bernoulli random variable. Suppose that we conduct successive trials of a Bernoulli experiment, recording at each trial the occurrence of 0's and 1's. Let X be the random variable representing the number of Bernoulli trials necessary until the rth 1 occurs. Then X is called a negative binomial random variable with probability mass function given by

$$p(x) = \binom{x-1}{r-1} p^r(1-p)^{x-r}, \quad x = r, r+1, \ldots \tag{1.75}$$

where r is integer valued and positive and $0 < p < 1$.

The negative binomial random variable may be characterized in another way. Suppose that a series of identical Bernoulli trials are conducted. The series is

1.3 PROBABILITY DISTRIBUTIONS

terminated when the first 1 is recorded. If Z_1 is the random variable representing the number of trials until termination, then Z_1 is geometrically distributed. Now a second series of Bernoulli trials is run, the number of trials necessary again recorded until the first 1 is detected, Z_2. Then Z_2 is also geometrically distributed. The series of Bernoulli trials is repeated r times where Z_i is the number of trials necessary before the first 1 is found in the ith series. If all r series are conducted under identical conditions, then $Z_1, Z_2, \ldots, Z_r$ are identically distributed independent geometric random variables. The total number of Bernoulli trials necessary to find the first r 1's, X, is then given by

$$X = \sum_{i=1}^{r} Z_i \qquad (1.76)$$

Hence, the negative binomial random variable with parameters r and p may be expressed as the sum of r geometric random variables each with parameter p.

(e) *Hypergeometric* Suppose that a lot contains N items of which M is defective. We randomly select a sample n items from the lot and record the number of defective items in the sample, X. Here X is called a hypergeometric random variable with parameters M, N, and n and may assume integer values between 0 and n inclusive.

Let us calculate the probability mass function, $p(x)$, of X. The number of samples of size n which may be randomly selected from a lot of size N is given by the combinations of N things taken n at a time or $\binom{N}{n}$. To obtain x defective items we must select x items from among the M defective items in the lot and $n - x$ from the remaining $N - M$ good items. The number of ways of randomly selecting x things from M is given by $\binom{M}{x}$, while $\binom{N-M}{n-x}$ is the number of ways of randomly selecting $n - x$ items from $N - M$. Hence, the total number of ways of selecting a sample containing exactly x defective items is $\binom{M}{x}\binom{N-M}{n-x}$. Now the probability of finding exactly x defects in a sample of size n is given by the ratio of the total number of samples which could be randomly selected and which contain exactly x defects to the total number of samples which could be drawn. Therefore

$$p(x) = \frac{\binom{M}{x}\binom{N-M}{n-x}}{\binom{N}{n}}, \qquad x = 0, 1, \ldots, n \qquad (1.77)$$

where $n < M \leq N$.

Because the combinatorials in Eq. (1.77) are cumbersome to work with in attempting to calculate probabilities for the hypergeometric random variable, the binomial probability mass function is frequently used as an approximation to the hypergeometric where the parameter p of the binomial is given by M/N. The accuracy of the approximation depends upon the magnitudes of the ratios x/M and $(n - x)/(N - M)$. One rule of thumb frequently used is that the binomial may be used to approximate the hypergeometric when x/M and $(n - x)/(N - M)$ are both less than 0.1.

(f) Poisson To characterize the Poisson random variable, consider the occurrence of events over some time interval of length t. If the number of events occurring in t, X satisfies the following conditions, then X is a Poisson random variable.

   1. The number of events in t, X, is independent of the number of events occurring in any other nonoverlapping time interval.
   2. The probability mass function of X is the same for all intervals of length t, no matter when the interval started.
   3. Two events cannot occur at the same point in time.
   4. $0 < P(X > 0 \,|\, t > 0) < 1$.
   5. $P(X > 0 \,|\, t = 0) = 0$.

The probability mass function of the Poisson random variable is defined by

$$p(x) = \frac{(\lambda t)^x}{x!} \exp[-\lambda t], \qquad x = 0, 1, 2, \ldots \tag{1.78}$$

It can be shown that if the time between the occurrence of successive events is exponentially distributed, then the number of events occurring in any fixed period of time is Poisson distributed.

The Poisson probability mass function is often used to approximate the binomial probability mass function. When the parameter, p, of the binomial distribution is less than 0.10, the Poisson distribution may be used to approximate the binomial by letting $\lambda t = np$, where n and p are the parameters of the binomial. Since the binomial can be used to approximate the hypergeometric under the conditions specified previously, the Poisson may also be used to approximate the hypergeometric.

(g) Rectangular Let X be an integer valued random variable with values $x = a$, $a + 1, \ldots, b$. If each value of the X is equally likely to occur, then X is called a rectangular random variable and has the probability mass function given by

$$p(x) = \frac{1}{b - a + 1}, \qquad x = a, a + 1, \ldots, b \tag{1.79}$$

where a and b are integers and $a < b$.

1.3.2 Continuous Random Variables

(a) Normal The probability density function of the normal random variable is given by

$$f(x) = \frac{1}{\sigma\sqrt{2\pi}} \exp\left[-\frac{(x - u)^2}{2\sigma^2}\right], \qquad -\infty < x < \infty \tag{1.80}$$

1.3 PROBABILITY DISTRIBUTIONS

where $\sigma > 0$. The normal density function is symmetric about u, called the mean, and has points of inflection at $u \pm \sigma$, where σ is referred to as the standard deviation. The normal random variable with $u = 0$ and $\sigma = 1$ is called the standard normal random variable and is of importance since the transformation

$$Z = \frac{X - u}{\sigma} \tag{1.81}$$

yields a standard normal random variable if X is normally distributed with parameters u and σ as given in Eq. (1.80). Tables of the distribution function of the standard normal random variable are readily available in texts on statistical methods.

(b) χ^2 Let $Y_1, Y_2, \ldots, Y_n$ be mutually independent standard normal random variables. Then

$$X = \sum_{i=1}^{n} Y_i^2 \tag{1.82}$$

is said to be a χ^2 random variable with n degrees of freedom and probability density function given by

$$f(x) = \frac{1}{2^{n/2} \Gamma\left(\dfrac{n}{2}\right)} x^{(n/2)-1} \exp[-x/2], \qquad 0 < x < \infty \tag{1.83}$$

where n is a positive integer.

(c) t The t random variable may be expressed as a function of a standard normal random variable and a χ^2 random variable. Let Z be a standard normal random variable and Y an independent χ^2 random variable with n degrees of freedom. Then

$$X = \frac{Z}{\sqrt{Y/n}} \tag{1.84}$$

is a t random variable with probability density function given by

$$f(x) = \frac{1}{\sqrt{n\pi}} \frac{\Gamma\left(\dfrac{n+1}{2}\right)}{\Gamma\left(\dfrac{n}{2}\right)} \left(1 + \frac{x^2}{n}\right)^{-(n+1)/2}, \qquad -\infty < x < \infty \tag{1.85}$$

(d) F Let Y_1 and Y_2 be independent χ^2 random variables with degrees of freedom n_1 and n_2 respectively. Further, define X as

$$X = \frac{Y_1/n_1}{Y_2/n_2} \tag{1.86}$$

Then X is an F random variable with n_1 and n_2 degrees of freedom and probability density function

$$f(x) = \frac{\Gamma\left(\dfrac{n_1 + n_2}{2}\right)}{\Gamma\left(\dfrac{n_1}{2}\right)\Gamma\left(\dfrac{n_2}{2}\right)} \left(\frac{n_1}{n_2}\right)^{n_1/2} \left(1 + \frac{n_1}{n_2}x\right)^{-(n_1+n_2)/2}, \qquad 0 < x < \infty \quad (1.87)$$

(e) *Exponential* As we have already mentioned, the exponential random variable is directly related to the Poisson random variable. That is, if the number of events occurring in a fixed time interval is Poisson distributed, then the time between the occurrence of successive events is exponentially distributed. We have also indicated that the exponential random variable is without memory. That is, if X is an exponential random variable, then

$$P(X > t_1 + t_2 \mid X > t_1) = P(X > t_2) \qquad (1.88)$$

We will derive the probability density function of the exponential random variable by noting its relationship to the Poisson. If Y is a Poisson random variable with parameter λ, then the probability of y events in time x is given by

$$p(y) = \frac{(\lambda x)^y}{y!} \exp[-\lambda x], \qquad y = 0, 1, 2, \ldots$$

Then

$$p(0) = \exp[-\lambda x]$$

But the probability of 0 events in time x is equal to the probability that the time until the first event is greater than x. Then

$$p(0) = 1 - F(x) \quad \text{or} \quad F(x) = 1 - \exp[-\lambda x] \qquad (1.89)$$

Since

$$f(x) = \frac{dF(x)}{dx}$$

we have for the probability density function of X

$$f(x) = \lambda \exp[-\lambda x], \qquad 0 < x < \infty \qquad (1.90)$$

(f) *Gamma* Let $Y_1, Y_2, \ldots, Y_n$ be a set of independent, identically distributed random variables each with parameter λ. Then

$$X = \sum_{i=1}^{n} Y_i \qquad (1.91)$$

has a gamma distribution with probability density function given by

$$f(x) = \frac{\lambda^n}{(n-1)!} x^{n-1} \exp[-\lambda x], \qquad 0 < x < \infty \qquad (1.92)$$

where $\lambda > 0$ and n is a positive integer.

1.3 PROBABILITY DISTRIBUTIONS

In the expression in Eq. (1.92) the parameter n is restricted to positive integer values. However, the gamma random variable may also be defined for noninteger n. When $n > 0$ is not an integer the probability density function of the gamma random variable, X, is defined by

$$f(x) = \frac{\lambda^n}{\Gamma(n)} x^{n-1} \exp[-\lambda x], \qquad 0 < x < \infty \tag{1.93}$$

In Eq. (1.93), if we let $\lambda = \frac{1}{2}$ and $n = m/2$, then

$$f(x) = \frac{1}{2^{m/2}\Gamma\left(\frac{m}{2}\right)} x^{(m/2)-1} \exp[-x/2], \qquad 0 < x < \infty \tag{1.94}$$

which is the density function of a χ^2 random variable with m degrees of freedom.

(g) *Beta* The beta random variable may be expressed as a function of independent gamma random variables. Let Y_1 and Y_2 be independent gamma random variables with probability density functions given by

$$f(y_1) = \frac{1}{\Gamma(a)} y_1^{a-1} \exp[-y_1], \qquad 0 < y_1 < \infty \tag{1.95}$$

and

$$g(y_2) = \frac{1}{\Gamma(b)} y_2^{b-1} \exp[-y_2], \qquad 0 < y_2 < \infty \tag{1.96}$$

Then

$$X = \frac{Y_1}{Y_1 + Y_2} \tag{1.97}$$

is a beta random variable with probability density function given by

$$f(x) = \frac{\Gamma(a+b)}{\Gamma(a)\Gamma(b)} x^{a-1}(1-x)^{b-1}, \qquad 0 < x < 1 \tag{1.98}$$

where $a > 0$ and $b > 0$.

(h) *Weibull* The probability density function of the Weibull random variable is defined as

$$f(x) = \frac{a}{b-c}\left(\frac{x-c}{b-c}\right)^{a-1} \exp\left[-\left(\frac{x-c}{b-c}\right)^a\right], \qquad c < x < \infty \tag{1.99}$$

where $a \geq 1$ and $b > c$. The Weibull density reduces to the exponential by letting $c = 0$ and $a = 1$.

(i) *Uniform* Let X be a continuous random variable with values defined between a and b, $a < b$. Now suppose that the probability that the value of X falls in the interval α to β, $\alpha < \beta$, is proportional to the width of the interval, $\beta - \alpha$, where $\alpha \geq a$ and $\beta \leq b$. Then

$$P(\alpha < x < \beta) = k(\beta - \alpha) = F(\beta) - F(\alpha) \tag{1.100}$$

Letting $\alpha = a$ and $\beta = x$

$$F(x) = k(x - a)$$

Since $F(b) = 1$, $k = 1/(b - a)$. Hence

$$F(x) = \frac{x - a}{b - a}$$

and from Eq. (1.46),

$$f(x) = \frac{1}{b - a}, \quad a < x < b, \quad \text{where} \quad a < b \tag{1.101}$$

Example 1.17 Show that the exponential random variable is without memory. From Eq. (1.88), if the random variable X is without memory, then

$$P(X > t_1 + t_2 | X > t_1) = P(X > t_2)$$

Now

$$P(X > t_1 + t_2 | X > t_1) = \frac{1 - F(t_1 + t_2)}{1 - F(t_1)}$$

If X is exponentially distributed with parameter λ, we have

$$1 - F(x) = \exp[-\lambda x]$$

Hence

$$P(X > t_1 + t_2 | X > t_1) = \frac{\exp[-\lambda(t_1 + t_2)]}{\exp[-\lambda t_1]} = \exp[-\lambda t_2]$$

$$= 1 - F(t_2) = P(X > t_2) \quad \blacksquare$$

Example 1.18 Find the distribution function of the Weibull random variable. The density function of the Weibull random variable is given by

$$f(x) = \frac{a}{b - c}\left(\frac{x - c}{b - c}\right)^{a-1} \exp\left[-\left(\frac{x - c}{b - c}\right)^a\right], \quad c < x < \infty$$

Now

$$F(x) = \int_{-\infty}^{x} f(y)\,dy = \int_{-\infty}^{c} f(y)\,dy + \int_{c}^{x} f(y)\,dy$$

$$= 0 + \int_{c}^{x} \frac{a}{b - c}\left(\frac{y - c}{b - c}\right)^{a-1} \exp\left[-\left(\frac{y - c}{b - c}\right)^a\right] dy$$

Let

$$z = \left(\frac{y - c}{b - c}\right)^a$$

Then
$$y = (b-c)z^{1/a} + c$$
$$dy = \frac{(b-c)}{a} z^{-(a-1)/a} dz, \quad \left(\frac{y-c}{b-c}\right)^{a-1} = z^{(a-1)/a}$$

and we are led to
$$F(x) = \int_0^{(x-c/b-c)^a} \exp[-z] \, dz = 1 - \exp\left[-\left(\frac{x-c}{b-c}\right)^a\right], \quad c < x < \infty$$

Example 1.19 The number of events, N, occurring in a time interval of length x has the probability mass function $p(n, x)$. The density function of the time elapsed, X, until n events occur is given by $f(x, n)$. Obtain an expression for $F(x, n)$ in terms of $P(n, x)$, where $F(x, n)$ is the distribution function of X and $P(n, x)$ is the distribution function of N.

From the definition of $F(x, n)$
$$F(x, n) = P(\text{time until occurrence of } n\text{th event} < x)$$

Now
$$P(n, x) = P(n + 1\text{st event occurs after } x) = 1 - F(x, n+1)$$

Hence
$$F(x, n) = 1 - P(n-1, x) \qquad (1.102)$$

Example 1.20 Using Eq. (1.102), define the distribution function of X, where X is gamma distributed with parameters λ and integer n.

Since n is an integer, X can be represented as the sum of independent exponential random variables each with parameter λ, $Y_1, Y_2, \ldots, Y_n$. Now consider Y_i to be the time between the occurrence of the $(i-1)$st and ith events. Since Y_i is exponentially distributed, the number of events occurring in a time period X is Poisson distributed with parameter λ. Therefore
$$p(n-1, x) = \frac{(\lambda x)^{n-1}}{(n-1)!} \exp[-\lambda x], \quad n = 1, 2, \ldots$$

and
$$F(x, n) = 1 - \sum_{k=1}^{n} \frac{(\lambda x)^{k-1}}{(k-1)!} \exp[-\lambda x]$$
$$= 1 - \sum_{k=0}^{n-1} \frac{(\lambda x)^k}{k!} \exp[-\lambda x], \quad 0 < x < \infty \qquad (1.103)$$

Example 1.21 A lot contains 10,000 units of product of which 100 are defective. A sample of 5 units are selected at random from the lot and inspected. Find the probability of finding 0, 1, 2, or 3 defective units in the sample using the hypergeometric, binomial, and Poisson probability mass functions.

Using Eq. (1.77) for the hypergeometric case, we have
$$N = 10{,}000, \quad M = 100, \quad n = 5$$

and

$$p(x) = \frac{\binom{100}{x}\binom{9900}{5-x}}{\binom{10{,}000}{5}}, \quad x = 0, 1, \ldots, 5$$

For the binomial case

$$p = \frac{100}{10{,}000} = 0.01$$

$$p(x) = \binom{5}{x}(0.01)^x(0.99)^{5-x}, \quad x = 0, 1, \ldots, 5$$

and for the Poisson case

$$\lambda t = (0.01)n = 0.05, \quad p(x) = \frac{(0.05)^x}{x!}\exp[-0.05], \quad x = 0, 1, 2, \ldots$$

The calculation of $p(x)$ for $x = 0, 1, 2, 3$ for the three distributions given is summarized in Table 1.6

TABLE 1.6
$p(x)$, $x = 0, 1, 2, 3$, FOR THE HYPERGEOMETRIC, BINOMIAL, AND POISSON DISTRIBUTIONS

x	Hypergeometric, $p(x)$	Binomial, $p(x)$	Poisson, $p(x)$
0	0.9510	0.9510	0.9512
1	0.0480	0.0480	0.0476
2	0.0010	0.0010	0.0011
3	0.0000	0.0000	0.0000

1.4 Joint Distributions

In our treatment of probability distributions we considered the distribution function and the probability density function or probability mass function of a single random variable. More often than not, physical systems include several random variables, of which all or part may be interdependent. As an illustration, suppose that manufactured lots are submitted for inspection. From each lot a sample of n units are selected at random, inspected, and the number of defective units in the sample, X, recorded. If the proportion of defective units per lot, P, is a random variable, then the inspection system includes two random variables, X and P, where X is certainly dependent upon the value of P for any given lot. Queueing systems also include several random variables such as the time between arrivals to the system, service time for each unit served, the number of units

1.4 JOINT DISTRIBUTIONS

entering the system at an arrival, and sometimes the number of service channels in operation. In operating an inventory system one must consider such random variables as the time between successive demands, the number of units of product requested per demand, and order lead time. In the analysis of vehicular traffic flow the analyst must take account of random variables such as vehicular velocity, the separation between vehicles, acceptable gaps in merging situations, and turning movements at intersections. An analysis of the operations in the vicinity of a metropolitan air terminal must consider aircraft velocity, the number of each class of aircraft in the terminal area, the separation between successive aircraft approaching the runway, and the time an aircraft spends on the runway.

We will denote the *joint* density, probability mass, and distribution functions of the random variables X and Y by $f(x, y)$, $p(x, y)$, and $F(x, y)$ respectively.

Definition 1.9 If $F(x, y)$, $f(x, y)$, and $p(x, y)$ are the distribution function, probability density function, and probability mass function respectively of the random variables X and Y, then

$$F(x, y) = P(X \le x \text{ and } Y \le y) \tag{1.104}$$

and

$$F(x, y) = \begin{cases} \int_{-\infty}^{x} \int_{-\infty}^{y} f(u, v) \, dv \, du, & \text{continuous } X \text{ and } Y \\ \sum_{u=-\infty}^{x} \sum_{v=-\infty}^{y} p(u, v) & \text{discrete } X \text{ and } Y \end{cases} \tag{1.105}$$

Theorem 1.8 If X and Y are independent random variables then

$$F(x, y) = F(x)F(y) \tag{1.106}$$

Proof From Theorem 1.4 if the events E_1 and E_2 are independent, then

$$P(E_1 E_2) = P(E_1)P(E_2)$$

If E_1 is the event that $X \le x$ and E_2 is the event that $Y \le y$, we have

$$P(E_1) = F(x), \quad P(E_2) = F(y), \quad \text{and} \quad P(E_1 E_2) = F(x, y)$$

Since

$$P(E_1 E_2) = P(E_1)P(E_2)$$

the desired result is obtained, and

$$F(x, y) = F(x)F(y) \qquad \blacksquare$$

From Eq. (1.105) it is easily seen that

$$f(x, y) = \frac{\partial^2}{\partial x \, \partial y} F(x, y) \tag{1.107}$$

Hence, if X and Y are independent, then

$$f(x, y) = f(x)f(y) \tag{1.108}$$

A similar result can be shown for discrete random variables. That is, if X and Y are discrete, independent random variables, then

$$p(x, y) = p(x)p(y) \tag{1.109}$$

Definition 1.10 Let $f(x, y)$, $p(x, y)$, and $F(x, y)$ be the joint probability density function, the joint probability mass function, and the joint distribution function of the random variables X and Y respectively. The *marginal density function*, the *marginal probability mass function*, and the *marginal distribution function* of X are given by

$$f(x) = \int_{-\infty}^{\infty} f(x, y) \, dy \tag{1.110}$$

$$p(x) = \sum_{y=-\infty}^{\infty} p(x, y) \tag{1.111}$$

$$F(x) = \begin{cases} \int_{-\infty}^{x} \int_{\infty}^{\infty} f(u, y) \, dy \, du, & \text{continuous } X \text{ and } Y \\ \sum_{u=-\infty}^{x} \sum_{y=-\infty}^{\infty} p(u, v) & \text{discrete } X \text{ and } Y \end{cases} \tag{1.112}$$

The expressions given in Eqs. (1.106), (1.108), and (1.109) are frequently used as tests for the *independence* of two random variables. That is, if $F(x, y), f(x, y)$, and $p(x, y)$ can be factored into the product of the marginal distribution, marginal density, or marginal probability mass functions, the random variables X and Y are independent.

Example 1.22 The joint density function of the random variables X and Y is given by

$$f(x, y) = \lambda u^2 y \exp[-\lambda x - uy], \quad 0 < x < \infty, \quad 0 < y < \infty$$

Find the marginal density functions of X and Y and determine whether X and Y are independent or dependent random variables.

From Eq. (1.110),

$$f(x) = \int_{-\infty}^{\infty} f(x, y) \, dy = \int_{0}^{\infty} \lambda u^2 y \exp[-\lambda x - uy] \, dy$$

$$= \lambda \exp[-\lambda x] \int_{0}^{\infty} u^2 y \exp[-uy] \, dy = \lambda \exp[-\lambda x], \quad 0 < x < \infty$$

The marginal density function of Y is

$$g(y) = \int_{0}^{\infty} \lambda u^2 y \exp[-\lambda x - uy] \, dx = u^2 y \exp[-uy] \int_{0}^{\infty} \lambda \exp[-\lambda x] \, dx$$

$$= u^2 y \exp[-uy], \quad 0 < y < \infty$$

Since $f(x, y) = f(x)g(y)$, X and Y are independent random variables.

1.4 JOINT DISTRIBUTIONS

Example 1.23 X and Y are jointly distributed normal random variables with joint density function given by

$$f(x, y) = \frac{1}{2\pi\sigma_1\sigma_2\sqrt{1-\rho^2}} \exp\left[-\frac{1}{2(1-\rho^2)}\left[\frac{x^2}{\sigma_1^2} - \frac{2\rho xy}{\sigma_1\sigma_2} + \frac{y^2}{\sigma_2^2}\right]\right],$$

$$-\infty < x, \ y < \infty$$

Determine the condition under which X and Y are independent random variables.

To determine whether X and Y are independent we must obtain the marginal density functions of X and Y.

$$f(x) = \frac{1}{\sigma_1\sqrt{2\pi}} \exp\left[-\frac{x^2}{2(1-\rho^2)\sigma_1^2}\right] \int_{-\infty}^{\infty} \frac{1}{\sqrt{2\pi}\sigma_2\sqrt{1-\rho^2}} \exp\left[-\frac{1}{2(1-\rho^2)}\left[\frac{y^2}{\sigma_2^2} - \frac{2\rho xy}{\sigma_1\sigma_2}\right]\right] dy$$

We may express the exponent inside the integral as

$$\frac{1}{2(1-\rho^2)}\left[\frac{y^2}{\sigma_2^2} - \frac{2\rho xy}{\sigma_1\sigma_2}\right] = \frac{1}{2(1-\rho^2)}\left[\frac{y}{\sigma_2} - \frac{\rho x}{\sigma_1}\right]^2 - \frac{\rho^2 x^2}{2(1-\rho^2)\sigma_1^2}$$

$$= \frac{1}{2(1-\rho^2)\sigma_2^2}\left[y - \frac{\sigma_2 \rho x}{\sigma_1}\right]^2 - \frac{\rho^2 x^2}{2(1-\rho^2)\sigma_1^2}$$

and

$$f(x) = \frac{1}{\sigma_1\sqrt{2\pi}} \exp\left[-\frac{x^2}{2\sigma_1^2}\right] \int_{-\infty}^{\infty} \frac{1}{\sqrt{2\pi}\sigma_2\sqrt{1-\rho^2}} \exp\left[-\frac{1}{2(1-\rho^2)\sigma_2^2}\left[y - \frac{\sigma_2 \rho x}{\sigma_1}\right]^2\right] dy$$

Now let

$$z = \frac{1}{\sqrt{(1-\rho^2)\sigma_2^2}}\left[y - \frac{\sigma_2 \rho x}{\sigma_1}\right]$$

Then

$$y = \sigma_2\sqrt{(1-\rho^2)}z + \frac{\sigma_2 \rho x}{\sigma_1} \quad \text{and} \quad dy = \sigma_2\sqrt{(1-\rho^2)}\,dz$$

Therefore

$$f(x) = \frac{1}{\sigma_1\sqrt{2\pi}} \exp\left[-\frac{x^2}{2\sigma_1^2}\right] \int_{-\infty}^{\infty} \frac{1}{\sqrt{2\pi}} \exp\left[-\frac{z^2}{2}\right] dz$$

$$= \frac{1}{\sigma_1\sqrt{2\pi}} \exp\left[-\frac{x^2}{2\sigma_1^2}\right], \quad -\infty < x < \infty$$

since

$$\int_{-\infty}^{\infty} \frac{1}{\sqrt{2\pi}} \exp\left[-\frac{z^2}{2}\right] dz = 1$$

By a similar argument, we may show that the marginal density function of y is given by

$$g(y) = \frac{1}{\sigma_2 \sqrt{2\pi}} \exp\left[-\frac{y^2}{2\sigma_2^2}\right], \qquad -\infty < y < \infty$$

Since $f(x, y) \neq f(x)g(y)$, X and Y are not independent random variables. However, if we let $\rho = 0$, we have

$$f(x, y) = \frac{1}{2\pi\sigma_1 \sigma_2} \exp\left[-\frac{1}{2}\left[\frac{x^2}{\sigma_1^2} + \frac{y^2}{\sigma_2^2}\right]\right], \qquad -\infty < x, \; y < \infty$$

and, in this case, $f(x, y) = f(x)g(y)$.

In summary then, jointly distributed normal random variables are independent if $\rho = 0$ and dependent if $\rho \neq 0$. The constant ρ is called the *coefficient of correlation* and is used as a measure of the association or dependence of X and Y.

Definition 1.11 The *conditional distribution function* of the random variable X for a given value y of the random variable Y, $F(x \mid y)$ is given by

$$F(x \mid y) = P(X \leq x \mid y) \tag{1.113}$$

and the joint distribution function of X and Y is

$$F(x, y) = \begin{cases} \displaystyle\int_{-\infty}^{y} F(x \mid u) f(u) \, du, & \text{continuous } X \text{ and } Y \\[1em] \displaystyle\sum_{u=-\infty}^{y} F(x \mid u) p(u), & \text{discrete } X \text{ and } Y \end{cases} \tag{1.114}$$

From Eq. (1.114) we are able to obtain the following important relationships for the conditional probability density function, $f(x \mid y)$.

Theorem 1.9 If $f(x \mid y)$ is the *conditional density function* of the random variable X given a value, y, of the random variable Y, then

$$f(x \mid y) = \frac{g(x, y)}{h(y)} \tag{1.115}$$

where $g(x, y)$ is the joint density function of X and Y, and $h(y)$ is the marginal density function of Y.

Proof From Eq. (1.114), we have

$$F(x, y) = \int_{-\infty}^{y} F(x \mid u) h(u) \, du$$

and from Eq. (1.107),

$$g(x, y) = \frac{\partial^2}{\partial x \, \partial y} F(x, y)$$

1.4 JOINT DISTRIBUTIONS

Now

$$\frac{\partial}{\partial y} F(x, y) = F(x \mid y)h(y) = \int_{-\infty}^{x} f(v \mid y)h(y)\, dv$$

and

$$\frac{\partial}{\partial x\, \partial y} F(x, y) = f(x \mid y)h(y)$$

Thus

$$g(x, y) = f(x \mid y)h(y) \quad \text{and} \quad f(x \mid y) = \frac{g(x, y)}{h(y)}$$

The reader will notice the similarity between the relationships given by Eqs. (1.115) and (1.24). It is not surprising, then, that a similar relationship holds for *conditional probability mass functions*. That is, if $p(x \mid y)$ is the conditional probability mass function of X given a value, y, of the random variable Y, then

$$p(x \mid y) = \frac{g(x, y)}{s(y)} \quad (1.116)$$

where $g(x, y)$ is the joint probability mass function of X and Y and $s(y)$ is the marginal probability mass function of Y.

Example 1.24 The conditional density function of X given a value, y, of Y is given by

$$f(x \mid y) = y \exp[-yx], \quad 0 < x < \infty$$

The marginal density function of Y is

$$g(y) = \frac{\lambda^n}{\Gamma(n)} y^{n-1} \exp[-\lambda y], \quad 0 < y < \infty$$

a. Find the joint density function of X and Y, $h(x, y)$.
b. Find the marginal density function of X, $s(x)$.
c. Find the conditional density function of Y for a given value, x, of X, $t(y \mid x)$.
d. Determine whether or not X and Y are independent random variables.

From Eq. (1.115) we have for part a

$$h(x, y) = f(x \mid y)g(y) = \frac{\lambda^n}{\Gamma(n)} y^n \exp[-y(\lambda + x)], \quad 0 < x < \infty, \quad 0 < y < \infty$$

In part b, we may obtain $s(x)$ from $h(x, y)$. That is

$$s(x) = \int_{-\infty}^{\infty} h(x, y)\, dy$$

$$= \int_{0}^{\infty} \frac{\lambda^n}{\Gamma(n)} y^n \exp[-y(\lambda + x)]\, dy = \frac{\lambda^n}{\Gamma(n)} \int_{0}^{\infty} y^n \exp[-y(\lambda + x)]\, dy$$

Now

$$\int_0^\infty v^n e^{-\alpha v} \, dv = \frac{\Gamma(n+1)}{\alpha^{n+1}} \tag{1.117}$$

Therefore

$$s(x) = \frac{\lambda^n}{\Gamma(n)} \frac{\Gamma(n+1)}{(\lambda + x)^{n+1}}$$

$$= \frac{n\lambda^n}{(\lambda + x)^{n+1}}, \quad 0 < x < \infty, \quad \text{since} \quad \Gamma(n+1) = n\Gamma(n)$$

In part c we may find $t(y \mid x)$ from

$$t(y \mid x) = \frac{h(x, y)}{s(x)} = \frac{\dfrac{\lambda^n}{\Gamma(n)} y^n (\lambda + x)^{n+1}}{n\lambda^n} \exp[-y(\lambda + x)]$$

$$= \frac{y^n}{\Gamma(n+1)} (\lambda + x)^{n+1} \exp[-y(\lambda + x)], \quad 0 < y < \infty$$

Since

$$s(x)g(y) = \frac{n\lambda^{2n}}{\Gamma(n)} \frac{y^{n-1}}{(\lambda + x)^{n+1}} \exp[-\lambda y]$$

we have $h(x, y) \neq s(x)g(y)$ and the random variables X and Y are not independent. ◻

Example 1.25 Manufactured lots of product of size L are received for inspection. Let X be a random variable representing the dimension of a randomly selected unit. The mean value of the dimension for a given lot is U. However, the mean is known to vary in a random manner from lot to lot. Therefore U is a random variable. In order to estimate the value of U for a given lot, a sample of n units is selected at random from the lot and inspected. Let X_i be the dimension of the ith unit inspected. The estimated value of U is then the sample mean, $\bar{X}$, where

$$\bar{X} = \frac{1}{n} \sum_{i=1}^n X_i$$

The density function of $\bar{X}$ for a given value of U, u, can be represented as

$$f(\bar{x} \mid u) = \frac{\sqrt{n}}{\sigma \sqrt{2\pi}} \exp\left[-\frac{n(\bar{x} - u)^2}{2\sigma^2}\right], \quad -\infty < \bar{x} < \infty$$

and the marginal density function of U is

$$g(u) = \frac{\sqrt{L}}{\sigma \sqrt{2\pi}} \exp\left[-\frac{L(u - u_0)^2}{2\sigma^2}\right], \quad -\infty < u < \infty$$

1.4 JOINT DISTRIBUTIONS

Find

a. the joint density function of $\bar{X}$ and U, $h(\bar{x}, u)$
b. the marginal density function of $\bar{X}$, $s(\bar{x})$
c. the conditional density function of U for a given value of $\bar{X}$, $\bar{x}$, $t(u \mid \bar{x})$.

The joint density function of $\bar{X}$ and U is given by $f(\bar{x} \mid u)g(u)$. Hence

$$h(\bar{x}, u) = \frac{\sqrt{nL}}{2\pi\sigma^2} \exp\left[-\frac{n(\bar{x} - u)^2 + L(u - u_0)^2}{2\sigma^2}\right], \qquad -\infty < \bar{x}, \ u < \infty$$

From Eq. (1.110), the marginal density function of $\bar{X}$ is

$$s(\bar{x}) = \int_{-\infty}^{\infty} \frac{\sqrt{nL}}{2\pi\sigma^2} \exp\left[-\frac{n(\bar{x} - u)^2 + L(u - u_0)^2}{2\sigma^2}\right] du, \qquad -\infty < \bar{x} < \infty$$

To evaluate the last integral, let

$$n(\bar{x} - u)^2 + L(u - u_0)^2 = n(\bar{x}^2 - 2\bar{x}u + u^2) + L(u^2 - 2uu_0 + u_0^2)$$
$$= n\bar{x}^2 + Lu_0^2 + (n + L)u^2 - 2(n\bar{x} + Lu_0)u$$

Completing the square in u yields

$$n(\bar{x} - u)^2 + L(u - u_0)^2 = n\bar{x}^2 + Lu_0^2 + (n + L)\left(u - \frac{n\bar{x} + Lu_0}{n + L}\right)^2$$
$$- \frac{(n\bar{x} + Lu_0)^2}{n + L}$$

$$= (n + L)\left(u - \frac{n\bar{x} + Lu_0}{n + L}\right)^2 + \frac{nL(\bar{x} - u_0)^2}{n + L}$$

and

$$s(\bar{x}) = \frac{\sqrt{nL}}{2\pi\sigma^2} \exp\left[-\frac{nL(\bar{x} - u_0)^2}{2(n + L)\sigma^2}\right] \int_{-\infty}^{\infty} \exp\left[-\frac{(n + L)\left(u - \frac{n\bar{x} + Lu_0}{n + L}\right)^2}{2\sigma^2}\right] du$$

Now multiply the right side of this equation by $\sqrt{n + L}/\sqrt{n + L}$ with the result

$$s(\bar{x}) = \frac{\sqrt{\frac{nL}{n + L}}}{\sigma\sqrt{2\pi}} \exp\left[-\frac{nL}{n + L} \frac{(\bar{x} - u_0)^2}{2\sigma^2}\right] \int_{-\infty}^{\infty} \frac{\sqrt{n + L}}{\sigma\sqrt{2\pi}} \exp\left[-\frac{\left(u - \frac{n\bar{x} + Lu_0}{n + L}\right)^2}{2\sigma^2/(n + L)}\right] du$$

The function inside the integral is recognized as the normal density function with mean $(n\bar{x} + Lu_0)/(n + L)$ and standard deviation $\sigma/\sqrt{n + L}$. Hence the integral in the above expression is unity, and

$$s(\bar{x}) = \frac{\sqrt{\frac{nL}{n + L}}}{\sigma\sqrt{2\pi}} \exp\left[-\frac{(\bar{x} - u_0)^2}{2\sigma^2 / \left(\frac{nL}{n + L}\right)}\right], \qquad -\infty < \bar{x} < \infty$$

Thus the marginal distribution of $\bar{X}$ is normal with mean u_0 and standard deviation $\sigma/\sqrt{nL/(n+L)}$.

For the conditional density function of U for a given value of $\bar{X}, \bar{x}$, we have

$$t(u|\bar{x}) = \frac{h(\bar{x}, u)}{s(\bar{x})} = \frac{\dfrac{\sqrt{nL}}{2\pi\sigma^2} \exp\left[-\dfrac{n(\bar{x}-u)^2 + L(u-u_0)^2}{2\sigma^2}\right]}{\dfrac{\sqrt{\dfrac{nL}{n+L}}}{\sigma\sqrt{2\pi}} \exp\left[-\dfrac{\left(\dfrac{nL}{n+L}\right)(\bar{x}-u_0)^2}{2\sigma^2}\right]}$$

$$= \frac{1}{\sigma\sqrt{2\pi(n+L)}} \exp\left[-\frac{n(\bar{x}-u)^2 + L(u-u_0)^2 - \left(\dfrac{nL}{n+L}\right)(\bar{x}-u_0)^2}{2\sigma^2}\right],$$

$$-\infty < u < \infty$$

We simplify the term in the exponent as follows. Let

$$\alpha = n(\bar{x}-u)^2 + L(u-u_0)^2 - \left(\frac{nL}{n+L}\right)(\bar{x}-u_0)^2$$

Then

$$\alpha = n(\bar{x}^2 - 2\bar{x}u + u^2) + L(u^2 - 2uu_0 + u_0^2) - \left(\frac{nL}{n+L}\right)(\bar{x}^2 - 2\bar{x}u_0 + u_0^2)$$

$$= \frac{[(n+L)u - (n\bar{x} + Lu_0)]^2}{n+L}$$

and

$$t(u|\bar{x}) = \frac{1}{\sigma\sqrt{2\pi(n+L)}} \exp\left[-\frac{[(n+L)u - (n\bar{x} + Lu_0)]^2}{2(n+L)\sigma^2}\right], \quad -\infty < u < \infty$$

with the result that the conditional distribution of U for a given value of $\bar{X}, \bar{x}$, is normal with mean $n\bar{x} + Lu_0$ and standard deviation $\sigma\sqrt{n+L}$.

The results of Example 1.25 are of importance in statistical theory. To summarize these results, if the conditional distribution of X for a given value of Y is normal and if the marginal distribution of Y is normal, then the marginal distribution of X is also normal, as is the conditional distribution of Y for a given value of X.

1.5 Derived Distributions

In an earlier section of this chapter we discussed several random variables which could be represented as functions of other random variables. For example, the χ^2 random variable can be expressed as the sum of the squares of indepen-

1.5 DERIVED DISTRIBUTIONS

dently distributed standard normal random variables. The binomial random variable was defined as the sum of independent Bernoulli random variables. Further, we described the *F* random variable as proportional to the ratio of two χ^2 random variables. These, and other random variables, which were defined as functions of one or more other random variables, are of importance from a statistical point of view.

While the relationships mentioned above are well known, we are frequently faced with the problem of finding the distribution of a function of one or more random variables in practical problems where the desired distribution is not known *a priori*. Let us consider a few examples. In the analysis of waiting line or queueing problems the analyst is often concerned with the time a unit spends waiting for service (Cooper, 1972; Saaty, 1961). Waiting time is a function of the number of units in the system when the unit in question arrives at the queueing system and the time required to service a unit. Therefore, waiting time is a function of two random variables, the number of units in the system and service time per unit. In simulation studies the analyst is faced with the problem of defining methods through which the random variables affecting the system studied can be generated (Schmidt and Taylor, 1970). Such a method is often called a process generator. To develop a process generator we normally relate the random variable to be generated to a uniformly distributed random variable with values lying between 0 and 1. The analysis of maintenance problems usually requires knowledge of the distribution of downtime per period of operation on the equipment to be maintained. Downtime, in this context, is a function of the time between equipment failures and the time required to repair the equipment, each of which is a random variable.

Throughout the remainder of this section we shall discuss methods through which the probability density functions or probability mass function of a random variable, which is a function of other random variables, can be developed. We shall first treat functions of a single random variable and then treat the general case of functions of several random variables. It should be mentioned that treatment of the latter case requires an elementary knowledge of determinants. Therefore, the reader may wish to refer to Chapter 2 for a discussion of determinants and their calculation.

1.5.1 Functions of a Single Random Variable

Consider the continuous random variable X and the relationship

$$Y = a + bX \tag{1.118}$$

Since X is a random variable, then so is Y. Now suppose that we wish to find the density function of Y if X has the density function given by $f(x)$, where $f(x) > 0$ for $\alpha < x < \beta$ and $f(x) = 0$ otherwise. If $b > 0$, then Y may assume values between $a + \alpha b$ and $a + \beta b$. Hence

$$P(X \le x) = P(Y \le a + bx)$$

or

$$P(Y \le y) = P\left(X \le \frac{y-a}{b}\right) \tag{1.119}$$

If $F(x)$ and $G(y)$ are the distribution functions of X and Y respectively, we have

$$G(y) = F\left(\frac{y-a}{b}\right) \tag{1.120}$$

Since the density function of Y, $g(y)$, is given by $g(y) = dG(y)/dy$ we have the result that

$$\begin{aligned} g(y) &= \frac{d}{dy} F\left(\frac{y-a}{b}\right) \\ &= \frac{d}{dy} \int_{\alpha}^{y-a/b} f(x)\, dx = \frac{1}{b} f\left(\frac{y-a}{b}\right), \quad a + \alpha b < y < a + \beta b \end{aligned} \tag{1.121}$$

The reader should note that if

$$H(x) = \int_{\phi(x)}^{\psi(x)} h(x, y)\, dy \tag{1.122}$$

then

$$\begin{aligned} \frac{d}{dx} H(x) = \int_{\phi(x)}^{\psi(x)} \frac{\partial}{\partial x} h(x, y)\, dy &+ \left[\frac{d}{dx} \psi(x)\right] h[x, \psi(x)] \\ &- \left[\frac{d}{dx} \phi(x)\right] h[x, \phi(x)] \end{aligned} \tag{1.123}$$

The transformation given in Eq. (1.118) is called a *one-to-one transformation*. To generalize, let

$$Y = \phi(X) \tag{1.124}$$

Since Y is a function of X, if we solve Eq. (1.124) for X, we find that X is a function of Y, or

$$X = \psi(Y) \tag{1.125}$$

The transformations $\phi(X)$ and $\psi(Y)$ are said to be one-to-one if for any value of X, $\phi(X)$ yields one and only one value of Y, and if for any value of Y, $\psi(Y)$ yields one and only one value of X.

To illustrate a transformation which is not one-to-one, let

$$Y = \phi(X) = X^2$$

1.5 DERIVED DISTRIBUTIONS

Solving for Y in terms of X yields

$$X = \psi(Y) = \sqrt{Y}$$

For each value of X, there is only one and only one value of Y. However, for each value of Y, y, there are two values of X, namely $-\sqrt{y}$ and $+\sqrt{y}$.

Theorem 1.10 Let X and Y be continuous random variables defined by the transformations

$$Y = \phi(X) \tag{1.126}$$

and

$$X = \psi(Y) \tag{1.127}$$

If these transformations are either increasing or decreasing functions of X and Y and are one-to-one and if $f(x)$ is the probability density function of X where $f(x) > 0$ for $\alpha < x < \beta$ and $f(x) = 0$ otherwise, then the probability density function of Y, $g(y)$, is given by

$$g(y) = \left|\frac{d}{dy}\psi(y)\right| f[\psi(y)], \qquad \gamma_1 < y < \gamma_2 \tag{1.128}$$

where

$$\gamma_1 = \min[\phi(\alpha), \phi(\beta)] \tag{1.129}$$

$$\gamma_2 = \max[\phi(\alpha), \phi(\beta)] \tag{1.130}$$

Proof First assume

$$\gamma_1 = \phi(\alpha), \qquad \gamma_2 = \phi(\beta)$$

In this case $\phi(X)$ is an increasing function of X, and

$$G(y) = P(Y \le y) = P[\phi(X) \le \phi(x)] = P[X \le \psi(y)]$$

$$= F[\psi(y)] = \int_{\alpha}^{\psi(y)} f(x)\, dx \tag{1.131}$$

The density function of Y is then given by

$$g(y) = \frac{d}{dy} G(y) = \frac{d}{dy} \int_{\alpha}^{\psi(y)} f(x)\, dx \tag{1.132}$$

or

$$g(y) = \frac{d}{dy}\psi(y) f[\psi(y)], \qquad \phi(\alpha) < y < \phi(\beta) \tag{1.133}$$

Since $\phi(X)$ is an increasing function of X, that is, Y increases with increasing X, X increases with Y, and $\psi(Y)$ is an increasing function of Y. Hence $d\psi(y)/dy > 0$.

Now suppose that $\phi(X)$ is a decreasing function of X. In this case, as X increases, $\phi(X)$ or Y decreases. Thus, the minimum value of Y is $\phi(\beta)$ and the maximum value of Y is $\phi(\alpha)$. Proceeding in a manner similar to that presented where $\phi(X)$ was an increasing function of X, we have

$$P(Y \le y) = P[\phi(X) \le \phi(x)] = P[X \ge \psi(y)] = 1 - F[\psi(y)] \quad (1.134)$$

Hence

$$G(y) = 1 - F[\psi(y)] \quad (1.135)$$

or

$$g(y) = -\frac{d}{dy}\psi(y) f[\psi(y)], \quad \phi(\beta) < y < \phi(\alpha) \quad (1.136)$$

Now $\phi(X)$ is a decreasing function of X, or Y decreases with X. Thus, X decreases with Y and $\psi(Y)$ is a decreasing function of Y and $d\psi(y)/dy < 0$. Therefore, $g(y)$ is positive valued. Equations (1.133) and (1.136) can be combined as

$$g(y) = \left|\frac{d}{dy}\psi(y)\right| f[\psi(y)], \quad \gamma_1 < y < \gamma_2$$

yielding the desired result.

Let us now consider transformations of discrete random variables. If X and Y are integer valued discrete random variables defined by the one-to-one transformations $Y = \phi(X)$ and $X = \psi(Y)$, then the probability mass function of Y, $g(y)$, is given by

$$g(y) = p[\psi(y)], \quad y = \gamma_1, \gamma_1 + 1, \ldots, \gamma_2 \quad (1.137)$$

where γ_1 and γ_2 are as defined in Eqs. 1.129 and 1.130 and where $p(x)$ is the probability mass function of X.

The following examples will illustrate the use of Eqs. 1.128 and 1.137.

Example 1.26 The random variable X is exponentially distributed with parameter λ. Let $Y = 3 + 2X$. Find the density function of Y.

Since X is exponentially distributed

$$f(x) = \lambda \exp[-\lambda x], \quad 0 < x < \infty$$

Now

$$\phi(X) = 3 + 2X \quad \text{and} \quad \psi(Y) = \frac{Y-3}{2}$$

From Eq. (1.128),

$$g(y) = \left|\frac{d}{dy}\psi(y)\right| f[\psi(y)], \quad \text{where} \quad \frac{d}{dy}\psi(y) = \frac{1}{2}$$

1.5 DERIVED DISTRIBUTIONS

Thus
$$g(y) = \frac{\lambda}{2} \exp\left[-\frac{\lambda}{2}(y-3)\right]$$

Since $\phi(X)$ is an increasing function of X
$$\gamma_1 = \phi(\alpha) = 3 \quad \text{and} \quad \gamma_2 = \phi(\beta) = \infty$$

Finally, then,
$$g(y) = \frac{\lambda}{2} \exp\left[-\frac{1}{2}(y-3)\right], \quad 3 < y < \infty$$

Example 1.27 X is a binomial random variable with probability mass function
$$p(x) = \binom{n}{x} p^x (1-p)^{n-x}, \quad x = 0, 1, 2, \ldots, n$$

Find the probability mass function of Y, $g(y)$, where
$$Y = -4X$$

From Eq. (1.137)
$$q(Y) = p[\psi(Y)], \quad \text{where} \quad \psi(Y) = -\frac{Y}{4} = X$$

Hence
$$q(y) = \binom{n}{-y/4} p^{-y/4}(1-p)^{(4n+y)/4}$$

Since $\phi(X)$ and $\psi(Y)$ are decreasing functions of X and Y
$$\gamma_1 = -4n, \quad \gamma_2 = 0$$

Therefore, $q(y)$ is defined by
$$q(y) = \binom{n}{-y/4} p^{-y/4}(1-p)^{(4n+y)/4}, \quad y = -4n, -4n+1, \ldots, 0$$

Example 1.28 X is a continuous random variable with density function $f(x)$. Let
$$Y = \int_{-\infty}^{X} f(v)\, dv$$

Find the density function of Y.

For the density function specified, let
$$X = \psi(Y)$$

Then
$$\frac{dy}{dx} = f(x) \quad \text{and} \quad \frac{dx}{dy} = \frac{1}{f(x)}$$

From Eq. (1.128)

$$g(y) = \left|\frac{d}{dy}\psi(y)\right| f[\psi(y)] = \left|\frac{dx}{dy}\right| f(x) = 1$$

Since Y defines the distribution function of X, it is obvious that $0 < y < 1$ and

$$g(y) = 1, \quad 0 < y < 1 \quad \blacksquare \quad (1.138)$$

The result of Example 1.28 has important applications in Monte Carlo simulation. If X is a random variable and Y is the corresponding value of the distribution function of X, then Y is a random variable that is uniformly distributed on the interval 0 to 1. By defining X as a function of Y we are able to relate X to a uniformly distributed random variable. The relationship between X and the uniform random variable Y is called a process generator. The development of a process generator is illustrated in the following example.

Example 1.29 The random variable X has a Weibull distribution with density function

$$f(x) = \frac{a}{b-c}\left(\frac{x-c}{b-c}\right)^{a-1} \exp\left[-\left(\frac{x-c}{b-c}\right)^a\right], \quad c < x < \infty$$

where $a \geq 1$ and $b > c$. Define X in terms of Y, where $g(y) = 1, 0 < y < 1$.

We know that the distribution function of X is uniformly distributed on the interval 0 to 1. Now

$$F(x) = 1 - \exp\left[-\left(\frac{x-c}{b-c}\right)^a\right], \quad c < x < \infty$$

from Example 1.18. Since $F(x)$ is uniformly distributed

$$Y = 1 - \exp\left[-\left(\frac{X-c}{b-c}\right)^a\right]$$

Solving for X, we have

$$X = c + (b-c)[-\ln(1-Y)]^{1/a} \quad \blacksquare \quad (1.139)$$

Thus for any value of the uniform random variable Y, we may obtain the corresponding value of the Weibull random variable X. To generate values of the Weibull random variable then, we first generate a value of the uniform random variable, y, on the interval 0 to 1. Substituting this value into Eq. (1.139) yields a value of the Weibull random variable.

There are several algorithmic methods available for generating values of the uniform random variable on the interval 0 to 1. Since these methods are algorithmic in nature, the numbers generated are usually referred to as pseudorandom numbers. However, the generated numbers may be considered random in the sense that they are indistinguishable from truely random numbers given that the observer is not aware of the algorithm used.

We will now consider a specific case of a transformation which is not

1.5 DERIVED DISTRIBUTIONS

one-to-one. Specifically, we are concerned with the transformation where X is both positive and negative valued, where $f(x)$ is the density function of X, and where $F(x)$ is its distribution function. Now

$$P(Y \le y) = P(X^2 \le y) = P(-\sqrt{y} \le X \le \sqrt{y})$$
$$= F(\sqrt{y}) - F(-\sqrt{y}) \qquad (1.140)$$

If $G(y)$ and $g(y)$ are the distribution and density functions of Y respectively, then

$$G(y) = F(\sqrt{y}) - F(-\sqrt{y}) = \int_{-\sqrt{y}}^{\sqrt{y}} f(x)\,dx \qquad (1.141)$$

Using Eq. (1.123) to differentiate $G(y)$ with respect to y yields

$$g(y) = \frac{d}{dy}G(y) = \frac{1}{2\sqrt{y}}f(\sqrt{y}) + \frac{1}{2\sqrt{y}}f(-\sqrt{y})$$
$$= \frac{f(\sqrt{y}) + f(-\sqrt{y})}{2\sqrt{y}} \qquad (1.142)$$

where the limits on y will depend upon the range of permissible values of X for which $f(x) > 0$.

Example 1.30 (χ^2 *Random Variable*) The continuous random variable X is normally distributed with density function

$$f(x) = \frac{1}{\sqrt{2\pi}}\exp\left[-\frac{x^2}{2}\right], \qquad -\infty < x < \infty$$

Find the density function of Y, $g(y)$, where

$$Y = X^2$$

From Eq. (1.142)

$$g(y) = \frac{f(\sqrt{y}) + f(-\sqrt{y})}{2\sqrt{y}} = \frac{1}{\sqrt{2\pi y}}\exp\left[-\frac{y}{2}\right]$$

Now

$$\sqrt{\pi} = \Gamma\left(\frac{1}{2}\right)$$

and

$$g(y) = \frac{y^{-\frac{1}{2}}}{2^{\frac{1}{2}}\Gamma\left(\frac{1}{2}\right)}\exp\left[-\frac{y}{2}\right], \qquad 0 < y < \infty \qquad (1.143)$$

which is the density function of the χ^2 random variable with one degree of freedom. The reader may recall that we characterized the χ^2 random variable with n degrees of freedom as the sum of the squares of n independent standard normal random variables. We have established this assertion where $n = 1$. ▨

Example 1.31 X is a continuous random variable with density function given by $f(x) = \frac{1}{4}$, $-\frac{3}{2} < x < \frac{5}{2}$. Let $Y = X^2$. Find the density function of Y, $g(y)$.

For $-\frac{3}{2} < x < \frac{3}{2}$, the same value of Y arises for $X = x$ and $X = -x$. However, for $\frac{3}{2} \leq x < \frac{5}{2}$, each value of X yields a unique value of Y. On the interval $-\frac{3}{2} < x < \frac{3}{2}$, $0 < y < \frac{9}{4}$ and on the interval $\frac{3}{2} \leq x < \frac{5}{2}$, $\frac{9}{4} \leq y < \frac{25}{4}$. Now

$$g(y) = \frac{f(\sqrt{y}) + f(-\sqrt{y})}{2\sqrt{y}}$$

Now $f(\sqrt{y})$ and $f(-\sqrt{y})$ are positive valued for $0 < y < \frac{9}{4}$, but $f(-\sqrt{y}) = 0$ for $\frac{9}{4} \leq y < \frac{25}{4}$, although $f(\sqrt{y}) > 0$ on this interval. Hence

$$g(y) = \begin{cases} \dfrac{\frac{1}{4} + \frac{1}{4}}{2\sqrt{y}}, & 0 < y < \dfrac{9}{4} \\ \dfrac{\frac{1}{4} + 0}{2\sqrt{y}}, & \dfrac{9}{4} \leq y < \dfrac{25}{4} \end{cases}$$

$$= \begin{cases} \dfrac{1}{4\sqrt{y}}, & 0 < y < \dfrac{9}{4} \\ \dfrac{1}{8\sqrt{y}}, & \dfrac{9}{4} \leq y < \dfrac{25}{4} \end{cases}$$

1.5.2 Functions of Several Random Variables

In this section we will consider one-to-one transformations of several random variables. Let $f(x_1, x_2, \ldots, x_n)$ be the joint probability density function of the random variables $X_1, X_2, \ldots, X_n$. Let

$$\begin{aligned} Y_1 &= \phi_1(X_1, X_2, \ldots, X_n) \\ Y_2 &= \phi_2(X_1, X_2, \ldots, X_n) \\ &\vdots \\ Y_n &= \phi_n(X_1, X_2, \ldots, X_n) \end{aligned} \quad (1.144)$$

and define the inverse transformations

$$\begin{aligned} X_1 &= \psi_1(Y_1, Y_2, \ldots, Y_n) \\ X_2 &= \psi_2(Y_1, Y_2, \ldots, Y_n) \\ &\vdots \\ X_n &= \psi_n(Y_1, Y_2, \ldots, Y_n) \end{aligned} \quad (1.145)$$

As in the preceding section, if $\phi_i(X_1, X_2, \ldots, X_n)$ and $\psi_i(Y_1, Y_2, \ldots, Y_n)$, $i = 1, 2, \ldots, n$, form a set of one-to-one transformations, then for any set of values of $X_1, X_2, \ldots, X_n$, there is one and only one corresponding value of Y_i given by $\phi_i(X_1, X_2, \ldots, X_n)$, $i = 1, 2, \ldots, n$. Similarly, for any set of values of $Y_1, Y_2, \ldots, Y_n$, there is one and only one value of X_i given by $\psi_i(Y_1, Y_2, \ldots, Y_n)$, $i = 1, 2, \ldots, n$. Our objective is to find the joint probability density function of $Y_1, Y_2, \ldots, Y_n$, $g(y_1, y_2, \ldots, y_n)$.

1.5 DERIVED DISTRIBUTIONS

Theorem 1.11 Let $X_1, X_2, \ldots, X_n$ be jointly distributed continuous random variables with joint probability density function given by $f(x_1, x_2, \ldots, x_n)$. Further, let

$$Y_i = \phi_i(X_1, X_2, \ldots, X_n), \quad i = 1, 2, \ldots, n$$
$$X_i = \psi_i(Y_1, Y_2, \ldots, Y_n), \quad i = 1, 2, \ldots, n$$

be one-to-one transformations. Then the joint probability density function of $Y_1, Y_2, \ldots, Y_n, g(y_1, y_2, \ldots, y_n)$, is given by

$$g(y_1, y_2, \ldots, y_n) = |J| f(\psi_1, \psi_2, \ldots, \psi_n) \tag{1.146}$$

where J is the *Jacobian* of the transformation and is defined by the determinant

$$J = \begin{vmatrix} \dfrac{\partial \psi_1}{\partial y_1} & \dfrac{\partial \psi_1}{\partial y_2} & \cdots & \dfrac{\partial \psi_1}{\partial y_n} \\ \dfrac{\partial \psi_2}{\partial y_1} & \dfrac{\partial \psi_2}{\partial y_2} & \cdots & \dfrac{\partial \psi_2}{\partial y_n} \\ \vdots & \vdots & & \vdots \\ \dfrac{\partial \psi_n}{\partial y_1} & \dfrac{\partial \psi_n}{\partial y_2} & \cdots & \dfrac{\partial \psi_n}{\partial y_n} \end{vmatrix} \tag{1.147}$$

Now let us consider the case of discrete random variables $X_1, X_2, \ldots, X_n$ with joint probability mass function $p(x_1, x_2, \ldots, x_n)$. If the transformations given in Eq. (1.144) and (1.145) are one-to-one, then the joint probability mass function of $Y_1, Y_2, \ldots, Y_n, g(y_1, y_2, \ldots, y_n)$, is given by

$$g(y_1, y_2, \ldots, y_n) = p(\psi_1, \psi_2, \ldots, \psi_n) \tag{1.148}$$

For both continuous and discrete random variable transformations, the limits of the random variables obtained from the transformation cannot be specified in a general manner. The problem of finding the required limits can be quite tedious. As a check on the limits defined and the joint density or probability mass function of the new random variables it is often useful to integrate the joint density function or sum the joint probability mass function over the defined limits. If the joint density or probability mass function and the limits of the new random variables are properly specified, then the result of either operation should be unity.

Example 1.32 (*t Random Variable*) Let Z be a standard normal random variable with density function

$$f_1(z) = \frac{1}{\sqrt{2\pi}} \exp\left[-\frac{z^2}{2}\right], \quad -\infty < z < \infty$$

and let X be a χ^2 random variable with n degrees of freedom and density function

$$f_2(x) = \frac{1}{2^{n/2}\Gamma\left(\frac{n}{2}\right)} x^{n/2-1} \exp[-x/2], \quad 0 < x < \infty$$

where Z and X are assumed to be independent. Let

$$Y_1 = \frac{Z}{\sqrt{X/n}}$$

Find the density function of Y_1.

Since Z and X are independent random variables, the joint density function of Z and X, $f(z, x)$, is given by

$$f(z, x) = f_1(z) f_2(x)$$

or

$$f(z, x) = \frac{1}{\sqrt{2\pi}\, 2^{n/2}\Gamma\left(\frac{n}{2}\right)} x^{n/2-1} \exp\left[-\frac{1}{2}(x + z^2)\right],$$

$$-\infty < Z < \infty, \quad 0 < x < \infty$$

As defined above

$$Y_1 = \frac{Z}{\sqrt{X/n}} = \phi_1(Z, X)$$

Further let

$$Y_2 = X = \phi_2(Z, X)$$

To determine the density function of Y_1, we shall first find the joint density function of Y_1 and Y_2, $g(y_1, y_2)$. The density function of Y_1 is then obtained by integrating $g(y_1, y_2)$ over y_2.

From the previous expressions for Y_1 and Y_2 we have

$$Z = Y_1\sqrt{Y_2/n} = \psi_1(Y_1, Y_2) \quad \text{and} \quad X = Y_2 = \psi_2(Y_1, Y_2)$$

The Jacobian of the transformation is given by Eq. 1.47 and is

$$J = \begin{vmatrix} \dfrac{\partial \psi_1}{\partial y_1} & \dfrac{\partial \psi_1}{\partial y_2} \\ \dfrac{\partial \psi_2}{\partial y_1} & \dfrac{\partial \psi_2}{\partial y_2} \end{vmatrix} = \begin{vmatrix} \sqrt{y_2/n} & \dfrac{y_1}{2\sqrt{ny_2}} \\ 0 & 1 \end{vmatrix} = \sqrt{y_2/n}$$

1.5 DERIVED DISTRIBUTIONS

Hence

$$g(y_1, y_2) = \frac{\sqrt{y_2}}{\sqrt{2\pi n}\, 2^{n/2} \Gamma\left(\frac{n}{2}\right)} y_2^{n/2-1} \exp\left[-\frac{1}{2}\left(y_2 + \frac{y_1^2 y_2}{n}\right)\right]$$

$$= \frac{1}{\sqrt{2\pi n}\, 2^{n/2} \Gamma\left(\frac{n}{2}\right)} y_2^{(n-1)/2} \exp\left[-\frac{y_2}{2}\left(1 + \frac{y_1^2}{n}\right)\right]$$

Since $-\infty < z < \infty$, we have $-\infty < y_1 < \infty$, and since $0 < x < \infty$, $0 < y_2 < \infty$. The density function of Y_1 is given by

$$g_1(y_1) = \int_0^\infty g(y_1, y_2)\, dy_2$$

$$= \frac{1}{\sqrt{2\pi n}\, 2^{n/2} \Gamma\left(\frac{n}{2}\right)} \int_0^\infty y_2^{(n-1)/2} \exp\left[-\frac{y_2}{2}\left(1 + \frac{y_1^2}{n}\right)\right] dy_2$$

From Example 1.24,

$$\int_0^\infty v^n \exp[-\alpha v]\, dv = \frac{\Gamma(n+1)}{\alpha^{n+1}}$$

Therefore

$$\int_0^\infty y_2^{(n-1)/2} \exp\left[-\frac{y_2}{2}\left(1 + \frac{y_1^2}{n}\right)\right] dy_2 = \frac{2^{(n+1)/2} \Gamma\left(\frac{n+1}{2}\right)}{\left(1 + \frac{y_1^2}{n}\right)^{(n+1)/2}}$$

and

$$g(y_1) = \frac{\Gamma\left(\frac{n+1}{2}\right)}{\sqrt{\pi n}\, \Gamma\left(\frac{n}{2}\right)} \left(1 + \frac{y_1^2}{n}\right)^{-(n+1)/2}, \qquad -\infty < y_1 < \infty$$

From Eq. (1.85) we see that Y_1 has a t distribution with n degrees of freedom.

Example 1.33 *(Gamma Random Variable)* $X_1, X_2, \ldots, X_n$ are independent exponential random variables, each with parameter λ. Let

$$Y_n = \sum_{i=1}^n X_i$$

Show that Y_n has a gamma distribution with parameters λ and n.

We shall obtain the density function of Y_n by first finding the joint density function of $Y_1, Y_2, \ldots, Y_n$, $g(y_1, y_2, \ldots, y_n)$, where

$$Y_i = \sum_{j=1}^{i} X_j, \quad i = 1, 2, \ldots, n$$

Hence
$$X_1 = Y_1$$
$$X_2 = Y_2 - Y_1$$
$$X_3 = Y_3 - Y_2$$
$$\vdots$$
$$X_n = Y_n - Y_{n-1}$$

The joint density function of $X_1, X_2, \ldots, X_n$ is given by

$$f(x_1, x_2, \ldots, x_n) = \begin{cases} \lambda^n \exp\left[-\lambda \sum_{i=1}^{n} x_i\right], & 0 < x_1, x_2, \ldots, x_n < \infty \\ 0, & \text{otherwise} \end{cases}$$

The Jacobian of the transformation is given by

$$J = \begin{vmatrix} 1 & 0 & 0 & \cdots & 0 \\ -1 & 1 & 0 & \cdots & 0 \\ 0 & -1 & 1 & \cdots & 0 \\ \cdot & \cdot & \cdot & \cdots & \cdot \\ \cdot & \cdot & \cdot & \cdots & \cdot \\ \cdot & \cdot & \cdot & \cdots & \cdot \\ 0 & 0 & 0 & \cdots & 1 \end{vmatrix}$$

and $|J| = 1$. Hence
$$g(y_1, y_2, \ldots, y_n) = \lambda^n \exp[-\lambda y_n]$$

Since $Y_i = \sum_{j=1}^{i} X_j$, the limits on y_i are y_{i-1} to y_{i+1} for $1 < i < n$, 0 to y_2 for y_1, and y_{n-1} to ∞ for y_n. Therefore $g(y, y_2, \ldots, y_n)$ is specified by

$$g(y_1, y_2, \ldots, y_n) = \lambda^n \exp[-\lambda y_n], \quad 0 \leq y_1 \leq y_2 \leq \cdots \leq y_n < \infty$$

The density function of Y_n is then

$$g_n(y_n) = \int_0^{y_n} \int_0^{y_{n-1}} \cdots \int_0^{y_3} \int_0^{y_2} \lambda^n \exp[-\lambda y_n] \, dy_1, dy_2 \cdots dy_{n-2} \, dy_{n-1}$$

$$= \int_0^{y_n} \int_0^{y_{n-1}} \cdots \int_0^{y_3} y_2 \lambda^n \exp[-\lambda y_n] \, dy_2 \cdots dy_{n-2} \, dy_{n-1}$$

$$= \int_0^{y_n} \int_0^{y_{n-1}} \cdots \int_0^{y_4} \frac{y_3^2}{2} \lambda^n \exp[-\lambda y_n] \, dy_3 \cdots dy_{n-2} \, dy_{n-1}$$

$$= \int_0^{y_n} \int_0^{y_{n-1}} \cdots \int_0^{y_5} \frac{y_4^3}{3!} \lambda^n \exp[-\lambda y_n] \, dy_4 \cdots dy_{n-2} \, dy_{n-1}$$

$$= \frac{y_n^{n-1}}{(n-1)!} \lambda^n \exp[-\lambda y_n], \quad 0 < y_n < \infty$$

and from Eq. (1.92), Y_n is gamma distributed with parameters λ and n.

1.5 DERIVED DISTRIBUTIONS

Example 1.34 (***Sum of Poisson Random Variables***) Show that the sum of independent Poisson random variables is Poisson distributed.

Let $X_1, X_2, \ldots, X_n$ be independent Poisson random variables with parameters $\lambda_1, \lambda_2, \ldots, \lambda_n$ and let $Y_i = \sum_{j=1}^{i} X_j$. Then

$$X_1 = Y_1$$
$$X_2 = Y_2 - Y_1$$
$$\vdots$$
$$X_n = Y_n - Y_{n-1}$$

Since $X_1, X_2, \ldots, X_n$ are independent Poisson random variables,

$$p(x_1, x_2, \ldots, x_n) = \frac{\lambda_1^{x_1} \lambda_2^{x_2} \cdots \lambda_n^{x_n}}{x_1! x_2! \cdots x_n!} \exp\left[\sum_{i=1}^{n} \lambda_i\right], \quad x_i = 0, 1, \ldots, \quad i = 1, 2, \ldots, n$$

By Eq. (1.148),

$$q(y_1, y_2, \ldots, y_n) = \frac{\lambda_1^{y_1} \lambda_2^{y_2 - y_1} \cdots \lambda_n^{y_n - y_{n-1}}}{y_1! (y_2 - y_1)! \cdots (y_n - y_{n-1})!} \exp\left[\sum_{i=1}^{n} \lambda_i\right]$$

$$= \frac{\left(\frac{\lambda_1}{\lambda_2}\right)^{y_1} \left(\frac{\lambda_2}{\lambda_3}\right)^{y_2} \cdots \left(\frac{\lambda_{n-1}}{\lambda_n}\right)^{y_{n-1}} \lambda_n^{y_n}}{y_1! (y_2 - y_1)! \cdots (y_n - y_{n-1})!} \exp\left[-\sum_{i=1}^{n} \lambda_i\right]$$

where $q(y_1, y_2, \ldots, y_n)$ is nonzero for integer valued $y_1, y_2, \ldots, y_n$ such that $0 \leq y_1 \leq y_2 \leq y_3 \leq \cdots \leq y_{n-1} \leq y_n$, and $q(y_1, y_2, \ldots, y_n) = 0$ otherwise. Then

$$q_n(y_n) = \sum_{y_{n-1}=0}^{y_n} \cdots \sum_{y_2=0}^{y_3} \sum_{y_1=0}^{y_2} q(y_1, y_2, \ldots, y_n)$$

Now

$$q_2(y_2, y_3, \ldots, y_n) = \sum_{y_1=0}^{y_2} \frac{\left(\frac{\lambda_1}{\lambda_2}\right)^{y_1} \cdots \left(\frac{\lambda_{n-1}}{\lambda_n}\right)^{y_{n-1}} \lambda_n^{y_n}}{y_1! (y_2 - y_1)! \cdots (y_n - y_{n-1})!} \exp\left[-\sum_{i=1}^{n} \lambda_i\right]$$

Multiplying and dividing the expression for $q(y_2, y_3, \ldots, y_n)$ by $y_2!$ yields

$$q_2(y_2, y_3, \ldots, y_n) = \frac{\lambda_n^{y_n} \exp\left[-\sum_{i=1}^{n} \lambda_i\right]}{y_2!} \left[\prod_{i=2}^{n-1} \frac{\left(\frac{\lambda_i}{\lambda_{i+1}}\right)^{y_i}}{(y_{i+1} - y_i)!}\right] \sum_{y_1=0}^{y_2} \frac{y_2!}{y_1!(y_2 - y_1)!} \left(\frac{\lambda_1}{\lambda_2}\right)^{y_1}$$

From the binomial expansion

$$\sum_{k=0}^{m} \frac{m!}{k!(m-k)!} a^k b^{m-k} = (a+b)^m$$

Hence

$$\sum_{y_1=0}^{y_2} \frac{y_2!}{y_1!(y_2 - y_1)!} \left(\frac{\lambda_1}{\lambda_2}\right)^{y_1} = \left(1 + \frac{\lambda_1}{\lambda_2}\right)^{y_2} = \left(\frac{\lambda_1 + \lambda_2}{\lambda_2}\right)^{y_2}$$

and
$$q_2(y_2, y_3, \ldots, y_n) = \frac{\left(\dfrac{\lambda_1 + \lambda_2}{\lambda_3}\right)^{y_2}\left(\dfrac{\lambda_3}{\lambda_4}\right)^{y_3} \cdots \left(\dfrac{\lambda_{n-1}}{\lambda_n}\right)^{y_{n-1}} \lambda_n^{y_n}}{y_2!(y_3 - y_2)! \cdots (y_n - y_{n-1})!} \exp\left[-\sum_{i=1}^{n} \lambda_i\right]$$

We now find $q_3(y_3, y_4, \ldots, y_n)$.

$$q_3(y_3, y_4, \ldots, y_n) = \sum_{y_2=0}^{y_3} q_3(y_3, y_4, \ldots, y_n)$$

Again multiplying and dividing by $y_3!$ yields us

$q_3(y_3, y_4, \ldots, y_n)$

$$= \frac{\lambda_n^{y_n} \exp\left[-\sum_{i=1}^{n} \lambda_i\right]}{y_3!} \left[\prod_{i=3}^{n-1} \frac{\left(\dfrac{\lambda_i}{\lambda_{i+1}}\right)^{y_i}}{(y_{i+1} - y_i)!}\right] \sum_{y_2=0}^{y_3} \frac{y_3!}{y_2!(y_3 - y_2)!} \left(\frac{\lambda_1 + \lambda_2}{\lambda_3}\right)^{y_2}$$

$$= \frac{\left(\dfrac{\lambda_1 + \lambda_2 + \lambda_3}{\lambda_4}\right)^{y_3}\left(\dfrac{\lambda_4}{\lambda_5}\right)^{y_4} \cdots \left(\dfrac{\lambda_{n-1}}{\lambda_n}\right)^{y_{n-1}} \lambda_n^{y_n}}{y_3!(y_4 - y_3)! \cdots (y_n - y_{n-1})!} \exp\left[-\sum_{i=1}^{n} \lambda_i\right]$$

We continue the derivation of $q_n(y_n)$ by induction. Assume

$$q_j(y_j, y_{j+1}, \ldots, y_n) = \frac{\left(\dfrac{\sum_{i=1}^{j} \lambda_i}{\lambda_{j+1}}\right)^{y_j}\left(\dfrac{\lambda_{j+1}}{\lambda_{j+2}}\right)^{y_{j+1}} \cdots \left(\dfrac{\lambda_{n-1}}{\lambda_n}\right)^{y_{n-1}} \lambda_n^{y_n}}{y_j!(y_{j+1} - y_j)! \cdots (y_n - y_{n-1})!} \exp\left[-\sum_{i=1}^{n} \lambda_i\right]$$

Then, using the above analysis, we have

$q_{j+1}(y_{j+1}, y_{j+2}, \ldots, y_n)$

$$= \frac{\lambda_n^{y_n} \exp\left[-\sum_{i=1}^{n} y_i\right]}{y_{j+1}!} \left[\prod_{i=j+1}^{n-1} \frac{\left(\dfrac{\lambda_i}{\lambda_{i+1}}\right)^{y_i}}{(y_{i+1} - y_i)!}\right] \sum_{y_j=0}^{y_{j+1}} \frac{y_{j+1}!}{y_j!(y_{j+1} - y_j)!} \left[\frac{\sum_{i=1}^{j} \lambda_i}{\lambda_{j+1}}\right]^{y_j}$$

$$= \frac{\left[\dfrac{\sum_{i=1}^{j+1} \lambda_i}{\lambda_{j+2}}\right]^{y_{j+1}}\left(\dfrac{\lambda_{j+2}}{\lambda_{j+3}}\right)^{y_{j+2}} \cdots \left(\dfrac{\lambda_{n-1}}{\lambda_n}\right)^{y_{n-1}} \lambda_n^{y_n}}{y_{j+1}!(y_{j+2} - y_{j+1})! \cdots (y_n - y_{n-1})!} \exp\left[-\sum_{i=1}^{n} \lambda_i\right]$$

Thus the expression for $q_j(y_j, y_{j+1}, \ldots, y_n)$ holds for all $j \leq n$. Hence

$$q_n(y_n) = \frac{\left[\dfrac{\sum_{i=1}^{n} \lambda_i}{\lambda_n}\right]^{y_n}}{y_n!} \lambda_n^{y_n} \exp\left[-\sum_{i=1}^{n} \lambda_i\right] = \frac{(\sum_{i=1}^{n} \lambda_i)^{y_n}}{y_n!} \exp\left[-\sum_{i=1}^{n} \lambda_i\right],$$

$$y_n = 0, 1, 2, \ldots \quad (1.149)$$

Comparing Eq. 1.149 with Eq. 1.78, we see that Y_n has a Poisson distribution with parameter $\sum_{i=1}^{n} \lambda_i$.

1.5 DERIVED DISTRIBUTIONS

Example 1.35 (*Noncentral χ^2 Random Variable*) Let $Z_1, Z_2, \ldots, Z_n$ be independent normal random variables with mean λ_i, $i = 1, 2, \ldots, n$, and standard deviation unity. Find the density function of Y_n where

$$Y_n = \sum_{i=1}^{n} Z_i^2 \tag{1.150}$$

To find the density function of Y_n, we will first find the density function of Z_i^2, $i = 1, 2, \ldots, n$. Letting

$$X_i = Z_i^2 \tag{1.151}$$

we have

$$Y_j = \sum_{i=1}^{j} X_i, \quad j = 1, 2, \ldots, n \tag{1.152}$$

We will then be in a position to find the density function of Y_j using the joint density of $X_1, X_2, \ldots, X_j$, $g(x_1, x_2, \ldots, x_j)$, as a starting point. Let

$$f(z_i) = \frac{1}{\sqrt{2\pi}} \exp\left[-\frac{(z_i - \lambda_i)^2}{2}\right], \quad -\infty < z_i < \infty, \quad i = 1, 2, \ldots, n \tag{1.153}$$

From Eq. (1.142)

$$\begin{aligned} g_i(x_i) &= \frac{f(\sqrt{x_i}) + f(-\sqrt{x_i})}{2\sqrt{x_i}} \\ &= \frac{x_i^{-\frac{1}{2}}}{2\sqrt{2\pi}} \left[\exp\left[-\frac{x_i - 2\sqrt{x_i}\lambda_i + \lambda_i^2}{2}\right] + \exp\left[-\frac{x_i + 2\sqrt{x_i}\lambda_i + \lambda_i^2}{2}\right]\right] \\ &= \frac{x_i^{-\frac{1}{2}}}{2\sqrt{2\pi}} \exp\left[-\frac{x_i + \lambda_i^2}{2}\right] [\exp[\sqrt{x_i}\lambda_i] + \exp[-\sqrt{x_i}\lambda_i]] \end{aligned} \tag{1.154}$$

Now, we may express $\exp[a]$ as

$$\exp[a] = \sum_{k=0}^{\infty} \frac{a^k}{k!}$$

Therefore,

$$\begin{aligned} \exp[\sqrt{x_i}\lambda_i] + \exp[-\sqrt{x_i}\lambda_i] &= \sum_{k=0}^{\infty} \left[\frac{(\sqrt{x_i}\lambda_i)^k}{k!} + \frac{(-\sqrt{x_i}\lambda_i)^k}{k!}\right] \\ &= \sum_{k=0}^{\infty} \left[\frac{2(\sqrt{x_i}\lambda_i)^{2k}}{(2k)!}\right] \end{aligned} \tag{1.155}$$

since $(\sqrt{x_i}\lambda_i)^k/k! + (-\sqrt{x_i}\lambda_i)^k/k! = 0$ for odd k. Now, one may show that (Abramowitz and Stegun, 1964)

$$\begin{aligned} (2k)! &= 2k(2k-1)! = 2k\Gamma(2k) \\ &= 2k(2\pi)^{-\frac{1}{2}} 2^{2k-\frac{1}{2}} \Gamma(k)\Gamma(k+\tfrac{1}{2}) = (2\pi)^{-\frac{1}{2}} 2^{2k+\frac{1}{2}} \Gamma(k+\tfrac{1}{2})k! \end{aligned} \tag{1.156}$$

Hence

$$\exp[\sqrt{x_i}\lambda_i] + \exp[-\sqrt{x_i}\lambda_i] = \sum_{k=0}^{\infty} \frac{\sqrt{2\pi}x_i^k \lambda_i^{2k}}{2^{2k-\frac{1}{2}}\Gamma(k+\frac{1}{2})k!} \quad (1.157)$$

Letting $\delta_i = \lambda_i^2/2$, we have

$$\exp[\sqrt{x_i}\lambda_i] + \exp[-\sqrt{x_i}\lambda_i] = \sum_{k=0}^{\infty} \frac{\sqrt{2\pi}x_i^k \delta_i^k}{2^{k-\frac{1}{2}}\Gamma(k+\frac{1}{2})k!} \quad (1.158)$$

and

$$g_i(x_i) = \exp\left[-\left(\delta_i + \frac{x_i}{2}\right)\right] \sum_{k=0}^{\infty} \frac{x_i^{k-\frac{1}{2}} \delta_i^k}{2^{k+\frac{1}{2}}\Gamma(k+\frac{1}{2})k!} \quad (1.159)$$

Since Z_1 and Z_2 are independently distributed, X_1 and X_2 are also independently distributed and

$$g_i(x_1, x_2) = \exp\left[-\left(\delta_1 + \delta_2 + \frac{x_1 + x_2}{2}\right)\right] \prod_{i=1}^{2} \sum_{k_i=0}^{\infty} \frac{x_i^{k_i-\frac{1}{2}} \delta_i^{k_i}}{2^{k_i+\frac{1}{2}}\Gamma(k_i+\frac{1}{2})k_i!},$$
$$0 < x_1, x_2 < \infty$$

Let
$$Y_2 = X_1 + X_2, \quad Y_1 = X_1$$

Therefore
$$X_1 = Y_1, \quad X_2 = Y_2 - Y_1$$

and the Jacobian of the transformation is

$$J = \begin{vmatrix} 1 & 0 \\ -1 & 1 \end{vmatrix} = 1$$

Thus

$$h(y_1, y_2) = \exp\left[-\left(\delta_1 + \delta_2 + \frac{y_2}{2}\right)\right] \sum_{k_1=0}^{\infty} \sum_{k_2=0}^{\infty} \frac{y_1^{k_1-\frac{1}{2}}(y_2-y_1)^{k_2-\frac{1}{2}}\delta_1^{k_1}\delta_2^{k_2}}{2^{k_1+k_2+1}\Gamma(k_1+\frac{1}{2})\Gamma(k_2+\frac{1}{2})k_1!k_2!},$$
$$0 < y_1 < y_2 < \infty$$

The marginal density function of Y_2 is then given by

$$h_2(y_2) = \exp\left[-\left(\delta_1 + \delta_2 + \frac{y_2}{2}\right)\right] \sum_{k_1=0}^{\infty} \sum_{k_2=0}^{\infty} \frac{\delta_1^{k_1} \delta_2^{k_2}}{2^{k_1+k_2+1}\Gamma(k_1+\frac{1}{2})\Gamma(k_2+\frac{1}{2})k_1!k_2!}$$
$$\times \int_0^{y_2} y_1^{k_1-\frac{1}{2}}(y_2-y_1)^{k_2-\frac{1}{2}} dy_1$$

Now,

$$\int_0^{y_2} y_1^{k_1-\frac{1}{2}}(y_2-y_1)^{k_2-\frac{1}{2}} dy_1 = y_2^{k_1+k_2-1}\int_0^{y_2}\left(\frac{y_1}{y_2}\right)^{k_1-\frac{1}{2}}\left(1-\frac{y_1}{y_2}\right)^{k_2-\frac{1}{2}} dy_1$$

1.5 DERIVED DISTRIBUTIONS

Let $v = (y_1/y_2)$. Then
$$y_1 = y_2 v, \qquad dy_1 = y_2\, dv$$

and

$$\int_0^{y_2} y_1^{k_1-\frac{1}{2}}(y_2 - y_1)^{k_2-\frac{1}{2}}\, dy_1 = y_2^{k_1+k_2}\int_0^1 v^{k_1-\frac{1}{2}}(1-v)^{k_2-\frac{1}{2}}\, dv$$

$$= y_2^{k_1+k_2}\frac{\Gamma(k_1+\frac{1}{2})\Gamma(k_2+\frac{1}{2})}{\Gamma(k_1+k_2+1)} \qquad (1.160)$$

Therefore

$$h_2(y_2) = \exp\left[-\left(\delta_1 + \delta_2 + \frac{y_2}{2}\right)\right]\sum_{k_1=0}^{\infty}\sum_{k_2=0}^{\infty}\frac{y_2^{k_1+k_2}\,\delta_1^{k_1}\,\delta_2^{k_2}}{2^{k_1+k_2+1}\Gamma(k_1+k_2+1)k_1!\,k_2!}$$

The double summation in the expression for $h_2(y_2)$ may be reduced to a single summation as follows. Let $m = k_1 + k_2$ and $l = k_1$. Then

$$\sum_{k_1=0}^{\infty}\sum_{k_2=0}^{\infty}\frac{\delta_1^{k_1}\delta_2^{k_2}y_2^{k_1+k_2}}{2^{k_1+k_2+1}\Gamma(k_1+k_2+1)k_1!\,k_2!}$$

$$= \sum_{m=0}^{\infty}\sum_{l=0}^{m}\frac{y_2^m\,\delta_1^l\,\delta_2^{m-l}}{2^{m+1}\Gamma(m+1)l!\,(m-l)!}$$

$$= \sum_{m=0}^{\infty}\frac{y_2^m}{2^{m+1}\Gamma(m+1)m!}\sum_{l=0}^{m}\frac{m!}{l!\,(m-l)!}\delta_1^l\,\delta_2^{m-l}$$

$$= \sum_{m=0}^{\infty}\frac{y_2^m(\delta_1+\delta_2)^m}{2^{m+1}\Gamma(m+1)m!} \qquad (1.161)$$

and

$$h_2(y_2) = \exp\left[-\left(\delta_1+\delta_2+\frac{y_2}{2}\right)\right]\sum_{m=0}^{\infty}\frac{(\delta_1+\delta_2)^m y_2^m}{2^{m+1}\Gamma(m+1)m!}, \qquad 0 < y_2 < \infty \qquad (1.162)$$

Let us now find the density function of Y_3 in an attempt to define a general form for $g_i(y_i)$ and complete the proof by induction.

$$Y_3 = Y_2 + X_3$$

From an argument identical to that given above

$$g(y_2, x_3) = \exp\left[-\left(\delta_1+\delta_2+\delta_3+\frac{y_2+x_3}{2}\right)\right]\sum_{m=0}^{\infty}\frac{(\delta_1+\delta_2)^m y_2^m}{2^{m+1}\Gamma(m+1)m!}$$

$$\times \sum_{k=0}^{\infty}\frac{x_3^{k-\frac{1}{2}}\delta_3^k}{2^{k+\frac{1}{2}}\Gamma(k+\frac{1}{2})k!}, \qquad 0 < x_3,\ y_2 < \infty$$

If $V = Y_2$, $X_3 = Y_3 - V$, then

$$h(v, y_3) = \exp\left[-\left(\sum_{i=1}^{3} \delta_i + \frac{y_3}{2}\right)\right] \sum_{m=0}^{\infty} \frac{(\delta_1 + \delta_2)^m v^m}{2^{m+1}\Gamma(m+1)m!}$$

$$\times \sum_{k=0}^{\infty} \frac{\delta_3^k (y_3 - v)^{k-\frac{1}{2}}}{2^{k+\frac{1}{2}}\Gamma(k+\frac{1}{2})k!}, \qquad 0 < v < y_3 < \infty$$

Finally

$$h_3(y_3) = \exp\left[-\left(\sum_{i=1}^{3} \delta_i + \frac{y_3}{2}\right)\right] \sum_{m=0}^{\infty} \frac{(\sum_{i=1}^{3} \delta_i)^m y_3^{m+\frac{1}{2}}}{2^{m+\frac{3}{2}}\Gamma(m+\frac{3}{2})m!},$$

$$0 < y_3 < \infty \qquad (1.163)$$

From Eqs. (1.159), (1.162), and (1.163) let us assume that

$$h_{j-1}(y_{j-1}) = \exp\left[-\left(\sum_{i=1}^{j-1} \delta_i + \frac{y_{j-1}}{2}\right)\right] \sum_{m=0}^{\infty} \frac{(\sum_{i=1}^{j-1} \delta_i)^m y_{j-1}^{m+(j-3)/2}}{2^{m+(j-1)/2}\Gamma\left(m + \dfrac{j-1}{2}\right)m!},$$

$$0 < y_{j-1} < \infty \qquad (1.164)$$

We will now attempt to complete the derivation by induction by developing $h_j(y_j)$ using the arguments already presented. Letting

$$Y_j = Y_{j-1} + X_j, \qquad V = X_j$$

we have

$$Y_{j-1} = Y_j - V, \qquad X_j = V$$

and

$$h(v, y_j) = \exp\left[-\left(\sum_{i=1}^{j} \delta_i + \frac{y_j}{2}\right)\right] \sum_{m=0}^{\infty} \frac{(\sum_{i=1}^{j-1} \delta_i)^m (y_j - v)^{m+(j-3)/2}}{2^{m+(j-1)/2}\Gamma\left(m + \dfrac{j-1}{2}\right)m!}$$

$$\times \sum_{k=0}^{\infty} \frac{v^{k-\frac{1}{2}} \delta_j^k}{2^{k+\frac{1}{2}}\Gamma(k+\frac{1}{2})k!} \qquad 0 < v < y_j < \infty$$

The density function of Y_j is given by

$$h_j(y_j) = \exp\left[-\left(\sum_{i=1}^{j} \delta_i + \frac{y_j}{2}\right)\right] \sum_{m=0}^{\infty} \sum_{k=0}^{\infty} \frac{(\sum_{i=1}^{j-1} \delta_i)^m \delta_j^k}{2^{m+k+(j/2)}\Gamma\left(m + \dfrac{j-1}{2}\right)\Gamma(k+\frac{1}{2})m!\,k!}$$

$$\times \int_0^{y_j} v^{k-\frac{1}{2}}(y_j - v)^{m+(j-3)/2}\,dv$$

$$= \exp\left[-\left(\sum_{i=1}^{j} \delta_i + \frac{y_j}{2}\right)\right] \sum_{m=0}^{\infty} \sum_{k=0}^{\infty} \frac{(\sum_{i=1}^{j-1} \delta_i)^m \delta_j^k y_j^{m+k+(j/2)-1}}{2^{m+k+(j/2)}\Gamma\left(m + k + \dfrac{j}{2}\right)m!\,k!}$$

1.5 DERIVED DISTRIBUTIONS

by Eq. (1.160). Applying Eq. (1.161) we have

$$\sum_{m=0}^{\infty}\sum_{k=0}^{\infty} \frac{(\sum_{i=1}^{j-1}\delta_i)^m \delta_j^k y_j^{m+k+(j/2)-1}}{2^{m+k+(j/2)}\Gamma\left(m+k+\frac{j}{2}\right)m!k!} = \sum_{t=0}^{\infty}\sum_{l=0}^{t} \frac{(\sum_{i=1}^{j-1}\delta_i)^{t-l}\delta_j^l y_j^{t+(j/2)-1}}{2^{t+(j/2)}\Gamma\left(t+\frac{j}{2}\right)(t-l)!l!}$$

$$= \sum_{t=0}^{\infty} \frac{y_j^{t+(j/2)-1}}{2^{t+(j/2)}\Gamma\left(t+\frac{j}{2}\right)t!} \sum_{l=0}^{t} \frac{t!}{l!(t-l)!}$$

$$\times \left(\sum_{i=1}^{j-1}\delta_i\right)^{t-l}\delta_j^l$$

$$\sum_{m=0}^{\infty}\sum_{k=0}^{\infty} \frac{(\sum_{i=1}^{j-1}\delta_i)^m \delta_j^k y_j^{m+k+(j/2)-1}}{2^{m+k+(j/2)}\Gamma\left(m+k+\frac{j}{2}\right)m!k!} = \sum_{t=0}^{\infty} \frac{(\sum_{i=1}^{j}\delta_i)^t y_j^{t+(j/2)-1}}{2^{t+(j/2)}\Gamma\left(t+\frac{j}{2}\right)t!}$$

and

$$h_j(y_j) = \exp\left[-\left(\sum_{i=1}^{j}\delta_i + \frac{y_j}{2}\right)\right] \sum_{m=0}^{\infty} \frac{(\sum_{i=1}^{j}\delta_i)^m y_j^{m+(j/2)-1}}{2^{m+(j/2)}\Gamma\left(m+\frac{j}{2}\right)m!}, \qquad 0 < y_j < \infty$$

(1.165)

Since Eq. (1.165) reduces to Eq. (1.164) by replacing j by $j-1$, the proof by induction is complete and

$$h_n(y_n) = \exp\left[-\left(\sum_{i=1}^{n}\delta_i + \frac{y_n}{2}\right)\right] \sum_{m=0}^{\infty} \frac{(\sum_{i=1}^{n}\delta_i)^m y_n^{m+(n/2)-1}}{2^{m+(n/2)}\Gamma\left(m+\frac{n}{2}\right)m!}, \qquad 0 < y_n < \infty$$

(1.166)

Equation (1.166) gives the probability density function of the noncentral χ^2 random variable with noncentrality parameter $\sum_{i=1}^{n}\delta_i$ and n degrees of freedom where

$$\sum_{i=1}^{n}\delta_i = \frac{1}{2}\sum_{i=1}^{n}\lambda_i^2 \qquad (1.167)$$

When $\lambda_i = 0$, $i = 1, 2, \ldots, n$, $Z_1, Z_2, \ldots, Z_n$ are independent standard normal random variables and

$$Y_n = \sum_{i=1}^{n} Z_i^2$$

then y_n has a central χ^2 (or simply χ^2) distribution with density function

$$h_n(y_n) = \frac{1}{2^{(n/2)}\Gamma\left(\dfrac{n}{2}\right)} y_n^{(n/2)-1} \exp\left[-\frac{y_n}{2}\right], \quad 0 < y_n < \infty$$

Example 1.36 (*Distribution of the Sample Mean*) A random sample of n units is drawn from a production process and a critical dimension of each is measured. Let $X_1, X_2, \ldots, X_n$ be the set of measurements. If $X_1, X_2, \ldots, X_n$ are independent, identically distributed normal random variables with mean u and standard deviation σ, show that the sample mean given by

$$\bar{X} = \frac{1}{n}\sum_{i=1}^{n} X_i$$

is also normally distributed with mean u and standard deviation $\sigma/\sqrt{n}$. Since $X_1, X_2, \ldots, X_n$ are identically distributed independent normal random variables

$$f(x_1, x_2, \ldots, x_n) = \left(\frac{1}{\sigma\sqrt{2\pi}}\right)^n \exp\left[-\frac{\sum_{i=1}^{n}(x_i - u)^2}{2\sigma^2}\right],$$
$$-\infty < x_i < \infty, \quad i = 1, 2, \ldots, n$$

We will first find the density function of $\sum_{i=1}^{n} X_i$. Let

$$Y_n = \sum_{i=1}^{n} X_i$$

$$Y_{n-1} = \sum_{i=1}^{n-1} X_i$$

$$\vdots$$

$$Y_1 = X_1$$

Then

$$X_1 = Y_1$$
$$X_2 = Y_2 - Y_1$$
$$\vdots$$
$$X_n = Y_n - Y_{n-1}$$

From Example 1.33, $|J| = 1$. Hence

$$g(y_1, y_2, \ldots, y_n) = \left(\frac{1}{\sigma\sqrt{2\pi}}\right)^n \exp\left[-\frac{\sum_{i=1}^{n}[(y_i - y_{i-1}) - u]^2}{2\sigma^2}\right],$$
$$-\infty < y_i < \infty, \quad i = 1, 2, \ldots, n$$

where $y_0 = 0$. The reader will note that the limits on y_i are minus and plus infinity since X_i, $i = 1, 2, \ldots, n$, may assume both negative and positive values. Now

1.5 DERIVED DISTRIBUTIONS

$$\sum_{i=1}^{n}[(y_i - y_{i-1}) - u]^2 = \sum_{i=1}^{n}[(y_i - y_{i-1})^2 - 2u(y_i - y_{i-1}) + u^2]$$

$$= \sum_{i=1}^{n}[y_i^2 - 2y_i y_{i-1} + y_{i-1}^2 - 2u(y_i - y_{i-1}) + u^2]$$

$$= \sum_{i=1}^{n}(y_i - y_{i-1})^2 + nu^2 - 2uy_n$$

or

$$g(y_1, y_2, \ldots, y_n) = \left(\frac{1}{\sigma\sqrt{2\pi}}\right)^n \exp\left[-\frac{(nu^2 - 2uy_n)}{2\sigma^2}\right] \exp\left[-\frac{\sum_{i=1}^{n}(y_i - y_{i-1})^2}{2\sigma^2}\right],$$

$$-\infty < y_i < \infty, \quad i = 1, 2, \ldots, n$$

We obtain the density function of Y_n by successive integration of $g(y_1, y_2, \ldots, y_n)$ over $y_1, y_2, \ldots, y_{n-1}$. Now,

$$g(y_2, y_3, \ldots, y_n)$$

$$= \left(\frac{1}{\sigma\sqrt{2\pi}}\right)^n \exp\left[-\frac{(nu^2 - 2uy_n)}{2\sigma^2}\right] \int_{-\infty}^{\infty} \exp\left[-\frac{\sum_{i=1}^{n}(y_i - y_{i-1})^2}{2\sigma^2}\right] dy_1$$

But

$$\int_{-\infty}^{\infty} \exp\left[-\frac{\sum_{i=1}^{n}(y_i - y_{i-1})^2}{2\sigma^2}\right] dy_1$$

$$= \exp\left[-\frac{\sum_{i=3}^{n}(y_i - y_{i-1})^2}{2\sigma^2}\right] \int_{-\infty}^{\infty} \exp\left[-\frac{y_1^2 + (y_2 - y_1)^2}{2\sigma^2}\right] dy_1$$

$$= \exp\left[-\frac{\sum_{i=3}^{n}(y_i - y_{i-1})^2}{2\sigma^2}\right] \int_{-\infty}^{\infty} \exp\left[-\frac{2\left(y_1 - \frac{y_2}{2}\right) + \frac{y_2^2}{2}}{2\sigma^2}\right] dy_1$$

$$= \exp\left[-\frac{\frac{y_2^2}{2} + \sum_{i=3}^{n}(y_i - y_{i-1})^2}{2\sigma^2}\right] \int_{-\infty}^{\infty} \exp\left[-\frac{\left(y_1 - \frac{y_2}{2}\right)^2}{\sigma^2}\right] dy_1$$

Let

$$z = \sqrt{2}\,\frac{y_1 - \frac{y_2}{2}}{\sigma}$$

Then

$$y_1 = \frac{\sigma z}{\sqrt{2}} + \frac{y_2}{2}, \quad dy_1 = \frac{\sigma}{\sqrt{2}} dz$$

and

$$\int_{-\infty}^{\infty} \exp\left[-\frac{\left(y_1 - \frac{y_2}{2}\right)^2}{2\sigma^2}\right] dy_1 = \frac{\sigma}{\sqrt{2}} \int_{-\infty}^{\infty} \exp\left[-\frac{z^2}{2}\right] dz = \sqrt{\pi}\sigma$$

which leads to

$$g(y_2, y_3, \ldots, y_n)$$
$$= \frac{1}{\sqrt{2}}\left(\frac{1}{\sigma\sqrt{2\pi}}\right)^{n-1} \exp\left[-\frac{(nu^2 - 2uy_n)}{2\sigma^2}\right] \exp\left[-\frac{\frac{y_2^2}{2} + \sum_{i=3}^{n}(y_i - y_{i-1})^2}{2\sigma^2}\right],$$
$$-\infty < y_i < \infty, \quad i = 2, 3, \ldots, n$$

Integrating $g(y_2, y_3, \ldots, y_n)$ over y_2

$$g(y_3, y_4, \ldots, y_n)$$
$$= \frac{1}{\sqrt{2}}\left(\frac{1}{\sigma\sqrt{2\pi}}\right)^{n-1} \exp\left[-\frac{nu^2 - 2uy_n}{2\sigma^2}\right] \exp\left[-\frac{\sum_{i=4}^{n}(y_i - y_{i-1})^2}{2\sigma^2}\right]$$
$$\times \int_{-\infty}^{\infty} \exp\left[-\frac{(y_3 - y_2)^2 + \frac{y_2^2}{2}}{2\sigma^2}\right] dy_2$$

where

$$\int_{-\infty}^{\infty} \exp\left[-\frac{(y_3 - y_2)^2 + \frac{y_2^2}{2}}{2\sigma^2}\right] dy_2 = \int_{-\infty}^{\infty} \exp\left[-\frac{\frac{3}{2}y_2^2 - 2y_2 y_3 + y_3^2}{2\sigma^2}\right] dy_2$$
$$= \exp\left[-\frac{y_3^2}{3}\right] \int_{-\infty}^{\infty} \exp\left[-\frac{\frac{3}{2}\left(y_2 - \frac{2}{3}y_3\right)^2}{2\sigma^2}\right] dy_2$$

Let

$$z = \frac{\frac{3}{2}(y_2 - \frac{2}{3}y_3)}{\sigma}$$

Then

$$y_2 = \tfrac{2}{3}\sigma z + \tfrac{2}{3}y_3, \qquad dy_2 = \tfrac{2}{3}\sigma \, dz$$

1.5 DERIVED DISTRIBUTIONS

and

$$\int_{-\infty}^{\infty} \exp\left[-\frac{(y_3 - y_2)^2 + \frac{y_2^2}{2}}{2\sigma^2}\right] dy_2 = \sqrt{\frac{2}{3}} \sigma \exp\left[-\frac{y_3^2}{3}\right] \int_{-\infty}^{\infty} \exp\left[-\frac{z^2}{2}\right] dz$$

$$= 2\sqrt{\frac{\pi}{3}} \sigma \exp\left[-\frac{y_3^2}{3}\right]$$

Thus

$$g(y_3, y_4, \ldots, y_n) = \frac{1}{\sqrt{3}} \left[\frac{1}{\sigma\sqrt{2\pi}}\right]^{n-2} \exp\left[-\frac{nu^2 - 2uy_n}{2\sigma^2}\right]$$

$$\times \exp\left[-\frac{\frac{y_3^2}{3} + \sum_{i=4}^{n}(y_i - y_{i-1})^2}{2\sigma^2}\right],$$

$$-\infty < y_i < \infty, \quad i = 3, 4, \ldots, n$$

We complete the derivation of the density function of Y_n by induction. From the expressions for $g(y_1, y_2, \ldots, y_n)$, $g(y_2, y_3, \ldots, y_n)$, and $g(y_3, y_4, \ldots, y_n)$, we assume

$$g(y_j, y_{j+1}, \ldots, y_n) = \frac{1}{\sqrt{j}} \left[\frac{1}{\sigma\sqrt{2\pi}}\right]^{n-j+1} \exp\left[-\frac{nu^2 - 2uy_n}{2\sigma^2}\right]$$

$$\times \exp\left[-\frac{\frac{y_j^2}{j} + \sum_{i=j+1}^{n}(y_i - y_{i-1})^2}{2\sigma^2}\right],$$

$$-\infty < y_i < \infty, \quad i = j, j+1, \ldots, n$$

Then

$$g(y_{j+1}, y_{j+2}, \ldots, y_n) = \frac{1}{\sqrt{j}} \left[\frac{1}{\sigma\sqrt{2\pi}}\right]^{n-j+1} \exp\left[-\frac{nu^2 - 2uy_n}{2\sigma^2}\right]$$

$$\times \int_{-\infty}^{\infty} \exp\left[-\frac{\frac{y_j^2}{j} + \sum_{i=j+1}^{n}(y_i - y_{i-1})^2}{2\sigma^2}\right] dy_j$$

and

$$\int_{-\infty}^{\infty} \exp\left[-\frac{\frac{y_j^2}{j} + \sum_{i=j+1}^{n}(y_j - y_{j-1})^2}{2\sigma^2}\right] dy_j = \exp\left[-\frac{\sum_{i=j+2}^{n}(y_i - y_{i-1})^2}{2\sigma^2}\right]$$

$$\times \int_{-\infty}^{\infty} \exp\left[-\frac{\frac{y_j^2}{j} + (y_{j+1} - y_j)^2}{2\sigma^2}\right] dy_j$$

$$= \exp\left[-\frac{\sum_{i=j+2}^{n}(y_i - y_{i-1})^2}{2\sigma^2}\right]$$

$$\times \int_{-\infty}^{\infty} \exp\left[-\frac{\left(\frac{j+1}{j}\right)\left[y_j - \left(\frac{j}{j+1}\right)y_{j+1}\right]^2 + \frac{y_{j+1}^2}{j+1}}{2\sigma^2}\right] dy_j$$

$$= \exp\left[-\frac{\frac{y_{j+1}^2}{j+1} + \sum_{i=j+2}^{n}(y_i - y_{i-1})^2}{2\sigma^2}\right]$$

$$\times \int_{-\infty}^{\infty} \exp\left[-\frac{\left(\frac{j+1}{j}\right)\left[y_j - \left(\frac{j}{j+1}\right)y_{j+1}\right]^2}{2\sigma^2}\right] dy_j$$

Let

$$z = \frac{j+1}{j} \frac{\left[y_j - \left(\frac{j}{j+1}\right)y_{j+1}\right]}{\sigma}$$

Then

$$y_j = \sqrt{\frac{j}{j+1}}\, \sigma z + \left(\frac{j}{j+1}\right)y_{j+1}, \qquad dy_j = \sqrt{\frac{j}{j+1}}\, \sigma\, dz$$

and

$$\int_{-\infty}^{\infty} \exp\left[-\frac{\frac{y_j^2}{j} + \sum_{i=j+1}^{n}(y_j - y_{j-1})^2}{2\sigma^2}\right] dy_j$$

$$= \sqrt{2\pi}\sqrt{\frac{j}{j+1}}\, \sigma \exp\left[-\frac{\frac{y_{j+1}^2}{j+1} + \sum_{i=j+2}^{n}(y_i - y_{i-1})^2}{2\sigma^2}\right]$$

1.5 DERIVED DISTRIBUTIONS

Hence

$$g(y_{j+1}, y_{j+2}, \ldots, y_n) = \frac{1}{\sqrt{j+1}} \left[\frac{1}{\sigma\sqrt{2\pi}}\right]^{n-j} \exp\left[-\frac{nu^2 - 2uy_n}{2\sigma^2}\right]$$

$$\times \exp\left[-\frac{\dfrac{y_{j+1}^2}{j+1} + \sum_{i=j+2}^{n}(y_i - y_{i-1})^2}{2\sigma^2}\right]$$

$$-\infty < y_i < \infty, \quad i = j+1, j+2, \ldots, n$$

Since $g(y_{j+1}, y_{j+2}, \ldots, y_n)$ and $g(y_j, y_{j+1}, \ldots, y_n)$ are equivalent forms, the expression for $g(y_j, y_{j+1}, \ldots, y_n)$ holds for all j. Hence

$$g(y_{n-1}, y_n) = \frac{1}{\sqrt{n-1}} \left[\frac{1}{\sigma\sqrt{2\pi}}\right]^2 \exp\left[-\frac{nu^2 - 2uy_n}{2\sigma^2}\right]$$

$$\times \exp\left[-\frac{\dfrac{y_{n-1}^2}{n-1} + (y_n - y_{n-1})^2}{2\sigma^2}\right], \quad -\infty < y_{n-1}, y_n < \infty$$

Applying the argument already presented, we have

$$g(y_n) = \frac{1}{\sqrt{n-1}} \left(\frac{1}{\sigma\sqrt{2\pi}}\right)^2 \exp\left[-\frac{nu^2 - 2uy_n}{2\sigma^2}\right]$$

$$\times \int_{-\infty}^{\infty} \exp\left[-\frac{\dfrac{y_{n-1}^2}{n-1} + (y_n - y_{n-1})^2}{2\sigma^2}\right] dy_{n-1}$$

$$= \frac{1}{\sigma\sqrt{2\pi n}} \exp\left[-\frac{nu^2 - 2uy_n + \dfrac{y_n^2}{n}}{2\sigma^2}\right]$$

$$= \frac{1}{\sigma\sqrt{2\pi n}} \exp\left[-\frac{(y_n - nu)^2}{2n\sigma^2}\right], \quad -\infty < y_n < \infty \quad (1.168)$$

Now let

$$\bar{X} = \frac{1}{n} Y_n = \phi(Y_n)$$

Then

$$Y_n = n\bar{X} = \psi(\bar{X}), \quad \frac{d}{d\bar{x}}\psi(\bar{x}) = n$$

Thus

$$h(\bar{x}) = \frac{\sqrt{n}}{\sigma\sqrt{2\pi}} \exp\left[-\frac{n(\bar{x} - u)^2}{2\sigma^2}\right], \quad -\infty < \bar{x} < \infty \quad (1.169)$$

which is the probability density function of a normal random variable with mean u and standard deviation $\sigma/\sqrt{n}$.

Example 1.37 X_1 and X_2 are independent random variables with joint probability density function

$$f(x_1, x_2) = \exp[-(x_1 + x_2)], \qquad 0 < x_1, \; x_2 < \infty$$

Find the joint density function of Y_1 and Y_2 where

$$Y_1 = X_1 + X_2, \qquad Y_2 = \frac{X_1}{X_1 + X_2}$$

From the definition of Y_1 and Y_2

$$X_1 = Y_1 Y_2, \qquad X_2 = Y_1(1 - Y_2)$$

Thus

$$J = \begin{vmatrix} Y_2 & Y_1 \\ (1 - Y_2) & -Y_1 \end{vmatrix} = -Y_1$$

and $|J| = Y_1$. Since $0 < x_1, x_2 < \infty$, $0 < x_1 + x_2 < \infty$ and $0 < y_1 < \infty$. Now $0 < x_1 < \infty$ and $0 < y_1 < \infty$. Thus $x_1/(x_1 + x_2) > 0$. Since $0 < x_2 < \infty$, $0 < y_1(1 - y_2) < \infty$, and $0 < y_2 < 1$. The joint density function of Y_1 and Y_2 is then

$$g(y_1, y_2) = y_1 \exp[-y_1], \qquad 0 < y_1 < \infty, \; 0 < y_2 < 1$$

1.6 Expectation

In the area of system modeling the concept of *mathematical expectation* plays a particularly significant role. Expected value theory is also used extensively in estimating the parameters of probability distributions such as those already discussed and in defining measures of central tendency and variability for random variables. Finally, the expected value of certain functions of a random variable leads to transforms which may be used to characterize the random variable in a fashion sometimes more useful than the probability mass and density functions. Among these transforms are the moment generating function, the characteristic function, the Fourier transform, the Laplace transform, and the Z transform. The moment generating function will be discussed in this section. Treatment of the remaining transforms will be delayed until Chapter 6.

Let X be a random variable and $\theta(X)$ some function of X. The *expected value* of $\theta(X)$ will be denoted by $E[\theta(X)]$. To give some intuitive meaning to $E[\theta(X)]$, let us consider the case where $\theta(X) = X$. That is, we wish to find the expected value of X, $E(X)$. Suppose that N values of the random variable X are selected, $x_1, x_2, \ldots, x_N$. The sample mean or average, $\bar{x}$, would then be given by

$$\bar{x} = \frac{1}{N} \sum_{i=1}^{N} x_i \qquad (1.170)$$

1.6 EXPECTATION

If we allow N to increase indefinitely then $\bar{x}$ approaches the mean of the random variable which is defined as $E(X)$.

Definition 1.12 Let $\theta(X)$ be a function of the random variable X. Then

$$E[\theta(X)] = \begin{cases} \sum_{x=-\infty}^{\infty} \theta(x)p(x), & \text{discrete case} \\ \int_{-\infty}^{\infty} \theta(x)f(x)\,dx, & \text{continuous case} \end{cases} \quad (1.171)$$

where $p(x)$ and $f(x)$ are the probability mass and probability density functions of X. If either the sum or the integral in Eq. (1.171) diverge $E[\theta(X)]$ does not exist.

As already indicated, when $\theta(X) = X$, $E[\theta(X)]$ is called the *mean*, m, of the random variable X. When $\theta(X) = [X - E(X)]^2$, $E[\theta(X)]$ is called the *variance* σ^2, or $\text{Var}(X)$, of X and $\sqrt{E[\theta(X)]}$ is called the *standard deviation*, σ, of X. The variance and standard deviation are used to measure the variability of the random variable.

The concept of expectation may be extended to functions of several random variables. If $\theta(X_1, X_2, \ldots, X_n)$ is a function of the random variables $X_1, X_2, \ldots, X_n$, then $E[\theta(X_1, X_2, \ldots, X_n)]$ is the expected value of $\theta(X_1, X_2, \ldots, X_n)$.

Definition 1.13 Let $\theta(X_1, X_2, \ldots, X_n)$ be a function of the random variables $X_1, X_2, \ldots, X_n$. Then

$$E[\theta(X_1, X_2, \ldots, X_n)] = \begin{cases} \sum_{x_n=-\infty}^{\infty} \cdots \sum_{x_2=-\infty}^{\infty} \sum_{x_1=-\infty}^{\infty} \theta(x_1, x_2, \ldots, x_n)p(x_1, x_2, \ldots, x_n), \\ \qquad\qquad \text{discrete case} \\ \int_{-\infty}^{\infty} \cdots \int_{-\infty}^{\infty} \int_{-\infty}^{\infty} \theta(x_1, x_2, \ldots, x_n) \\ \qquad\qquad \times f(x_1, x_2, \ldots, x_n)dx_1\,dx_2\cdots dx_n, \\ \qquad\qquad \text{continuous case} \end{cases} \quad (1.172)$$

where $p(x_1, x_2, \ldots, x_n)$ is the joint probability mass function of $X_1, X_2, \ldots, X_n$ and $f(x_1, x_2, \ldots, x_n)$ is the joint probability density function of $X_1, X_2, \ldots, X_n$.

Theorem 1.12 Let $X_1, X_2, \ldots, X_n$ be discrete or continuous random variables. Then, if $a_1, a_2, \ldots, a_n$ are real constants

a. $E\left(\sum_{i=1}^{n} a_i X_i\right) = \sum_{i=1}^{n} a_i E(X_i)$ \hfill (1.173)

b. $E(X_i, X_j) = E(X_i)E(X_j)$ \quad if X_i and X_j are independent \hfill (1.174)

c. $\operatorname{Var}\left(\sum_{i=1}^{n} a_i X_i\right)$

$$= \sum_{i=1}^{n} a_i^2 \operatorname{Var}(X_i), \quad \text{if } X_1, X_2, \ldots, X_n \text{ are independent} \quad (1.175)$$

d. $\operatorname{Var}\left(\sum_{i=1}^{n} a_i X_i\right) = \sum_{i=1}^{n} a_i^2 \operatorname{Var}(X_i) + 2 \sum_{i=1}^{n-1} \sum_{j=i+1}^{n} a_i a_j \operatorname{Cov}(X_i, X_j),$

$$\text{if } X_1 \text{ and } X_2 \text{ are not independent} \quad (1.176)$$

where

$$\operatorname{Cov}(X_i, X_j) = E\{[X_i - E(X_i)][X_j - E(X_j)]\} \quad (1.177)$$

Proof The proof of this theorem will be given for the case of continuous $X_1, X_2, \ldots, X_n$ only. For the proof of part a, let

$$\theta(X_1, X_2, \ldots, X_n) = \sum_{i=1}^{n} a_i X_i$$

Then, from Eq. (1.172),

$E[\theta(X_1, X_2, \ldots, X_n)]$

$$= \int_{-\infty}^{\infty} \cdots \int_{-\infty}^{\infty} \int_{-\infty}^{\infty} \sum_{i=1}^{n} a_i x_i f(x_1, x_2, \ldots, x_n) \, dx_1 \, dx_2 \cdots dx_n$$

$$= \sum_{i=1}^{n} a_i \int_{-\infty}^{\infty} \cdots \int_{-\infty}^{\infty} \int_{-\infty}^{\infty} x_i f(x_1, x_2, \ldots, x_n) \, dx_1 \, dx_2 \cdots dx_n$$

Now

$$\int_{-\infty}^{\infty} \cdots \int_{-\infty}^{\infty} \int_{-\infty}^{\infty} x_i f(x_1, x_2, \ldots, x_n) \, dx_1 \, dx_2 \cdots dx_n$$

$$= \int_{-\infty}^{\infty} x_i \int_{-\infty}^{\infty} \cdots \int_{-\infty}^{\infty} f(x_1, x_2, \ldots, x_n) \, dx_1 \cdots dx_n \, dx_i$$

$$= \int_{-\infty}^{\infty} x_i f_i(x_i) \, dx_i = E(X_i)$$

from Eq. (1.171) where $f_i(x_i)$ is the marginal probability density function of X_i. Finally then

$$E\left(\sum_{i=1}^{n} a_i X_i\right) = \sum_{i=1}^{n} a_i E(X_i)$$

For part b, let

$$\theta(X_1, X_2) = X_1 X_2$$

1.6 EXPECTATION

Then

$$E(X_i X_j) = \int_{-\infty}^{\infty} \int_{-\infty}^{\infty} x_i x_j f(x_i, x_j) \, dx_i \, dx_j$$

since X_i and X_j are independent random variables

$$f(x_i, x_j) = f_i(x_i) f_j(x_j)$$

where $f_i(x_i)$ and $f_j(x_j)$ are the marginal probability density functions of X_i and X_j. Thus

$$E(X_i X_j) = \int_{-\infty}^{\infty} x_i f_i(x_i) \, dx_i \int_{-\infty}^{\infty} x_j f_j(x_j) \, dx_j = E(X_i) E(X_j)$$

In part c,

$$\text{Var}\left(\sum_{i=1}^{n} a_i X_i\right) = E\left[\sum_{i=1}^{n} a_i X_i - E\left(\sum_{i=1}^{n} a_i X_i\right)\right]^2$$

$$= E\left[\sum_{i=1}^{n} a_i X_i - \sum_{i=1}^{n} a_i E(X_i)\right]^2$$

$$= E\left\{\sum_{i=1}^{n} a_i [X_i - E(X_i)]\right\}^2$$

$$= E \sum_{i=1}^{n} a_i^2 [X_i - E(X_i)]^2$$

$$+ 2 \sum_{i=1}^{n-1} \sum_{j=i+1}^{n} a_i a_j E\{[X_i - E(X_i)][X_j - E(X_j)]\} \quad (1.178)$$

from Eq. (1.173). Now

$$E\{[X_i - E(X_i)][X_j - E(X_j)]\} = E[X_i X_j - X_i E(X_j) - X_j E(X_i) + E(X_i) E(X_j)]$$

Since X_i and X_j are independent and $E(X_i)$ and $E(X_j)$ are constants

$$E\{[X_i - E(X_i)][X_j - E(X_j)]\} = E(X_i X_j) - E(X_i) E(X_j)$$
$$= E(X_i) E(X_j) - E(X_i) E(X_j) = 0$$

Therefore

$$\text{Var}\left[\sum_{i=1}^{n} a_i X_i\right] = \sum_{i=1}^{n} a_i^2 E[X_i - E(X_i)]^2 = \sum_{i=1}^{n} a_i^2 \text{Var}(X_i)$$

The expression given in part d is a direct result of Eq. (1.178). When $X_1, X_2, \ldots, X_n$ are dependent random variables

$$E\{[X_i - E(X_i)][X_j - E(X_j)]\} \neq 0$$

and since

$$\text{Cov}(X_i, X_j) = E\{[X_i - E(X_i)][X_j - E(X_j)]\}$$

the desired result is obtained.

Example 1.38 Find the mean, m, and variance, σ^2, of the following random variables: a. Poisson; b. gamma; c. binomial.

To find the mean and variance of the Poisson random variable we note that

$$\sum_{x=0}^{\infty} \frac{\lambda^x}{x!} \exp[-\lambda] = 1$$

Now

$$m = E(x) = \sum_{x=0}^{\infty} x \frac{\lambda^x}{x!} \exp[-\lambda] = \sum_{x=1}^{\infty} \frac{\lambda^x}{(x-1)!} \exp[-\lambda]$$

$$= \lambda \sum_{x=1}^{\infty} \frac{\lambda^{x-1}}{(x-1)!} \exp[-\lambda] = \lambda \sum_{y=0}^{\infty} \frac{\lambda^y}{y!} \exp[-\lambda] = \lambda \quad (1.179)$$

For the variance we note that

$$\text{Var}(X) = E(X^2) - [E(X)]^2 \quad (1.180)$$

Hence

$$\sigma^2 = \sum_{x=0}^{\infty} x^2 \frac{\lambda^x}{x!} \exp[-\lambda] - \lambda^2$$

Let

$$x^2 = x(x-1) + x$$

Then

$$\sigma^2 = \sum_{x=0}^{\infty} x(x-1) \frac{\lambda^x}{x!} \exp[-\lambda] + \sum_{x=0}^{\infty} x \frac{\lambda^x}{x!} \exp[-\lambda] - \lambda^2$$

$$= \sum_{x=2}^{\infty} \frac{\lambda^x}{(x-2)!} \exp[-\lambda] + \lambda - \lambda^2 = \lambda^2 \sum_{y=0}^{\infty} \frac{\lambda^y}{y!} \exp[-\lambda] + \lambda - \lambda^2 = \lambda$$

$$(1.181)$$

Hence the mean and variance of a Poisson random variable are equal.

For the gamma random variable

$$m = \int_0^{\infty} x \frac{\lambda^n}{\Gamma(n)} x^{n-1} \exp[-\lambda x] \, dx = \int_0^{\infty} \frac{\lambda^n}{\Gamma(n)} x^n \exp[-\lambda x] \, dx = \frac{\lambda^n}{\Gamma(n)} \frac{\Gamma(n+1)}{\lambda^{n+1}}$$

from Eq. (1.117). Hence

$$m = \frac{n}{\lambda} \quad (1.182)$$

For the variance

$$\sigma^2 = \int_0^{\infty} \frac{\lambda^n}{\Gamma(n)} x^{n+1} \exp[-\lambda x] \, dx - \left(\frac{n}{\lambda}\right)^2 = \frac{\lambda^n}{\Gamma(n)} \frac{\Gamma(n+2)}{\lambda^{n+2}} - \left(\frac{n}{\lambda}\right)^2$$

and

$$\sigma^2 = \frac{(n+1)n}{\lambda^2} - \frac{n^2}{\lambda^2} = \frac{n}{\lambda^2} \quad (1.183)$$

1.6 EXPECTATION

The probability mass function of the binomial random variable is given by

$$p(x) = \frac{n!}{x!(n-x)!} p^x (1-p)^{n-x}, \quad x = 0, 1, \ldots, n$$

Thus

$$m = \sum_{x=0}^{n} x \frac{n!}{x!(n-x)!} p^x (1-p)^{n-x} = \sum_{x=1}^{n} \frac{n!}{(x-1)!(n-x)!} p^x (1-p)^{n-x}$$

Let $y = x - 1$. Then

$$m = \sum_{y=0}^{n-1} \frac{n!}{y!(n-y-1)!} p^{y+1} (1-p)^{n-y-1}$$

$$= np \sum_{y=0}^{n-1} \frac{(n-1)!}{y!(n-y-1)!} p^y (1-p)^{n-y-1} = np \quad (1.184)$$

since

$$\sum_{i=0}^{k} \frac{k!}{i!(k-i)!} a^i b^{k-i} = (a+b)^k$$

The variance of the binomial random variable is

$$\sigma^2 = \sum_{x=0}^{n} x^2 \frac{n!}{x!(n-x)!} p^x (1-p)^{n-y} - (np)^2$$

Letting $x = x(x-1) + x$ yields

$$\sigma^2 = \sum_{x=0}^{n} x(x-1) \frac{n!}{x!(n-x)!} p^x (1-p)^{n-x}$$

$$+ \sum_{x=0}^{n} x \frac{n!}{x!(n-x)!} p^x (1-p)^{n-x} - (np)^2$$

$$= \sum_{x=2}^{n} \frac{n!}{(x-2)!(n-x)!} p^x (1-p)^{n-x} + np - (np)^2$$

$$= n(n-1)p^2 \sum_{x=2}^{n} \frac{(n-2)!}{(x-2)!(n-x)!} p^{x-2} (1-p)^{n-x}$$

$$+ np - (np)^2$$

$$= n(n-1)p^2 + np - (np)^2 = np(1-p) \quad \text{◪} \quad (1.185)$$

Example 1.39 X is a uniform random variable with parameters a and b $(a < b)$. Let

$$\theta(x) = \begin{cases} ax, & a \le x < (a+b)/2 \\ bx^2, & (a+b)/2 \le x \le b \end{cases}$$

Find $E[\theta(X)]$.

From Eq. (1.171)

$$E[\theta(X)] = \int_a^b \frac{\theta(x)}{b-a} dx$$

But $\theta(X)$ changes definition at $x = (a+b)/2$. Hence

$$E[\theta(X)] = \int_a^{(a+b)/2} \frac{ax}{a+b} dx + \int_{(a+b)/2}^b \frac{bx^2}{a+b} dx$$

$$= \frac{ax^2}{2(a+b)} \bigg|_a^{(a+b)/2} + \frac{bx^3}{3(a+b)} \bigg|_{(a+b)/2}^b$$

$$= \frac{8b^4 - 12a^3 - b(a+b)^3 + 3a(a+b)^2}{24(a+b)}$$

Example 1.40 A manufacturer places an order for Q units of raw materials T days prior to the start of production. Lead time, the time required for delivery of the order after it is placed, X, is an exponential random variable with parameter λ. If the order arrives before the production start up it must be stored at a cost of C_I per unit per day stored. If the order does not arrive until after the scheduled production start up, production is delayed at a cost of C_D per day delayed. Find the expected cost of the ordering policy.

We wish to find $E[\theta(X)]$ where

$$\theta(x) = \begin{cases} C_I Q(T-x), & 0 < x \le T \\ C_D(x-T), & T < x < \infty \end{cases}$$

Then

$$E[\theta(X)] = \int_0^T C_I Q(T-x)\lambda \exp[-\lambda x] dx$$

$$+ \int_T^\infty C_D(x-T)\lambda \exp[-\lambda x] dx$$

$$= C_I QT \int_0^T \lambda \exp[-\lambda x] dx - C_I Q \int_0^T \lambda x \exp[-\lambda x] dx$$

$$+ C_D \int_T^\infty \lambda x \exp[-\lambda x] dx - C_D T \int_T^\infty \lambda \exp[-\lambda x] dx$$

$$= C_I QT(1 - \exp[-\lambda T]) - \frac{C_I Q}{\lambda}(1 - \lambda T \exp[-\lambda T] - \exp[-\lambda T])$$

$$+ \frac{C_D}{\lambda}(1 + \lambda T)\exp[-\lambda T] - C_D T \exp[-\lambda T]$$

$$= C_I Q\left(\frac{\lambda T - 1}{\lambda}\right) + \frac{C_I Q + C_D}{\lambda} \exp[-\lambda T]$$

1.6 EXPECTATION

1.6.1 Moment Generating Functions

In several examples in the preceding section, we determined the distribution of the sum of independent identically distributed random variables. The reader will recall that this was a rather tedious task using the methods described. In certain cases, this process can be achieved by use of the *moment generating function* with much less computational effort. In addition, the moment generating function can be used to determine the moments of a random variable. In general, when it exists, the moment generating function uniquely characterizes a random variable as do the probability density, probability mass, and distribution functions.

Definition 1.14 If X is a random variable, then $E(\exp[tX])$ is the moment generating function, denoted $M(t, X)$, of X if the expectation exists for all values of t on some interval containing zero.

Definition 1.15 If $X_1, X_2, \ldots, X_n$ are random variables, then $E(\exp[t_1 X_1 + t_2 X_2 + \cdots + t_n X_n])$ is the joint moment generating function of $X_1, X_2, \ldots, X_n$, denoted $M(t_1, t_2, \ldots, t_n; X_1, X_2, \ldots, X_n)$ if the expectation exists for all values of t_i on some interval containing zero, $-h_i < t_i < h_i$, $h_i > 0$, $i = 1, 2, \ldots, n$.

To clarify the above definitions, let X be a random variable with probability density function $f(x)$ or probability mass function $p(x)$. Then the moment generating function of X is given by

$$M(t, X) = \begin{cases} \int_{-\infty}^{\infty} \exp[tx] f(x)\, dx, & \text{continuous case} \\ \sum_{x=-\infty}^{\infty} \exp[tx] p(x), & \text{discrete case} \end{cases} \quad (1.186)$$

Now, if an interval can be found, $-h < t < h$, $h > 0$, such that the expectation defining $M(t, X)$ exists for all t in this interval, then $M(t, X)$ exists. That is, $M(t, X)$ exists if the integral, in the continuous case, or the sum, in the discrete case, in Eq. (1.186) converges for all $-h < t < h$ for at least some $h > 0$.

To illustrate, let X be an exponential random variable with parameter λ. Then

$$f(x) = \lambda \exp[-\lambda x], \quad 0 < x < \infty$$

The moment generating function of X is given by

$$M(t, X) = \int_0^\infty \lambda \exp[-x(\lambda - t)]\, dx = \frac{\lambda}{\lambda - t}$$

The reader will note that $M(t, X)$ goes to infinity for $t = \lambda$. However, it is not necessary that $M(t, X)$ exist for all t. For $-\lambda < t < \lambda$, $M(t, X)$ does exist. Therefore, an interval for t containing zero does exist for which $M(t, X)$ exists and the moment generating function of the exponential random variable is defined.

Now let us illustrate the case of a random variable which has no moment generating function. Let X be a random variable with probability mass function

$$p(x) = \frac{6}{\pi^2 x^2}, \quad x = 1, 2, 3, \ldots$$

Then

$$M(t, X) = \sum_{x=1}^{\infty} \frac{6}{\pi^2 x^2} \exp[tx]$$

$$M(t, X) = \frac{6}{\pi^2} e^t + \frac{6}{4\pi^2} e^{2t} + \frac{6}{9\pi^2} \exp[3t] + \cdots$$

Now, as we will show in Chapter 3, the above sum converges only if the last term in the sum approaches zero. However, in this case the last term approaches zero only if $t \leq 0$ and therefore the series converges only if $t \leq 0$. Hence, no real number $h > 0$ exists such that $M(t, X)$ exists for $-h < t < h$. Thus the moment generating function of X does not exist.

Example 1.41 Find the moment generating function of the following random variables: a. normal; b. gamma; c. binomial; d. Poisson.

The density function of the normal random variable is given by

$$f(x) = \frac{1}{\sigma \sqrt{2\pi}} \exp\left[-\frac{(x-u)^2}{2\sigma^2}\right], \quad -\infty < x < \infty$$

Then

$$M(t, X) = \int_{-\infty}^{\infty} \frac{1}{\sigma \sqrt{2\pi}} \exp\left[tx - \frac{(x-u)^2}{2\sigma^2}\right] dx$$

The exponent in the integral may be expressed as

$$tx - \frac{(x-u)^2}{2\sigma^2} = -\frac{1}{2\sigma^2}[x^2 - 2x(\sigma^2 t + u) + u^2]$$

$$= -\frac{1}{2\sigma^2}[x - (\sigma^2 t + u)]^2 + tu + \frac{\sigma^2 t^2}{2}$$

Hence

$$M(t, X) = \exp\left[ut + \frac{\sigma^2 u^2}{2}\right] \int_{-\infty}^{\infty} \frac{1}{\sigma \sqrt{2\pi}} \exp\left[-\frac{1}{2\sigma^2}[x - (\sigma^2 t + u)]^2\right] dx$$

$$= \exp\left[ut + \frac{\sigma^2 t^2}{2}\right], \quad -\infty < t < \infty$$

The moment generating function of the gamma random variable with density function,

$$f(x) = \frac{\lambda^n}{\Gamma(n)} x^{n-1} \exp[-\lambda x], \quad 0 < x < \infty$$

1.6 EXPECTATION

is given by

$$M(t, X) = \int_0^\infty \frac{\lambda^n}{\Gamma(n)} x^{n-1} \exp[-(\lambda - t)x] \, dx = \frac{\lambda^n}{\Gamma(n)} \frac{\Gamma(n)}{(\lambda + t)^n} = \left(1 - \frac{t}{\lambda}\right)^{-n}, \quad t < \lambda$$

For the binomial random variable with parameters $0 < p < 1$ and integer $n > 0$, we have

$$M(t, X) = \sum_{x=0}^{n} \exp[tx] \frac{n!}{x!(n-x)!} p^x (1-p)^{n-x}$$

$$= \sum_{x=0}^{n} \frac{n!}{x!(n-x)!} (p \exp[t])^x (1-p)^{n-x}$$

$$= [p \exp[t] + (1-p)]^n, \quad -\infty < t < \infty$$

For the Poisson random variable with parameter λ,

$$M(t, X) = \sum_{x=0}^{\infty} \frac{\lambda^x}{x!} \exp[tx] \exp[-\lambda] = \sum_{x=0}^{\infty} \frac{(\lambda \exp[t])^x}{x!} \exp[-\lambda]$$

Since

$$\sum_{x=0}^{\infty} \frac{a^x}{x!} = \exp[a]$$

we have

$$M(t, X) = \exp[-\lambda] \exp[\lambda \exp[t]] = \exp[-\lambda(1 - \exp[t])], \quad -\infty < t < \infty \quad \blacksquare$$

Theorem 1.13 If the moment generating function, $M(t, X)$, of the random variable X exists, then

$$E(X^m) = \frac{d^m}{dt^m} M(t, X) \bigg|_{t=0} \tag{1.187}$$

where $E(X^m)$ is sometimes called the *mth central moment* of the random variable X.

Proof The mth derivative of $M(t, X)$ with respect to t is given by

$$\frac{d^m}{dt^m} M(t, X) = \begin{cases} \int_{-\infty}^{\infty} x^m \exp[tx] f(x) \, dx, & \text{continuous case} \\ \sum_{x=-\infty}^{\infty} x^m \exp[tx] p(x), & \text{discrete case} \end{cases}$$

Evaluating $(d^m/dt^m)M(t, X)$ at $t = 0$, yields

$$\frac{d^m}{dt^m} M(t, x) \bigg|_{t=0} = \begin{cases} \int_{-\infty}^{\infty} x^m f(x) \, dx, & \text{continuous case} \\ \sum_{x=-\infty}^{\infty} x^m p(x), & \text{discrete case} \end{cases}$$

since $\exp[(0)x] = 1$. Hence

$$\left.\frac{d^m}{dt^m} M(t, X)\right|_{t=0} = E(X^m)$$

Theorem 1.13 gives us a method for developing the central moments, $m = 1, 2, \ldots$, of any random variable provided that the moment generating function exists, and it is for this reason that $M(t, X)$ is called a moment generating function. The first central moment, $E(x)$, is called the mean. Hence

$$m = \left.\frac{d}{dt} M(t, X)\right|_{t=0} \tag{1.188}$$

We may also use the moment generating function to determine the variance of a random variable since the variance of a random variable is a function of the first two central moments. From Eq. (1.180)

$$\sigma^2 = E(X^2) - [E(X)]^2 = \left.\frac{d^2}{dt^2} M(t, X)\right|_{t=0} - \left[\left.\frac{d}{dt} M(t, X)\right|_{t=0}\right]^2 \tag{1.189}$$

Other moments and functions of moments may be calculated in a similar manner.

Example 1.42 Using the moment generating function, calculate the mean and variance of the following random variables: a. Poisson; b. gamma; c. binomial.

For part a, the moment generating function of the Poisson random variable is

$$M(t, X) = \exp[-\lambda(1 - \exp[t])], \quad -\infty < t < \infty$$

from Example 1.41. The first and second derivatives of $M(t, X)$ are determined as follows.

$$\frac{d}{dt} M(t, X) = \frac{d}{dt} \exp[-\lambda(1 - \exp[t])] = \lambda \exp[t]\exp[-\lambda(1 - \exp[t])]$$

and

$$\frac{d^2}{dt^2} M(t, X) = \frac{d}{dt}\left[\frac{d}{dt} M(t, X)\right]$$

$$= \frac{d}{dt} \lambda \exp[t]\exp[-\lambda(1 - \exp[t])]$$

$$= \lambda \exp[t]\exp[-\lambda(1 - \exp[t])] + \lambda^2 \exp[2t]\exp[-\lambda(1 - \exp[t])]$$

Evaluating $(d/dt)M(t, X)$ and $(d^2/dt^2)M(t, X)$ at $t = 0$, we have

$$\left.\frac{d}{dt} M(t, X)\right|_{t=0} = \lambda$$

and

$$\left.\frac{d^2}{dt^2} M(t, X)\right|_{t=0} = \lambda + \lambda^2$$

1.6 EXPECTATION

Hence, from Eqs. (1.188) and (1.189)
$$m = \lambda \quad \text{and} \quad \sigma^2 = \lambda^2 + \lambda - \lambda^2 = \lambda$$

For the gamma random variable,
$$M(t, X) = \left(1 - \frac{t}{\lambda}\right)^{-n}, \quad t < \lambda$$

from Example 1.41. Now
$$\frac{d}{dt} M(t, X) = (-n)\left(1 - \frac{t}{\lambda}\right)^{-(n+1)}\left(-\frac{1}{\lambda}\right) = \frac{n}{\lambda}\left(1 - \frac{t}{\lambda}\right)^{-(n+1)}$$

and
$$\frac{d^2}{dt^2} M(t, X) = \frac{d}{dt}\left[\frac{d}{dt} M(t, X)\right]$$
$$= -(n+1)\frac{n}{\lambda}\left(1 - \frac{t}{\lambda}\right)^{-(n+2)}\left(-\frac{1}{\lambda}\right) = \frac{n(n+1)}{\lambda^2}\left(1 - \frac{t}{\lambda}\right)^{-(n+2)}$$

Evaluating the first and second derivatives at $t = 0$ leads us to
$$\left.\frac{d}{dt} M(t, X)\right|_{t=0} = \frac{n}{\lambda} \quad \text{and} \quad \left.\frac{d^2}{dt^2} M(t, X)\right|_{t=0} = \frac{n(n+1)}{\lambda^2}$$

Finally, from Eqs. (1.188) and (1.189)
$$m = \frac{n}{\lambda} \quad \text{and} \quad \sigma^2 = \frac{n(n+1)}{\lambda^2} - \frac{n^2}{\lambda^2} = \frac{n}{\lambda^2}$$

For the binomial random variable,
$$M(t, X) = [p \exp[t] + (1 - p)]^n, \quad -\infty < t < \infty$$

from Example 1.41. Now
$$\frac{d}{dt} M(t, X) = n[p \exp[t] + (1 - p)]^{n-1} p \exp[t]$$

and
$$\frac{d^2}{dt^2} M(t, X) = \frac{d}{dt}\left[\frac{d}{dt} M(t, X)\right]$$
$$= n(n-1)[p \exp[t] + (1 - p)]^{n-2} p^2 \exp[2t]$$

Hence
$$\left.\frac{d}{dt} M(t, X)\right|_{t=0} = np \quad \text{and} \quad \left.\frac{d^2}{dt^2} M(t, X)\right|_{t=0} = n(n-1)p^2 + np$$

and the mean and variance of the binomial random variable are

$$m = np$$
$$\sigma^2 = n(n-1)p^2 + np - (np)^2 = -np^2 + np = np(1-p)$$ ∎

Theorem 1.14 If $X_1, X_2, \ldots, X_n$ are independent random variables whose marginal or individual moment generating functions, $M(t_i, X_i)$, $i = 1, 2, \ldots, n$, exist, then the joint moment generating function of $X_1, X_2, \ldots, X_n$ is given by

$$M(t_1, t_2, \ldots, t_n; X_1, X_2, \ldots, X_n) = M(t_1, X_1)M(t_2, X_2) \cdots M(t_n, X_n) \quad (1.190)$$

That is, the joint moment generating function of n independent random variables can be expressed as the product of the individual or marginal moment generating functions of each random variable.

Proof The proof will be given for the case of continuous $X_1, X_2, \ldots, X_n$ only. From Definition 1.15,

$$M(t_1, t_2, \ldots, t_n; X_1, X_2, \ldots, X_n)$$
$$= \int_{-\infty}^{\infty} \cdots \int_{-\infty}^{\infty} \exp\left[\sum_{i=1}^{n} t_i x_i\right] f(x_1, \ldots, x_n) \, dx_1 \cdots dx_n$$

Since $X_1, X_2, \ldots, X_n$ are assumed to be independently distributed,

$$f(x_1, x_2, \ldots, x_n) = f_1(x_1) f_2(x_2) \cdots f_n(x_n)$$

where $f_i(x_i)$ is the marginal probability density function of X_i, $i = 1, 2, \ldots, n$. Thus

$$M(t_1, t_2, \ldots, t_n; X_1, X_2, \ldots, X_n)$$
$$= \int_{-\infty}^{\infty} \cdots \int_{-\infty}^{\infty} \exp[t_1 x_1] f_1(x_1) \cdots \exp[t_n x_n] f_n(x_n) \, dx_1 \cdots dx_n$$

or

$$M(t_1, t_2, \ldots, t_n; X_1, X_2, \ldots, X_n)$$
$$= \left[\int_{-\infty}^{\infty} \exp[t_1 x_1] f_1(x_1) \, dx\right] \cdots \left[\int_{-\infty}^{\infty} \exp[t_n x_n] f_n(x_n) \, dx_n\right]$$

Since

$$\int_{-\infty}^{\infty} \exp[t_i x_i] f_i(x_i) \, dx_i = M(t_i X_i), \quad i = 1, 2, \ldots, n$$

we have

$$M(t_1, t_2, \ldots, t_n; X_1, X_2, \ldots, X_n) = M(t_1, X_1) M(t_2, X_2) \cdots M(t_n, X_n)$$ ∎

As mentioned earlier, the moment generating function is frequently used to determine the distribution of the sum of independent random variables and occassionally to determine the distribution of other functions of independent

1.6 EXPECTATION

random variables. The following theorem provides the basis for using the moment generating function to define the distribution of the sum of independent random variables.

Theorem 1.15 Let $X_1, X_2, \ldots, X_n$ be a set of independent random variables with marginal moment generating functions given by $M(t_i, X_i)$, $i = 1, 2, \ldots, n$. If

$$Y = \sum_{i=1}^{n} X_i \qquad (1.191)$$

then

$$M(t, Y) = M(t, X_1)M(t, X_2) \cdots M(t, X_n) \qquad (1.192)$$

if $M(t_i, X_i)$ exist for $i = 1, 2, \ldots, n$.

Proof The proof will be given for the case of continuous $X_1, X_2, \ldots, X_n$ only. If

$$Y = \sum_{i=1}^{n} X_i$$

we have

$$M(t, Y) = E[\exp[tY]] = E\left[\exp\left[t \sum_{i=1}^{n} X_i\right]\right] \qquad (1.193)$$

$$M(t, y) = \int_{-\infty}^{\infty} \cdots \int_{-\infty}^{\infty} \int_{-\infty}^{\infty} \exp\left[t \sum_{i=1}^{n} X_i\right] f_1(x_1) f_2(x_2) \cdots f_n(x_n) \, dx_1 \, dx_2 \cdots dx_n$$

since $X_1, X_2, \ldots, X_n$ are assumed to be independently distributed. Thus

$$M(t, Y) = \left[\int_{-\infty}^{\infty} \exp[tx_1] f_1(x_1) \, dx_1\right]\left[\int_{-\infty}^{\infty} \exp[tx_2] f_2(x_2) \, dx_2\right]$$

$$\cdots \left[\int_{-\infty}^{\infty} \exp[tx_n] f_n(x_n) \, dx_n\right]$$

$$= M(t, X_1)M(t, X_2) \cdots M(t, X_n) \qquad \blacksquare$$

Example 1.43 $X_1, X_2, \ldots, X_n$ are independently distributed normal random variables with mean m_i and variance σ_i^2, $i = 1, 2, \ldots, n$. Find the distribution of Y, where

$$Y = \sum_{i=1}^{n} X_i$$

The moment generating function of X_i is given by

$$M(t, X_i) = \exp\left[m_i t + \frac{\sigma_i^2 t^2}{2}\right], \qquad -\infty < t < \infty, \quad i = 1, 2, \ldots, n$$

Since $X_1, X_2, \ldots, X_n$ are independently distributed we may use the results of Theorem 1.15. Hence

$$M(t, Y) = \exp\left[m_1 t + \frac{\sigma_1^2 t^2}{2}\right] \exp\left[m_2 t + \frac{\sigma_2^2 t^2}{2}\right] \cdots \exp\left[m_n t + \frac{\sigma_n^2 t^2}{2}\right]$$

$$= \exp\left[t \sum_{i=1}^{n} m_i + \frac{t}{2} \sum_{i=1}^{n} \sigma_i^2\right], \quad \infty < t < \infty$$

Let

$$\alpha = \sum_{i=1}^{n} m_i, \quad \beta = \sum_{i=1}^{n} \sigma_i^2$$

Then

$$M(t, Y) = \exp\left[\alpha t + \frac{\beta t^2}{2}\right]$$

But $M(t, Y)$ is the moment generating function of a normal random variable with mean α and variance β. Thus, Y is normally distributed with mean $\sum_{i=1}^{n} m_i$ and variance $\sum_{i=1}^{n} \sigma_i^2$.

Example 1.44 Show that the sum, Y, of n independent identically distributed Bernoulli random variables each with parameter p is binomially distributed with probability mass function given by

$$p(y) = \frac{n!}{y!(n-y)!} p^y (1-p)^{n-y}, \quad y = 0, 1, \ldots, n$$

Let $X_1, X_2, \ldots, X_n$ be n independent Bernoulli random variables with probability mass functions

$$p(x_i) = p^{x_i}(1-p)^{1-x_i}, \quad x_i = 0, 1, \quad i = 1, 2, \ldots, n$$

The moment generating function of X_i is

$$M(t, X_i) = \sum_{x_i=0}^{1} \exp[tx_i] p^{x_i}(1-p)^{1-x_i}, \quad i = 1, 2, \ldots, n$$

$$= (1-p) + p \exp[t]$$

Now let

$$Y = \sum_{i=1}^{n} X_i$$

Since $X_1, X_2, \ldots, X_n$ are independently distributed

$$M(t, Y) = M(t, X_1) M(t, X_2) \cdots M(t, X_n)$$

$$M(t, y) = [(1-p) + p \exp[t]]^n$$

1.6 EXPECTATION

But $M(t, Y)$ is the moment generating function of a binomial random variable with parameters n and p from Example 1.41. Since the moment generating function is unique, the probability mass function of Y is

$$p(y) = \frac{n!}{y!(n-y)!} p^y (1-p)^{n-y}, \quad y = 0, 1, \ldots, n$$

Example 1.45 If the moment generating function of the χ^2 random variable, X, is given by

$$M(t, X) = (1 - 2t)^{r/2}, \quad t < \tfrac{1}{2} \quad (1.194)$$

where r is the degrees of freedom, show that the sum of the squares of independent, identically distributed standard normally random variables is χ^2 distributed.

Let $Z_1, Z_2, \ldots, Z_n$ be independent standard normal random variables. Then

$$M(t, Z_i) = \exp\left[\frac{t^2}{2}\right], \quad -\infty < t < \infty, \quad i = 1, 2, \ldots, n$$

since $m_i = 0$ and $\sigma_i^2 = 1$ for a standard normal random variable. Now let

$$Y_i = Z_i^2, \quad i = 1, 2, \ldots, n$$

The moment generating function of Y is given by

$$M(t, Y_i) = E(\exp[tY_i]) = E(\exp[tZ_i^2]) = \int_{-\infty}^{\infty} \frac{1}{\sqrt{2\pi}} \exp[tz_i^2] \exp\left[\frac{z_i^2}{2}\right] dz_i$$

Letting $v = \sqrt{1 - 2t}\, z_i$ we have

$$dz_i = \frac{dv}{\sqrt{1 - 2t}}$$

and

$$M(t, Y_i) = \frac{1}{\sqrt{1 - 2t}} \int_{-\infty}^{\infty} \frac{1}{\sqrt{2\pi}} \exp\left[-\frac{v^2}{2}\right] dv = (1 - 2t)^{-\tfrac{1}{2}}$$

Since $Z_1, Z_2, \ldots, Z_n$ are independently distributed, so are $Y_1, Y_2, \ldots, Y_n$. If

$$X = \sum_{i=1}^{n} Z_i^2 = \sum_{i=1}^{n} Y_i$$

then

$$M(t, X) = M(t, Y_1) M(t, Y_2) \cdots M(t, Y_n) = (1 - 2t)^{n/2}$$

Thus X is χ^2 distributed with n degrees of freedom.

1.7 Applications

1.7.1 Statistical Methods

Point Estimation—Method of Moments

In our discussion of probability distributions we found that the probability mass and density functions for the random variables discussed were functions of one or more constants or parameters. Suppose that we have data available which indicates that a random variable is of a given type: Poisson, binomial, normal, χ^2, etc. In order to completely specify the probability mass function or the probability density function of the random variable it is necessary to determine the numerical values of the parameters of the distribution. For example, suppose that we have reason to believe that X is an exponential random variable. Then

$$f(x) = \lambda \exp[-\lambda x], \qquad 0 < x < \infty$$

However, each value of λ defines a different density function. We wish to know the value of λ associated with the random variable of current interest.

Suppose that the probability density or mass function of X is a function of k parameters, $\theta_1, \theta_2, \ldots, \theta_k$, such that

$$E(X^i) = g_i(\theta_1, \theta_2, \ldots, \theta_k), \qquad i = 1, 2, \ldots, k \qquad (1.195)$$

where each parameter θ_j, $j = 1, 2, \ldots, k$, is included in at least one $g_i(\theta_1, \theta_2, \ldots, \theta_k)$, $i = 1, 2, \ldots, k$, and where $E(X^i)$ is the ith central moment of X. We may estimate $E(X^i)$ by $(1/n)\sum_{j=1}^{n} X_j^i$, where X_j is the jth sample value of X, $j = 1, 2, \ldots, n$. If we designate an estimate of $E(X^i)$ by $\hat{E}(X^i)$, then

$$\hat{E}(X^i) = \frac{1}{n}\sum_{j=1}^{n} X_j^i, \qquad i = 1, 2, \ldots, k \qquad (1.196)$$

Now equating $\hat{E}(X^i)$ and $g_i(\theta_1, \theta_2, \ldots, \theta_k)$ we have k equations and k unknown values of θ_i, $i = 1, 2, \ldots, k$. Hence, we may solve for $\hat{\theta}_1, \hat{\theta}_2, \ldots, \hat{\theta}_k$, the estimated values of $\theta_1, \theta_2, \ldots, \theta_k$.

Example 1.46 Using the method of moments develop estimates for the parameters of

    a. the normal probability density function;
    b. the binomial probability mass function.
    For part a

$$f(x) = \frac{1}{\sigma\sqrt{2\pi}} \exp\left[-\frac{(x-m)^2}{2\sigma^2}\right], \qquad -\infty < x < \infty$$

1.7 APPLICATIONS

Since $f(x)$ is a function of two parameters, m and σ^2, we must determine the first two central moments of X. From Eq. (1.187)

$$E(X) = \frac{d}{dt} M(t, X) \bigg|_{t=0}$$

and

$$E(X^2) = \frac{d}{dt} M(t, X) \bigg|_{t=0}$$

From Example 1.41, the moment generating function of the normal random variable is given by

$$M(t, X) = \exp\left[mt + \frac{\sigma^2 t^2}{2}\right], \quad -\infty < t < \infty$$

Thus

$$\frac{d}{dt} M(t, X) \bigg|_{t=0} = (m + \sigma^2 t) \exp\left[mt + \frac{\sigma^2 t^2}{2}\right]\bigg|_{t=0} = m$$

and

$$\frac{d^2}{dt^2} M(t, X) \bigg|_{t=0} = \sigma^2 \exp\left[mt + \frac{\sigma^2 t^2}{2}\right]\bigg|_{t=0} + (m + \sigma^2 t)^2 \exp\left[mt + \frac{\sigma^2 t^2}{2}\right]\bigg|_{t=0}$$

$$= \sigma^2 + m^2$$

Equating $\hat{E}(X)$ and $E(X)$ and $\hat{E}(X^2)$ and $E(X^2)$ yields

$$\frac{1}{n} \sum_{j=1}^{n} X_j = m \quad \text{and} \quad \frac{1}{n} \sum_{j=1}^{n} X_j^2 = \sigma^2 + m^2$$

Therefore

$$\hat{m} = \frac{1}{n} \sum_{j=1}^{n} X_j \tag{1.197}$$

and

$$\hat{\sigma}^2 = \frac{1}{n} \left[\sum_{j=1}^{n} X_j^2 - \frac{1}{n}\left(\sum_{j=1}^{n} X_j\right)^2\right]$$

The probability mass function of the binomial random variable with parameters m and p is

$$p(x) = \frac{m!}{x!(m-x)!} p^x (1-p)^{m-x}, \quad x = 0, 1, \ldots, m$$

and

$$M(t, X) = [p \exp[t] + (1-p)]^m, \quad -\infty < t < \infty$$

from Example 1.41. Thus

$$E(X) = \frac{d}{dt} M(t, X)\bigg|_{t=0} = mp$$

and

$$E(X^2) = \frac{d^2}{dt^2} M(t, X)\bigg|_{t=0} = m(m-1)p^2 + mp$$

from Example 1.42. Hence

$$\frac{1}{n} \sum_{j=1}^{n} X_j = mp \qquad (1.198)$$

and

$$\frac{1}{n} \sum_{j=1}^{n} X_j^2 = m(m-1)p^2 + mp \qquad (1.199)$$

Solving Eqs. (1.198) and (1.199) for $\hat{m}$ and $\hat{p}$ leads to

$$\hat{p} = 1 - \frac{\sum_{j=1}^{n} X_j^2 - \frac{1}{n}(\sum_{j=1}^{n} X_j)^2}{\sum_{j=1}^{n} X_j} \qquad (1.200)$$

and

$$\hat{m} = \frac{\sum_{j=1}^{n} X_j}{n\hat{p}} \qquad \text{⑤} \qquad (1.201)$$

Example 1.47 (*Unbiased Estimators*) A point estimate $\hat{\theta}$ for a parameter θ is said to be an unbiased estimate if

$$E(\hat{\theta}) = \theta \qquad (1.202)$$

Find unbiased estimators for the parameters m and σ^2 of the normal probability density function.

From Example 1.46, estimates of m and σ^2, $\hat{m}$ and $\hat{\sigma}^2$, are given by

$$\hat{m} = \frac{1}{n} \sum_{j=1}^{n} X_j \quad \text{and} \quad \hat{\sigma}^2 = \frac{1}{n}\left[\sum_{j=1}^{n} X_j^2 - \frac{1}{n}\left(\sum_{j=1}^{n} X_j\right)^2\right]$$

We will first determine whether $\hat{m}$ and $\hat{\sigma}^2$ are unbiased estimates of m and σ^2. It will be assumed that the observed values of X, X_j, $j = 1, 2, \ldots, n$, are randomly selected and therefore independent normal random variables. Now

$$E(\hat{m}) = E\left(\frac{1}{n} \sum_{j=1}^{n} X_j\right) = \frac{1}{n} \sum_{j=1}^{n} E(X_j)$$

Since X_j is normally distributed with parameters m and σ^2 we have

$$E(X_j) = m \quad \text{and} \quad E(\hat{m}) = \frac{1}{n} \sum_{j=1}^{n} m = m$$

Thus, $\hat{m} = (1/n) \sum_{j=1}^{n} X_j$ is an unbiased estimate of m.

1.7 APPLICATIONS

We now consider $E(\hat{\sigma}^2)$.

$$E(\hat{\sigma}^2) = E\frac{1}{n}\left[\sum_{j=1}^{n} X_j^2 - \frac{1}{n}\left(\sum_{j=1}^{n} X_j\right)^2\right] = \frac{1}{n}\sum_{j=1}^{n} E(X_j^2) - \frac{1}{n^2} E\left(\sum_{j=1}^{n} X_j\right)^2$$

Now $E(X_j^2) = \sigma^2 + m^2$ from Example 1.46.

$$E\left(\sum_{j=1}^{n} X_j\right)^2 = E\left[\sum_{j=1}^{n} X_j^2 + 2\sum_{j=1}^{n-1}\sum_{k=j+1}^{n} X_j X_k\right] = \sum_{j=1}^{n} E(X_j^2) + 2\sum_{j=1}^{n-1}\sum_{k=j+1}^{n} E(X_j X_k)$$

Since X_j and X_k are independently distributed for $j \neq k$

$$E(X_j X_k) = m^2$$

and

$$E\left(\sum_{j=1}^{n} X_j\right)^2 = \sum_{j=1}^{n} (\sigma^2 + m^2) + 2\sum_{j=1}^{n-1}\sum_{k=j+1}^{n} m^2 = n(\sigma^2 + m^2) + n(n-1)m^2$$

Therefore

$$E(\hat{\sigma}^2) = (\sigma^2 + m^2) - \frac{\sigma^2 + m^2}{n} - \frac{n-1}{n}m^2 = \frac{(n-1)}{n}\sigma^2$$

Since

$$E(\hat{\sigma}^2) \neq \sigma^2$$

$\hat{\sigma}^2$ is not an unbiased estimate for σ^2. However, if we multiply $\hat{\sigma}^2$ by $n/(n-1)$ such that a new estimator, $\hat{\sigma}_1^2$ is defined, then

$$E(\hat{\sigma}_1^2) = \sigma^2$$

Hence

$$\hat{\sigma}_1^2 = \frac{1}{n-1}\left[\sum_{j=1}^{n} X_j^2 - \frac{1}{n}\left(\sum_{j=1}^{n} X_j\right)^2\right] \tag{1.203}$$

is an unbiased estimate of σ^2.

Example 1.48 **(Hypothesis Testing)** A large lot of manufactured items is to be submitted for inspection. The mean dimension of the units in the lot is supposed to be m_0. If the mean dimension varies from m_0 by $\pm \delta$ the lot should be rejected where $\delta > 0$. Thus we wish to test the hypothesis that $m = m_0$ against the alternative that $m \neq m_0$. In order to determine whether the true lot mean, m, varies significantly from m_0, m is to be estimated using the sample mean $\bar{X}$. To calculate $\bar{X}$ a sample of n items is selected at random from the lot and inspected. Let $X_1, X_2, \ldots, X_n$ be the dimension of each of the n items. Then

$$\bar{X} = \frac{1}{n}\sum_{i=1}^{n} X_i$$

is an unbiased estimate of m. A history of collected data on this item indicates that X_i is normally distributed with variance σ^2, although the lot mean, m, is known to vary from lot to lot.

A statistical test is to be designed which can be used to assess the quality of the lot where the quality of the lot is measured by the deviation of m from m_0. That is, if $m = m_0$ the lot is considered to be of high quality and thus acceptable. On the other hand if $|m - m_0| \geq \delta$, the lot is considered to be of poor quality and therefore rejectable. The test is to be designed such that it will be rejected with probability α, $0 < \alpha < 1$, if $m = m_0$. The error of incorrectly rejecting a lot is usually called a *Type I error*. On the other hand, if $|m - m_0| = \delta$, the lot is to be accepted with probability β, $0 < \beta < 1$. The error of incorrectly accepting a lot is usually called a *Type II error*. The criteria for rejection or acceptance are to be control limits on $\bar{X}$, C_L and C_U, such that if $C_L \leq \bar{x} \leq C_U$ the lot is accepted and is rejected if $\bar{x} < C_L$ or $\bar{x} > C_U$ where C_L and C_U are to be equidistant from m_0. That is, $m_0 - C_L = C_U - m_0$. Determine the control limits, C_L and C_U, and the sample size n such that the criteria for the test are met.

From Example 1.36, the distribution of the sample mean, $\bar{X}$, is normal with mean m and standard deviation $\sigma/\sqrt{n}$. Therefore

$$f(\bar{x}) = \frac{\sqrt{n}}{\sigma\sqrt{2\pi}} \exp\left[-\frac{n(\bar{x} - m)^2}{2\sigma^2}\right], \quad -\infty < \bar{x} < \infty$$

Now the test is to be designed such that

$$P(\text{rejection} \,|\, m = m_0) = P(\text{Type I error}) = \alpha \quad (1.204)$$

and

$$P(\text{acceptance} \,|\, |m - m_0| = \delta) = P(\text{Type II error}) = \beta \quad (1.205)$$

But

$$P(\text{rejection} \,|\, m = m_0) = 1 - P(\text{acceptance} \,|\, m = m_0)$$
$$= 1 - P(C_L \leq \bar{x} \leq C_U \,|\, m = m_0)$$
$$= 1 - \int_{C_L}^{C_U} \frac{\sqrt{n}}{\sigma\sqrt{2\pi}} \exp\left[-\frac{n(\bar{x} - m_0)^2}{2\sigma^2}\right] d\bar{x} \quad (1.206)$$

Since $C_U - m_0 = m_0 - C_L$ let

$$\Delta = C_U - m_0 = m_0 - C_L$$

Then

$$P(\text{rejection} \,|\, m = m_0) = 1 - \int_{m_0 - \Delta}^{m_0 + \Delta} \frac{\sqrt{n}}{\sigma\sqrt{2\pi}} \exp\left[-\frac{n(\bar{x} - m_0)^2}{2\sigma^2}\right] d\bar{x}$$

Let

$$z = \frac{\bar{x} - m_0}{\sigma/\sqrt{n}} \quad \text{and} \quad d\bar{x} = \frac{\sigma}{\sqrt{n}} dz + m_0$$

1.7 APPLICATIONS

Thus

$$P(\text{rejection} \mid m = m_0) = 1 - \int_{-\sqrt{n}\Delta/\sigma}^{\sqrt{n}\Delta/\sigma} \frac{1}{\sqrt{2\pi}} \exp\left[-\frac{z^2}{2}\right] dz = \alpha$$

and

$$\int_{-\sqrt{n}\Delta/\sigma}^{\sqrt{n}\Delta/\sigma} \frac{1}{\sqrt{2\pi}} \exp\left[-\frac{z^2}{2}\right] dz = 1 - \alpha \qquad (1.207)$$

or

$$F\left(\frac{\sqrt{n}\Delta}{\sigma}\right) - F\left(-\frac{\sqrt{n}\Delta}{\sigma}\right) = 1 - \alpha$$

Since the standard normal probability density function is symmetric about zero,

$$F\left(-\frac{\sqrt{n}\Delta}{\sigma}\right) = 1 - F\left(\frac{\sqrt{n}\Delta}{\sigma}\right), \qquad 2F\left(\frac{\sqrt{n}\Delta}{\sigma}\right) = 2 - \alpha$$

or

$$F\left(\frac{\sqrt{n}\Delta}{\sigma}\right) = 1 - \frac{\alpha}{2} \qquad (1.208)$$

That is, $\sqrt{n}\Delta/\sigma$ is the value of the standard normal random variable, Z, such that

$$P\left(Z \leq \frac{\sqrt{n}\Delta}{\sigma}\right) = 1 - \frac{\alpha}{2}$$

Let us designate this value by $Z_{1-\alpha/2}$. Hence

$$\Delta = \frac{\sigma}{\sqrt{n}} Z_{1-\alpha/2} \qquad (1.209)$$

The cumulative distribution function, $F(z)$, of the standard normal random variable is given in Table 1 in the Appendix such that for any value of α the corresponding value of $Z_{1-\alpha/2}$ can be easily found. Thus the control limits C_L and C_U are given by

$$C_U = m_0 + \frac{\sigma}{\sqrt{n}} Z_{1-\alpha/2} \qquad (1.210)$$

$$C_L = m_0 - \frac{\sigma}{\sqrt{n}} Z_{1-\alpha/2} \qquad (1.211)$$

To determine the sample size, n, we use Eq. (1.205).

$$P(\text{acceptance} \mid |m - m_0| = \delta) = \beta$$

or

$$P(\text{acceptance} \mid m = m_0 + \delta) = P(\text{acceptance} \mid m = m_0 - \delta) = \beta$$

Now,
$$P(\text{acceptance} \mid m = m_0 + \delta) = \int_{C_L}^{C_U} \frac{\sqrt{n}}{\sigma\sqrt{2\pi}} \exp\left[-\frac{n(\bar{x} - m_0 - \delta)^2}{2\sigma^2}\right] d\bar{x} \quad (1.212)$$

Unless δ is quite small,
$$\int_{-\infty}^{C_L} \frac{\sqrt{n}}{\sigma\sqrt{2\pi}} \exp\left[-\frac{n(x - m_0 - \delta)^2}{2\sigma^2}\right] d\bar{x} \simeq 0 \quad (1.213)$$

Assuming that Eq. 1.213 holds, we have
$$P(\text{acceptance} \mid m = m_0 + \delta) = \int_{-\infty}^{C_U} \frac{\sqrt{n}}{\sigma\sqrt{2\pi}} \exp\left[-\frac{n(\bar{x} - m_0 - \delta)^2}{2\sigma^2}\right] d\bar{x} \quad (1.214)$$

Letting
$$z = \frac{\bar{x} - m_0 - \delta}{\sigma/\sqrt{n}} \quad \text{and} \quad d\bar{x} = \frac{\sigma}{\sqrt{n}} dz + m_0 + \delta$$

yields us
$$P(\text{acceptance} \mid m = m_0 + \delta) = \int_{-\infty}^{\sqrt{n}(\Delta - \delta)/\sigma} \frac{1}{\sqrt{2\pi}} \exp\left[-\frac{z^2}{2}\right] dz$$

or
$$F\left[\frac{\sqrt{n}(\Delta - \delta)}{\sigma}\right] = \beta \quad (1.215)$$

and
$$\frac{\sqrt{n}(\Delta - \delta)}{\sigma} = Z_\beta \quad (1.216)$$

Combining Eqs. (1.209) and (1.216) gives us
$$\Delta = \frac{\sigma}{\sqrt{n}} Z_{1-\alpha/2} \quad \text{and} \quad \Delta = \frac{\sigma}{\sqrt{n}} Z_\beta + \delta$$

Thus
$$\frac{\sigma}{\sqrt{n}} Z_{1-\alpha/2} = \frac{\sigma}{\sqrt{n}} Z_\beta + \delta \quad (1.217)$$

Solving for n we have
$$n = \left(\frac{\sigma}{\delta}\right)^2 (Z_{1-\alpha/2} - Z_\beta)^2 \quad (1.218)$$

and the complete solution to the problem is determined and given by Eqs. (1.210), (1.211), and (1.218).

1.7 APPLICATIONS

Example 1.49 A certain container is produced in lots of size 100,000. The lot mean inside diameter of the containers is to be one inch. If the lot mean is as large as 1.01 inches or as small as 0.99 inches the lot should be rejected with probability 0.90 $(1 - \beta)$. If the lot mean is in fact one inch the probability of rejecting the lot is to be 0.05 (α). The inside diameter of the containers is known to be normally distributed with standard deviation 0.007 inches. Find the control limits, C_L and C_U, and the sample size n for the test.

From Table 1 of the Appendix

$$Z_{1-\alpha/2} = Z_{0.975} = 1.96 \quad \text{and} \quad Z_\beta = Z_{0.10} = -1.28$$

Since $m_0 = 1$, we have for the control limits,

$$C_U = 1 + \frac{0.007}{\sqrt{n}}(1.96) \quad \text{and} \quad C_L = 1 - \frac{0.007}{\sqrt{n}}(1.96)$$

From Eq. (1.209),

$$n = \left(\frac{0.007}{0.01}\right)^2 (1.96 + 1.28)^2 = (0.49)(10.50) = 5.15$$

Hence $n = 5$ and the control limits for the test are

$$C_U = 1 + \frac{0.007}{\sqrt{5}}(1.96) = 1.0064$$

and

$$C_L = 1 - \frac{0.007}{\sqrt{5}}(1.96) = 0.9936$$

Example 1.50 (*Confidence Interval Estimation*) The mean, m, of a given measurable characteristic is to be estimated by means of the sample mean, $\bar{X}$, based upon n observations of the characteristic. However, $\bar{X}$ is a random variable and therefore contains error. For small n, $\bar{X}$ might prove to be a poor estimate of m. Thus we seek a method whereby the precision with which $\bar{X}$ estimates m can be determined. To accomplish this we will develop a confidence interval estimate for m.

If the interval L to U ($L < U$) is a $(1 - \alpha)$ 100% confidence interval for m, where $0 < \alpha < 1$, and where L and U are functions of $\bar{x}$, α, and n, then the probability that the interval *will* contain the true mean m is $1 - \alpha$. The reader should note that this probability statement is made prior to the collection of data. That is, once the data is collected and a value of $\bar{X}$ calculated, numerical values of L and U are determined and m either lies within the interval or it does not, since m is a constant. However, if we were to repeat the experiment over and over again we would find that $(1 - \alpha)100\%$ of the calculated intervals would contain the true mean, m. Hence, we say that the probability that the confidence interval *will* contain the true mean is $1 - \alpha$, while there is a $100\alpha\%$ chance that the interval *will not* contain the true mean. Assume that the measurable characteristic is normally distributed with standard deviation σ. Define the $(1 - \alpha)\, 100\%$ confidence limits, L and U, for m if $U - m = m - L$.

Since $\bar{X}$ is a normally distributed random variable with mean m and standard deviation $\sigma/\sqrt{n}$, we have

$$f(\bar{x}) = \frac{\sqrt{n}}{\sigma\sqrt{2\pi}} \exp\left[-\frac{n(\bar{x} - m)^2}{2\sigma^2}\right], \qquad -\infty < \bar{x} < \infty$$

Hence

$$z = \frac{\bar{x} - m}{\sigma/\sqrt{n}}$$

is a standard normal random variable and

$$P\left(Z_{\alpha/2} \leq \frac{\bar{x} - m}{\sigma/\sqrt{n}} \leq Z_{1-\alpha/2}\right) = 1 - \alpha$$

or

$$P\left(\frac{\sigma}{\sqrt{n}} Z_{\alpha/2} - \bar{x} \leq -m \leq \frac{\sigma}{\sqrt{n}} Z_{1-\alpha/2} - \bar{x}\right) = 1 - \alpha$$

Multiplying throughout the inequality by -1 gives us

$$P\left(\bar{x} - \frac{\sigma}{\sqrt{n}} Z_{1-\alpha/2} \leq m \leq \bar{x} - \frac{\sigma}{\sqrt{n}} Z_{\alpha/2}\right) = 1 - \alpha$$

Since

$$Z_{\alpha/2} = -Z_{1-\alpha/2}$$

we have for the confidence limits for m,

$$L = \bar{x} - \frac{\sigma}{\sqrt{n}} Z_{1-\alpha/2} \qquad (1.219)$$

$$U = \bar{x} + \frac{\sigma}{\sqrt{n}} Z_{1-\alpha/2} \qquad (1.220)$$

1.7.2 Mathematical Modeling

Example 1.51 (*Inventory*) A store has a standing order for delivery of Q units of a perishable product at the beginning of each day. The price paid by the store for the units is C per unit. The selling price per unit is S $(C < S)$. If any unit is not sold by the end of the day it is discarded at a loss C per unsold unit. Daily demand for the product, X, is Poisson distributed with parameter λ. Find the expected profit per day from the policy calling for the delivery of Q units per day.

Let $P(X)$ be the profit derived if X units of product are demanded in a given day. Then

$$P(x) = \begin{cases} (S - C)x - C(Q - x), & x = 0, 1, 2, \ldots, Q \\ (S - C)Q, & x = Q + 1, Q + 2, \ldots \end{cases}$$

Then the expected profit per day is given by

$$E[P(X)] = \sum_{x=0}^{Q}(Sx - CQ)\frac{\lambda^x}{x!}\exp[-\lambda] + \sum_{x=Q+1}^{\infty}(S - C)Q\frac{\lambda^x}{x!}\exp[-\lambda]$$

$$= S\sum_{x=0}^{Q} x\frac{\lambda^x}{x!}\exp[-\lambda] + SQ\sum_{x=Q+1}^{\infty}\frac{\lambda^x}{x!}\exp[-\lambda] - CQ \qquad \square$$

Example 1.52 **(Reliability)** A company manufactures a product which it guarantees for S years. If the product fails within the guarantee period it is replaced at a cost C_r. The time until failure of the product, T, is exponentially distributed with parameter λ, where λ is the failure rate. The cost of manufacturing a product with failure rate λ is C_m/λ. That is, as the failure rate is reduced the cost of producing items with the reduced failure rate increases. Determine the expected cost of manufacturing and replacement of units.

As given above, the cost of manufacture is C_m/λ. The expected cost of replacement is given by

$$E(\text{replacement cost per unit}) = C_r P(\text{unit is replaced})$$

$$= C_r P(T \le S) = C_r \int_0^S \lambda \exp[-\lambda t]\, dt$$

Hence the expected total cost of manufacture and replacement per unit $C_T(\lambda)$ is

$$C_T(\lambda) = \frac{C_m}{\lambda} + C_r\int_0^S \lambda \exp[-\lambda T]\, dt = \frac{C_m}{\lambda} + C_r(1 - \exp[-\lambda S]) \qquad \square$$

Example 1.53 **(Quality Control)** The critical dimension, X, of a particular product is known to be normally distributed. The product is manufactured in large lots of size L units. Let S_L and S_U ($S_L < S_U$) be the specification limits for the critical dimension, X, of a unit. If $S_L \le x \le S_U$ the unit is considered acceptable and is defective if $x < S_L$ or $x > S_U$. The variance of the critical dimension per lot is σ^2 and is constant from one lot to another. However, the lot mean varies from lot to lot in a random fashion. Specifically, the lot mean M is normally distributed with mean m_0 and variance σ^2/L.

From each lot a sample of n items are selected at random, inspected and the sample mean $\bar{X}$ calculated. If the value of $\bar{X}$, $\bar{x}$, lies inside the control limits C_L and C_U ($C_L < C_U$) the lot is accepted. If $\bar{x} < C_L$ or $\bar{x} > C_U$ the lot is rejected and scrapped. The inspection process is destructive and inspected units are scrapped.

The cost of inspecting any unit of product is C_I and includes the cost of scrapping the unit. The cost of scrapping a unit in a rejected lot is C_r. If the lot is accepted it is used in its entirety in a subsequent manufacturing process. Each defective unit in an accepted lot results in a cost C_a. Thus each unit in an accepted lot such that $x < S_L$ or $x > S_U$ is defective and results in a cost C_a. Find the expected total cost of the quality control per lot.

There are three random variables to be considered in this system; the lot mean, M, the sample mean, $\bar{X}$, and the critical dimension X. Let $C_T(C_L, C_U, n|m)$

be the expected total cost per lot given that the lot mean, M, is m. If $h(m)$ is the density function of M, then the unconditional expected total cost, $C_T(C_L, C_U, n)$, is given by

$$C_T(C_L, C_U, n) = \int_{-\infty}^{\infty} C_T(C_L, C_U, n \mid m) h(m) \, dm \qquad (1.221)$$

The cost of inspection per lot is $C_I n$ regardless of whether the lot is accepted or rejected. Let $C_R(m)$ and $C_A(m)$ be the expected costs of rejection and acceptance given $M = m$. Then

$$C_T(C_L, C_U, n \mid m) = C_I n + C_R(m) + C_A(m) \qquad (1.222)$$

After the inspection process is completed there are $(L - n)$ units remaining in the lot. If the lot is rejected all $(L - n)$ units are scrapped at a cost $C_r(L - n)$. If the lot is accepted then the cost of the lot is $C_a(L - n)[1 - \int_{S_L}^{S_U} f(x \mid m) \, dx]$, where

$$\left[1 - \int_{S_L}^{S_U} f(x \mid m) \, dx\right] = P(\text{any unit is defective}) \qquad (1.223)$$

and $f(x \mid m)$ is the conditional density function of X given $M = m$. Thus

$$C_R(m) = C_r(L - n) P(\text{lot is rejected} \mid m)$$

$$= C_r(L - n)\left[1 - \int_{C_L}^{C_U} \frac{\sqrt{n}}{\sigma \sqrt{2\pi}} \exp\left[-\frac{n(\bar{x} - m)^2}{2\sigma^2}\right] d\bar{x}\right] \qquad (1.224)$$

and

$$C_A(m) = C_a(L - n)\left[1 - \int_{S_L}^{S_U} f(x) \, dx\right] P(\text{lot is accepted} \mid m)$$

$$= C_a(L - n)\left[1 - \int_{S_L}^{S_U} \frac{1}{\sigma \sqrt{2\pi}} \exp\left[-\frac{(x - m)^2}{2\sigma^2}\right] dx\right]$$

$$\times \left[\int_{C_L}^{C_U} \frac{\sqrt{n}}{\sigma \sqrt{2\pi}} \exp\left[-\frac{n(\bar{x} - m)^2}{2\sigma^2}\right] d\bar{x}\right] \qquad (1.225)$$

Let

$$F(y \mid m) = \int_{-\infty}^{y} \frac{1}{\sigma \sqrt{\pi}} \exp\left[-\frac{(x - m)^2}{2\sigma^2}\right] dx$$

and

$$G(y \mid m) = \int_{-\infty}^{y} \frac{\sqrt{n}}{\sigma \sqrt{2\pi}} \exp\left[-\frac{n(\bar{x} - m)^2}{2\sigma^2}\right] d\bar{x}$$

Then

$$C_R(m) = C_r(L - n)[1 - G(C_U \mid m) + G(C_L \mid m)]$$

1.7 APPLICATIONS

and

$$C_A(m) = C_a(L - n)[1 - F(S_U | m) + F(S_L | m)][G(C_U | m) - G(C_L | m)]$$

Hence

$$C_T(C_L, C_U, n | m) = C_I n + (L - n)\{C_r + (C_a - C_r)[G(C_U | m) - G(C_L | m)] \\ - C_a[F(S_U | m) - F(S_L | m)][G(C_U | m) - G(C_L | m)]\} \quad (1.226)$$

Since M is normally distributed with mean m_0 and variance $\sigma^2/\sqrt{L}$, we have

$$C_T(C_L, C_U, n) = C_I n + C_r(L - n) + \int_{-\infty}^{\infty} [G(C_U | m) - G(C_L | m)]\{(C_a - C_r) \\ - C_a[F(S_U | m) - F(S_L | m)]\} \frac{\sqrt{L}}{\sigma\sqrt{2\pi}} \exp\left[-\frac{L(m - m_0)^2}{2\sigma^2}\right] dm \quad (1.227)$$

1.7.3 Engineering Applications

Example 1.54 An electrical device must operate for T hours to fulfill its intended purpose. However, there is a particular circuit in the device which was proved unreliable in the past. Specifically, the life of this circuit is exponentially distributed with parameter λ, where $1/\lambda < T$. To circumvent this problem, n of these circuits are to be arranged in parallel. Let t_i be the operating life of the ith parallel circuit in hours. If $t_1 < T$, then the second circuit takes over. If $t_1 + t_2 < T$, then the third circuit takes over, and so forth. If $\sum_{i=1}^{n} t_i < T$, the device fails. If

$$\lambda = 5.00, \quad T = 1.00$$

what is the maximum number of parallel circuits necessary such that the probability of successful operation of the device for T hours or more is at least 0.97?

Let τ be the operating life of the device in hours. Then

$$P(\tau \geq T) = 1 - P(\tau < T) = 1 - P(t_1 + t_2 + \cdots + t_n < T)$$

Since

$$f(t_i) = \lambda \exp[-\lambda t_i], \quad 0 < t_i < \infty, \quad i = 1, 2, \ldots, n$$

from Example 1.33,

$$g(\tau) = \frac{\tau^{n-1}}{(n-1)!} \lambda^n \exp[-\lambda \tau], \quad 0 < \tau < \infty$$

Now,

$$P(\tau \geq T) = 1 - \int_0^T g(\tau) \, d\tau$$

One may show [see Eq. (3.139) and the development following] that

$$P(\tau \geq T) = \sum_{i=0}^{n-1} \frac{(\lambda T)^i}{i!} \exp[-\lambda T] \quad \text{or} \quad P(\tau \geq 1) = \sum_{i=0}^{n-1} \frac{(5)^i}{i!} \exp[-5]$$

Evaluation of $P(\tau \geq T)$ for successive values of n leads to the results shown in Table 1.7. As shown in this table, 11 parallel circuits are necessary to satisfy the condition that $P(\tau \geq T) \geq 0.97$.

TABLE 1.7

n	$P(\tau \geq T)$	n	$P(\tau \geq T)$
1	0.007	7	0.762
2	0.040	8	0.867
3	0.125	9	0.932
4	0.265	10	0.968
5	0.440	11	0.986
6	0.616		

Example 1.55 In Example 1.54, the exact distribution of τ was determined to find the number of parallel circuits necessary such that $P(\tau \geq T) \geq 0.97$ where

$$E(\tau) = \frac{n}{\lambda}, \quad \text{Var}(\tau) = \frac{n}{\lambda^2}$$

Suppose that we assume that τ has an approximate normal distribution with mean n/λ and variance n/λ^2. Determine the number of parallel circuits necessary under the assumption that τ is normally distributed.

From Example 1.54,

$$P(\tau \geq T) = \int_T^\infty g(\tau) \, d\tau$$

Under the assumption of normality, this leads to

$$P(\tau \geq T) = 1 - \int_{-\infty}^T \frac{\lambda}{\sqrt{2\pi n}} \exp\left[-\frac{\lambda^2 \left(\tau - \frac{n}{\lambda}\right)^2}{2n}\right] d\tau$$

Letting

$$z = \frac{\lambda\left(\tau - \frac{n}{\lambda}\right)}{\sqrt{n}}$$

we have

$$P(\tau > T) = 1 - \int_{-\infty}^{\lambda(T - n/\lambda)/\sqrt{n}} \frac{1}{\sqrt{2\pi}} \exp\left[-\frac{z^2}{2}\right] dz$$

1.7 APPLICATIONS

Hence we wish to find n such that

$$\int_{-\infty}^{\lambda(T-n/\lambda)/\sqrt{n}} \frac{1}{\sqrt{2\pi}} \exp\left[-\frac{z^2}{2}\right] dz = 0.03$$

That is

$$F\left[\frac{\lambda\left(T-\frac{n}{\lambda}\right)}{\sqrt{n}}\right] = 0.03$$

or from Table 1 of the Appendix

$$\frac{\lambda\left(T-\frac{n}{\lambda}\right)}{\sqrt{n}} = -1.88$$

Letting $\lambda = 5$ and $T = 1$, gives us

$$5 - n = -1.88\sqrt{n}$$

Squaring both sides of this equation and rearranging terms we have

$$n^2 - 13.53n + 25 = 0$$

Solving for n, we have

$$n = 11.32, 2.21$$

Since n must be integer valued,

$$n = 2, 3, 11, 12$$

Now for $n = 2$,

$$F\left(\frac{5-n}{\sqrt{n}}\right) = F(2.12) = 0.983 > 0.030$$

For $n = 3$,

$$F\left(\frac{5-n}{\sqrt{n}}\right) = F(1.16) = 0.877 > 0.030$$

For $n = 11$,

$$F\left(\frac{5-n}{\sqrt{n}}\right) = F(-1.81) = 0.035 > 0.030$$

Finally, for $n = 12$,

$$F\left(\frac{5-n}{\sqrt{n}}\right) = F(-2.02) = 0.022 < 0.030$$

Hence, the assumption of normality leads to the conclusion that 12 parallel circuits are necessary if the criterion $P(\tau \geq T) \geq 0.97$ is to be satisfied.

Example 1.56 If E is the electromotive force and R is the resistance in the circuit shown (Fig. 1.4), the power, P, is given by $P = E^2/R$.

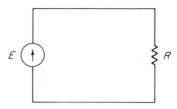

Figure 1.4

If E is a uniformly distributed random variable on the interval (a, b), $0 < a < b$, find the density function of P.

The density function of E is given by

$$f(e) = \frac{1}{b-a}, \quad a < e < b$$

Let $p = e^2/R$. Then

$$e = \sqrt{pR}, \quad de = \sqrt{\frac{R}{p}}\frac{dp}{2}$$

and $a^2/R < p < b^2/R$. Therefore,

$$g(p) = \frac{\sqrt{R/p}}{2(b-a)}, \quad \frac{a^2}{R} < p < \frac{b^2}{R}$$

Example 1.57 A shell is fired at a target which is a distance ρ from the point of firing. The shell is fired at an initial velocity V_0 and at an angle of $\theta°$ with the horizontal. The angle θ is a random variable which is uniformly distributed on the interval (a, b). That is

$$f(\theta) = \frac{1}{b-a}, \quad a < \theta < b$$

Neglecting atmospheric resistance and error in the horizontal plane, the range, R, of the shell is given by

$$R = \frac{V_0^2 \sin(2\theta)}{g}$$

where g is the acceleration due to gravity. The shell will hit the target if $\theta = (a+b)/2$. Hence

$$\rho = \frac{V_0^2 \sin(a+b)}{g}$$

Find the density function of $S = R - \rho$.

1.7 APPLICATIONS

Let $y = \sin(2\theta)$. Then

$$\theta = \frac{\sin^{-1}(y)}{2}, \qquad d\theta = \frac{dy}{2\sqrt{1-y^2}}$$

and

$$g(y) = \frac{1}{2(b-a)\sqrt{1-y^2}}, \qquad \sin(2a) < y < \sin(2b)$$

Now

$$r = \frac{V_0^2 y}{g} \quad \text{and} \quad y = \frac{rg}{V_0^2}, \qquad dy = \frac{g}{V_0^2} dr$$

Hence

$$h(r) = \frac{g}{2V_0^2(b-a)\sqrt{1-(rg/V_0^2)^2}}, \qquad \frac{V_0^2}{g}\sin(2a) < r < \frac{V_0^2}{g}\sin(2b)$$

Let $s = r - \rho$. Then

$$r = \rho + s, \qquad dr = ds$$

and the density function S is given by

$$q(s) = \frac{g}{2V_0^2(b-a)\sqrt{1-[g(s+\rho)/V_0^2]^2}}, \qquad A < s < B$$

where

$$A = \frac{V_0^2}{g}[\sin(2a) - \sin(a+b)], \qquad B = \frac{V_0^2}{g}[\sin(2b) - \sin(a+b)] \qquad \blacksquare$$

Example 1.58 Vehicles arrive at a toll booth at a rate of λ per hour. The rate at which tolls are paid is μ per hour. However, λ and μ are independent random variables with probability density functions given by

$$f_1(\lambda) = k_1 \exp[-k_1 \lambda], \quad 0 < \lambda < \infty, \qquad f_2(\mu) = k_2 \exp[-k_2], \quad 0 < \mu < \infty$$

The ratio, defined by $\rho = \lambda/\mu$ is called the traffic intensity factor and an unstable condition arises if $\rho > 1$. Find the probability density function of ρ, $g(\rho)$, and determine the value of k_2 such that

$$P(\rho \geq 1) = 0.05$$

Let

$$\rho = \lambda/\mu, \qquad y = \mu$$

Then

$$\lambda = y\rho, \qquad \mu = y, \qquad J = \begin{vmatrix} \rho & y \\ 1 & 0 \end{vmatrix} = -y$$

Hence, $|J| = y$. Since λ and μ are independently distributed, we have
$$f(\lambda, \mu) = k_1 k_2 \exp[-k_1\lambda - k_2\mu], \qquad 0 < \lambda < \infty, \quad 0 < \mu < \infty$$
and
$$h(\rho, y) = k_1 k_2 y \exp[-y(k_1\rho + k_2)], \qquad 0 < y < \infty, \quad 0 < \rho < \infty$$
Thus $g(\rho)$ is given by
$$g(\rho) = \int_0^\infty h(\rho, y)\, dy = \frac{k_1 k_2}{(k_1\rho + k_2)^2}, \qquad 0 < \rho < \infty$$

Now
$$P(\rho \geq 1) = 1 - P(\rho < 1) = 1 - \int_0^1 \frac{k_1 k_2}{(k_1\rho + k_2)^2}\, d\rho$$
$$= 1 - \left[-\frac{k_1 k_2}{k_1(k_1\rho + k_2)}\right]\Big|_0^1 = \frac{k_1 k_2}{k_1(k_1 + k_2)}$$

Thus
$$\frac{k_1 k_2}{k_1(k_1 + k_2)} = 0.05$$
or
$$k_1 k_2 = 0.05\, k_1(k_1 + k_2)$$
and
$$k_2 = \frac{1}{19} k_1$$

Example 1.59 The random variables X and Y are statistically dependent. The conditional variance of Y given X is given by
$$\mathrm{Var}(Y\,|\,X) = E\{[Y - E(Y\,|\,X)]\,|\,X\}$$
Show that
$$\mathrm{Var}(Y) = E[\mathrm{Var}(Y\,|\,X)] + \mathrm{Var}[E(Y\,|\,X)]$$

The variance of Y may be expressed as
$$\mathrm{Var}(Y) = E[Y - E(Y)]^2 = E\{E[Y - E(Y)]^2\,|\,X\}$$
Then $E\{E[Y - E(Y)]^2\,|\,X\}$ may be expressed as
$$E\{E[Y - E(Y)]^2\,|\,X\} = E\{E[Y - E(Y\,|\,X) + E(Y\,|\,X) - E(Y)]^2\,|\,X\}$$
$$= E\{E[Y - E(Y\,|\,X)]^2\,|\,X\} + E\{E[E(Y\,|\,X) - E(Y)]^2\,|\,X\}$$
$$+ 2E\{E[Y - E(Y\,|\,X)][E(Y\,|\,X) - E(Y)]\,|\,X\}$$

1.7 APPLICATIONS

Now

$$E\{E[Y - E(Y|X)][E(Y|X) - E(Y)]|X\} = E\{E[YE(Y|X) - YE(Y) \\ - E^2(Y|X) + E(Y|X)E(Y)]|X\}$$

Since

$$\{E[YE(Y|X)]|X\} = E(Y|X)E(Y|X) = E^2(Y|X)$$
$$\{E[YE(Y)]|X\} = E(Y)E(Y|X)$$

and

$$\{E(Y|X)E(Y)|X\} = E(Y|X)E(Y)$$

we have

$$E\{E[Y - E(Y|X)][E(Y|X) - E(Y)]|X\} = 0$$

Therefore

$$\text{Var}(Y) = E\{E[Y - E(Y|X)]^2|X\} + E\{E[E(Y|X) - E(Y)]^2|X\}$$

Thus $\{E[Y - E(Y|X)]^2|X\}$ is, by definition, the conditional variance of Y given X. Similarly, $\{E[E(Y|X) - E(Y)]^2|X\}$ is the variance of $E(Y|X)$. Hence

$$\text{Var}(Y) = E[\text{Var}(Y|X)] + \text{Var}[E(Y|X)] \qquad \text{▣}$$

Example 1.60 Let X_i be the random variable representing the number of passengers carried by a landing aircraft at an air terminal and let N be the number of aircraft landing per hour. Then Y, the total number of passengers landing per hour, is given by

$$Y = \sum_{i=1}^{N} X_i$$

If X_i and X_j are independently distributed as the random variable X for $i \neq j$ and if N is a random variable, find the mean and variance of Y.

For given N,

$$E(Y|N) = \sum_{i=1}^{N} E(X_i) = \sum_{i=1}^{N} E(X) = NE(X)$$

Now

$$E(Y) = E[E(Y|N)] = E(N)E(X)$$

From Example 1.59,

$$\text{Var}(Y) = E[\text{Var}(Y|N)] + \text{Var}[E(Y|N)]$$

Since X_i, $i = 1, 2, \ldots, N$, are independently distributed,

$$E[\text{Var}(Y|N)] = E\left[\text{Var}\left(\sum_{i=1}^{N} X_i\right)\right] = E\left[\sum_{i=1}^{N} \text{Var}(X_i)\right] = E[N\,\text{Var}(X)]$$
$$= E(N)\text{Var}(X)$$

Further,
$$\text{Var}[E(Y\mid N)] = \text{Var}[NE(X)] = \text{Var}(N)\text{Var}[E(X)] = \text{Var}(N)E^2(X)$$
since $E(X)$ is a constant. Hence
$$\text{Var}(Y) = E(N)\text{Var}(X) + \text{Var}(N)E^2(X)$$

Example 1.61 A measure of uncertainty associated with a random variable, X, which is often used in *information theory* is defined by
$$H(X) = E\{-\ln[f(x)]\}$$
and is called *entropy*. Show that the entropy of a normal random variable is an increasing function of its standard deviation, σ.

If X is normally distributed with mean μ and variance σ^2, then
$$H(X) = E\left\{-\ln\left[\frac{1}{\sigma\sqrt{2\pi}}\exp\left[-\frac{(x-\mu)^2}{2\sigma^2}\right]\right]\right\} = E\left[\ln(\sigma\sqrt{2\pi}) + \frac{(x-\mu)^2}{2\sigma^2}\right]$$
$$= \ln(\sigma\sqrt{2\pi}) + \frac{1}{2\sigma^2}E(x-\mu)^2 = \ln(\sigma\sqrt{2\pi}) + \frac{1}{2}$$

Hence, for two normal random variables, X_1 and X_2, with associated variances σ_1^2 and σ_2^2 such that $\sigma_1^2 > \sigma_2^2$, we have
$$H(X_1) > H(X_2)$$

Example 1.62 (*Continuation of Example 1.61*) Show that two random variables, X_1 and X_2, which have equal variances, σ^2, do not necessarily have equal entropy.

Let X_1 be a normally distributed random variable with mean μ and variance σ^2 and let X_2 be exponentially distributed with parameter $1/\sigma$. Then
$$f(x_1) = \frac{1}{\sigma\sqrt{2\pi}}\exp\left[-\frac{(x-\mu)^2}{2\sigma^2}\right], \quad -\infty < x < \infty$$
and
$$g(x_2) = \frac{1}{\sigma}\exp[-x/\sigma], \quad 0 < x < \infty$$

Hence
$$\text{Var}(X_1) = \sigma^2, \quad \text{Var}(X_2) = \sigma^2$$

From Example 1.61,
$$H(X_1) = \ln(\sigma\sqrt{2\pi}) + \tfrac{1}{2} = \ln(\sigma) + \ln(\sqrt{2\pi}) + \tfrac{1}{2}$$

For X_2,
$$H(X_2) = E\{-\ln[g(x_2)]\} = E\left[-\ln(1/\sigma) + \left(\frac{x}{\sigma}\right)\right] = \ln\sigma + 1$$

Since
$$\ln(\sqrt{2\pi}) = 0.92$$
we have
$$H(X_1) = \ln(\sigma) + 1.42 \quad \text{and} \quad H(X_2) = \ln(\sigma) + 1.00$$
Hence
$$H(X_1) \neq H(X_2)$$

Problems

1. There are 26 people at a party. What is the probability that two or more people were born on the same day of the year? Assume 365 days per year and that
$$P(\text{birth on } i\text{th day of the year}) = \tfrac{1}{365}, \quad i = 1, 2, \ldots, 365$$

2. Five cards are drawn from a standard deck of playing cards one after another without replacement. Find
   a. The probability of drawing 2 aces and 3 jacks;
   b. The probability of drawing 4 aces;
   c. The probability of drawing an ace, a king, a queen, a jack, and a ten.

3. A set of dice are tossed. Find
   a. The probability of a 7 or 11;
   b. The probability that the same number of points show on each die;
   c. The probability that the same number of points do not appear on each die.

4. Prove Theorem 1.7.

5. Prove Eqs. (1.60), (1.62), and (1.63) if x is an integer valued random variable.

6. X is a gamma random variable with probability density function
$$f(x) = \lambda^2 x \exp[-\lambda x], \quad 0 < x < \infty$$
Find the distribution function of X.

7. Define the constant k such that
$$f(x) = kx, \quad 0 < x < 5$$
is a probability density function.

8. Find the constant k such that
$$p(x) = k(\tfrac{1}{2})^x, \quad x = 1, 2, \ldots$$
is a probability mass function.

9. The random variable X has the density function
$$f(x) = kx^3, \quad 0 < x < 1$$
Find the value of k.

10. Show that the geometric random variable is without memory.

11. Let X and Y be independent random variables. Show that
$$f(x, y) = f(x)f(y)$$

if X and Y are continuous, and

$$p(x, y) = p(x)p(y)$$

if X and Y are discrete.

12. Prove Eq. 1.116.
13. Prove Eq. 1.137.
14. Let

$$p(x) = \tfrac{1}{10}, \quad x = 0, 1, 2, \ldots, 9$$

If $Y = 2X + 6$, find the probability mass function of Y.

15. Two tanks, A and B, are engaged in a battle. Each fires in turn with A firing first. The probability that A hits B with a single shot is 0.20 and the probability that B hits A with a single shot is 0.10. Each tank has an inexhaustable supply of shells. The battle is over when either tank hits the other. What is the probability that
    a. A wins?
    b. B wins?

16. X is a Poisson random variable. Find the probability mass function of $X + a$, where a is a constant.

17. Let X be an exponential random variable with parameter $\lambda > 0$ and let R be a random variable that is uniformly distributed on the interval 0 to 1. Express X as a function of R.

18. Let

$$Y = |X|$$

If $f(x)$ is the density function of X, find the density function of Y, $g(y)$. Note that the transformation is not one-to-one.

19. Z is a standard normal random variable. Find the probability density function of $|Z|$.

20. The joint probability density function of X and Y is given by

$$f(x, y) = 4x(y - x)\exp[-(x + y)], \quad 0 < y < \infty, \ 0 < x < y$$

Are X and Y independent random variables?

21. The joint probability mass function of X and Y is

$$p(x, y) = \frac{1}{y!(x-y)!} p^y (1-p)^{x-y} \lambda^x \exp[-\lambda x], \quad y = 0, 1, \ldots, x, \ x = 0, 1, 2, \ldots$$

Are X and Y independent random variables?

22. The joint probability density function of X and Y is

$$f(x, y) = \frac{1}{xT}, \quad 0 < x < T, \ 0 < y < x$$

Find
 a. $f(x)$ b. $g(y)$
 c. $f(x|y)$ d. $g(y|x)$

23. The dimension of a product is normally distributed with mean μ and unit variance ($\sigma^2 = 1$). If the dimension, X, of a unit deviates from μ_0, a cost C is incurred depending upon the magnitude of deviation of X from μ_0. That is

$$D = C|X - \mu_0|$$

Find the distribution of cost per unit, D.

PROBLEMS

24. A manufacturing line produces units in lots of size L. The proportion, P, of defective units in each lot is beta distributed with parameters a and b. The cost of each defective unit in the lot is d. Thus the total cost per lot, C, is

$$C = dLP$$

Find the distribution of C and E(C).

25. Determine whether the random variables Y_1 and Y_2 given in Example 1.37 are independently distributed.

26. Show that the sum of independent, identically distributed geometric random variables has a negative binomial distribution.

27. Show that the sum of independent, identically distributed binomial random variables is also binomially distributed.

28. Let X_1 and X_2 be independent χ^2 random variables with n_1 and n_2 degrees of freedom. If

$$y = \frac{X_1/n_1}{X_2/n_2}$$

Find the density function of Y. The random variable Y is the F random variable.

29. If X_1 is a noncentral χ^2 random variable with noncentrality parameter γ and n_1 degrees of freedom and X_2 is a χ^2 random variable with n_2 degrees of freedom, find the density function of the noncentral F random variable defined by

$$Y = \frac{X_1/n_1}{X_2/n_2}$$

30. X_1 and X_2 are independent Cauchy random variables with joint probability density function

$$f(x_1, x_2) = \frac{1}{\pi^2[1 + (x_1 - \lambda)^2][1 + (x_2 - \lambda)^2]}, \quad -\infty < x_1, x_2 < \infty$$

Find the probability density function of $\bar{X}$ where

$$\bar{X} = \tfrac{1}{2}(X_1 + X_2)$$

31. X_1 and X_2 are independent gamma random variables with joint probability density function given by

$$f(x_1, x_2) = \frac{1}{\Gamma(a)\Gamma(b)} x_1^{a-1} x_2^{b-1} \exp[-x_1 - x_2], \quad 0 < x_1, x_2 < \infty$$

Let

$$Y = \frac{X_1}{X_1 + X_2}$$

Show that Y is a beta random variable with probability density function

$$g(y) = \frac{\Gamma(a+b)}{\Gamma(a)\Gamma(b)} y^{a-1}(1 - y)^{b-1}, \quad 0 < y < 1$$

32. $X_1, X_2, \ldots, X_n$ are independent uniform random variables with marginal probability density functions

$$f(x_i) = 1, \quad 0 < x_i < 1, \quad i = 1, 2, \ldots, n$$

Let

$$Y = -\frac{1}{\lambda} \sum_{i=1}^{n} \ln(X_i)$$

Find the probability density function of Y.

33. X is a Laplace random variable with probability density function
$$f(x) = \tfrac{1}{2} \exp[-|x - \lambda|], \qquad -\infty < x < \infty$$
Find the mean and variance of X.
34. Find the mean and variance of the Weibull random variable.
35. Find the mean and variance of the Beta random variable.
36. Find the mean of the F random variable.
37. Find the mean of the noncentral χ^2 random variable.
38. The proportion, P, of defective items in a lot of size L has the probability density function given by
$$f(p) = b(1 - p)^{b-1}, \qquad 0 < p < 1$$
From each lot a sample of size n is selected at random and inspected. The number of defective items, X, appearing in the sample has the probability mass function
$$p(x \mid p) = \binom{n}{x} p^x (1 - p)^{n-x}, \qquad x = 0, 1, \ldots, n$$
Find
   a. The joint distribution of X and P;
   b. The conditional density function of P given X;
   c. The marginal probability density function of X;
   d. $E(X)$.
39. Derive the moment generating function of the uniform random variable.
40. Derive the moment generating function of the rectangular random variable.
41. The probability density function of the hyperexponential random variable, X, is given by
$$f(x) = p\lambda_1 \exp[-\lambda_1 x] + (1 - p)\lambda_2 \exp[-\lambda_2 x], \qquad 0 < x < \infty, \ 0 < p < 1, \ \lambda_1, \lambda_2 > 0$$
Find the mean, variance, and moment generating function of X.
42. Derive the moment generating function of the noncentral χ^2 random variable.
43. Prove Theorem 1.12 for the case where $X_1, X_2, \ldots, X_n$ are discrete random variables.
44. Using the moment generating function, find the mean and variance of the uniform random variable.
45. Using the moment generating function, find the mean and variance of the negative binomial random variable.
46. Using the moment generating function, show that the sum of independent, identically distributed exponential random variables is a gamma random variable.
47. Prove Theorem 1.14, where $X_1, X_2, \ldots, X_n$ are discrete random variables.
48. Prove Theorem 1.15 where $X_1, X_2, \ldots, X_n$ are discrete random variables.
49. Show that
$$\mathrm{Var}(X) = E(X^2) - [E(X)]^2$$
50. The characteristic function of a random variable is defined as $E[E^{iuX}]$ where $i = \sqrt{-1}$. Find the characteristic function of
   a. The normal random variable;
   b. The binomial random variable.

References

Abramowitz, M., and Stegun, I. A., *Handbook of Mathematical Functions*. Washington: National Bureau of Standards, 1964.

Bowker, A. H., and Lieberman, G. J., *Engineering Statistics*. Englewood Cliffs, New Jersey: Prentice-Hall, 1959.

Brownlee, K. A., *Statistical Theory and Methodology in Science and Engineering*. New York: Wiley (Interscience), 1960.

REFERENCES

Cooper, R. B., *Introduction to Queueing Theory*. New York: The Macmillan Co., 1972.
Draper, N., and Smith, H., *Applied Regression Analysis*. New York: Wiley (Interscience), 1966.
Duncan, A. J., *Quality Control and Industrial Statistics*. Homewood, Illinois: Irwin, 1965.
Freeman, H., *Introduction to Statistical Inference*. Reading, Massachusetts: Addison-Wesley, 1963.
Graybill, F. A., *An Introduction to Linear Statistical Models*. New York: McGraw-Hill, 1961.
Hadley, G., and Whitin, T. M., *Analysis of Inventory Systems*. Englewood Cliffs, New Jersey: Prentice-Hall, 1963.
Karlin, S., *A First Course in Stochastic Processes*. New York: Academic Press, 1966.
Morse, P. M., *Queues, Inventories, and Maintenance*. New York: Wiley (Interscience), 1962.
Ostle, B., *Statistics in Research*. Ames, Iowa: The Iowa State University Press, 1963.
Parzen, E., *Modern Probability Theory and Its Applications*. New York: Wiley (Interscience), 1960.
Parzen, E., *Stochastic Processes*. San Francisco: Holden-Day, 1962.
Saaty, T. L., *Elements of Queueing Theory*. New York: McGraw-Hill, 1961.
Schmidt, J. W., and Taylor, R. E., *Simulation and Analysis of Industrial Systems*. Homewood, Illinois: Irwin, 1970.
Shooman, M. L., *Probabilistic Reliability, An Engineering Approach*. New York: McGraw-Hill, 1968.
Wilks, S. S., *Mathematical Statistics*. New York: Wiley (Interscience), 1962.

CHAPTER 2

MATRIX ALGEBRA

2.1 Introduction

There are many areas of operations research in which the analyst must deal with arrays of numbers or functions. For example, the transition probabilities of a Markov chain are usually expressed in a transition matrix. In linear and nonlinear programming, the objective function and the restrictions on the solution space are frequently expressed in matrix form. As will be demonstrated in Chapter 3, matrix theory plays an important role in classical optimization theory. There are many fields allied to operations research in which a knowledge of matrix theory is fundamental. In the area of statistical design and analysis, one would have difficulty using the method of least squares, which forms the foundation of regression analysis, without being able to find the inverse of a matrix. In fact, matrix algebra is a basic requirement for study in the field of design of experiments in general. The analyst frequently encounters optimization problems where the classical methods presented in Chapter 4 are impractical. In such cases he may resort to an iterative optimum seeking technique. However, an understanding of many of these techniques requires a basic knowledge of matrix algebra.

2.2 Definitions and Basic Operations

Definition 2.1 A *matrix* is defined as an array of numbers or functions.

In this chapter a matrix will be denoted by a capital letter and the elements of the matrix by lower case letters. Let A be a matrix and let a_{ij} be the *element* in the ith row and jth column of the matrix. Therefore

$$A = \begin{bmatrix} a_{11} & a_{12} & \cdots & a_{1j} & \cdots & a_{1n} \\ a_{21} & a_{22} & \cdots & a_{2j} & \cdots & a_{2n} \\ \vdots & \vdots & & \vdots & & \vdots \\ a_{i1} & a_{i2} & \cdots & a_{ij} & \cdots & a_{in} \\ \vdots & \vdots & & \vdots & & \vdots \\ a_{m1} & a_{m2} & \cdots & a_{mj} & \cdots & a_{mn} \end{bmatrix} \qquad (2.1)$$

2.2 DEFINITIONS AND BASIC OPERATIONS

The matrix A is said to be an $m \times n$ matrix since it has m rows and n columns. In words $m \times n$ means m by n. An alternative method of defining the number of rows and columns is to define its order.

Definition 2.2 An $m \times n$ matrix is said to be of *order* (m, n).

When the number of rows and columns are equal the matrix is called *square*. The order of a square matrix is defined by the number of rows in the matrix. That is, if A is an $m \times m$ matrix, then it is said to be of order m. If A is a square $n \times n$ matrix, then the elements $a_{11}, a_{22}, \ldots, a_{nn}$ constitute the main diagonal of the matrix.

Definition 2.3 Two matrices, A and B, are *equal* if and only if both are of the same order and $a_{ij} = b_{ij}$ for all i and j.

That is, for two matrices to be equal, the number of rows in each must be the same, the number of columns in each must be the same, and each element in one must be identical to the corresponding element in the other.

Matrices may be added or subtracted provided they are of the same order. Such matrices are said to be conformable for addition and subtraction. For example, let A, B, C, and D be defined as follows.

$$A = \begin{bmatrix} 4 & 3 \\ 1 & 0 \end{bmatrix}, \quad B = \begin{bmatrix} 1 & 4 & 1 \\ 2 & 1 & 0 \end{bmatrix}, \quad C = \begin{bmatrix} 0 & 7 \\ 1 & 1 \end{bmatrix}, \quad D = \begin{bmatrix} 4 & 10 \\ 2 & 1 \end{bmatrix}$$

Any combination of the matrices A, C, and D may be added or subtracted from one another. However, an operation involving the addition or subtraction of B with A, C, or D or any combination thereof is not permitted since B is not of the same order as the remaining matrices. Note that although A, C, and D are of the same order, no two are equal to one another.

Two matrices of the same order may be added (subtracted) by adding (subtracting) corresponding elements. Thus the matrix representing the sum (difference) is of the same order as the matrices summed (subtracted). For example, let A and B be matrices of order (m, n) and let C be the matrix representing their sum or difference. Then

$$C = A \pm B \tag{2.2}$$

and the elements of C, are given by

$$c_{ij} = a_{ij} \pm b_{ij}, \quad i = 1, 2, \ldots, m \quad \text{and} \quad j = 1, 2, \ldots, n \tag{2.3}$$

Example 2.1 Let the matrices A, B, and C be defined as

$$A = \begin{bmatrix} 1 & 1 \\ 4 & 1 \\ 2 & 3 \end{bmatrix}, \quad B = \begin{bmatrix} 2 & 0 \\ 1 & 1 \\ 3 & 0 \end{bmatrix}, \quad C = \begin{bmatrix} 3 & 1 \\ 5 & 2 \\ 5 & 3 \end{bmatrix}$$

Define the matrices D, E, and F where

$$D = A + B, \quad E = A + B - C, \quad F = A + B + C$$

To define D, we simply add together the corresponding elements of A and B.

$$D = \begin{bmatrix} 3 & 1 \\ 5 & 2 \\ 5 & 3 \end{bmatrix}$$

Since $D = A + B$, $E = D - C$ and

$$E = \begin{bmatrix} 0 & 0 \\ 0 & 0 \\ 0 & 0 \end{bmatrix}$$

To define F, the corresponding elements of A, B, and C are summed.

$$F = \begin{bmatrix} 6 & 2 \\ 10 & 4 \\ 10 & 6 \end{bmatrix}$$

In the above example the matrix E contains elements all of which are zero. Such a matrix is called a *zero or null matrix* and will hereafter be denoted by $\emptyset$.

Multiplication of a matrix, A, by a scalar, k, corresponds to summing A, k times. That is,

$$kA = \sum_{i=1}^{k} A \qquad (2.4)$$

Therefore, if $B = kA$, then

$$b_{ij} = ka_{ij} \quad \text{for all } i \text{ and } j \qquad (2.5)$$

and B is of the same order as A. In other words, multiplication of a matrix by a scalar results in multiplication of each element of the matrix by the scalar.

2.2.1 Matrix Multiplication

Two matrices may be multiplied together if the number of columns in one is equal to the number of rows in the other. Let A be a matrix of order (m, n) and B a matrix of order (n, q). Then the product AB is defined and A is said to be *conformable* to B for multiplication. However, for $m \neq q$ the product BA is not defined. Let $Q = AB$ where A is $m \times n$ and B is $n \times p$. The product matrix, Q, is of order (m, p). Let A and B be given by

2.2 DEFINITIONS AND BASIC OPERATIONS

$$A = \begin{bmatrix} a_{11} & a_{12} & \cdots & a_{1j} & \cdots & a_{1n} \\ a_{21} & a_{22} & \cdots & a_{2j} & \cdots & a_{2n} \\ \vdots & \vdots & & \vdots & & \vdots \\ a_{i1} & a_{i2} & \cdots & a_{ij} & \cdots & a_{in} \\ \vdots & \vdots & & \vdots & & \vdots \\ a_{m1} & a_{m2} & \cdots & a_{mj} & \cdots & a_{mn} \end{bmatrix} \quad (2.6)$$

$$B = \begin{bmatrix} b_{11} & b_{12} & \cdots & b_{1k} & \cdots & b_{1p} \\ b_{21} & b_{22} & \cdots & b_{2k} & \cdots & b_{2p} \\ \vdots & \vdots & & \vdots & & \vdots \\ b_{j1} & b_{j2} & \cdots & b_{jk} & \cdots & b_{jp} \\ \vdots & \vdots & & \vdots & & \vdots \\ b_{n1} & b_{n2} & \cdots & b_{nk} & \cdots & b_{np} \end{bmatrix} \quad (2.7)$$

The element in the ith row, kth column of the product matrix Q, q_{ik}, is given by

$$q_{ik} = \sum_{j=1}^{n} a_{ij} b_{jk}, \quad i = 1, 2, \ldots, m \quad \text{and} \quad k = 1, 2, \ldots, p \quad (2.8)$$

Therefore, the element in the ith row, kth column of the product matrix is obtained by multiplying each element in the ith row of matrix A by its corresponding element in the kth column of matrix B and summing the resulting n products.

Example 2.2 Let A and B be defined by

$$A = \begin{bmatrix} 1 & 4 & 0 \\ 2 & 1 & 3 \end{bmatrix}, \quad B = \begin{bmatrix} 1 & 0 & 1 \\ 5 & 1 & 1 \\ 1 & 2 & 1 \end{bmatrix}$$

Define Q and P as the products AB and BA respectively.

a. Are P and Q defined?
b. Define the order of those product matrices which are defined.
c. Evaluate those product matrices which are defined.

Part a Matrix A is of order (2, 3) and B is a square matrix of order 3. Therefore AB is defined while BA is not.

Part b The product matrix Q is of order (2, 3).

Part c

$$Q = AB = \begin{bmatrix} 1 & 4 & 0 \\ 2 & 1 & 3 \end{bmatrix} \begin{bmatrix} 1 & 0 & 1 \\ 5 & 1 & 1 \\ 1 & 2 & 1 \end{bmatrix} = \begin{bmatrix} q_{11} & q_{12} & q_{13} \\ q_{21} & q_{22} & q_{23} \end{bmatrix}$$

From Eq. (2.8)

$$q_{11} = (1)(1) + (4)(5) + (0)(1) = 21$$
$$q_{12} = (1)(0) + (4)(1) + (0)(2) = 4$$
$$q_{13} = (1)(1) + (4)(1) + (0)(1) = 5$$
$$q_{21} = (2)(1) + (1)(5) + (3)(1) = 10$$
$$q_{22} = (2)(0) + (1)(1) + (3)(2) = 7$$
$$q_{23} = (2)(1) + (1)(1) + (3)(1) = 6$$

Therefore

$$Q = \begin{bmatrix} 21 & 4 & 5 \\ 10 & 7 & 6 \end{bmatrix}$$

Example 2.3 Let A and B be defined by

$$A = \begin{bmatrix} 1 & 2 \\ 1 & 1 \\ 4 & 1 \end{bmatrix}, \quad B = \begin{bmatrix} 1 & 4 & 0 \\ 1 & 0 & 1 \end{bmatrix}$$

Evaluate the product matrices P and Q where

$$P = AB, \quad Q = BA$$

Since A is of order (3, 2) and B is of order (2, 3), both product matrices, P and Q, are defined. P is of order 3 and Q is of order 2.

$$P = \begin{bmatrix} 1 & 2 \\ 1 & 1 \\ 4 & 1 \end{bmatrix} \begin{bmatrix} 1 & 4 & 0 \\ 1 & 0 & 1 \end{bmatrix}$$

$$= \begin{bmatrix} (1)(1) + (2)(1) & (1)(4) + (2)(0) & (1)(0) + (2)(1) \\ (1)(1) + (1)(1) & (1)(4) + (1)(0) & (1)(0) + (1)(1) \\ (4)(1) + (1)(1) & (4)(4) + (1)(0) & (4)(0) + (1)(1) \end{bmatrix} = \begin{bmatrix} 3 & 4 & 2 \\ 2 & 4 & 1 \\ 5 & 16 & 1 \end{bmatrix}$$

$$Q = \begin{bmatrix} 1 & 4 & 0 \\ 1 & 0 & 1 \end{bmatrix} \begin{bmatrix} 1 & 2 \\ 1 & 1 \\ 4 & 1 \end{bmatrix}$$

$$= \begin{bmatrix} (1)(1) + (4)(1) + (0)(4) & (1)(2) + (4)(1) + (0)(1) \\ (1)(1) + (0)(1) + (1)(4) & (1)(2) + (0)(1) + (1)(1) \end{bmatrix} = \begin{bmatrix} 5 & 6 \\ 5 & 3 \end{bmatrix}$$

Example 2.4 Let

$$A = \begin{bmatrix} 1 & 0 \\ 4 & 1 \end{bmatrix}, \quad B = \begin{bmatrix} 1 & 1 \\ 2 & 4 \end{bmatrix}, \quad C = \begin{bmatrix} 1 & 0 \\ 0 & 1 \end{bmatrix}$$

Evaluate AB, BA, AC, and CA.

2.2 DEFINITIONS AND BASIC OPERATIONS

Since A and B are both of order 2, AB and BA are both defined.

$$AB = \begin{bmatrix} 1 & 0 \\ 4 & 1 \end{bmatrix} \begin{bmatrix} 1 & 1 \\ 2 & 4 \end{bmatrix}$$

$$AB = \begin{bmatrix} (1)(1)+(0)(2) & (1)(1)+(0)(4) \\ (4)(1)+(1)(2) & (4)(1)+(1)(4) \end{bmatrix} = \begin{bmatrix} 1 & 1 \\ 6 & 8 \end{bmatrix}$$

$$BA = \begin{bmatrix} 1 & 1 \\ 2 & 4 \end{bmatrix} \begin{bmatrix} 1 & 0 \\ 4 & 1 \end{bmatrix} = \begin{bmatrix} (1)(1)+(1)(4) & (1)(0)+(1)(1) \\ (2)(1)+(4)(4) & (2)(0)+(4)(1) \end{bmatrix} = \begin{bmatrix} 5 & 1 \\ 18 & 4 \end{bmatrix}$$

$$AC = \begin{bmatrix} 1 & 0 \\ 4 & 1 \end{bmatrix} \begin{bmatrix} 1 & 0 \\ 0 & 1 \end{bmatrix} = \begin{bmatrix} (1)(1)+(0)(0) & (1)(0)+(0)(1) \\ (4)(1)+(1)(0) & (4)(0)+(1)(1) \end{bmatrix} = \begin{bmatrix} 1 & 0 \\ 4 & 1 \end{bmatrix}$$

$$CA = \begin{bmatrix} 1 & 0 \\ 0 & 1 \end{bmatrix} \begin{bmatrix} 1 & 0 \\ 4 & 1 \end{bmatrix} = \begin{bmatrix} (1)(1)+(0)(4) & (1)(0)+(0)(1) \\ (0)(1)+(1)(4) & (0)(0)+(1)(1) \end{bmatrix} = \begin{bmatrix} 1 & 0 \\ 4 & 1 \end{bmatrix}$$

Examples 2.3 and 2.4 illustrate an important property of matrix multiplication. In general $AB \neq BA$ even though both products may be defined. However, this statement does not imply that the products AB and BA are never equal. These two points are illustrated in Example 2.4 where

$$AB \neq BA$$

but

$$AC = CA$$

The matrices A and C are said to commute.

Theorem 2.1 Matrix multiplication is *associative* or

$$(AB)C = A(BC) \tag{2.9}$$

where the matrices are conformable for the operations indicated.

Proof Let A be of order (p, n), B of order (n, m), and C of order (m, q). Let the elements of A, B, and C be given by a_{ij}, b_{jk}, c_{kl}, where $i = 1, 2, \ldots, p, j = 1, 2, \ldots, n, k = 1, 2, \ldots, m$, and $l = 1, 2, \ldots, q$. Let

$$R = (AB)C, \quad S = A(BC)$$

From Eq. (2.8), the element in the ith row, kth column of the product AB is $\sum_{j=1}^{n} a_{ij} b_{jk}$. Then

$$R = \begin{bmatrix} \sum_{j=1}^{n} a_{1j}b_{j1} & \sum_{j=1}^{n} a_{1j}b_{j2} & \cdots & \sum_{j=1}^{n} a_{1j}b_{jm} \\ \sum_{j=1}^{n} a_{2j}b_{j1} & \sum_{j=1}^{n} a_{2j}b_{j2} & \cdots & \sum_{j=1}^{n} a_{2j}b_{jm} \\ \vdots & \vdots & & \vdots \\ \sum_{j=1}^{n} a_{pj}b_{j1} & \sum_{j=1}^{n} a_{pj}b_{j2} & \cdots & \sum_{j=1}^{n} a_{pj}b_{jm} \end{bmatrix} \begin{bmatrix} c_{11} & c_{12} & \cdots & c_{1q} \\ c_{21} & c_{22} & \cdots & c_{2q} \\ \vdots & \vdots & & \vdots \\ c_{m1} & c_{m2} & \cdots & c_{mq} \end{bmatrix}$$

The element in the ith row, lth column of R is then

$$r_{il} = \sum_{k=1}^{m} \sum_{j=1}^{n} a_{ij} b_{jk} c_{kl}, \quad i = 1, 2, \ldots, p, \quad l = 1, 2, \ldots, q$$

The element in the jth row, lth column of the matrix BC is $\sum_{k=1}^{m} b_{jk} c_{kl}$.

$$S = \begin{bmatrix} a_{11} & a_{12} & \cdots & a_{1n} \\ a_{21} & a_{22} & \cdots & a_{2n} \\ \vdots & \vdots & & \vdots \\ a_{p1} & a_{p2} & \cdots & a_{pn} \end{bmatrix} \begin{bmatrix} \sum_{k=1}^{m} b_{1k} c_{k1} & \sum_{k=1}^{m} b_{1k} c_{k2} & \cdots & \sum_{k=1}^{m} b_{1k} c_{kq} \\ \sum_{k=1}^{m} b_{2k} c_{k1} & \sum_{k=1}^{m} b_{2k} c_{k2} & \cdots & \sum_{k=1}^{m} b_{2k} c_{kq} \\ \vdots & \vdots & & \vdots \\ \sum_{k=1}^{m} b_{nk} c_{k1} & \sum_{k=1}^{m} b_{nk} c_{k2} & \cdots & \sum_{k=1}^{m} b_{nk} c_{kq} \end{bmatrix}$$

and s_{il} is given by

$$s_{il} = \sum_{j=1}^{n} a_{ij} \sum_{k=1}^{m} b_{jk} c_{kl}, \quad i = 1, 2, \ldots, p, \quad l = 1, 2, \ldots, q$$

It is obvious that R and S are of the same order, (p, q). Since

$$s_{il} = \sum_{j=1}^{n} \sum_{k=1}^{m} a_{ij} b_{jk} c_{kl} = \sum_{k=1}^{m} \sum_{j=1}^{n} a_{ij} b_{jk} c_{kl}$$

each element of S is equal to the corresponding element of R, and $R = S$. ∎

Theorem 2.2 Matrix multiplication is *distributive* with respect to addition. That is,

$$A(B + C) = AB + AC$$

where the matrices are conformable for the operations indicated.

Proof Let B and C be of order (m, n) and let A be of order (p, m). The element in the ith row, kth column of $A(B + C)$ is

$$\sum_{j=1}^{m} a_{ij}(b_{jk} + c_{jk}), \quad i = 1, 2, \ldots, p \text{ and } k = 1, 2, \ldots, n$$

Let R be defined as

$$R = A(B + C) - AC$$

Then

$$r_{ik} = \sum_{j=1}^{m} a_{ij}(b_{jk} + c_{jk}) - \sum_{j=1}^{m} a_{ij} c_{jk} = \sum_{j=1}^{m} a_{ij} b_{jk}$$

for all i and k. Therefore $R = AB$. But

$$A(B + C) = A(B + C) - AC + AC = R + AC = AB + AC \quad ∎$$

2.3 The Transpose of a Matrix

Definition 2.4 The *transpose* A^T, of the matrix A, is obtained by interchanging the rows and columns of A.

Thus if A is of order (m, n), then A^T is of order (n, m). That is, if

$$A = \begin{bmatrix} a_{11} & a_{12} & \cdots & a_{1j} & \cdots & a_{1n} \\ a_{21} & a_{22} & \cdots & a_{2j} & \cdots & a_{2n} \\ \vdots & \vdots & & \vdots & & \vdots \\ a_{i1} & a_{i2} & \cdots & a_{ij} & \cdots & a_{in} \\ \vdots & \vdots & & \vdots & & \vdots \\ a_{m1} & a_{m2} & \cdots & a_{mj} & \cdots & a_{mn} \end{bmatrix} \quad (2.10)$$

then

$$A^T = \begin{bmatrix} a_{11} & a_{21} & \cdots & a_{i1} & \cdots & a_{m1} \\ a_{12} & a_{22} & \cdots & a_{i2} & \cdots & a_{m2} \\ \vdots & \vdots & & \vdots & & \vdots \\ a_{1j} & a_{2j} & \cdots & a_{ij} & \cdots & a_{mj} \\ \vdots & \vdots & & \vdots & & \vdots \\ a_{1n} & a_{2n} & \cdots & a_{in} & \cdots & a_{mn} \end{bmatrix} \quad (2.11)$$

Example 2.5 Let the matrices A, B, and C be defined as follows.

$$A = \begin{bmatrix} 1 \\ 2 \\ 4 \end{bmatrix}, \quad B = \begin{bmatrix} 1 & 0 \\ 4 & 1 \\ 2 & 0 \end{bmatrix}, \quad C = \begin{bmatrix} 1 & 9 & 2 \\ 9 & 2 & 5 \\ 2 & 5 & 4 \end{bmatrix}$$

Find the transpose of each matrix.

Since A, B, and C are of order (3, 1), (3, 2), and 3 respectively, A^T, B^T, and C^T are of order (1, 3), (2, 3), and 3 respectively.

$$A^T = [1 \; 2 \; 4], \quad B^T = \begin{bmatrix} 1 & 4 & 2 \\ 0 & 1 & 0 \end{bmatrix}, \quad C^T = \begin{bmatrix} 1 & 9 & 2 \\ 9 & 2 & 5 \\ 2 & 5 & 4 \end{bmatrix}$$

In Example 2.5, A has only one column while A^T has one row. A matrix having only one row or column is usually referred to as a *vector*. A matrix with one row is called a *row vector* while a matrix having one column is called a *column vector*.

Theorem 2.3 If the matrices A and B are conformable for addition, then

$$(A + B)^T = A^T + B^T \quad (2.12)$$

where the matrices are conformable for the operations indicated.

Proof Assume A and B are of order (m, n). Then the element in the ith row, jth column of $(A + B)$ is $a_{ij} + b_{ij}$. Let

$$C = (A + B)^T$$

Then

$$c_{ji} = a_{ij} + b_{ij}, \quad i = 1, 2, \ldots, m, \quad j = 1, 2, \ldots, n$$

Let R and S be defined such that

$$r_{ji} = a_{ij}, \quad s_{ji} = b_{ij}$$

where i and j assume the same values as above. Then

$$C = R + S$$

But $R = A^T$ and $S = B^T$. Thus

$$(A + B)^T = A^T + B^T$$

Theorem 2.4 Let A be conformable to B for multiplication, then

$$(AB)^T = B^T A^T \tag{2.13}$$

Proof Let A be of order (p, n) and B of order (n, q). Then the elements of AB are given by $\sum_{j=1}^{n} a_{ij} b_{jk}$, $i = 1, 2, \ldots, p$ and $k = 1, 2, \ldots, q$. Let $R = (AB)^T$. Therefore R is of order (q, p) and

$$r_{kl} = \sum_{j=1}^{n} a_{ij} b_{jk}$$

where the permissible values of k and i are as defined above. Since A and B are of order (p, n) and (n, q) respectively, A^T and B^T are of order (n, p) and (q, n) respectively and B^T is conformable to A^T for multiplication. Let $S = B^T A^T$ or

$$S = \begin{bmatrix} b_{11} & b_{21} & \cdots & b_{n1} \\ b_{12} & b_{22} & \cdots & b_{n2} \\ \vdots & \vdots & & \vdots \\ b_{1q} & b_{2q} & \cdots & b_{nq} \end{bmatrix} \begin{bmatrix} a_{11} & a_{21} & \cdots & a_{p1} \\ a_{12} & a_{22} & \cdots & a_{p2} \\ \vdots & \vdots & & \vdots \\ a_{1n} & a_{2n} & \cdots & a_{pn} \end{bmatrix}$$

Therefore S is of order (q, p) and the element in the kth row, ith column of S is $\sum_{j=1}^{n} b_{jk} a_{ij}$. Since R and S are of the same order and $r_{ki} = s_{ki}$ for all permissible k and i, hence

$$(AB)^T = B^T A^T$$

Example 2.6 Let A, B, and C be defined as

$$A = \begin{bmatrix} 1 & 4 & 2 \\ 1 & 5 & 3 \end{bmatrix}, \quad B = \begin{bmatrix} 1 & 5 & 0 \\ 0 & 1 & 0 \end{bmatrix}, \quad C = \begin{bmatrix} 1 \\ 5 \\ 1 \end{bmatrix}$$

2.3 THE TRANSPOSE OF A MATRIX

Show that
$$(A + B)^T = A^T + B^T, \qquad (AC)^T = C^T A^T$$

$$(A + B)^T = \begin{bmatrix} 2 & 9 & 2 \\ 1 & 6 & 3 \end{bmatrix}^T = \begin{bmatrix} 2 & 1 \\ 9 & 6 \\ 2 & 3 \end{bmatrix}$$

$$A^T = \begin{bmatrix} 1 & 1 \\ 4 & 5 \\ 2 & 3 \end{bmatrix}, \qquad B^T = \begin{bmatrix} 1 & 0 \\ 5 & 1 \\ 0 & 0 \end{bmatrix}$$

Therefore
$$A^T + B^T = \begin{bmatrix} 2 & 1 \\ 9 & 6 \\ 2 & 3 \end{bmatrix}$$

and $(A + B)^T = A^T + B^T$

$$(AC)^T = \begin{bmatrix} 23 \\ 29 \end{bmatrix}^T = [23 \quad 29]$$

$$C^T = [1 \quad 5 \quad 1], \qquad A^T = \begin{bmatrix} 1 & 1 \\ 4 & 5 \\ 2 & 3 \end{bmatrix}$$

$$C^T A^T = [23 \quad 29] \qquad \text{and} \qquad (AC)^T = C^T A^T$$

2.3.1 Symmetric, Skew-Symmetric, Scalar, Identity, and Periodic Matrices

There are several matrices which are repeatedly encountered in operations research, statistics, and applied mathematics in general as well as matrix theory itself. Although the application of these special matrices to problems in operations research may not be immediately evident their importance will be demonstrated by example in this and succeeding sections of this chapter as well as in later chapters.

Definition 2.5. A matrix A is said to be *symmetric* if $A = A^T$.

Definition 2.6 A matrix A is said to be *skew-symmetric* if $A = -A^T$.

By definition of the equivalence of two matrices, if A is either symmetric or skew-symmetric it must be square. However, the converse is not necessarily true. If A is symmetric then

$$a_{ij} = a_{ji} \qquad \text{for all } i \text{ and } j \tag{2.14}$$

and if A is skew-symmetric

$$a_{ij} = -a_{ji} \quad \text{for all } i \text{ and } j$$

Theorem 2.5 The diagonal elements of a skew-symmetric matrix must be zero.

Proof Let A be skew-symmetric and therefore a square matrix. If A is skew-symmetric

$$a_{ij} = -a_{ji} \quad \text{for all } i \text{ and } j$$

Therefore

$$a_{ii} = -a_{ii}$$

However, if this relationship holds then a_{ii} must be zero for all i. ▨

Theorem 2.6 Let

$$B = A^T A$$

The product matrix B is always defined and is symmetric.

Proof Let A be of order (n, m). Then A^T is of order (m, n) and A^T is conformable to A for multiplication. Therefore B is a square matrix of order m. The elements b_{ik} and b_{ki} of B are given by

$$b_{ik} = \sum_{j=1}^{n} a_{ji} a_{jk}, \quad b_{ki} = \sum_{j=1}^{n} a_{jk} a_{ji}$$

By definition of the transpose of a matrix, a_{kj} of A is equal to a_{jk} of A^T and a_{ji} of A^T is equal to a_{ij} of A. Therefore

$$b_{ik} = b_{ki}$$ ▨

Example 2.7 Let A be a square matrix. Show that A can be expressed by

$$A = B + C$$

where B is a symmetric matrix and C is a skew-symmetric matrix.

Let A, B, and C be square matrices of order m. Then

$$a_{ij} = b_{ij} + c_{ij}, \quad a_{ji} = b_{ji} + c_{ji}$$

for all i and j. Since C is presumed to be skew-symmetric, $c_{ij} = 0$ for $i = j$ and $c_{ij} = -c_{ji}$ for $i \neq j$. Since B is symmetric $b_{ij} = b_{ji}$ for all i and j. Therefore

$$a_{ji} = b_{ij} - c_{ij}$$

and

$$b_{ij} = \frac{a_{ij} + a_{ji}}{2}, \quad c_{ij} = \frac{a_{ij} - a_{ji}}{2} \quad \text{for all } i \text{ and } j$$ ▨

2.3 THE TRANSPOSE OF A MATRIX

Definition 2.7 A *scalar matrix* is a square matrix with off-diagonal elements equal to zero and diagonal elements equal to a constant k.

Therefore, if A is a scalar matrix, it can be expressed by the general form

$$A = \begin{bmatrix} k & 0 & 0 & \cdots & 0 \\ 0 & k & 0 & \cdots & 0 \\ 0 & 0 & k & \cdots & 0 \\ \vdots & \vdots & \vdots & & \vdots \\ 0 & 0 & 0 & \cdots & k \end{bmatrix} \tag{2.15}$$

The term scalar matrix derives from the fact that if a scalar matrix A is conformable to a matrix B for multiplication, then

$$AB = kB \tag{2.16}$$

That is, multiplication of a matrix B by a scalar matrix A is equivalent to multiplication of B by the scalar quantity k.

The identity matrix, usually denoted by I, is a special scalar matrix for which the diagonal elements are equal to unity. That is,

$$I = \begin{bmatrix} 1 & 0 & 0 & \cdots & 0 \\ 0 & 1 & 0 & \cdots & 0 \\ 0 & 0 & 1 & \cdots & 0 \\ \vdots & \vdots & \vdots & & \vdots \\ 0 & 0 & 0 & \cdots & 1 \end{bmatrix} \tag{2.17}$$

Since I is a scalar matrix

$$AI = A \tag{2.18}$$

where A is assumed conformable to I for multiplication.

Definition 2.8 The square matrix A is *periodic* and of *period* k if k is the least integer such that

$$A^{k+1} = A \tag{2.19}$$

In the special case where $k = 1$, the matrix A is said to be *idempotent*.

Example 2.8 Determine whether the following matrices are idempotent.

$$A = \begin{bmatrix} 1 & 1 & 1 \\ 0 & 0 & 0 \\ 0 & 0 & 0 \end{bmatrix}, \quad B = \begin{bmatrix} k & 0 & 0 \\ 0 & k & 0 \\ 0 & 0 & k \end{bmatrix}, \quad C = \begin{bmatrix} 1 & 0 & 0 \\ 0 & 1 & 0 \\ 0 & 1 & 0 \end{bmatrix}$$

We have

$$A^2 = \begin{bmatrix} 1 & 1 & 1 \\ 0 & 0 & 0 \\ 0 & 0 & 0 \end{bmatrix}$$

Therefore A is idempotent. Also,

$$B^2 = \begin{bmatrix} k^2 & 0 & 0 \\ 0 & k^2 & 0 \\ 0 & 0 & k^2 \end{bmatrix}$$

Therefore B is symmetric idempotent if and only if $k = 1$. Finally,

$$C^2 = \begin{bmatrix} 1 & 0 & 0 \\ 0 & 1 & 0 \\ 0 & 1 & 0 \end{bmatrix}$$

and C is idempotent.

2.4 The Determinant of a Matrix

Before discussing the determinant of a matrix, a brief discussion of permutations and inversions is in order. Consider the following arrangements of the integers 1, 2, and 3.

$$123 \quad 132 \quad 213 \quad 231 \quad 312 \quad 321$$

Each of these arrangements is called a *permutation* of the integers. To generalize, let $k_1, k_2, \ldots, k_n$ be some permutation of the integers 1 to n, where each integer is included once and only once. If $k_i = i$ for $i = 1, 2, \ldots, n$, the integers are arranged in their natural order and this arrangement constitutes one of $n!$ possible permutations of the integers. Another permutation is defined by

$$k_1 = n$$

$$k_i = i - 1, \quad i = 2, 3, \ldots, n$$

Definition 2.9 If, in a given permutation, an integer is preceded by a larger integer an *inversion* is said to exist.

For example, in the permutation 1243, one inversion exists since 3 is preceded by 4. However, in the permutation 4123 there are 3 inversions since 4 precedes 1, 2, and 3. In a given permutation if the number of inversions is odd the permutation is said to be of *odd parity*. If the number of inversions is even, the permutation is of *even parity*. To illustrate, consider the following permutation of the integers 1 to 4 (Table 2.1).

2.4 THE DETERMINANT OF A MATRIX

TABLE 2.1
PERMUTATIONS OF THE INTEGERS 1, 2, 3, 4

Permutation	No. Inversions	Parity	Permutation	No. Inversions	Parity
1234	0	Even	3124	2	Even
1243	1	Odd	3142	3	Odd
1324	1	Odd	3214	3	Odd
1342	2	Even	3241	4	Even
1423	2	Even	3412	4	Even
1432	3	Odd	3421	5	Odd
2134	1	Odd	4123	3	Odd
2143	2	Even	4132	4	Even
2314	2	Even	4213	4	Even
2341	3	Odd	4231	5	Odd
2413	3	Odd	4312	5	Odd
2431	4	Even	4321	6	Even

To obtain the number of inversions in a given permutation, one simply counts the number of smaller integers following each integer in the permutation. For example, consider the permutation 1,3,5,7,9,2,4,6,8. 1 is followed by no integer less than itself. 3 is followed by one integer less than itself, 2. 5 is followed by 2 and 4; 7 by 2, 4, and 6; and 9 is followed by 2, 4, 6, and 8. The remaining integers, 2, 4, 6, and 8, are not followed by smaller integers and the number of inversions is 10 and the parity of the permutation is even. Hereafter a permutation will simply be referred to as even or odd to denote even parity or odd parity.

Theorem 2.7 If any two adjacent terms of a permutation of distinct integers are interchanged the parity of the permutation is reversed.

Proof Let k_i and k_{i+1} be the terms interchanged. If $k_i < k_{i+1}$, the number of inversions is increased by one by the interchange. If $k_i > k_{i+1}$ the number of inversions is reduced by one. Therefore, in either case, the parity of the permutation is reversed. ▩

Let

$$\epsilon_{k_1 k_2 \cdots k_m} = \begin{cases} +1, & \text{if } k_1 k_2 \cdots k_m \text{ is even} \\ -1, & \text{if } k_1 k_2 \cdots k_m \text{ is odd} \end{cases} \quad (2.20)$$

where $k_1 k_2 \cdots k_m$ is a permutation of the integers 1 to m inclusive. Let A be a square matrix of order m with elements a_{ij}. The *determinant* of A will be denoted by $|A|$ or $\det(A)$ and is defined by

$$|A| = \sum \epsilon_{k_1 k_2 \cdots k_m} a_{1k_1} a_{2k_2} \cdots a_{mk_m} \quad (2.21)$$

where the sum is over all permutations of $k_1 k_2 \cdots k_m$. Therefore, each term in the summation of Eq. (2.21) contains one and only one element from each row and column of A. The sign of the term is determined by the parity of $k_1 k_2 \cdots k_m$.

Definition 2.10 If A is a square matrix such that $|A| = 0$, then A is said to be a *singular* matrix. If $|A| \neq 0$, then A is nonsingular.

Example 2.9 Evaluate the determinant of the following matrix

$$A = \begin{bmatrix} 3 & 1 & 3 \\ 0 & 4 & 1 \\ 1 & 1 & 0 \end{bmatrix}$$

$k_1 k_2 k_3$	$\epsilon_{k_1 k_2 k_3}$	$a_{1k_1} a_{2k_2} a_{3k_3}$	$\epsilon_{k_1 k_2 k_3} a_{1k_1} a_{2k_2} a_{3k_3}$
123	+1	$a_{11} a_{22} a_{33} = (3)(4)(0)$	0
132	−1	$a_{11} a_{23} a_{32} = (3)(1)(1)$	−3
213	−1	$a_{12} a_{21} a_{33} = (1)(0)(0)$	0
231	+1	$a_{12} a_{23} a_{31} = (1)(1)(1)$	+1
312	+1	$a_{13} a_{21} a_{32} = (3)(0)(0)$	0
321	−1	$a_{13} a_{22} a_{31} = (3)(4)(1)$	−12

$$|A| = -14$$

Example 2.10 Evaluate the determinant of the following matrix.

$$A = \begin{bmatrix} 1 & 1 & 2 & 3 \\ 0 & 3 & 1 & 0 \\ 1 & 1 & 0 & 0 \\ 4 & 2 & 1 & 1 \end{bmatrix}$$

See Table 2.2.

Theorem 2.8 Interchanging any two rows of a square matrix reverses the sign of the determinant of the matrix.

Proof Let the matrices A and B be defined as

$$A = \begin{bmatrix} a_{11} & a_{12} & \cdots & a_{1j} & \cdots & a_{1m} \\ a_{21} & a_{22} & \cdots & a_{2j} & \cdots & a_{2m} \\ \vdots & \vdots & & \vdots & & \vdots \\ a_{i1} & a_{i2} & \cdots & a_{ij} & \cdots & a_{im} \\ \vdots & \vdots & & \vdots & & \vdots \\ a_{k1} & a_{k2} & \cdots & a_{kj} & \cdots & a_{km} \\ \vdots & \vdots & & \vdots & & \vdots \\ a_{m1} & a_{m2} & \cdots & a_{mj} & \cdots & a_{mm} \end{bmatrix}, \quad B = \begin{bmatrix} a_{11} & a_{12} & \cdots & a_{1j} & \cdots & a_{1m} \\ a_{21} & a_{22} & \cdots & a_{2j} & \cdots & a_{2m} \\ \vdots & \vdots & & \vdots & & \vdots \\ a_{k1} & a_{k2} & \cdots & a_{kj} & \cdots & a_{km} \\ \vdots & \vdots & & \vdots & & \vdots \\ a_{i1} & a_{i2} & \cdots & a_{ij} & \cdots & a_{im} \\ \vdots & \vdots & & \vdots & & \vdots \\ a_{m1} & a_{m2} & \cdots & a_{mj} & \cdots & a_{mm} \end{bmatrix}$$

That is, the matrix B is obtained by interchanging the ith and kth rows of matrix A:

$$|A| = \sum \epsilon_{p_1 p_2 \cdots p_i \cdots p_k \cdots p_m} a_{1p_1} a_{2p_2} \cdots a_{ip_i} \cdots a_{kp_k} \cdots a_{mp_m}$$

$$|B| = \sum \epsilon_{p_1 p_2 \cdots p_k \cdots p_i \cdots p_m} a_{1p_1} a_{2p_2} \cdots a_{kp_k} \cdots a_{ip_i} \cdots a_{mp_m}$$

TABLE 2.2

$k_1 k_2 k_3 k_4$	$\epsilon_{k_1 k_2 k_3 k_4}$	$a_{1k_1} a_{2k_2} a_{3k_3} a_{4k_4}$		$\epsilon_{k_1 k_2 k_3 k_4} a_{1k_1} a_{2k_2} a_{3k_3} a_{4k_4}$
1234	+1	$a_{11} a_{22} a_{33} a_{44}$	= (1)(3)(0)(1)	0
1243	−1	$a_{11} a_{22} a_{34} a_{43}$	= (1)(3)(0)(1)	0
1324	−1	$a_{11} a_{23} a_{32} a_{44}$	= (1)(1)(1)(1)	−1
1342	+1	$a_{11} a_{23} a_{34} a_{42}$	= (1)(1)(0)(2)	0
1423	+1	$a_{11} a_{24} a_{32} a_{43}$	= (1)(0)(1)(1)	0
1432	−1	$a_{11} a_{24} a_{33} a_{42}$	= (1)(0)(0)(2)	0
2134	−1	$a_{12} a_{21} a_{33} a_{44}$	= (1)(0)(0)(1)	0
2143	+1	$a_{12} a_{21} a_{34} a_{43}$	= (1)(0)(0)(1)	0
2314	+1	$a_{12} a_{23} a_{31} a_{44}$	= (1)(1)(1)(1)	+1
2341	−1	$a_{12} a_{23} a_{34} a_{41}$	= (1)(1)(0)(4)	0
2413	−1	$a_{12} a_{24} a_{31} a_{43}$	= (1)(0)(1)(1)	0
2431	+1	$a_{12} a_{24} a_{33} a_{41}$	= (1)(0)(0)(4)	0
3124	+1	$a_{13} a_{21} a_{32} a_{44}$	= (2)(0)(1)(1)	0
3142	−1	$a_{13} a_{21} a_{34} a_{42}$	= (2)(0)(0)(2)	0
3214	−1	$a_{13} a_{22} a_{31} a_{44}$	= (2)(3)(1)(1)	−6
3241	+1	$a_{13} a_{22} a_{34} a_{41}$	= (2)(3)(0)(4)	0
3412	+1	$a_{13} a_{24} a_{31} a_{42}$	= (2)(0)(1)(2)	0
3421	−1	$a_{13} a_{24} a_{32} a_{41}$	= (2)(0)(1)(4)	0
4123	−1	$a_{14} a_{21} a_{32} a_{43}$	= (3)(0)(1)(1)	0
4132	+1	$a_{14} a_{21} a_{33} a_{42}$	= (3)(0)(0)(2)	0
4213	+1	$a_{14} a_{22} a_{31} a_{43}$	= (3)(3)(1)(1)	+9
4231	−1	$a_{14} a_{22} a_{33} a_{41}$	= (3)(3)(0)(4)	0
4312	−1	$a_{14} a_{23} a_{31} a_{42}$	= (3)(1)(1)(2)	−6
4321	+1	$a_{14} a_{23} a_{32} a_{41}$	= (3)(1)(1)(4)	+12
			$\lvert A \rvert =$	+9

$\lvert B \rvert$ is obtained by interchanging p_i and p_k in $\epsilon_{p_1 p_2 \cdots p_m}$ of $\lvert A \rvert$ since interchanging $a_{i p_i}$ and $a_{k p_k}$ has no effect on the value of the determinant. Assume $i < k$. The permutation $p_1 p_2 \cdots p_i \cdots p_k \cdots p_m$ can be changed to $p_1 p_2 \cdots p_k \cdots p_i \cdots p_m$ by first interchanging p_k and p_{k-1}, then interchanging p_k and p_{k-2}, etc., until p_k lies between p_{i-1} and p_i. Next p_i is interchanged with p_{i+1}, then interchanged with p_{i+2}, etc., until p_i lies between p_{k-1} and p_{k+1}. The succession of interchanges which places p_k between p_{i-1} and p_i requires $(k-1)-(i-1)$ adjacent transpositions. The succession of interchanges which places p_i between p_{k-1} and p_{k+1} requires $(k-1)-i$ adjacent transpositions. Therefore the total number of adjacent transpositions is $2(k-i-1)+1$, which results in $2(k-i-1)+1$ inversions. Since the number of inversions is odd, the sign of $\epsilon_{p_1 p_2 \cdots p_m}$ is reversed after the interchange of p_i and p_k. Therefore

$$\lvert B \rvert = -\lvert A \rvert$$

It is worth noting that the result proven here also applies when two columns of a square matrix are interchanged.

Theorem 2.9 Let A be a square matrix of order m. If any two rows of A are identical, then $|A| = 0$.

Proof Assume that $a_{ij} = a_{kj}$ for $j = 1, 2, \ldots, m$ and $i < k$.

$$|A| = \sum \epsilon_{p_1 p_2 \cdots p_i \cdots p_k \cdots p_m} a_{1 p_1} a_{2 p_2} \cdots a_{i p_i} \cdots a_{k p_k} \cdots a_{m p_m}$$

For each term of the summation of the form

$$\epsilon_{p_1 p_2 \cdots p_i \cdots p_k \cdots p_m} a_{1 p_1} a_{2 p_2} \cdots a_{i p_i} \cdots a_{k p_k} \cdots a_{m p_m}$$

there is a corresponding term

$$\epsilon_{p_1 p_2 \cdots p_k \cdots p_i \cdots p_m} a_{1 p_1} a_{2 p_2} \cdots a_{k p_k} \cdots a_{i p_i} \cdots a_{m p_m}.$$

Since $a_{ij} = a_{kj}$ for all j, these two terms differ only in their signs. The change in signs results from the transposition of p_i and p_k in $\epsilon_{p_1 p_2 \cdots p_m}$. Therefore, for each term of the summation there is another term of equal magnitude but of opposite sign. The determinant of A can then be expressed by

$$A = \sum [\epsilon_{p_1 p_2 \cdots p_i \cdots p_k \cdots p_m} + \epsilon_{p_1 p_2 \cdots p_k \cdots p_i \cdots p_m}] a_{1 p_1} a_{2 p_2} \cdots a_{i p_i} \cdots a_{k p_k} \cdots a_{m p_m} = 0$$

where the sum is taken over all permutations for which $i < k$. ∎

Theorem 2.10 If any row of a square matrix A is zero, then $|A| = 0$.

Proof Assume $a_{ij} = 0$ for all j and any i. The determinant of A is given by

$$|A| = \sum \epsilon_{k_1 k_2 \cdots k_i \cdots k_m} a_{1 k_1} a_{2 k_2} \cdots a_{i k_i} \cdots a_{m k_m}$$

Each term of the summation is zero since each term, $a_{1 k_1} a_{2 k_2} \cdots a_{i k_i} \cdots a_{m k_m}$, contains an element which is zero, $a_{i k_i}$. Therefore $|A| = 0$. ∎

The following properties of determinants are given without proof.

Property 1 If A is a square matrix of order m, the determinant of A can be defined as

$$|A| = \sum \epsilon_{k_1 k_2 \cdots k_m} a_{k_1 1} a_{k_2 2} \cdots a_{k_m m} \tag{2.22}$$

where the summation is over the $m!$ permutations of $k_1 k_2 \cdots k_m$.

Property 2 If A is a square matrix, then

$$|A^T| = |A| \tag{2.23}$$

Property 3 If A is a square matrix such that one row is a common multiple of another row or such that one column is a common multiple of another column then

$$|A| = 0$$

Property 4 Let A be a square matrix. If to any row of A a constant multiple of any other row is added or if to any column a constant multiple of any other

2.4 THE DETERMINANT OF A MATRIX

column is added, the determinant of the resulting matrix is equal to the determinant of A.

Property 5 Let A and B be square matrices of order m with identical corresponding elements except that one row of B is k times the corresponding row of A or one column of B is k times the corresponding column of A then

$$|B| = k|A|$$

Property 6 If A and B are square matrices of order m, then

$$|AB| = |A| \, |B| \qquad (2.24)$$

Property 7 Let A be a scalar matrix of order m with diagonal elements equal to k. Then

$$|A| = k^m \qquad (2.25)$$

Property 8

$$|I| = 1 \qquad (2.26)$$

Property 9 Let A be a symmetric idempotent matrix. Then either

$$|A| = 0 \qquad \text{or} \qquad A = I$$

Theorem 2.11 If A is a square matrix of order m, then

$$|kA| = k^m |A|$$

where k is a scalar.

Proof Let B be a scalar matrix of order m with diagonal elements equal to k. By Eq. (2.16),

$$BA = kA$$

By Property 6,

$$|BA| = |B| \, |A|$$

By Property 7,

$$|B| = k^m$$

Therefore

$$|kA| = k^m |A| \qquad \blacksquare$$

Example 2.11 Let A be a skew-symmetric matrix of order m. Evaluate the determinant of A for odd m.

Since A is skew-symmetric

$$A^T = -A$$

Let B be a scalar matrix of order m with diagonal elements equal to -1. Then

$$A^T = BA$$

and
$$|A^T| = (-1)^m |A| = -|A|$$
since m is odd. But by Property 2,
$$|A^T| = |A| \quad \text{which implies} \quad |A| = -|A|$$
Therefore
$$|A| = 0$$

2.4.1 Cofactor Method

If A is a square matrix of order 2 or 3 then the determinant of A can be computed by application of the following equations.

$$|A| = a_{11}a_{22} - a_{12}a_{21} \qquad [A \text{ is of order } 2] \qquad (2.27)$$

$$\begin{aligned}|A| = &a_{11}a_{22}a_{33} + a_{12}a_{23}a_{31} + a_{13}a_{21}a_{32} \\ &- a_{13}a_{22}a_{31} - a_{11}a_{23}a_{32} - a_{12}a_{21}a_{33}\end{aligned}$$
$$[A \text{ is of order } 3] \qquad (2.28)$$

When the order of the square matrix under consideration is greater than 3, evaluation of the determinant becomes more complicated and techniques less complicated than those given in the previous section would be welcomed. Of course any determinant can be evaluated by complete enumeration of the permutations involved and determination of the parity associated with each permutation as illustrated in Examples 2.9 and 2.10. However, the tedium involved in enumerating the permutations alone increases significantly as the order of the matrix is increased. Fortunately there are computational methods available which reduce the complexity of this process considerably.

Let A be a square matrix of order m with elements a_{ij}. From A delete the ith row and the jth column leaving a square matrix of order $m - 1$. This $(m - 1)$ square matrix will be denoted by M_{ij} and is taken as the matrix A after the ith row and jth column have been removed. The determinant of M_{ij} is called the *minor* of a_{ij}. Therefore, each element a_{ij}, of a square matrix, has an associated minor $|M_{ij}|$.

Example 2.12 Evaluate the minors of the following matrix.

$$A = \begin{bmatrix} 1 & 4 & 2 \\ 3 & 1 & 1 \\ 4 & 1 & 0 \end{bmatrix}$$

$$|M_{11}| = \begin{vmatrix} 1 & 1 \\ 1 & 0 \end{vmatrix} = -1, \quad |M_{12}| = \begin{vmatrix} 3 & 1 \\ 4 & 0 \end{vmatrix} = -4, \quad |M_{13}| = \begin{vmatrix} 3 & 1 \\ 4 & 1 \end{vmatrix} = -1$$

2.4 THE DETERMINANT OF A MATRIX

$$|M_{21}| = \begin{vmatrix} 4 & 2 \\ 1 & 0 \end{vmatrix} = -2, \quad |M_{22}| = \begin{vmatrix} 1 & 2 \\ 4 & 0 \end{vmatrix} = -8, \quad |M_{23}| = \begin{vmatrix} 1 & 4 \\ 4 & 1 \end{vmatrix} = -15$$

$$|M_{31}| = \begin{vmatrix} 4 & 2 \\ 1 & 1 \end{vmatrix} = 2, \quad |M_{32}| = \begin{vmatrix} 1 & 2 \\ 3 & 1 \end{vmatrix} = -5, \quad |M_{33}| = \begin{vmatrix} 1 & 4 \\ 3 & 1 \end{vmatrix} = -11$$

In addition to its minor, each element of a square matrix has an associated cofactor.

Definition 2.11 The *cofactor*, A_{ij}, of an element, a_{ij}, is obtained by multiplying its minor, $|M_{ij}|$, by $(-1)^{i+j}$, or

$$A_{ij} = (-1)^{i+j} |M_{ij}| \tag{2.29}$$

The cofactor of an element is sometimes referred to as a signed minor, since it is obtained by multiplying the minor by $+1$ or -1 depending upon whether $i + j$ is even or odd.

Example 2.13 Calculate the cofactors of the elements in the following matrix.

$$A = \begin{bmatrix} 1 & 1 & 3 \\ 2 & 1 & 0 \\ 1 & 5 & 1 \end{bmatrix}$$

$\|M_{11}\| = 1,$	$(-1)^{1+1} = 1,$	$A_{11} = 1$
$\|M_{12}\| = 2,$	$(-1)^{1+2} = -1,$	$A_{12} = -2$
$\|M_{13}\| = 9,$	$(-1)^{1+3} = 1,$	$A_{13} = 9$
$\|M_{21}\| = -14,$	$(-1)^{2+1} = -1,$	$A_{21} = 14$
$\|M_{22}\| = -2,$	$(-1)^{2+2} = 1,$	$A_{22} = -2$
$\|M_{23}\| = 4,$	$(-1)^{2+3} = -1,$	$A_{23} = -4$
$\|M_{31}\| = -3,$	$(-1)^{3+1} = 1,$	$A_{31} = -3$
$\|M_{32}\| = -6,$	$(-1)^{3+2} = -1,$	$A_{32} = 6$
$\|M_{33}\| = -1,$	$(-1)^{3+3} = 1,$	$A_{33} = -1$

Cofactors are useful in computing the determinant of a square matrix, particularly when the matrix is of order greater than 3. Let A be a square matrix of order m with elements a_{ij} and cofactors A_{ij}. Then

$$|A| = \sum_{i=1}^{m} a_{ij} A_{ij}, \quad j = 1, 2, 3, \ldots, m \tag{2.30}$$

or

$$|A| = \sum_{j=1}^{m} a_{ij} A_{ij}, \quad i = 1, 2, 3, \ldots, m \tag{2.31}$$

That is, the determinant of a square matrix may be evaluated by selecting any column of the matrix, j, multiplying each element in that column by its cofactor, and taking the sum of the resulting products. Equation (2.31) implies that the same result can be achieved by selecting a row, i, instead of a column and carrying out the indicated operations on the elements in the selected row.

Example 2.14 Evaluate the determinant of the following matrix using

a. Eq. (2.30) where $j = 1$;
b. Eq. (2.31) where $i = 3$

$$A = \begin{bmatrix} 1 & 1 & 1 & 2 \\ 4 & 0 & 2 & 0 \\ 2 & 0 & 1 & 2 \\ 0 & 1 & 1 & 1 \end{bmatrix}$$

Part a

$$|A| = (1)(-1)^{1+1}\begin{vmatrix} 0 & 2 & 0 \\ 0 & 1 & 2 \\ 1 & 1 & 1 \end{vmatrix} + (4)(-1)^{2+1}\begin{vmatrix} 1 & 1 & 2 \\ 0 & 1 & 2 \\ 1 & 1 & 1 \end{vmatrix}$$

$$+ (2)(-1)^{1+3}\begin{vmatrix} 1 & 1 & 2 \\ 0 & 2 & 0 \\ 1 & 1 & 1 \end{vmatrix} + (0)(-1)^{1+4}\begin{vmatrix} 1 & 1 & 2 \\ 0 & 2 & 0 \\ 0 & 1 & 2 \end{vmatrix}$$

$$= (1)(1)(4) + (4)(-1)(-1) + (2)(1)(-2) + (0)(-1)(4)$$
$$= 4 + 4 - 4 + 0 = 4$$

Part b

$$|A| = (2)(-1)^{3+1}\begin{vmatrix} 1 & 1 & 2 \\ 0 & 2 & 0 \\ 1 & 1 & 1 \end{vmatrix} + (0)(-1)^{3+2}\begin{vmatrix} 1 & 1 & 2 \\ 4 & 2 & 0 \\ 0 & 1 & 1 \end{vmatrix}$$

$$+ (1)(-1)^{3+3}\begin{vmatrix} 1 & 1 & 2 \\ 4 & 0 & 0 \\ 0 & 1 & 1 \end{vmatrix} + (2)(-1)^{3+4}\begin{vmatrix} 1 & 1 & 1 \\ 4 & 0 & 2 \\ 0 & 1 & 1 \end{vmatrix}$$

$$= (2)(1)(-2) + (0)(-1)(6) + (1)(1)(4) + (2)(-1)(-2)$$
$$= -4 + 0 + 4 + 4 = 4$$

Example 2.15 Using the cofactor method, evaluate the determinant of the following matrix

$$A = \begin{bmatrix} 1 & 0 & 1 & 2 & 0 \\ 0 & 1 & 1 & 2 & 0 \\ 0 & 1 & 1 & 1 & 0 \\ 0 & 0 & 1 & 2 & 1 \\ 1 & 2 & 1 & 1 & 1 \end{bmatrix}$$

2.4 THE DETERMINANT OF A MATRIX

Applying Eq. (2.31) to the first row leads to

$$|A| = (1)(-1)^{1+1}\begin{vmatrix} 1 & 1 & 2 & 0 \\ 1 & 1 & 1 & 0 \\ 0 & 1 & 2 & 1 \\ 2 & 1 & 1 & 1 \end{vmatrix} + (0)(-1)^{1+2}\begin{vmatrix} 0 & 1 & 2 & 0 \\ 0 & 1 & 1 & 0 \\ 0 & 1 & 2 & 1 \\ 1 & 1 & 1 & 1 \end{vmatrix}$$

$$+ (1)(-1)^{1+3}\begin{vmatrix} 0 & 1 & 2 & 0 \\ 0 & 1 & 1 & 0 \\ 0 & 0 & 2 & 1 \\ 1 & 2 & 1 & 1 \end{vmatrix} + (2)(-1)^{1+4}\begin{vmatrix} 0 & 1 & 1 & 0 \\ 0 & 1 & 1 & 0 \\ 0 & 0 & 1 & 1 \\ 1 & 2 & 1 & 1 \end{vmatrix}$$

$$+ (0)(-1)^{1+5}\begin{vmatrix} 0 & 1 & 1 & 2 \\ 0 & 1 & 1 & 1 \\ 0 & 0 & 1 & 2 \\ 1 & 2 & 1 & 1 \end{vmatrix}$$

Since $a_{12} = a_{15} = 0$, the second and fifth terms in the above sum are immediately eliminated. The fourth term of this is also zero by Property 3 of determinants since rows one and two of the determinant in this term are identical and therefore constant multiples of one another. Thus the determinant is reduced to the sum of the first and third terms only.

$$|A| = (1)(-1)^{1+1}\begin{vmatrix} 1 & 1 & 2 & 0 \\ 1 & 1 & 1 & 0 \\ 0 & 1 & 2 & 1 \\ 2 & 1 & 1 & 1 \end{vmatrix} + (1)(-1)^{1+3}\begin{vmatrix} 0 & 1 & 2 & 0 \\ 0 & 1 & 1 & 0 \\ 0 & 0 & 2 & 1 \\ 1 & 2 & 1 & 1 \end{vmatrix}$$

The cofactor method is now reapplied to the first row of each of the 4 × 4 determinants in the above expression.

$$|A| = (1)(-1)^{1+1}\left[(1)(-1)^{1+1}\begin{vmatrix} 1 & 1 & 0 \\ 1 & 2 & 1 \\ 1 & 1 & 1 \end{vmatrix} + (1)(-1)^{1+2}\begin{vmatrix} 1 & 1 & 0 \\ 0 & 2 & 1 \\ 2 & 1 & 1 \end{vmatrix}\right.$$

$$+ (2)(-1)^{1+3}\begin{vmatrix} 1 & 1 & 0 \\ 0 & 1 & 1 \\ 2 & 1 & 1 \end{vmatrix} + (0)(-1)^{1+4}\begin{vmatrix} 1 & 1 & 1 \\ 0 & 1 & 2 \\ 2 & 1 & 1 \end{vmatrix}\Bigg]$$

$$+ (1)(-1)^{1+3}\left[(0)(-1)^{1+1}\begin{vmatrix} 1 & 1 & 0 \\ 0 & 2 & 1 \\ 2 & 1 & 1 \end{vmatrix} + (1)(-1)^{1+2}\begin{vmatrix} 0 & 1 & 0 \\ 0 & 2 & 1 \\ 1 & 1 & 1 \end{vmatrix}\right.$$

$$+ (2)(-1)^{1+3}\begin{vmatrix} 0 & 1 & 0 \\ 0 & 0 & 1 \\ 1 & 2 & 1 \end{vmatrix} + (0)(-1)^{1+4}\begin{vmatrix} 0 & 1 & 1 \\ 0 & 0 & 2 \\ 1 & 2 & 1 \end{vmatrix}\Bigg]$$

Evaluating each of the 3 × 3 determinants

$$|A| = (1)(1)[(1)(1)(1) + (1)(-1)(3) + (2)(1)(2) + (0)(-1)(1)]$$
$$+ (1)(1)[(0)(1)(3) + (1)(-1)(1) + (2)(1)(1) + (0)(-1)(2)]$$
$$= (1 - 3 + 4 + 0) + (0 - 1 + 2 + 0) = 3$$

Example 2.16 Show that the following matrix is singular.

$$A = \begin{bmatrix} 1 & 1 & 0 & 0 & -2 \\ -2 & 4 & 0 & 0 & 1 \\ 0 & 0 & 1 & 0 & 0 \\ 9 & 1 & 6 & 1 & 3 \\ -1 & 1 & 0 & 0 & 1 \end{bmatrix}$$

If A is singular, then $|A| = 0$. Applying the cofactor method to the fourth column of A yields

$$|A| = a_{44} A_{44}$$

since $a_{14} = a_{24} = a_{34} = a_{54} = 0$.

$$|A| = (1)(-1)^{4+4} \begin{vmatrix} 1 & 1 & 0 & -2 \\ -2 & 4 & 0 & 1 \\ 0 & 0 & 1 & 0 \\ -1 & 1 & 0 & 1 \end{vmatrix}$$

Applying the cofactor method to the third column of the above 4 × 4 determinant,

$$|A| = (1)(-1)^{4+4}(1)(-1)^{3+3} \begin{vmatrix} 1 & 1 & -2 \\ -2 & 4 & 1 \\ -1 & 1 & 1 \end{vmatrix} = (1)(1)(1)(1)(0) = 0$$

2.5 Rank of a Matrix

Let A be a matrix of order (m, n), where $m < n$. To evaluate the rank of A, each square submatrix within A is examined.

Definition 2.12 A is said to be of *rank k* if there exists at least one nonsingular square submatrix in A of order (k, k), and if all square submatrices of order $(k + 1, k + 1)$ or greater, if any, are singular.

Therefore, the rank of a matrix A is determined by evaluating the determinants of the square submatrices within A. If A is of order (m, n) and $m < n$, then A may not have rank greater than m.

2.6 INVERSE OF A SQUARE MATRIX

Example 2.17 Find the rank of the following matrices.

$$A = \begin{bmatrix} 1 & 0 & 1 & 1 \\ 3 & 2 & 4 & 2 \\ 1 & 0 & 1 & 6 \end{bmatrix}, \quad B = \begin{bmatrix} 1 & 1 & 2 \\ 1 & 4 & 2 \\ 0 & 3 & 0 \end{bmatrix}, \quad C = \begin{bmatrix} 1 & 0 & 0 & 0 \\ 4 & 1 & 0 & 1 \\ 1 & 0 & 1 & 0 \\ 1 & 1 & 0 & 1 \end{bmatrix}$$

Since A is not a square matrix, it has no determinant. Examining the submatrices of A of order $(3, 3)$ yields

$$\begin{vmatrix} 1 & 0 & 1 \\ 3 & 2 & 4 \\ 1 & 0 & 1 \end{vmatrix} = 0 \quad \begin{vmatrix} 0 & 1 & 1 \\ 2 & 4 & 2 \\ 0 & 1 & 6 \end{vmatrix} = -10 \quad \begin{vmatrix} 1 & 1 & 1 \\ 3 & 4 & 2 \\ 1 & 1 & 6 \end{vmatrix} = 5 \quad \begin{vmatrix} 1 & 0 & 1 \\ 3 & 2 & 2 \\ 1 & 0 & 6 \end{vmatrix} = 10$$

Since A contains a nonsingular square submatrix of order $(3, 3)$, A is of rank 3.
Since

$$|B| = 0$$

B is of rank 2 or less. Since

$$\begin{vmatrix} 1 & 1 \\ 1 & 4 \end{vmatrix} = 3$$

B is of rank 2.

Using the cofactor method on the first row of C to evaluate the determinant leads to the conclusion that C is a singular matrix. Since

$$\begin{vmatrix} 1 & 0 & 0 \\ 4 & 1 & 0 \\ 1 & 0 & 1 \end{vmatrix} = 1$$

C is of rank 3.

2.6 Inverse of a Square Matrix

Definition 2.13 Let A be a square matrix of order m. The *inverse matrix* of A, denoted A^{-1}, is a square matrix of order m such that

$$A^{-1}A = AA^{-1} = I \tag{2.32}$$

where I is also of order m.

As will be shown below, the inverse of a square matrix may not exist. In particular, the inverse is defined only if the matrix under consideration is *nonsingular*; that is, it has a nonzero determinant.

The inverse of a nonsingular square matrix, A, is given by

$$A^{-1} = \frac{\text{adj}(A)}{|A|} \tag{2.33}$$

where $\text{adj}(A)$ is the adjoint of the matrix A.

Definition 2.14 The *adjoint* of a square matrix is defined as the transpose of the matrix of cofactors. Therefore

$$\text{adj}(A) = \begin{bmatrix} A_{11} & A_{21} & \cdots & A_{i1} & \cdots & A_{m1} \\ A_{12} & A_{22} & \cdots & A_{i2} & \cdots & A_{m2} \\ \vdots & \vdots & & \vdots & & \vdots \\ A_{1j} & A_{2j} & \cdots & A_{ij} & \cdots & A_{mj} \\ \vdots & \vdots & & \vdots & & \vdots \\ A_{1m} & A_{2m} & \cdots & A_{im} & \cdots & A_{mm} \end{bmatrix} \qquad (2.34)$$

From Eq. (2.33) one can easily see why the inverse is not defined if $|A| = 0$.

Example 2.18 Find the inverse of the matrices A and B and show that $A^{-1}A = I$ and $B^{-1}B = I$.

$$A = \begin{bmatrix} 1 & 1 & 0 \\ 2 & 1 & 3 \\ 1 & 1 & 4 \end{bmatrix}, \quad B = \begin{bmatrix} 1 & 0 & 1 \\ 0 & 4 & 0 \\ 1 & 0 & 2 \end{bmatrix}$$

The cofactors of the elements of A are

$$\begin{aligned} A_{11} &= 1, & A_{21} &= -4, & A_{31} &= 3 \\ A_{12} &= -5, & A_{22} &= 4, & A_{32} &= -3 \\ A_{13} &= 1, & A_{23} &= 0, & A_{33} &= -1 \end{aligned}$$

The matrix of cofactors is given by

$$\begin{bmatrix} 1 & -5 & 1 \\ -4 & 4 & 0 \\ 3 & -3 & -1 \end{bmatrix}$$

Taking the transpose of this matrix yields the adjoint of A:

$$\text{adj}(A) = \begin{bmatrix} 1 & -4 & 3 \\ -5 & 4 & -3 \\ 1 & 0 & -1 \end{bmatrix}$$

The determinant of A is evaluated by the cofactor method as follows.

$$|A| = a_{11}A_{11} + a_{12}A_{12} + a_{13}A_{13} = (1)(1) + (1)(-5) + (0)(1) = -4$$

Using Eq. (2.33) the inverse of A is given by

$$A^{-1} = -\frac{1}{4}\begin{bmatrix} 1 & -4 & 3 \\ -5 & 4 & -3 \\ 1 & 0 & -1 \end{bmatrix} = \begin{bmatrix} -\frac{1}{4} & 1 & -\frac{3}{4} \\ \frac{5}{4} & -1 & \frac{3}{4} \\ -\frac{1}{4} & 0 & \frac{1}{4} \end{bmatrix}$$

$$A^{-1}A = \begin{bmatrix} -\frac{1}{4} & 1 & -\frac{3}{4} \\ \frac{5}{4} & -1 & \frac{3}{4} \\ -\frac{1}{4} & 0 & \frac{1}{4} \end{bmatrix} \begin{bmatrix} 1 & 1 & 0 \\ 2 & 1 & 3 \\ 1 & 1 & 4 \end{bmatrix} = \begin{bmatrix} 1 & 0 & 0 \\ 0 & 1 & 0 \\ 0 & 0 & 1 \end{bmatrix} = I$$

2.6 INVERSE OF A SQUARE MATRIX

The cofactors of the elements f matrix B are

$$B_{11} = 8, \quad B_{21} = 0, \quad B_{31} = -4$$
$$B_{12} = 0, \quad B_{22} = 1, \quad B_{32} = 0$$
$$B_{13} = -4, \quad B_{23} = 0, \quad B_{33} = 4$$

yielding the following matrix of cofactors

$$\begin{bmatrix} 8 & 0 & -4 \\ 0 & 1 & 0 \\ -4 & 0 & 4 \end{bmatrix}$$

Since the matrix of cofactors is symmetric in this case, the matrix of cofactors is also the adjoint of B. The determinant of B is given by

$$|B| = b_{11}B_{11} + b_{12}B_{12} + b_{13}B_{13} = (1)(8) + (0)(0) + (1)(-4) = 4$$

and

$$B^{-1} = \frac{1}{4}\begin{bmatrix} 8 & 0 & -4 \\ 0 & 1 & 0 \\ -4 & 0 & 4 \end{bmatrix} = \begin{bmatrix} 2 & 0 & -1 \\ 0 & \frac{1}{4} & 0 \\ -1 & 0 & 1 \end{bmatrix}$$

$$B^{-1}B = \begin{bmatrix} 2 & 0 & -1 \\ 0 & \frac{1}{4} & 0 \\ -1 & 0 & 1 \end{bmatrix}\begin{bmatrix} 1 & 0 & 1 \\ 0 & 4 & 0 \\ 1 & 0 & 2 \end{bmatrix} = \begin{bmatrix} 1 & 0 & 0 \\ 0 & 1 & 0 \\ 0 & 0 & 1 \end{bmatrix} = I$$

The following properties concern the adjoint of a matrix:

Property 1 The adjoint of a scalar matrix is a scalar matrix.

Property 2 If A is a symmetric square matrix, $\text{adj}(A)$ is also a symmetric matrix.

Property 3 If A is a square matrix of order m, then

$$A[\text{adj}(A)] = |A|I \tag{2.35}$$

and

$$A[\text{adj}(A)] = [\text{adj}(A)]A \tag{2.36}$$

where I is of order m.

Theorem 2.12 If A and B are nonsingular square matrices of the same order, then

$$\text{adj}(AB) = \text{adj}(B) \cdot \text{adj}(A) \tag{2.37}$$

Proof From Property 3,

$$AB[\text{adj}(AB)] = |AB|I$$

Now

$$AB[\text{adj}(B)][\text{adj}(A)] = A[B\,\text{adj}(B)][\text{adj}(A)] = A(|B|I)[\text{adj}(A)] = |B|AI[\text{adj}(A)]$$

since $|B|$ is a scalar. From Eq. (2.18), $AI = A$. Therefore
$$AB[\mathrm{adj}(B)][\mathrm{adj}(A)] = |B|A[\mathrm{adj}(A)] = |B||A|I = |AB|I$$
from Eq. (2.24).
$$AB[\mathrm{adj}(AB)] = AB[\mathrm{adj}(B)][\mathrm{adj}(A)]$$
Multiplying both sides by $(AB)^{-1}$ and recalling that $(AB)^{-1}(AB) = I$
$$\mathrm{adj}(AB) = \mathrm{adj}(B) \cdot \mathrm{adj}(A)$$

Theorem 2.13 If A is a square matrix of order m, then
$$|\mathrm{adj}(A)| = |A|^{m-1} \tag{2.38}$$

Proof From Property 3,
$$A\,\mathrm{adj}(A) = |A|I \quad \text{and} \quad |A||\mathrm{adj}(A)| = ||A|I| = |A|^m$$
since
$$B = |A|I = \begin{bmatrix} |A| & 0 & \cdots & 0 \\ 0 & |A| & \cdots & 0 \\ \vdots & \vdots & & \vdots \\ 0 & 0 & \cdots & |A| \end{bmatrix}$$
From Property 7 of determinants, Eq. (2.25),
$$|B| = |A|^m$$
Therefore
$$|\mathrm{adj}(A)| = |A|^{m-1}$$

The following properties concern the inverse of a matrix:

Property 1 If A is a nonsingular square matrix of order m with off-diagonal elements equal to zero, then
$$A^{-1} = \begin{bmatrix} 1/a_{11} & 0 & \cdots & 0 & \cdots & 0 \\ 0 & 1/a_{22} & \cdots & 0 & \cdots & 0 \\ \vdots & \vdots & & \vdots & & \vdots \\ 0 & 0 & \cdots & 1/a_{ii} & \cdots & 0 \\ \vdots & \vdots & & \vdots & & \vdots \\ 0 & 0 & \cdots & 0 & \cdots & 1/a_{mm} \end{bmatrix} \tag{2.39}$$

Property 2 If a nonsingular matrix, A, is symmetric then A^{-1} is symmetric.

Property 3 If A is a nonsingular square matrix, then
$$(A^{-1})^{-1} = A \tag{2.40}$$

Property 4 If A is a nonsingular square matrix, then
$$(A^{\mathrm{T}})^{-1} = (A^{-1})^{\mathrm{T}} \tag{2.41}$$

2.6 INVERSE OF A SQUARE MATRIX

Property 5 If A and B are nonsingular square matrices of the same order, then
$$(AB)^{-1} = B^{-1}A^{-1} \tag{2.42}$$

Example 2.19 Show that
$$(AB)^{-1} = B^{-1}A^{-1}$$
where A and B are as defined in Property 5.
$$(AB)^{-1} = \frac{1}{|AB|}\operatorname{adj}(AB) = \frac{1}{|A||B|}\operatorname{adj}(B)\operatorname{adj}(A)$$
from Property 6 of determinants and Theorem 2.12. Therefore
$$(AB)^{-1} = \frac{\operatorname{adj}(B)}{|B|}\frac{\operatorname{adj}(A)}{|A|} = B^{-1}A^{-1}$$

Example 2.20 Let
$$\sum_{j=1}^{m} a_{1j}x_j = b_1$$
$$\sum_{j=1}^{m} a_{2j}x_j = b_2$$
$$\vdots$$
$$\sum_{j=1}^{m} a_{mj}x_j = b_m$$

Using matrix methods, derive the values of $x_1, x_2, \ldots, x_m$ satisfying these equations.

Let
$$X = \begin{bmatrix} x_1 \\ x_2 \\ \vdots \\ x_m \end{bmatrix}, \quad B = \begin{bmatrix} b_1 \\ b_2 \\ \vdots \\ b_m \end{bmatrix}, \quad A = \begin{bmatrix} a_{11} & a_{12} & \cdots & a_{1m} \\ a_{21} & a_{22} & \cdots & a_{2m} \\ \vdots & \vdots & & \vdots \\ a_{m1} & a_{m2} & \cdots & a_{mm} \end{bmatrix}$$

The system of equations given in the statement of the problem can be expressed by
$$AX = B \tag{2.43}$$
Multiplying both sides by A^{-1}
$$A^{-1}AX = A^{-1}B$$
Since $A^{-1}A = I$, then
$$X = A^{-1}B \tag{2.44}$$
provided that A is a nonsingular matrix.

Example 2.21 Solve the following system of equations for $x_1, x_2,$ and x_3.
$$3x_1 + x_2 + x_3 = 1, \quad x_1 + x_2 + x_3 = 0, \quad x_1 + 3x_2 + 2x_3 = 1$$

Let
$$X = \begin{bmatrix} x_1 \\ x_2 \\ x_3 \end{bmatrix}, \quad B = \begin{bmatrix} 1 \\ 0 \\ 1 \end{bmatrix}, \quad A = \begin{bmatrix} 3 & 1 & 1 \\ 1 & 1 & 1 \\ 1 & 3 & 2 \end{bmatrix}$$

From Eq. (2.43)
$$AX = B$$

The adjoint of A is given by
$$\text{adj}(A) = \begin{bmatrix} -1 & -1 & 2 \\ 1 & 5 & -8 \\ 0 & -2 & 2 \end{bmatrix}^T = \begin{bmatrix} -1 & 1 & 0 \\ -1 & 5 & -2 \\ 2 & -8 & 2 \end{bmatrix}$$

$$|A| = -2, \quad A^{-1} = \begin{bmatrix} \frac{1}{2} & -\frac{1}{2} & 0 \\ \frac{1}{2} & -\frac{5}{2} & 1 \\ -1 & 4 & -1 \end{bmatrix}$$

From Eq. (2.44)
$$\begin{bmatrix} x_1 \\ x_2 \\ x_3 \end{bmatrix} = \begin{bmatrix} \frac{1}{2} & -\frac{1}{2} & 0 \\ \frac{1}{2} & -\frac{5}{2} & 1 \\ -1 & 4 & -1 \end{bmatrix} \begin{bmatrix} 1 \\ 0 \\ 1 \end{bmatrix} = \begin{bmatrix} \frac{1}{2} \\ \frac{3}{2} \\ -2 \end{bmatrix}$$

2.7 Vectors

As already indicated a matrix having either one row or one column is called a vector. Therefore, the properties discussed with reference to matrices of order (m, n) also apply to vectors. Unless otherwise specified, column vectors will be used. That is, if X is an n-dimensional vector, it can be defined by a matrix with n rows and one column. Therefore

$$X = \begin{bmatrix} x_1 \\ x_2 \\ \vdots \\ x_n \end{bmatrix} \quad (2.45)$$

where the domain of definition of x_i will be assumed to be the real numbers.

Vectors are frequently used to define points in space. For example, let the vectors X_1, X_2, and X_3 be given by

$$X_1 = \begin{bmatrix} 1 \\ 1 \end{bmatrix}, \quad X_2 = \begin{bmatrix} 2 \\ 4 \end{bmatrix}, \quad X_3 = \begin{bmatrix} 4 \\ 3 \end{bmatrix}$$

2.7 VECTORS

X_1, X_2, and X_3 are then vectors or points in 2-space and are shown graphically in Fig. 2.1 where the coordinates of any point in the space are x_1 and x_2 respectively. Thus if X is a vector in this space then

$$X = \begin{bmatrix} x_1 \\ x_2 \end{bmatrix}$$

If $X_1, X_2, \ldots, X_m$ are vectors which are conformable for addition, then any linear combination of these vectors yields a vector or point in the space of

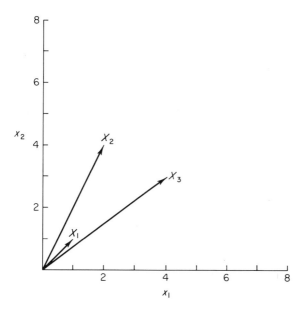

Figure 2.1 Vectors X_1, X_2, and X_3

definition of the original vectors. Therefore, if $X_1, X_2, \ldots, X_m$ are n-dimensional vectors, $\alpha_1, \alpha_2, \ldots, \alpha_m$ are scalars, and

$$X_{m+1} = \alpha_1 X_1 + \alpha_2 X_2 + \cdots + \alpha_m X_m \tag{2.46}$$

then X_{m+1} is an n-dimensional vector.

Example 2.22 Let X_1, X_2, and X_3 be defined as follows.

$$X_1 = \begin{bmatrix} 1 \\ 3 \end{bmatrix}, \quad X_2 = \begin{bmatrix} 3 \\ 2 \end{bmatrix}, \quad X_3 = \begin{bmatrix} 8 \\ 5 \end{bmatrix}$$

Define and plot the following vectors.

$$X_4 = X_1 + X_2, \quad X_5 = X_3 - 2X_2, \quad X_6 = 2X_3 - 3X_2 - 2X_1$$

$$X_4 = \begin{bmatrix} 1 \\ 3 \end{bmatrix} + \begin{bmatrix} 3 \\ 2 \end{bmatrix} = \begin{bmatrix} 4 \\ 5 \end{bmatrix}$$

$$X_5 = \begin{bmatrix} 8 \\ 5 \end{bmatrix} - 2\begin{bmatrix} 3 \\ 2 \end{bmatrix} = \begin{bmatrix} 2 \\ 1 \end{bmatrix}$$

$$X_6 = 2\begin{bmatrix} 8 \\ 5 \end{bmatrix} - 3\begin{bmatrix} 3 \\ 2 \end{bmatrix} - 2\begin{bmatrix} 1 \\ 3 \end{bmatrix} = \begin{bmatrix} 5 \\ -2 \end{bmatrix}$$

See Fig. 2.2.

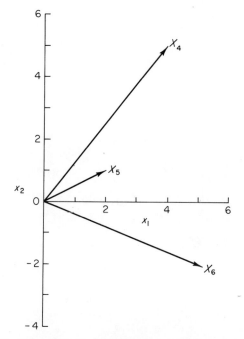

Figure 2.2 Vectors X_4, X_5, and X_6 in Example 2.22

Definition 2.15 The *length* of a vector, X, is the distance of the point defined by X from the origin, $\emptyset$, where

$$\emptyset = \begin{bmatrix} 0 \\ 0 \\ \vdots \\ 0 \end{bmatrix} \tag{2.47}$$

and will be referred to by $\|X\|$.

If X is an n-dimensional vector then

$$\|X\| = \sqrt{x_1^2 + x_2^2 + \cdots + x_n^2} = \sqrt{X^T X} \tag{2.48}$$

2.7 VECTORS

where the *positive* value of the square root is taken. The length of a vector is sometimes represented by $|X|$ instead of $\|X\|$. However, the notation used here avoids confusion with that for the determinant.

Let X be defined by

$$X = \begin{bmatrix} 2 \\ 2 \\ 1 \\ 0 \end{bmatrix}$$

then

$$\|X\| = \sqrt{(2)^2 + (2)^2 + (1)^2 + (0)^2} = 3$$

The notion of the length of a vector may be extended to define the distance between two vectors. Let X_1 and X_2 be two n-dimensional vectors. The distance between X_1 and X_2 is defined as the length of the vector $X_1 - X_2$. Therefore, the distance between X_1 and X_2 is given by

$$\|X_1 - X_2\| = \|X_2 - X_1\|$$
$$= \sqrt{(x_{11} - x_{21})^2 + (x_{12} - x_{22})^2 + \cdots + (x_{1n} - x_{2n})^2} \quad (2.49)$$

where x_{1j} and x_{2j} are the jth elements in X_1 and X_2 respectively. Since the positive value of the square root in Eq. (2.49) is taken, the distance between two vectors is always positive as is the length of a vector. An alternative expression for $\|X_1 - X_2\|$ is given by

$$\|X_1 - X_2\| = \sqrt{(X_1 - X_2)^T(X_1 - X_2)} \quad (2.50)$$

Example 2.23 Find the length of each of the following vectors and the distance between X_1 and X_2 and X_1 and X_3.

$$X_1 = \begin{bmatrix} 4 \\ 0 \\ 3 \end{bmatrix}, \quad X_2 = \begin{bmatrix} -4 \\ 0 \\ -3 \end{bmatrix}, \quad X_3 = \begin{bmatrix} 8 \\ 8 \\ 4 \end{bmatrix}$$

$$\|X_1\| = \sqrt{(4)^2 + (0)^2 + (3)^2} = 5$$
$$\|X_2\| = \sqrt{(-4)^2 + (0)^2 + (-3)^2} = 5$$
$$\|X_3\| = \sqrt{(8)^2 + (8)^2 + (4)^2} = 12$$
$$\|X_1 - X_2\| = \sqrt{(4+4)^2 + (0-0)^2 + (3+3)^2} = 10$$
$$\|X_1 - X_3\| = \sqrt{(4-8)^2 + (0-8)^2 + (3-4)^2} = 9$$

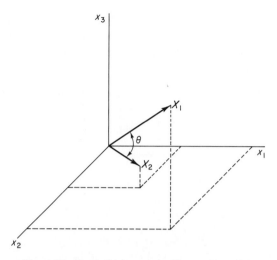

Figure 2.3 Angle θ between the Vectors X_1 and X_2

Let X_1 and X_2 be n-dimensional vectors and θ the angle between them. X_1, X_2, and θ are shown in Fig. 2.3 in three dimensions. The cosine of the angle θ is defined by

$$\cos(\theta) = \frac{X_1^T X_2}{\|X_1\| \, \|X_2\|} \tag{2.51}$$

If the vectors X_1 and X_2 are perpendicular to one another, or orthogonal, then $\theta = 90°$ and $\cos(\theta) = 0$. Equation (2.51) can be used to develop an alternative expression for the distance between two vectors. From Eq. (2.50)

$$\|X_1 - X_2\| = \sqrt{(X_1 - X_2)^T (X_1 - X_2)}$$

Expanding the term in the radical leads to

$$\begin{aligned}(X_1 - X_2)^T (X_1 - X_2) &= X_1^T X_1 - X_1^T X_2 - X_2^T X_1 + X_2^T X_2 \\ &= \|X_1\|^2 + \|X_2\|^2 - 2X_1^T X_2 \end{aligned} \tag{2.52}$$

where $X_1^T X_2 = X_2^T X_1$ since $X_1^T X_2$ and $X_2^T X_1$ are scalars.

From Eq. (2.51)

$$(X_1 - X_2)^T (X_1 - X_2) = \|X_1\|^2 + \|X_2\|^2 - 2\|X_1\| \, \|X_2\| \cos(\theta) \tag{2.53}$$

and

$$\|X_1 - X_2\|^2 = \|X_1\|^2 + \|X_2\|^2 - 2\|X_1\| \, \|X_2\| \cos(\theta) \tag{2.54}$$

Equation (2.54) leads to the following inequality

$$\|X_1 - X_2\| \leq \|X_1\| + \|X_2\| \tag{2.55}$$

Example 2.24 Determine whether or not the following vectors are orthogonal.

$$X_1 = \begin{bmatrix} 1 \\ 1 \\ 1 \end{bmatrix}, \quad X_2 = \begin{bmatrix} 1 \\ 0 \\ 1 \end{bmatrix}$$

2.8 VECTOR SPACES

If X_1 and X_2 are orthogonal then $\cos(\theta) = 0$.

$$\cos(\theta) = \frac{X_1^T X_2}{\|X_1\| \|X_2\|}$$

$\|X_1\| = \sqrt{3}, \qquad \|X_2\| = \sqrt{2}, \qquad X_1^T X_2 = 2, \qquad \cos(\theta) = \frac{2}{\sqrt{6}}$

Therefore X_1 and X_2 are not orthogonal.

Example 2.25 Show that the following vectors are mutually orthogonal (pairwise orthogonal).

$$X_1 = \begin{bmatrix} 2 \\ 0 \\ 0 \end{bmatrix}, \qquad X_2 = \begin{bmatrix} 0 \\ 3 \\ 0 \end{bmatrix}, \qquad X_3 = \begin{bmatrix} 0 \\ 0 \\ 4 \end{bmatrix}$$

Since $\|X_i\| > 0$ for $i = 1, 2, 3$, X_1, X_2, and X_3 are *mutually orthogonal* if $X_i^T X_j = 0$ for all i and j.

$$X_1^T X_2 = 0, \qquad X_1^T X_3 = 0, \qquad X_2^T X_3 = 0$$

Therefore, X_1, X_2, and X_3 form a set of mutually orthogonal vectors.

2.8 Vector Spaces

Definition 2.16 Let $\mathscr{X}_n$ be a set of n-dimensional vectors, X_i, such that kX_i and $X_i + X_j$ are defined and belong to $\mathscr{X}_n$, where k is a scalar. If $\mathscr{X}_n$ is not empty then $\mathscr{X}_n$ is called a *vector space* and has *dimension* equal to that of the vectors comprising the space.

Let $\mathscr{X}_n$ be the set of *all* n-dimensional vectors, X_i, the elements of which are the real numbers. That is,

$$X_i = \begin{bmatrix} x_{i1} \\ x_{i2} \\ \vdots \\ x_{ij} \\ \vdots \\ x_{in} \end{bmatrix} \qquad (2.56)$$

where x_{ij} is a real number, $j = 1, 2, \ldots, n$. If the scalar k is taken as a real number, then $\mathscr{X}_n$ is a vector space since kX_i and $X_i + X_j$ belong to $\mathscr{X}_n$. Now let $\mathscr{Y}_n$ be the set of n-dimensional vectors defined by

$$Y_i = \begin{bmatrix} x_{i1} \\ 0 \\ \vdots \\ 0 \\ \vdots \\ 0 \end{bmatrix} \qquad (2.57)$$

Since kY_i and $Y_i + Y_j$ belong to $\mathcal{Y}_n$, $\mathcal{Y}_n$ is a vector space. However, $\mathcal{Y}_n$ is a subset of $\mathcal{X}_n$ since every vector belonging to $\mathcal{Y}_n$ also belongs to $\mathcal{X}_n$. Therefore, $\mathcal{Y}_n$ is called a vector subspace of $\mathcal{X}_n$.

Definition 2.17 $\mathcal{Y}_n$ is a *vector subspace* of $\mathcal{X}_n$ if

1. $\mathcal{X}_n$ is a vector space;
2. $\mathcal{Y}_n$ is a vector space;
3. $Y_i \in \mathcal{Y}_n \Rightarrow Y_i \in \mathcal{X}_n$.

Example 2.26 Let $\mathcal{X}_3$ be the set of three-dimensional vectors with elements x_{ij}, $j = 1, 2, 3$, where x_{ij} may assume integer values only. Is $\mathcal{X}_3$ a vector space?

If $\mathcal{X}_3$ is a vector space then $X_i + X_j$ and kX_i belong to $\mathcal{X}_3$.

$$X_i + X_j = \begin{bmatrix} x_{i1} \\ x_{i2} \\ x_{i3} \end{bmatrix} + \begin{bmatrix} x_{j1} \\ x_{j2} \\ x_{j3} \end{bmatrix} = \begin{bmatrix} x_{i1} + x_{j1} \\ x_{i2} + x_{j2} \\ x_{i3} + x_{j3} \end{bmatrix}$$

Since the sum of any two integers is an integer, $X_i + X_j$ belongs to $\mathcal{X}_3$. Let $k = \frac{1}{2}$.

$$kX_i = \begin{bmatrix} x_{i1}/2 \\ x_{i2}/2 \\ x_{i3}/2 \end{bmatrix}$$

If x_{ij} is an odd integer then kX_i does not belong to $\mathcal{X}_3$. Therefore, $\mathcal{X}_3$ is not a vector space.

Theorem 2.14 *Every vector space contains the zero vector.*

Proof Let $\mathcal{X}_n$ be a set of n-dimensional vectors which does not contain the zero vector. Let X_i be a vector belonging to $\mathcal{X}_n$. If $\mathcal{X}_n$ is a vector space then $-X_i$ must belong to $\mathcal{X}_n$. Then $X_i - X_i$ also belongs to $\mathcal{X}_n$. But $X_i - X_i = 0$. However, by assumption, 0 does not belong to $\mathcal{X}_n$. Therefore, $\mathcal{X}_n$ is not a vector space since there exists a linear combination of vectors belonging to $\mathcal{X}_n$, which yields a vector not belonging to $\mathcal{X}_n$.

Definition 2.18 Let $\mathcal{X}_n$ be a vector space containing the vectors $X_1, X_2, \ldots, X_m$. If any vector, X_i, belonging to $\mathcal{X}_n$ can be expressed by a linear combination of $X_1, X_2, \ldots, X_m$,

$$X_i = \sum_{j=1}^{m} k_j X_j \tag{2.58}$$

where the k_j's are scalars, then $X_1, X_2, \ldots, X_m$ are said to *span* $\mathcal{X}_n$ or $\mathcal{X}_n$ is *spanned* by the vectors $X_1, X_2, \ldots, X_m$.

2.8 VECTOR SPACES

The set of vectors spanning a given vector space is not necessarily unique. That is, a vector space may be spanned by several sets of vectors. For example, let

$$X_1 = \begin{bmatrix} a_1 \\ 0 \end{bmatrix}, \quad X_2 = \begin{bmatrix} 0 \\ a_2 \end{bmatrix}, \quad X_3 = \begin{bmatrix} b_1 \\ 0 \end{bmatrix}, \quad X_4 = \begin{bmatrix} 0 \\ b_2 \end{bmatrix}$$

where $a_1 \neq b_1$ and $a_2 \neq b_2$. The vector space $\mathscr{X}_2$ is spanned by X_1, X_2 and by X_3, X_4 since scalars k_1, k_2, k_3, and k_4 exist such that any vector X_i, belonging to $\mathscr{X}_2$ can be represented by

$$X_i = k_1 X_1 + k_2 X_2 = k_3 X_3 + k_4 X_4$$

To prove this assertion, it is necessary to show that both of the two sets of vectors span $\mathscr{X}_2$. Let X_i be any vector in $\mathscr{X}_2$ and defined by

$$X_i = \begin{bmatrix} x_{i1} \\ x_{i2} \end{bmatrix}$$

By choosing $k_1 = x_{i1}/a_1$ and $k_2 = x_{i2}/a_2$

$$X_i = k_1 X_1 + k_2 X_2$$

Similarly, letting $k_3 = x_{i1}/b_1$ and $k_4 = x_{i2}/b_2$

$$X_i = k_3 X_3 + k_4 X_4$$

Example 2.27 Define the vector space spanned by the vectors X_1 and X_2, where

$$X_1 = \begin{bmatrix} 1 \\ 0 \\ 0 \end{bmatrix}, \quad X_2 = \begin{bmatrix} 0 \\ 0 \\ 1 \end{bmatrix}$$

The vector space spanned by X_1 and X_2 is defined by all linear combinations of X_1 and X_2. Let

$$X_i = k_1 X_1 + k_2 X_2$$

where k_1 and k_2 are arbitrary real scalars. The space spanned by X_1 and X_2, $\mathscr{X}_i$, is given by the set of vectors X_i, where

$$X_i = \begin{bmatrix} k_1 \\ 0 \\ k_2 \end{bmatrix}$$

and where k_1 and k_2 may assume any real values. Let

$$Y_1 = \begin{bmatrix} a_1 \\ 0 \\ a_2 \end{bmatrix}, \quad Y_2 = \begin{bmatrix} b_1 \\ 0 \\ b_2 \end{bmatrix}$$

where a_1, a_2, b_1, and b_2 are real numbers. Therefore, Y_1 and Y_2 belong to the set of vectors defined by X_i. If c_1 is an arbitrary scalar and if

$$Z_1 = c_1 Y_1 \quad \text{and} \quad Z_2 = Y_1 + Y_2$$

belong to the space defined by the set of vectors defined by X_i, then that space is a vector space. However, since c_1 and the elements of Y_1 and Y_2 are real numbers, Z_1 and Z_2 belong to $\mathscr{X}_i$ and X_1 and X_2 span $\mathscr{X}_i$.

Theorem 2.15 The space spanned by any set of n-dimensional vectors is a vector space.

Proof Let $X_1, X_2, \ldots, X_m$ be a set of n-dimensional vectors spanning $\mathscr{X}_i$ and let X_i be a vector defined by

$$X_i = \sum_{j=1}^{m} k_j X_j$$

where $k_1, k_2, \ldots, k_m$ are arbitrary scalars. Let

$$Z_1 = d_1 Y_1 \quad \text{and} \quad Z_2 = Y_2 + Y_3$$

where

$$Y_1 = \sum_{j=1}^{m} a_j X_j, \quad Y_2 = \sum_{j=1}^{m} b_j X_j, \quad Y_3 = \sum_{j=1}^{m} c_j X_j$$

and where a_j, b_j, c_j, and d_1 are scalars. By definition, Y_1, Y_2, and Y_3 belong to $\mathscr{X}_i$ and

$$Z_1 = \sum_{j=1}^{m} d_1 a_j X_j, \quad Z_2 = \sum_{j=1}^{m} (b_j + c_j) X_j$$

Since $d_1 a_j$ and $b_j + c_j$ are scalars, Z_1 and Z_2 belong to $\mathscr{X}_i$, and $\mathscr{X}_i$ is a vector space.

Definition 2.19 If $X_1, X_2, \ldots, X_m$ is a set of distinct vectors belonging to $\mathscr{X}_n$ such that

$$\sum_{j=1}^{m} k_j X_j = 0 \tag{2.59}$$

if and only if $k_j = 0, j = 1, 2, \ldots, m$, then the vectors $X_1, X_2, \ldots, X_m$ are said to be *linearly independent*. It should be noted that linearly independent vectors are not necessarily orthogonal, although *orthogonal vectors are always linearly independent*.

Example 2.28 Show that the following vectors are linearly independent but are not orthogonal.

$$X_1 = \begin{bmatrix} 1 \\ 0 \\ 0 \end{bmatrix}, \quad X_2 = \begin{bmatrix} 1 \\ 0 \\ 1 \end{bmatrix}$$

If X_1 and X_2 are linearly independent, then

$$k_1 X_1 + k_2 X_2 = 0$$

2.8 VECTOR SPACES

implies $k_1 = k_2 = 0$:

$$\begin{bmatrix} k_1 + k_2 \\ 0 + 0 \\ 0 + k_2 \end{bmatrix} = \begin{bmatrix} 0 \\ 0 \\ 0 \end{bmatrix}$$

Solving for k_1 and k_2 leads to the conclusion that X_1 and X_2 are linearly independent:

$$\|X_1\| = 1, \quad \|X_2\| = \sqrt{2}, \quad X_1^T X_2 = 1$$

Therefore

$$\cos(\theta) = \frac{1}{\sqrt{2}}$$

and X_1 and X_2 are not orthogonal.

Theorem 2.16 If two vectors are orthogonal they are linearly independent.

Proof Let X_1 and X_2 be two n-dimensional, nonzero, orthogonal vectors. If X_1 and X_2 are orthogonal then

$$X_1^T X_2 = 0 \tag{2.60}$$

To show that X_1 and X_2 are independent, it is necessary to show that

$$k_1 X_1 + k_2 X_2 = 0 \tag{2.61}$$

implies $k_1 = k_2 = 0$. Multiplying Eq. (2.61) by X_1^T

$$k_1 X_1^T X_1 + k_2 X_1^T X_2 = 0$$

By Eq. (2.60), $X_1^T X_2 = 0$. Therefore

$$k_1 X_1^T X_1 = 0$$

Since X_1 is assumed to be a nonzero vector, $k_1 = 0$. To show that k_2 is zero, Eq. (2.61) is multiplied by X_2^T.

$$k_1 X_2^T X_1 + k_2 X_2^T X_2 = 0$$

Since $X_2^T X_1 = X_1^T X_2$,

$$k_2 X_2^T X_2 = 0$$

and $k_2 = 0$ since X_2 is a nonzero vector. Therefore orthogonality implies linear independence.

Theorem 2.17 If $X_1, X_2, \ldots, X_m$ is a set of linearly independent vectors, then every subset of these vectors is also linearly independent.

Proof Since $X_1, X_2, \ldots, X_m$ are assumed to be linearly independent

$$\sum_{j=1}^{m} k_j X_j = 0 \Rightarrow k_j = 0, \quad j = 1, 2, \ldots, m$$

Assume that there are r vectors, $X_{i+1}, X_{i+2}, \ldots, X_{i+r}$, within this set which are linearly dependent. That is,

$$\sum_{j=1}^{r} k_{i+j} X_{i+j} = 0 \not\Rightarrow k_{i+j} = 0, \qquad j = 1, 2, \ldots, r$$

Then

$$\sum_{j=1}^{m} k_j X_j = 0$$

when $k_j = 0, j = 1, 2, \ldots, i, i + r + 1, \ldots, m$ and $k_j \neq 0$ for at least some j between $i + 1$ and $i + r$. Therefore, if $X_{i+1}, X_{i+2}, \ldots, X_{i+r}$ are linearly dependent, so are $X_1, X_2, \ldots, X_m$, which contradicts the original assumption of linear independence. Hence, if $X_1, X_2, \ldots, X_m$ is a set of linearly independent vectors, then any subset of these vectors is also linearly independent. ▫

Definition 2.20 If $X_1, X_2, \ldots, X_n$ are linearly independent and span the vector space $\mathscr{X}_n$, then $X_1, X_2, \ldots, X_n$ are said to form a *basis* for $\mathscr{X}_n$.

Although a set of vectors may span a given vector space, they are not necessarily independent and therefore do not necessarily form a basis for that vector space. For example, the vectors X_1, X_2, and X_3 defined by

$$X_1 = \begin{bmatrix} 1 \\ 0 \end{bmatrix}, \qquad X_2 = \begin{bmatrix} 0 \\ 1 \end{bmatrix}, \qquad X_3 = \begin{bmatrix} 1 \\ 1 \end{bmatrix}$$

span two-dimensional vector space, but they are not independent since

$$k_1 X_1 + k_2 X_2 + k_3 X_3 = 0$$

for $k_1 = k_2 = -k_3$. However, there is a subset of these three vectors, X_1 and X_2, which spans two-dimensional vector space and is independent. Therefore, X_1 and X_2 form a basis for the space.

Theorem 2.18 If $\mathscr{X}_n$ is a vector space with a basis of n vectors, then

a. No set of more than n vectors in $\mathscr{X}_n$ can be linearly independent;
b. No set of fewer than n vectors in $\mathscr{X}_n$ can span $\mathscr{X}_n$;
c. Every basis for $\mathscr{X}_n$ is composed of n vectors.

Proof *Part a* Let the set of vectors $X_1, X_2, \ldots, X_n$ be a basis for $\mathscr{X}_n$. Assume that there is a set of $n + r$ vectors in $\mathscr{X}_n$ that are linearly independent and are defined by $Y_1, Y_2, \ldots, Y_{n+r}$. Then

$$\sum_{j=1}^{n+r} k_j Y_j = 0$$

if and only if $k_j = 0, j = 1, 2, \ldots, n + r$. Since $X_1, X_2, \ldots, X_n$ is a basis for $\mathscr{X}_n$, this set spans $\mathscr{X}_n$ by the definition of the basis for a vector space. Therefore, every vector in $\mathscr{X}_n$ can be represented by a linear combination of $X_1, X_2, \ldots, X_n$. Let

2.8 VECTOR SPACES

$$Y_j = \sum_{i=1}^{n} a_{ij} X_i$$

where $a_{ij} \neq 0$ for some i. Then

$$\sum_{j=1}^{n+r} k_j Y_j = \sum_{j=1}^{n+r} k_j \sum_{i=1}^{n} a_{ij} X_i = \sum_{i=1}^{n} \sum_{j=1}^{n+r} k_j a_{ij} X_i = 0$$

holds if $\sum_{j=1}^{n+r} k_j a_{ij} = 0$ for all i. Assume $a_{ih} \neq 0$ and $a_{il} \neq 0$. For $k_h = 1/a_{ih}$, $k_l = -1/a_{il}$, and $k_j = 0$ for $j \neq h$, $j \neq l$, $\sum_{j=1}^{n+r} k_j a_{ij} = 0$ for some $k_j \neq 0$ and $\sum_{j=1}^{n+r} k_j Y_j = 0$ for some $k_j \neq 0$. Therefore, $Y_1, Y_2, \ldots, Y_{n+r}$ are not linearly independent and no set of more than n vectors belonging to $\mathscr{X}_n$ can be linearly independent if $X_1, X_2, \ldots, X_n$ span $\mathscr{X}_n$.

Proof Part b Again let $X_1, X_2, \ldots, X_n$ be a basis for $\mathscr{X}_n$. Assume that there exists a set of r vectors $(r < n)$ which spans $\mathscr{X}_n$, $Y_1, Y_2, \ldots, Y_r$. By part a, if $Y_1, Y_2, \ldots, Y_r$ span $\mathscr{X}_n$, then there exists no set of n vectors that are linearly independent for $n > r$. Therefore, either $X_1, X_2, \ldots, X_n$ are not linearly independent or the set of vectors $Y_1, Y_2, \ldots, Y_r$ does not span $\mathscr{X}_n$. But since $X_1, X_2, \ldots, X_n$ is a basis for $\mathscr{X}_n$, $Y_1, Y_2, \ldots, Y_r$ must not span $\mathscr{X}_n$. Therefore, no fewer than n vectors can span $\mathscr{X}_n$.

Proof Part c Let $X_1, X_2, \ldots, X_n$ be a basis for $\mathscr{X}_n$. Assume that there exists another basis for $\mathscr{X}_n$, $Y_1, Y_2, \ldots, Y_r$, where $r \neq n$. If $r > n$, then $Y_1, Y_2, \ldots, Y_r$ cannot be independent by part a and therefore no basis of $r > n$ vectors exists for $\mathscr{X}_n$. If $r < n$, then by part b $Y_1, Y_2, \ldots, Y_r$ cannot span $\mathscr{X}_n$ and therefore no basis of $r < n$ vectors exists for $\mathscr{X}_n$. Hence, every basis for $\mathscr{X}_n$ must be composed of n vectors.

Definition 2.21 Let $\mathscr{X}_n$ and $\mathscr{Y}_m$ be vector spaces composed of the n and m dimensional vectors X_i in Y_j respectively. A *linear transformation* from $\mathscr{X}_n$ to some subspace of $\mathscr{Y}_m$ can be defined by a matrix of order (m, n) and every linear transformation from $\mathscr{Y}_m$ to some subspace of $\mathscr{X}_n$ can be defined by a matrix of order (n, m).

A linear transformation is an operation which defines a vector in $\mathscr{Y}_m$ for each vector in $\mathscr{X}_n$. Let Y_i belong to $\mathscr{Y}_m$, X_j belong to $\mathscr{X}_n$ and let A be the *matrix of the transformation* from $\mathscr{X}_n$ to $\mathscr{Y}_m$ and therefore of order (m, n). Then

$$Y_i = AX_j \qquad (2.62)$$

Example 2.29 Let $\mathscr{X}_3$ be the vector space of all three-dimensional vectors X_j. Let A and B be transformations from $\mathscr{X}_3$ to $\mathscr{Y}_m$ and $\mathscr{Y}_n$ respectively.

$$A = \begin{bmatrix} 1 & 1 & 0 \\ 2 & 0 & 1 \end{bmatrix}, \qquad B = \begin{bmatrix} 1 & 1 & 2 \\ 1 & 0 & 1 \\ 0 & 0 & 0 \end{bmatrix}$$

Define the spaces $\mathscr{Y}_m$ and $\mathscr{Y}_n$. Are $\mathscr{Y}_m$ and $\mathscr{Y}_n$ vector spaces?

From Eq. (2.62),
$$AX_j = Y_1$$
Therefore, $\mathcal{Y}_m$ is a space defined by the set of vectors Y_1 where
$$Y_1 = \begin{bmatrix} x_{j1} + x_{j2} \\ 2x_{j1} + x_{j3} \end{bmatrix}$$
By the definition of Y_1, $\mathcal{Y}_m$ is the set of all two-dimensional vectors and is a vector space.
$$BX_j = Y_1$$
where
$$Y_1 = \begin{bmatrix} x_{j1} + x_{j2} + 2x_{j3} \\ x_{j1} \phantom{+ x_{j2}} + x_{j3} \\ 0 \end{bmatrix}$$
Therefore, $\mathcal{Y}_n$ is the set of all three-dimensional vectors for which $Y_{3i} = 0$. By an argument similar to that given in Example 2.27, $\mathcal{Y}_n$ is a vector space. ◙

Theorem 2.19 Let A be the matrix of a linear transformation from $\mathcal{X}_n$ to $\mathcal{Y}_m$. If $\mathcal{X}_n$ is a vector space then $\mathcal{Y}_m$ is a vector space.

Proof Let Y_i be the set of vectors defining $\mathcal{Y}_m$ and let X_j be any vector in $\mathcal{X}_n$. Then
$$Y_i = AX_j$$
If $\mathcal{Y}_m$ is a vector space, then $k_1 Y_1$ and $Y_1 + Y_2$ belong to $\mathcal{Y}_m$ for all Y_1 and Y_2 defined by the transformation A. Let
$$Z_1 = k_1 Y_1, \quad \text{where} \quad Y_1 = AX_1$$
Therefore
$$Z_1 = Ak_1 X_1$$
Since $\mathcal{X}_n$ is a vector space, $k_1 X_1$ belongs to $\mathcal{X}_n$ and Z_1 belongs to $\mathcal{Y}_m$. Let
$$Z_2 = Y_1 + Y_2$$
where Y_1 is defined above and
$$Y_2 = AX_2$$
Therefore,
$$Z_2 = AX_1 + AX_2 = A(X_1 + X_2)$$
Since $\mathcal{X}_n$ is a vector space, $X_1 + X_2$ belongs to $\mathcal{X}_n$ and Z_2 belongs $\mathcal{Y}_m$. Therefore $\mathcal{Y}_m$ is a vector space. ◙

2.9 CONVEX SETS

Definition 2.22 A linear transformation, A, from $\mathcal{X}_n$ to $\mathcal{Y}_m$ is *one-to-one* if corresponding to each vector in $\mathcal{X}_n$ there is one and only one vector in $\mathcal{Y}_m$ and corresponding to each vector in $\mathcal{Y}_m$ there is one and only one vector in $\mathcal{X}_n$. That is, if X_1 and X_2 belong to $\mathcal{X}_n$ and

$$Y_1 = AX_1, \qquad Y_2 = AX_2$$

then $Y_1 = Y_2$ if and only if $X_1 = X_2$.

Example 2.30 Show that the transformation B is not one-to-one.

$$B = \begin{bmatrix} 1 & 1 & 0 \\ 1 & 0 & 0 \\ 0 & 0 & 0 \end{bmatrix}$$

Let

$$X_1 = \begin{bmatrix} a_1 \\ a_2 \\ a_3 \end{bmatrix}, \qquad X_2 = \begin{bmatrix} a_1 \\ a_2 \\ b_3 \end{bmatrix}, \qquad \text{where } a_3 \neq b_3$$

$$Y_1 = \begin{bmatrix} a_1 + a_2 \\ a_1 \\ 0 \end{bmatrix}, \qquad Y_2 = \begin{bmatrix} a_1 + a_2 \\ a_1 \\ 0 \end{bmatrix}$$

Since $Y_1 = Y_2$ and $X_1 \neq X_2$, the transformation, B, is not one-to-one.

2.9 Convex Sets

Definition 2.23 Let $X_1, X_2, \ldots, X_m$ be a set of vectors belonging to the vector space $\mathcal{X}_n$. If

$$Y_1 = \sum_{j=1}^{m} \lambda_j X_j \qquad (2.63)$$

where λ_j is a nonnegative scalar for all j and $\sum_{j=1}^{m} \lambda_j = 1$, then Y_1 is said to be formed by a *convex combination* of $X_1, X_2, \ldots, X_m$.

Definition 2.24 The set of vectors comprising a space $\mathcal{X}_n$ (not necessarily a vector space) is a *convex set* if every convex combination of vectors in $\mathcal{X}_n$ yields another vector in $\mathcal{X}_n$. Although a convex set of vectors is not necessarily a vector space, a vector space is always convex.

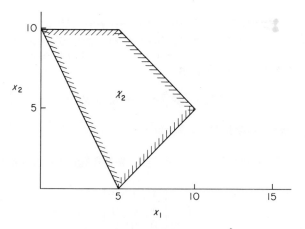

Figure 2.4   Convex Space $\mathscr{X}_2$

To illustrate these concepts consider the set of two-dimensional vectors belonging to the space, $\mathscr{X}_2$, shown in Fig. 2.4. The space $\mathscr{X}_2$ consists of all vectors of the form

$$X_i = \begin{bmatrix} x_1 \\ x_2 \end{bmatrix}$$

subject to the restriction that

$$x_1 + x_2 \leq 15$$
$$x_1 - x_2 \leq 5$$
$$2x_1 + x_2 \geq 10$$
$$x_2 \leq 10$$

In matrix form, $\mathscr{X}_2$ is composed of all vectors X_i such that

$$AX_i \leq B \tag{2.64}$$

where

$$A = \begin{bmatrix} 1 & 1 \\ 1 & -1 \\ -2 & -1 \\ 0 & 1 \end{bmatrix}, \quad B = \begin{bmatrix} 15 \\ 5 \\ -10 \\ 10 \end{bmatrix}$$

If $\mathscr{X}_2$ is a convex set, then a convex combination of n vectors belonging to $\mathscr{X}_2$ will yield another vector in $\mathscr{X}_2$. Let the following three vectors belonging to $\mathscr{X}_2$ be selected.

$$X_1 = \begin{bmatrix} 8 \\ 5 \end{bmatrix}, \quad X_2 = \begin{bmatrix} 3 \\ 8 \end{bmatrix}, \quad X_3 = \begin{bmatrix} 5 \\ 2 \end{bmatrix}$$

Let Y_1 be any convex combination of these vectors.

$$Y_1 = \lambda_1 X_1 + \lambda_2 X_2 + \lambda_3 X_3$$

2.9 CONVEX SETS

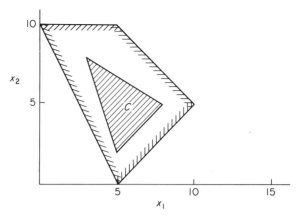

Figure 2.5 Convex space C

where $\lambda_j \geq 0$ and $\sum_{j=1}^{3} \lambda_j = 1$. The space defined by all such convex combinations, C, is shown in Fig. 2.5.

A convex combination of two vectors results in a point on the line segment joining the two vectors. This result is important since it leads to a simple means of defining a convex set. That is, if a line segment joining any two vectors in the space $\mathscr{X}_n$ lies entirely within $\mathscr{X}_n$, then $\mathscr{X}_n$ is convex. Referring to Fig. 2.4, it is obvious that $\mathscr{X}_2$ is convex since any line segment connecting any two points or vectors in $\mathscr{X}_2$ lies entirely in $\mathscr{X}_2$.

Definition 2.25 Let X_j be an n-dimensional column vector and B_j an n-dimensional row vector with elements the constants $b_{1j}, b_{2j}, \ldots, b_{nj}$. The relationship

$$B_j X_j = k \tag{2.65}$$

where k is a scalar, defines a *hyperplane* in n-dimensional space.

A hyperplane in two-dimensional space is a straight line and is a plane in three-dimensional space. A hyperplane divides a space into two half spaces. The *half space* defined by

$$B_j X_j \leq k \tag{2.66}$$

is composed of all vectors X_j satisfying Eq. (2.66). The remaining half space is defined by

$$B_j X_j > k \tag{2.67}$$

Example 2.31 Let the vector space $\mathscr{X}_2$ be composed of all two-dimensional vectors. Let $B_j X_j = 0$ be a hyperplane in $\mathscr{X}_2$ where

$$B_j = [-1, 2], \qquad X_j = \begin{bmatrix} x_{j1} \\ x_{j2} \end{bmatrix}$$

Graph the hyperplane and define the half spaces given by
$$B_j X_j < 0, \qquad B_j X_j \geq 0$$
The hyperplane defined by $B_j X_j = 0$ is the straight line defined by
$$-x_{j1} + 2x_{j2} = 0$$
and is shown graphically in Fig. 2.6 where $\mathscr{X}'_2$ is the set of vectors for which $B_j X_j < 0$ and $\mathscr{X}''_2$ is the set of vectors for which $B_j X_j \geq 0$.

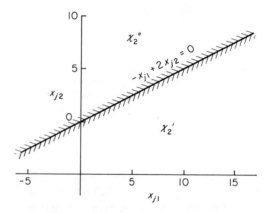

Figure 2.6 Half Spaces $\mathscr{X}'_2$ and $\mathscr{X}''_2$ Defined by the Hyperplane $-x_{j1} + 2x_{j2} = 0$

Theorem 2.20 A half space is convex.

Proof Let $\mathscr{X}'_n$ be the half space defined by
$$B_j X_j \leq k$$
and $\mathscr{X}''_n$ the half space defined by
$$B_j X_j > k$$
Let X_1 and X_2 be any two vectors belonging to $\mathscr{X}'_n$ and let
$$X_3 = \lambda X_1 + (1 - \lambda)X_2$$
where $0 \leq \lambda \leq 1$.
$$B_j X_3 = B_j[\lambda X_1 + (1 - \lambda)X_2] = \lambda B_j X_1 + (1 - \lambda)B_j X_2$$
Since $B_j X_1 \leq k$, $B_j X_2 \leq k$, and $\lambda \leq 1$
$$B_j X_3 \leq \lambda k + (1 - \lambda)k = k$$
Therefore, every convex combination of two points in $\mathscr{X}'_n$ yields another point in $\mathscr{X}'_n$ and $\mathscr{X}'_n$ is convex. By a similar argument $\mathscr{X}''_n$ can be shown to be convex.

2.9 CONVEX SETS

Let B_j be a $1 \times n$ row vector which defines a half space by

$$B_j X_i \leq k_j$$

where X_i is an $n \times 1$ column vector and k_j is a scalar. Let B be a matrix of the row vectors B_j and k a column vector of the scalars k_j. Then

$$B = \begin{bmatrix} B_1 \\ B_2 \\ \vdots \\ B_m \end{bmatrix}, \quad k = \begin{bmatrix} k_1 \\ k_2 \\ \vdots \\ k_m \end{bmatrix}$$

and

$$BX_i \leq k \tag{2.68}$$

Equation (2.68) defines the *intersection* of m half spaces in the vector space defined by the set of vectors X_i. When the intersection of half spaces is not empty, the resulting space is convex and is usually referred to as a *convex polyhedron* or a *polyhedral convex set*.

Theorem 2.21 If the intersection of m half spaces is not empty, then the intersection is a convex set.

Proof The intersection of m half spaces is the collection of vectors belonging to each of the m half spaces. Let X_1 and X_2 be any two vectors belonging to the intersection and let

$$Y = \lambda X_1 + (1 - \lambda) X_2$$

where $0 \leq \lambda \leq 1$. Therefore, Y lies on the line segment joining X_1 and X_2. Since X_1 and X_2 belong to each of the m half spaces, and since each half space is convex, Y must lie in each half space and therefore in the intersection. Therefore, the intersection of m half spaces is convex.

Example 2.32 Show graphically that the intersection of the following half spaces is empty.

$$\begin{aligned} x_{j1} - x_{j2} &\leq 5 \\ -x_{j1} - x_{j2} &\leq -10 \\ x_{j1} + 2x_{j2} &\leq 10 \\ -x_{j2} &\leq 0 \end{aligned}$$

From Fig. 2.7 it is apparent that there is no vector which lies in all four half spaces. Therefore, the intersection is empty.

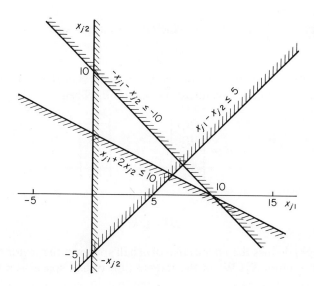

Figure 2.7 Half Spaces for Example 2.32

Example 2.33 Show the following convex polyhedron graphically.

$$\begin{bmatrix} -1 & -2 \\ 2 & 1 \\ 1 & -1 \\ -1 & 2 \end{bmatrix} \begin{bmatrix} x_{j1} \\ x_{j2} \end{bmatrix} \leq \begin{bmatrix} -4 \\ 8 \\ 0 \\ 6 \end{bmatrix}$$

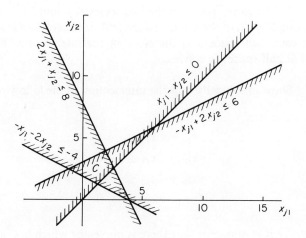

Figure 2.8 Convex Polyhedron C Defined in Example 2.33

The four half spaces are defined by

$$-x_{j1} - 2x_{j2} \leq -4$$
$$2x_{j1} + x_{j2} \leq 8$$
$$x_{j1} - x_{j2} \leq 0$$
$$-x_{j1} + 2x_{j2} \leq 6$$

The convex polyhedron defined by this set of half spaces is shown in Fig. 2.8 and is defined by C.

2.10 Characteristic Values and Quadratic Forms

Let A be a symmetric matrix of order n, X be an n-dimensional column vector and λ be a scalar such that

$$AX = \lambda X \qquad (2.69)$$

for some $X \neq 0$. The nonzero vectors, X_i, satisfying Eq. (2.69) are called *eigenvectors*, *characteristic vectors*, *latent vectors*, or *invariant vectors*. The scalars, λ_i, satisfying Eq. (2.69) are called *eigenvalues*, *characteristic roots*, or *latent roots*.

Equation (2.69) can also be expressed by

$$(\lambda I - A)X = 0 \qquad (2.70)$$

The solution vector, X, for (2.70) is *nonzero* if and only if

$$|\lambda I - A| = 0 \qquad (2.71)$$

Evaluation of $|\lambda I - A|$ results in a polynomial in λ of degree n and is known as the *characteristic equation* of A.

Example 2.34 Find the characteristic roots of the following matrices.

$$A = \begin{bmatrix} 2 & 1 \\ 1 & 2 \end{bmatrix}, \quad B = \begin{bmatrix} 2 & -3 \\ -3 & 1 \end{bmatrix}$$

From Eq. (2.71),

$$|\lambda I - A| = 0 = \begin{vmatrix} \lambda - 2 & -1 \\ -1 & \lambda - 2 \end{vmatrix} = \lambda^2 - 4\lambda + 3$$

Solving for the characteristic roots of A,

$$\lambda_1 = 3, \quad \lambda_2 = 1$$

By similar analysis,

$$|\lambda I - B| = 0 = \begin{vmatrix} \lambda - 2 & 3 \\ 3 & \lambda - 1 \end{vmatrix} = \lambda^2 - 3\lambda - 7$$

and the characteristic roots of B are

$$\lambda_1 = \frac{3 + \sqrt{37}}{2}, \quad \lambda_2 = \frac{3 - \sqrt{37}}{2}$$

Example 2.35 Find the eigenvectors, X_1 and X_2, corresponding to the characteristic roots for the matrix A in Example 2.34. Show that X_1 and X_2 are orthogonal.

From Example 2.34
$$\lambda_1 = 3, \qquad \lambda_2 = 1$$

For λ_1,
$$(\lambda_1 I - A)X_1 = \begin{bmatrix} 1 & -1 \\ -1 & 1 \end{bmatrix} \begin{bmatrix} x_{11} \\ x_{12} \end{bmatrix}$$

and
$$x_{11} - x_{12} = 0, \qquad -x_{11} + x_{12} = 0$$

Therefore any vector, X_1, of the form
$$X_1 = \begin{bmatrix} a \\ a \end{bmatrix}$$

is an eigenvector for A corresponding to the characteristic root λ_1.

For λ_2
$$(\lambda_2 I - A)X_2 = \begin{bmatrix} -1 & -1 \\ -1 & -1 \end{bmatrix} \begin{bmatrix} x_{21} \\ x_{22} \end{bmatrix}$$

and any vector, X_2, of the form
$$X_2 = \begin{bmatrix} b \\ -b \end{bmatrix}$$

is an eigenvector for A corresponding to the characteristic root λ_2.

To show that X_1 and X_2 are orthogonal
$$X_1^T X_2 = [a, a] \begin{bmatrix} b \\ -b \end{bmatrix} = 0$$

Theorem 2.22 If X_i and X_j are two eigenvectors corresponding to the distinct eigenvalues λ_i and λ_j for a symmetric matrix A, then X_i and X_j are mutually orthogonal.

Proof From Eq. (2.69)
$$AX_i = \lambda_i X_i \qquad \text{and} \qquad AX_j = \lambda_j X_j$$

Then
$$X_j^T A X_i = X_j^T \lambda_i X_i \qquad \text{and} \qquad X_i^T A X_j = X_i^T \lambda_j X_j$$

Since λ_i and λ_j are scalars, then
$$X_j^T \lambda_i X_i = \lambda_i X_j^T X_i \qquad \text{and} \qquad X_i^T \lambda_j X_j = \lambda_j X_i^T X_j$$

Since $X_i^T A X_j$ and $X_j^T A X_i$ are scalars, then

$$\lambda_i X_j^T X_i = \lambda_j X_i^T X_j$$

By the assumption that λ_i and λ_j are distinct, $\lambda_i \neq \lambda_j$, and since

$$X_j^T X_i = X_i^T X_j$$

then

$$(\lambda_i - \lambda_j) X_j^T X_i = 0$$

implies

$$X_j^T X_i = 0$$

and X_j and X_i are orthogonal vectors.

Definition 2.26 If A is a square matrix such that

$$A^T = A^{-1} \tag{2.72}$$

then A is said to be an *orthogonal matrix*.

Definition 2.27 If A is a symmetric matrix of order n with k ($\leq n$) identical eigenvalues each equal to the eigenvalue λ_j, then λ_j has *multiplicity k*.

Theorem 2.23 If A is a symmetric matrix of order n with eigenvalues $\lambda_1, \lambda_2, \ldots, \lambda_n$, distinct or not, then there exists a set of n mutually orthogonal eigenvectors associated with $\lambda_1, \lambda_2, \ldots, \lambda_n$.

Although a proof of Theorem 2.23 is beyond the scope of this text, some clarification of the implications of this theorem is in order. Suppose that A is a symmetric matrix of order n with eigenvalues $\lambda_1, \lambda_2, \ldots, \lambda_n$ and associated eigenvectors $X_1, X_2, \ldots, X_n$. By Theorem 2.23, the n eigenvectors should be mutually orthogonal. Suppose the eigenvalues $\lambda_{i+1}, \lambda_{i+2}, \ldots, \lambda_{i+k}$ are equal to the scalar α. It can be shown that there exists a set of k mutually orthogonal eigenvectors associated with the eigenvalue α. If the remaining $n - k$ eigenvalues $\lambda_1, \lambda_2, \ldots, \lambda_i, \lambda_{i+k+1}, \ldots, \lambda_n$ are distinct then $X_1, X_2, \ldots, X_i, X_{i+k+1}, \ldots, X_n$ are mutually orthogonal by Theorem 2.22. Also, by Theorem 2.22, any eigenvector associated with λ_{i+j}, $j = 1, 2, \ldots, k$, must be orthogonal to $X_1, X_2, \ldots, X_i$, $X_{i+k+1}, \ldots, X_n$. Therefore, all of the eigenvectors associated with λ_{i+j} must be orthogonal to $X_1, X_2, \ldots, X_i, X_{i+k+1}, \ldots, X_n$ as well as to each other. Thus there is a set of n mutually orthogonal eigenvectors for the symmetric matrix A. To illustrate these concepts consider the following examples.

Example 2.36 Find the eigenvalues and a set of three mutually orthogonal eigenvectors for the following matrix

$$A = \begin{bmatrix} 3 & 0 & -1 \\ 0 & 2 & 0 \\ -1 & 0 & 3 \end{bmatrix},$$

$$|\lambda I - A| = \begin{vmatrix} \lambda - 3 & 0 & 1 \\ 0 & \lambda - 2 & 0 \\ 1 & 0 & \lambda - 3 \end{vmatrix} = (\lambda - 2)[(\lambda - 3)^2 - 1] = 0$$

The eigenvalues for A are

$$\lambda_1 = 2, \quad \lambda_2 = 2, \quad \lambda_3 = 4$$

Since $\lambda_1 = \lambda_2$, then λ_1 has multiplicity 2. Therefore, there should be two distinct eigenvectors corresponding to λ_1 (or λ_2). Since

$$\begin{aligned} (\lambda I - A)X &= 0 \\ (\lambda_i - 3)x_{i1} \quad\quad\quad + \quad\quad x_{i3} &= 0 \\ (\lambda_i - 2)x_{i2} \quad\quad\quad &= 0 \\ x_{i1} \quad\quad + (\lambda_i - 3)x_{i3} &= 0 \end{aligned}$$

For λ_1

$$-x_{11} + x_{13} = 0, \quad x_{11} - x_{13} = 0$$

where x_{12} may assume an arbitrary real value. Letting $x_{12} = a$, X_1 corresponding to λ_1 is

$$X_1 = \begin{bmatrix} b \\ a \\ b \end{bmatrix}$$

Since λ_1 has multiplicity 2, there is another eigenvector, X_2, for $\lambda_2 = 2$ such that $X_1^T X_2 = 0$.

$$X_1^T X_2 = bx_{21} + ax_{22} + bx_{23} = 0$$

In addition, X_2 must satisfy the relationship $\lambda_2 X_2 - AX_2 = 0$ or

$$-x_{21} + x_{23} = 0, \quad x_{21} - x_{23} = 0$$

Therefore

$$X_2 = \begin{bmatrix} c \\ -2bc/a \\ c \end{bmatrix}$$

For λ_3,

$$\begin{aligned} x_{31} \quad\quad + x_{33} &= 0 \\ 2x_{32} \quad\quad &= 0 \\ x_{31} \quad\quad + x_{33} &= 0 \end{aligned}$$

2.10 CHARACTERISTIC VALUES AND QUADRATIC FORMS

and

$$X_3 = \begin{bmatrix} -d \\ 0 \\ d \end{bmatrix}$$

Example 2.37 For the problem given in Example 2.36, show that there is no vector $X_4 \neq 0$ associated with λ_3 which is orthogonal to X_3.

If there exists a vector X_4 which is orthogonal to X_3 and which is an eigenvector for λ_3 then

$$X_3^T X_4 = 0 \quad \text{and} \quad (\lambda_3 X_4 - AX_4) = 0$$

Therefore

$$-dx_{41} + dx_{43} = 0$$

and

$$x_{41} = x_{43}$$

However, this solution must satisfy $(\lambda_3 X_4 - AX_4) = 0$ or

$$\begin{aligned} x_{41} & & +x_{43} &= 0 \\ & 2x_{42} & &= 0 \\ x_{41} & & +x_{43} &= 0 \end{aligned}$$

which implies $x_{41} = x_{42} = x_{43} = 0$, and X_4 must be a zero vector.

Example 2.38 Determine the eigenvalues and eigenvectors for the following matrix.

$$A = \begin{bmatrix} 2 & 0 & 0 & 0 \\ 0 & 2 & 0 & 0 \\ 0 & 0 & 2 & 0 \\ 0 & 0 & 0 & 2 \end{bmatrix}$$

By inspection, the eigenvalues for A are all 2. That is

$$|\lambda I - A| = (\lambda - 2)^4$$

Since the eigenvalue 2 has multiplicity 4, there are four eigenvectors associated with this eigenvalue.

$$AX_i - 2X_i = 0, \quad i = 1, 2, 3, 4$$

Since AX_i is a column vector with elements $2x_{ij}, j = 1, 2, 3, 4$, any set of mutually orthogonal vectors satisfies the eigenvalue 2. Therefore

$$X_1 = \begin{bmatrix} 1 \\ 0 \\ 0 \\ 0 \end{bmatrix}, \quad X_2 = \begin{bmatrix} 0 \\ 1 \\ 0 \\ 0 \end{bmatrix}, \quad X_3 = \begin{bmatrix} 0 \\ 0 \\ 1 \\ 0 \end{bmatrix}, \quad X_4 = \begin{bmatrix} 0 \\ 0 \\ 0 \\ 1 \end{bmatrix}$$

are eigenvectors for A.

It is worth noting that $k_i X_i$ is also an eigenvector for A in Example 2.38, where k_i is a scalar. This implies that there is an infinite number of sets of mutually orthogonal eigenvectors for A. In general, if λ_j is an eigenvalue of multiplicity k ($\leq n$) for a symmetric matrix A of order n, then there is an infinite number of sets of k mutually orthogonal eigenvectors associated with λ_j. Further, since any given set of k such eigenvectors is mutually orthogonal, the set spans a vector space of dimension k, which is a subset of the vector space of dimension n, which is spanned by a set of n mutually orthogonal eigenvectors.

Definition 2.28 Two square matrices, A and R, are said to be *congruent* if there exists a nonsingular matrix B such that

$$R = B^T A B \qquad (2.73)$$

Definition 2.29 Two square matrices, A and R, are said to be *similar* if there exists a nonsingular matrix B such that

$$R = B^{-1} A B \qquad (2.74)$$

Theorem 2.24 If A is a symmetric matrix and B is an orthogonal matrix such that

$$R = B^T A B$$

then R and A have the same eigenvalues.

Proof The eigenvalues for R are those values of λ_i such that

$$|R - \lambda I| = 0$$

By definition of R

$$|R - \lambda I| = |B^T A B - \lambda I|$$

Since B is an orthogonal matrix, $B^T B = I$ and

$$B^T \lambda I B = \lambda B^T I B = \lambda I$$

Therefore

$$|R - \lambda I| = |B^T(A - \lambda I)B| = |B^T| |A - \lambda I| |B| = |B^T B| |A - \lambda I|$$
$$= |A - \lambda I|$$

Thus R and A have the same characteristic equation and therefore the same eigenvalues. ∎

Theorem 2.25 Let A be a symmetric matrix of order n with eigenvalues $\lambda_1, \lambda_2, \ldots, \lambda_n$ and let B be a matrix with columns a set of mutually orthogonal eigenvectors, X_i, associated with the eigenvalues, λ_i. Then the matrix R is a diagonal matrix where

$$R = B^T A B$$

2.10 CHARACTERISTIC VALUES AND QUADRATIC FORMS

Proof Since B is a matrix of eigenvectors such that X_i is associated with λ_i

$$B^T B = \begin{bmatrix} X^T_1 \\ X^T_2 \\ \vdots \\ X^T_n \end{bmatrix} [X_1, X_2, \ldots, X_n]$$

or

$$B^T B = \begin{bmatrix} X^T_1 X_1 & 0 & 0 & \cdots & 0 \\ 0 & X^T_2 X_2 & 0 & \cdots & 0 \\ 0 & 0 & X^T_3 X_3 & \cdots & 0 \\ \vdots & & & & \\ 0 & 0 & 0 & \cdots & X^T_n X_n \end{bmatrix}$$

since $X^T_i X_j = 0$ for $i \neq j$ due to the orthogonality of the eigenvectors X_i and X_j. By the same argument $X^T_i X_i \neq 0$.

$$R = \begin{bmatrix} X^T_1 \\ X^T_2 \\ \vdots \\ X^T_n \end{bmatrix} \begin{bmatrix} a_{11} & a_{12} & \cdots & a_{1n} \\ a_{12} & a_{22} & \cdots & a_{2n} \\ \vdots & \vdots & & \vdots \\ a_{1n} & a_{2n} & \cdots & a_{nn} \end{bmatrix} [X_1, X_2, \ldots, X_n]$$

Since X_i is an eigenvector for A,

$$AX_i = \lambda_i X_i$$

Therefore,

$$R = \begin{bmatrix} X^T_1 \\ X^T_2 \\ \vdots \\ X^T_n \end{bmatrix} [\lambda_1 X_1, \lambda_2 X_2, \ldots, \lambda_n X_n]$$

$$R^T = \begin{bmatrix} \lambda_1 X^T_1 \\ \lambda_2 X^T_2 \\ \vdots \\ \lambda_n X^T_n \end{bmatrix} [X_1, X_2, \ldots, X_n]$$

$$= \begin{bmatrix} \lambda_1 & 0 & \cdots & 0 \\ 0 & \lambda_2 & \cdots & 0 \\ \vdots & \vdots & & \vdots \\ 0 & 0 & \cdots & \lambda_n \end{bmatrix} \begin{bmatrix} X^T_1 \\ X^T_2 \\ \vdots \\ X^T_n \end{bmatrix} [X_1, X_2, \ldots, X_n] = L B^T B$$

where L is a diagonal matrix with elements the eigenvalues $\lambda_1, \lambda_2, \ldots, \lambda_n$. Since $L = L^T$ and $B^T = B$, $LB^T B$ is a diagonal matrix and, therefore, R is also a diagonal matrix.

Definition 2.30 Let X be an n-dimensional column vector with elements x_i and A a square matrix of order n with elements a_{ij}. The polynomial given by

$$q = X^T A X \tag{2.75}$$

or
$$q = \sum_{i=1}^{n} \sum_{j=1}^{n} a_{ij} x_i x_j \tag{2.76}$$

is called a *quadratic form*.

Quadratic forms play an important role in experimental design and, as will be demonstrated in a later chapter, are used extensively in classical optimization theory. The term quadratic refers to the fact that q, as defined in Eqs. (2.75) and (2.76) is a homogeneous quadratic equation in the variables $x_1, x_2, \ldots, x_n$. The matrix A in Eq. (2.75) is called the *matrix of the quadratic form*. If $|A|$ is zero, the quadratic form is said to be singular, and nonsingular otherwise. If A is a symmetric matrix, then q becomes

$$q = \sum_{i=1}^{n} a_{ii} x_i^2 + 2 \sum_{i=1}^{n-1} \sum_{j=i+1}^{n} a_{ij} x_i x_j \tag{2.77}$$

Example 2.39 Let X be a three-dimensional vector and let the matrices A and B be the matrices of two quadratic forms in X. Define the quadratic equations given by the forms

$$q_1 = X^T A X, \qquad q_2 = X^T B X$$

where

$$A = \begin{bmatrix} 2 & 0 & 0 \\ 0 & 3 & 0 \\ 0 & 0 & 1 \end{bmatrix}, \quad B = \begin{bmatrix} 0 & 1 & 1 \\ 1 & 1 & 2 \\ 1 & 2 & 0 \end{bmatrix}$$

$$q_1 = [x_1, x_2, x_3] \begin{bmatrix} 2 & 0 & 0 \\ 0 & 3 & 0 \\ 0 & 0 & 1 \end{bmatrix} \begin{bmatrix} x_1 \\ x_2 \\ x_3 \end{bmatrix}$$

$$= [x_1, x_2, x_3] \begin{bmatrix} 2x_1 \\ 3x_2 \\ x_3 \end{bmatrix} = 2x_1^2 + 3x_2^2 + x_3^2$$

$$q_2 = [x_1, x_2, x_3] \begin{bmatrix} 0 & 1 & 1 \\ 1 & 1 & 2 \\ 1 & 2 & 0 \end{bmatrix} \begin{bmatrix} x_1 \\ x_2 \\ x_3 \end{bmatrix} = [x_1, x_2, x_3] \begin{bmatrix} x_2 + x_3 \\ x_1 + x_2 + 2x_3 \\ x_1 + 2x_2 \end{bmatrix}$$

$$= 2x_1 x_2 + 2x_1 x_3 + x_2^2 + 4x_2 x_3 \qquad \blacksquare$$

Theorem 2.26 Let A be a symmetric matrix of order n and X a column vector of dimension n. The quadratic form

$$q = X^T A X$$

can be reduced to a *sum of squares* by the transformation

$$X = BY$$

2.10 CHARACTERISTIC VALUES AND QUADRATIC FORMS

where B is a matrix of order n with columns a set of mutually orthogonal eigenvectors for A.

Proof Replacing X by BY in the quadratic form, q, leads to

$$q = Y^T B^T A B Y$$

By Theorem 2.25, $B^T A B$ is a diagonal matrix with diagonal elements equal to $\lambda_i X_i^T X_i$. Then

$$q = \sum_{i=1}^{n} \lambda_i X_i^T X_i y_i^2$$

Example 2.40 Find a matrix, B, which reduces the quadratic form

$$q = X^T A X$$

to a sum of squares of the form

$$q = Y^T B^T A B Y$$

where

$$A = \begin{bmatrix} 4 & 0 & 3 \\ 0 & 2 & 0 \\ 3 & 0 & 4 \end{bmatrix}$$

By Theorem 2.25, B is a matrix with columns equal to a set of mutually orthogonal eigenvectors for A. The eigenvalues for A are obtained from

$$|\lambda I - A| = (\lambda - 2)[(\lambda - 4)^2 - 9] = 0$$

The eigenvalues for A are then

$$\lambda_1 = 2, \quad \lambda_2 = 7, \quad \lambda_3 = 1$$

Let X_i be an eigenvector for A associated with λ_i such that

$$(\lambda I - A)X_i = 0$$

Solving for X_1, X_2, and X_3 yields

$$X_1 = \begin{bmatrix} 0 \\ a \\ 0 \end{bmatrix}, \quad X_2 = \begin{bmatrix} b \\ 0 \\ b \end{bmatrix}, \quad X_3 = \begin{bmatrix} -c \\ 0 \\ c \end{bmatrix}$$

and B is given by

$$B = \begin{bmatrix} 0 & b & -c \\ a & 0 & 0 \\ 0 & b & c \end{bmatrix}$$

That B is a matrix which reduces the quadratic form q to a sum of squares is shown as follows.

$$q = Y^T B^T A B Y = Y^T \begin{bmatrix} 0 & 2a & 0 \\ 7b & 0 & 7b \\ -c & 0 & c \end{bmatrix} BY = Y^T \begin{bmatrix} 2a^2 & 0 & 0 \\ 0 & 14b^2 & 0 \\ 0 & 0 & 2c^2 \end{bmatrix} Y$$

Therefore

$$q = 2a^2 y_1^2 + 14b^2 y_2^2 + 2c^2 y_3^2$$

The reader should recall that the reduction of the original quadratic form to a sum of squares was accomplished by the transformation

$$X = BY$$

Solving for Y in terms of X yields

$$Y = B^{-1} X$$

where

$$|B| = -2abc$$

and

$$B^{-1} = -\frac{1}{2abc} \begin{bmatrix} 0 & -2bc & 0 \\ -ac & 0 & -ac \\ ab & 0 & -ab \end{bmatrix} = \begin{bmatrix} 0 & 1/a & 0 \\ 1/2b & 0 & 1/2b \\ -1/2c & 0 & 1/2c \end{bmatrix}$$

Therefore

$$Y = \begin{bmatrix} x_2/a \\ (1/2b)(x_1 + x_3) \\ -(1/2c)(x_1 - x_3) \end{bmatrix}$$

In the above example and throughout the discussion of characteristic roots and quadratic forms, diagonalization of a matrix has been accomplished through an orthogonal transformation. However, it is sometimes more convenient to employ other means to this end. Only one method will be discussed here. For a more complete discussion the reader should consult a more advanced treatment of matrix theory.

Let A be a symmetric matrix of order n and q the quadratic form

$$q = X^T A X$$

In this quadratic form let

$$y_1 = a_{11} x_1^2 + 2 \sum_{j=2}^{n} a_{1j} x_1 x_j = a_{11}\left(x_1^2 + 2 \sum_{j=2}^{n} \frac{a_{1j}}{a_{11}} x_1 x_j\right)$$

$$= a_{11}\left(x_1 + \sum_{j=2}^{n} \frac{a_{1j}}{a_{11}} x_j\right)^2 - \frac{1}{a_{11}}\left(\sum_{j=2}^{n} a_{1j} x_j\right)^2$$

2.10 CHARACTERISTIC VALUES AND QUADRATIC FORMS

Then

$$q = a_{11}z_{11}^2 - \frac{1}{a_{11}}\left[\sum_{j=2}^{n}(a_{1j}x_j)^2 + 2\sum_{j=2}^{n-1}\sum_{k=j+1}^{n}a_{1j}a_{1k}x_jx_k\right] + \sum_{j=2}^{n}\sum_{k=2}^{n}a_{jk}x_jx_k \tag{2.78}$$

where

$$z_{11} = x_1 + \sum_{j=2}^{n}\frac{a_{1j}}{a_{11}}x_j \tag{2.79}$$

Defining q in terms of the variables Z_i, where

$$z_{12} = x_2$$
$$\vdots$$
$$z_{1n} = x_n$$

leads to

$$q = p_{111}z_{11}^2 + \sum_{j=2}^{n}\sum_{k=2}^{n}p_{1jk}z_{1j}z_{1k} \tag{2.80}$$

Let

$$B_1 = \begin{bmatrix} 1 & a_{12}/a_{11} & a_{13}/a_{11} & \cdots & a_{1n}/a_{11} \\ 0 & 1 & 0 & \cdots & 0 \\ 0 & 0 & 1 & \cdots & 0 \\ \vdots & \vdots & \vdots & & \vdots \\ 0 & 0 & 0 & \cdots & 1 \end{bmatrix}, \quad Z_1 = \begin{bmatrix} z_{11} \\ z_{12} \\ z_{13} \\ \vdots \\ z_{1n} \end{bmatrix}$$

Then

$$q = Z_1^T(B_1^{-1})^T A B_1^{-1} Z_1 \tag{2.81}$$

where $(B_1^{-1})^T A B_1^{-1}$ has elements p_{1jk}. In Eq. (2.80) let

$$y_2 = p_{122}z_{12}^2 + 2\sum_{k=3}^{n}p_{12k}z_{12}z_{1k} = p_{122}\left(z_{12}^2 + 2\sum_{k=3}^{n}\frac{p_{12k}}{p_{122}}z_{12}z_{1k}\right)$$

$$= p_{122}\left(z_{12} + \sum_{k=3}^{n}\frac{p_{12k}}{p_{122}}z_{1k}\right)^2 - \frac{1}{p_{122}}\left(\sum_{k=3}^{n}p_{12k}z_{1k}\right)^2$$

Equation (2.80) becomes

$$q = p_{111}z_{11}^2 + p_{122}z_{22}^2 - \frac{1}{p_{122}}\left(\sum_{k=3}^{n}p_{12k}z_{1k}\right)^2 + \sum_{j=3}^{n}\sum_{k=3}^{n}p_{1jk}z_{1j}z_{1k}$$

where

$$z_{22} = z_{12} + \sum_{k=3}^{n}\frac{p_{12k}}{p_{122}}z_{1k}$$

Let

$$z_{21} = z_{11}$$

$$z_{22} = z_{12} + \sum_{k=3}^{n} \frac{p_{12k}}{p_{122}} z_{1k}$$

$$z_{23} = z_{13}$$

$$\vdots$$

$$z_{2n} = z_{1n}$$

and define the matrix B_2 and the column vector Z_2 as

$$B_2 = \begin{bmatrix} 1 & 0 & 0 & 0 & \cdots & 0 \\ 0 & 1 & p_{123}/p_{122} & p_{124}/p_{122} & \cdots & p_{12n}/p_{122} \\ 0 & 0 & 1 & 0 & \cdots & 0 \\ 0 & 0 & 0 & 1 & \cdots & 0 \\ \vdots & \vdots & \vdots & \vdots & & \vdots \\ 0 & 0 & 0 & 0 & \cdots & 1 \end{bmatrix}, \quad Z_2 = \begin{bmatrix} z_{21} \\ z_{22} \\ \vdots \\ z_{2n} \end{bmatrix}$$

The quadratic form q becomes

$$q = Z_2^{\mathrm{T}}(B_2^{-1})^{\mathrm{T}}(B_1^{-1})^{\mathrm{T}} A B_1^{-1} B_2^{-1} Z_2$$

$$= p_{211} z_{21}^2 + p_{222} z_{22}^2 + \sum_{j=3}^{n} \sum_{k=3}^{n} p_{2jk} z_{2j} z_{2k} \tag{2.82}$$

Continuing in this fashion, the quadratic form $X^{\mathrm{T}} A X$ is reduced to the sum of squares

$$q = Z_n^{\mathrm{T}} (B^{-1})^{\mathrm{T}} A B^{-1} Z_n \tag{2.83}$$

$$= \sum_{i=1}^{n} p_{nii} z_{ni}^2 \tag{2.84}$$

where

$$B = \prod_{i=1}^{n} B_{n-i+1} \tag{2.85}$$

It should be noted that each transformation, B_i, is nonsingular with $|B_i| = 1$ since all elements below the diagonal of B_i are zero. Hence

$$|B| = \prod_{i=1}^{n} |B_{n-i+1}| = 1 \tag{2.86}$$

Further, B is not, in general, an orthogonal transformation. This can be shown by the following simple counterexample. Let $n = 2$. Then

$$B_1 = \begin{bmatrix} 1 & b_{12} \\ 0 & 1 \end{bmatrix}, \quad B_2 = \begin{bmatrix} 1 & 0 \\ 0 & 1 \end{bmatrix}$$

2.10 CHARACTERISTIC VALUES AND QUADRATIC FORMS

and

$$B_2 B_1 = \begin{bmatrix} 1 & b_{12} \\ 0 & 1 \end{bmatrix}$$

From Eq. (2.86),

$$|B_2 B_1| = 1$$

If $B_2 B_1$ is in an orthogonal transformation, then

$$(B_2 B_1)^{-1} = (B_2 B_1)^{\mathrm{T}}$$

However

$$(B_2 B_1)^{\mathrm{T}} = \begin{bmatrix} 1 & 0 \\ b_{12} & 1 \end{bmatrix}, \quad (B_2 B_1)^{-1} = \begin{bmatrix} 1 & 0 \\ -b_{12} & 1 \end{bmatrix}$$

Therefore, $B_2 B_1$ is not an orthogonal transformation and B is not, in general, an orthogonal transformation.

Example 2.41 Reduce the following quadratic form to a sum of squares by a series of nonorthogonal transformations.

$$q = X^{\mathrm{T}} A X, \quad X = \begin{bmatrix} 1 \\ 4 \\ 1 \end{bmatrix}, \quad A = \begin{bmatrix} 1 & 0 & 2 \\ 0 & 2 & 3 \\ 2 & 3 & 1 \end{bmatrix}$$

Let

$$B_1 = \begin{bmatrix} 1 & 0 & 2 \\ 0 & 1 & 0 \\ 0 & 0 & 1 \end{bmatrix}, \quad B_1^{-1} = \begin{bmatrix} 1 & 0 & -2 \\ 0 & 1 & 0 \\ 0 & 0 & 1 \end{bmatrix}$$

Therefore

$$q = Z_1^{\mathrm{T}} (B_1^{-1})^{\mathrm{T}} A B_1^{-1} Z_1 = Z_1^{\mathrm{T}} \begin{bmatrix} 1 & 0 & 0 \\ 0 & 2 & 3 \\ 0 & 3 & -3 \end{bmatrix} Z_1 = z_{11}^2 + 2 z_{12}^2 + 6 z_{12} z_{13} - 3 z_{13}^2$$

Now let

$$B_2 = \begin{bmatrix} 1 & 0 & 0 \\ 0 & 1 & \frac{3}{2} \\ 0 & 0 & 1 \end{bmatrix}$$

from which

$$B_2^{-1} = \begin{bmatrix} 1 & 0 & 0 \\ 0 & 1 & -\frac{3}{2} \\ 0 & 0 & 1 \end{bmatrix}$$

and

$$q = Z_2^T(B_2^{-1})^T(B_1^{-1})^T A B_1^{-1} B_2^{-1} Z_2 = Z_2^T \begin{bmatrix} 1 & 0 & 0 \\ 0 & 2 & 0 \\ 0 & 0 & -\frac{15}{2} \end{bmatrix} Z_2$$

$$= z_{21}^2 + 2z_{22}^2 - \tfrac{15}{2} z_{23}^2 \qquad \text{⑤}$$

Definition 2.31 If A is a square matrix of order n, then $|A_i|$ is called the *leading principal minor* of order i ($i \leq n$), where

$$A_i = \begin{bmatrix} a_{11} & a_{12} & a_{13} & \cdots & a_{1i} \\ a_{21} & a_{22} & a_{23} & \cdots & a_{2i} \\ a_{31} & a_{32} & a_{33} & \cdots & a_{31} \\ \vdots & \vdots & \vdots & & \vdots \\ a_{i1} & a_{i2} & a_{i3} & \cdots & a_{ii} \end{bmatrix} \qquad (2.87)$$

By Definition 2.31, the leading principal minor of order i of a square matrix of order n is obtained by deleting the last $(n - i)$ rows and columns of A and taking the determinant of the resulting matrix. The leading principal minor of order zero is defined as unity. That is

$$|A_0| = 1 \qquad (2.88)$$

Theorem 2.27 If A is the nonsingular symmetric matrix of the quadratic form, q, with nonsingular leading principal minors,

$$q = X^T A X$$

then p_{nii} in the sum of squares

$$q = Z_n^T \prod_{i=1}^n (B_{n-i+1}^{-1})^T A \prod_{j=1}^n B_j^{-1} Z_n \qquad (2.89)$$

$$= \sum_{i=1}^n p_{nii} z_{ni}^2 \qquad (2.90)$$

can be expressed by

$$p_{nii} = \frac{|A_i|}{|A_{i-1}|} \qquad (2.91)$$

where

$$A_i = \begin{bmatrix} a_{11} & a_{12} & \cdots & a_{1i} \\ a_{12} & a_{22} & \cdots & a_{2i} \\ \vdots & & & \\ a_{1i} & a_{2i} & \cdots & a_{ii} \end{bmatrix} \qquad (2.92)$$

and

$$|A_0| = 1$$

2.10 CHARACTERISTIC VALUES AND QUADRATIC FORMS

Proof The proof given here is for the case where A is of order 3. Let

$$B_1 = \begin{bmatrix} 1 & a_{12}/a_{11} & a_{13}/a_{11} \\ 0 & 1 & 0 \\ 0 & 0 & 1 \end{bmatrix}$$

Then

$$B_1^{-1} = \begin{bmatrix} 1 & -a_{12}/a_{11} & -a_{13}/a_{11} \\ 0 & 1 & 0 \\ 0 & 0 & 1 \end{bmatrix}$$

and

$$(B_1^{-1})^T A B_1^{-1} = \begin{bmatrix} a_{11} & a_{12} & a_{13} \\ 0 & \dfrac{a_{11}a_{22} - a_{12}^2}{a_{11}} & \dfrac{a_{11}a_{23} - a_{12}a_{13}}{a_{11}} \\ 0 & \dfrac{a_{11}a_{23} - a_{12}a_{13}}{a_{11}} & \dfrac{a_{11}a_{33} - a_{13}^2}{a_{11}} \end{bmatrix}$$

Now let the transformation B_2 be given by

$$B_2 = \begin{bmatrix} 1 & 0 & 0 \\ 0 & 1 & \dfrac{a_{11}a_{23} - a_{12}a_{13}}{a_{11}a_{22} - a_{12}^2} \\ 0 & 0 & 1 \end{bmatrix}$$

and

$$B_2^{-1} = \begin{bmatrix} 1 & 0 & 0 \\ 0 & 1 & \dfrac{-a_{11}a_{23} - a_{12}a_{13}}{a_{11}a_{22} - a_{12}^2} \\ 0 & 0 & 1 \end{bmatrix}$$

The diagonal elements of $(B_2^{-1})^T(B_1^{-1})^T A B_1^{-1} B_2^{-1}$ are

$$a_{11}, \quad \dfrac{a_{11}a_{22} - a_{12}^2}{a_{11}}, \quad \dfrac{a_{11}a_{22}a_{33} + 2a_{12}a_{23}a_{13} - a_{11}a_{23}^2 - a_{22}a_{13}^2 - a_{33}a_{12}^2}{a_{11}a_{22} - a_{12}^2}$$

or

$$\dfrac{|A_1|}{|A_0|}, \quad \dfrac{|A_2|}{|A_1|}, \quad \text{and} \quad \dfrac{|A_3|}{|A_2|}$$

Since $(B_2^{-1})^T(B_1^{-1})^T A B_1^{-1} B_2^{-1}$ is a diagonal matrix,

$$(B_2^{-1})^T(B_1^{-1})^T A B_1^{-1} B_2^{-1} = \begin{bmatrix} \frac{|A_1|}{|A_0|} & 0 & 0 \\ 0 & \frac{|A_2|}{|A_1|} & 0 \\ 0 & 0 & \frac{|A_3|}{|A_2|} \end{bmatrix}$$

and

$$q = \frac{|A_1|}{|A_0|} z_{31}^2 + \frac{|A_2|}{|A_1|} z_{32}^2 + \frac{|A_3|}{|A_2|} z_{33}^2$$

Extending this result to a nonsingular symmetric matrix or order n,

$$q = \sum_{i=1}^{n} \frac{|A_i|}{|A_{i-1}|} z_{ni}^2 \qquad \qquad (2.93)$$

Definition 2.32 A quadratic form $X^T A X$ is said to be

1. *positive definite* if and only if $X^T A X > 0$ for all vectors $X \neq 0$,
2. *positive semidefinite* if and only if $X^T A X \geq 0$ for all vectors $X \neq 0$ and $X^T A X = 0$ for at least one $X \neq 0$,
3. *negative definite* if and only if $-X^T A X$ is positive definite,
4. *negative semidefinite* if and only if $-X^T A X$ is positive semidefinite.

Definition 2.33 If A is the matrix of a quadratic form, then A is positive definite, positive semidefinite, negative definite, negative semidefinite if the quadratic form is positive definite, positive semidefinite, negative definite, negative semidefinite.

Definition 2.34 A quadratic form which is neither positive definite, negative definite, positive semidefinite, nor negative semidefinite is said to be indefinite.

Theorem 2.28 The quadratic form $X^T A X$ is positive definite if and only if

$$|A_i| > 0, \qquad i = 1, 2, \ldots, n$$

where A is a symmetric matrix of order n.

Proof From Theorem 2.27, the quadratic form $X^T A X$ can be expressed by

$$q = \sum_{i=1}^{n} \frac{|A_i|}{|A_{i-1}|} z_{ni}^2 \qquad (2.94)$$

To prove that $|A_i| > 0$ is a necessary condition for $X^T A X$ to be positive definite one must show that

$$X^T A X > 0$$

for all $X \neq 0$ implies that

$$|A_i| > 0, \qquad i = 1, 2, \ldots, n$$

Suppose $|A_i|$ and $|A_{i-1}|$ are opposite in sign. Then
$$\frac{|A_i|}{|A_{i-1}|} < 0$$
However, if
$$Z_n = \begin{bmatrix} 0 \\ 0 \\ \vdots \\ z_{ni} \\ \vdots \\ 0 \end{bmatrix}$$

$X^T A X$ is negative. Therefore $|A_i|$ and $|A_{i-1}|$ must have like signs which implies either $|A_i| > 0$ for all i or $|A_i| < 0$ for all i. But, since $A_0 = 1$, $|A_i| > 0$ for $i = 1, 2, \ldots, n$. Thus if $X^T A X$ is positive definite,
$$|A_i| > 0, \quad i = 1, 2, \ldots, n$$

To prove sufficiency one must show that if $|A_i| > 0$ for all i, then $X^T A X$ is positive definite. If $|A_i| > 0$ for all i then
$$\frac{|A_i|}{|A_{i-1}|} z_{ni}^2 > 0, \quad i = 1, 2, \ldots, n$$
since z_{ni}^2 cannot be a negative number. Therefore
$$\sum_{i=1}^{n} \frac{|A_i|}{|A_{i-1}|} z_{ni}^2 > 0$$
since each component of the sum is nonnegative and at least one $z_{ni} \neq 0$. Hence, $X^T A X$ is positive definite.

Theorem 2.29 The quadratic form $X^T A X$ is negative definite if and only if
$$(-1)^i |A_i| > 0, \quad i = 1, 2, \ldots, n$$
where A is a symmetric matrix of order n.

Theorem 2.29 states that the quadratic form $X^T A X$ is negative definite if and only if the determinants of the leading principal minors alternate in sign with $|A_1| < 0$. That is,
$$|A_1| < 0, \quad |A_2| > 0, \quad |A_3| < 0, \ldots, (-1)^n |A_n| > 0$$

Example 2.42 Determine whether the quadratic forms $X^T A X$ and $X^T B X$ are positive or negative definite where
$$A = \begin{bmatrix} 2 & 1 & 1 \\ 1 & 3 & 0 \\ 1 & 0 & 3 \end{bmatrix}, \quad B = \begin{bmatrix} -4 & 0 & 0 \\ 0 & -1 & 1 \\ 0 & 1 & -2 \end{bmatrix}$$

Considering the quadratic form $X^T AX$,

$$|A_1| = 2, \quad |A_2| = 5, \quad |A_3| = 12$$

By Theorem 2.28, $X^T AX$ is positive definite. For $X^T BX$

$$|B_1| = -4, \quad |B_2| = 4, \quad |B_3| = -4$$

Since $|B_1| < 0$, $|B_2| > 0$, $|B_3| < 0$, $X^T BX$ is negative definite by Theorem 2.29.

The criteria for definiteness given in Theorems 2.28 and 2.29 are not the only criteria which may be used to determine whether a quadratic form is positive (negative) definite. The reader will recall that the criteria given in Theorems 2.28 and 2.29 were based upon diagonalization of the matrix of the quadratic form by a nonorthogonal transformation, reducing the quadratic form to a sum of squares. Therefore, it is not surprising that criteria for definiteness can also be developed when the matrix of the quadratic form is diagonalized by an orthogonal transformation. These criteria are given in Theorem 2.30.

Theorem 2.30 The quadratic form $X^T AX$, where A is symmetric, is

1. *positive definite* if and only if every eigenvalue for A is positive;
2. *negative definite* if and only if every eigenvalue for A is negative;
3. *positive semidefinite* if and only if every eigenvalue for A is nonnegative and at least one eigenvalue is zero;
4. *negative semidefinite* if and only if every eigenvalue for A is nonpositive and at least one eigenvalue is zero.

Example 2.43 Using Theorem 2.30, determine whether the quadratic forms given in Example 2.42 are positive or negative definite.

The characteristic equation for matrix A is given by

$$|\lambda I - A| = (\lambda - 3)[(\lambda - 2)(\lambda - 3) - 2] = 0$$

and the eigenvalues for A are

$$\lambda_1 = 3, \quad \lambda_2 = 4, \quad \lambda_3 = 1$$

Since the eigenvalues are all positive, $X^T AX$ is positive definite. The characteristic equation for matrix B is

$$|\lambda I - B| = (\lambda + 4)[(\lambda + 1)(\lambda + 2) - 1]$$

and the eigenvalues are

$$\lambda_1 = -4, \quad \lambda_2 = \frac{-3 + \sqrt{5}}{2}, \quad \lambda_3 = \frac{-3 - \sqrt{5}}{2}$$

Since the eigenvalues are all negative, $X^T BX$ is negative definite.

2.11 Applications

2.11.1 Finite Markov Chains

Suppose that a system or process can be characterized as being in any one of a finite number of states at a given point in time. To illustrate, suppose an inventory system is observed at the beginning of each workday and the inventory level at that time is recorded. Assume that the capacity of the system is 100 units of product. The states, levels of inventory, which the system may assume at the beginning of each workday are the integers 0 to 100 inclusive. Now, let it be assumed that units are manufactured and placed in inventory at a rate of 5 per day, that the probability of m demands for product in any one day is p_m, where m may vary between 0 and 25 units and that demand which occurs when the inventory level is zero results in a lost sale. Therefore

$$\sum_{m=0}^{25} p_m = 1 \tag{2.95}$$

Let the inventory level on the nth day be 30 units. Based upon the information given, one can calculate the probability that the inventory system will be in any given state on the $(n + 1)$st day. For example, to calculate the probability that the inventory level on the $(n + 1)$st day will be 15 units, the analyst must recognize that 5 units will be placed in stock before the start of the $(n + 1)$st day. Therefore, the inventory level will be 15 units if and only if demand between the beginning of the nth day and the $(n + 1)$st day is 20 units. Therefore, if i is the state of the system

$$P[i = 15 \text{ at } n + 1] = p_{20} \tag{2.96}$$

It should be noted the probability of observing any given inventory level at the beginning of any day is dependent only on the state of the system at the beginning of the previous day. For example, if the inventory level at the start of the $(n - 2)$nd day had been 37 units, 29 units at the start of the $(n - 1)$st day, and, as before, 30 units at the start of the nth day, the probability of an inventory level of 15 units at the start of the $(n + 1)$st day is still p_{20}. That is, this probability is independent of what happened prior to the nth day.

A process exhibiting the properties described for the above inventory system is called a finite Markov chain. In general a process may be characterized as a finite Markov chain if it exhibits the following properties.

    1. The process may be in one of a finite number of states at a given point in time.

    2. The probability that the process is in state i after n changes or transitions of the process is dependent only upon the state of the process after the $(n - 1)$st transition.

Associated with every finite Markov chain is a transition probability matrix, P, often simply referred to as the transition matrix. The element p_{ij} of P, $i \neq j$, represents the probability that the process will change from state i to state j in one transition or step. That is, p_{ij} is the probability that the state i of the system will be followed by a change to state j without passing through any other state $k \neq j$ before reaching j. The probability that the process remains in state i, given that i is the present state, for one transition is p_{ii}. Therefore, P can be defined by

$$P = \begin{bmatrix} p_{11} & p_{12} & \cdots & p_{1j} & \cdots & p_{1m} \\ p_{21} & p_{22} & \cdots & p_{2j} & \cdots & p_{2m} \\ \vdots & \vdots & & \vdots & & \vdots \\ p_{i1} & p_{i2} & \cdots & p_{ij} & \cdots & p_{im} \\ \vdots & \vdots & & \vdots & & \vdots \\ p_{m1} & p_{m2} & \cdots & p_{mj} & \cdots & p_{mm} \end{bmatrix} \quad (2.97)$$

where m is the number of states and $\sum_{i=1}^{m} p_{ij} = 1$.

Let $Q(n)$ be a column vector of probabilities that the system is in states 1 through m after n transitions. That is,

$$Q(n) = \begin{bmatrix} q_1(n) \\ q_2(n) \\ \vdots \\ q_j(n) \\ \vdots \\ q_m(n) \end{bmatrix} \quad (2.98)$$

where $q_j(n)$ is the probability that the process is in state j after n transitions or changes and $\sum_{j=1}^{m} q_j(n) = 1$. $Q(n)$ can be expressed by

$$Q(n) = P^T Q(n-1) \quad (2.99)$$

where the elements of $Q(n)$ are given as

$$q_j(n) = \sum_{i=1}^{m} P[i \text{ at } n-1] P[\text{transition from } i \text{ to } j] = \sum_{i=1}^{m} q_i(n-1) p_{ij} \quad (2.100)$$

Applying Eq. (2.99) to $Q(n-1)$, $n-1$ times leads to

$$Q(n) = P^T P^T Q(n-2) = (P^T)^n Q(0) \quad (2.101)$$

where $Q(0)$ is the column vector of probabilities that the system starts in the states 1 through m. Taking the transpose of both sides of Eq. (2.101) leads us to

$$Q(n)^T = Q(0)^T [(P^T)^n]^T \quad (2.102)$$

by Eq. (2.13). Applying (2.13) to $[(P^T)^n]^T$ yields

$$Q(n)^T = Q(0)^T \left[\prod_{k=1}^{n} P^T \right]^T = Q(0)^T \prod_{k=1}^{n} P = Q(0)^T P^n \quad (2.103)$$

The matrix P^n is called the n-step transition probability matrix. Let $p_{ij}(n)$ be the element in the ith row, jth column of P^n. Then $p_{ij}(n)$ represents the probability of

2.11 APPLICATIONS

passing from state i to state j in exactly n transitions. Based upon this definition of P^n, P is sometimes referred to as the one-step transition probability matrix.

Example 2.44 Given P as defined in Eq. (2.97), show that the n-step transition probability matrix is given by P^n.

Let the probability of passing from state i to state j in exactly n transitions be $p_{ij}(n)$. Now

$$p_{ij}(2) = \sum_{k=1}^{m} P[i \text{ to } k \text{ in one transition}]P[k \text{ to } j \text{ in one transition}]$$

$$= \sum_{k=1}^{m} p_{ik}p_{kj}, \quad i, j = 1, 2, \ldots, m \quad (2.104)$$

Noting that Eq. (2.104) is obtained by multiplying the ith row of P by the jth column of P, the two-step transition probability matrix is given by PP or P^2. Assume that the h-step transition probability matrix is given by P^h.

$$p_{ij}(h+1) = \sum_{k=1}^{m} P[i \text{ to } k \text{ in } h \text{ transitions}]P[k \text{ to } j \text{ in one transition}]$$

$$= \sum_{k=1}^{m} p_{ik}(h)p_{kj} \quad (2.105)$$

But $p_{ij}(h+1)$ is obtained by multiplying the ith row of P^h by the jth column of P. Therefore, the $h+1$-step transition probability matrix is given by P^{h+1} and, by induction, the proof is complete. ◻

Example 2.45 Given the following one-step transition probability matrix, find the two-and three-step transition probability matrices.

$$P = \begin{bmatrix} 0.0 & 0.4 & 0.6 \\ 0.5 & 0.0 & 0.5 \\ 0.2 & 0.8 & 0.0 \end{bmatrix}$$

The two-step transition matrix is given by P^2.

$$P^2 = \begin{bmatrix} 0.0 & 0.4 & 0.6 \\ 0.5 & 0.0 & 0.5 \\ 0.2 & 0.8 & 0.0 \end{bmatrix} \begin{bmatrix} 0.0 & 0.4 & 0.6 \\ 0.5 & 0.0 & 0.5 \\ 0.2 & 0.8 & 0.0 \end{bmatrix} = \begin{bmatrix} 0.32 & 0.48 & 0.20 \\ 0.10 & 0.60 & 0.30 \\ 0.40 & 0.08 & 0.52 \end{bmatrix}$$

The three-step transition matrix is given by P^3.

$$P^3 = \begin{bmatrix} 0.32 & 0.48 & 0.20 \\ 0.10 & 0.60 & 0.30 \\ 0.40 & 0.08 & 0.52 \end{bmatrix} \begin{bmatrix} 0.00 & 0.40 & 0.60 \\ 0.50 & 0.00 & 0.50 \\ 0.20 & 0.80 & 0.00 \end{bmatrix} = \begin{bmatrix} 0.2800 & 0.2880 & 0.4320 \\ 0.3600 & 0.2800 & 0.3600 \\ 0.1440 & 0.5760 & 0.2800 \end{bmatrix}$$

◻

Example 2.46 For the problem given in Example 2.45, suppose that the vector of starting probabilities, $Q(0)$ is given by

$$Q(0) = \begin{bmatrix} 0.4 \\ 0.3 \\ 0.3 \end{bmatrix}$$

Find $Q(1)^T$, $Q(2)^T$, and $Q(3)^T$.

From Eq. (2.103), $Q(n)^T = Q(0)^T p^n$. Therefore

$$Q(1)^T = Q(0)^T P = [0.4 \quad 0.3 \quad 0.3] \begin{bmatrix} 0.0 & 0.4 & 0.6 \\ 0.5 & 0.0 & 0.5 \\ 0.2 & 0.8 & 0.0 \end{bmatrix} = [0.21 \quad 0.40 \quad 0.39]$$

$$Q(2)^T = Q(0)^T P^2 = Q(1)^T P$$
$$= [0.21 \quad 0.40 \quad 0.39] \begin{bmatrix} 0.0 & 0.4 & 0.6 \\ 0.5 & 0.0 & 0.5 \\ 0.2 & 0.8 & 0.0 \end{bmatrix} = [0.2780 \quad 0.3960 \quad 0.3260]$$

$$Q(3)^T = Q(0)^T P^3 = Q(2)^T P$$
$$= [0.2780 \quad 0.3960 \quad 0.3260] \begin{bmatrix} 0.0 & 0.4 & 0.6 \\ 0.5 & 0.0 & 0.5 \\ 0.2 & 0.8 & 0.0 \end{bmatrix}$$
$$= [0.2632 \quad 0.3720 \quad 0.3648]$$

Example 2.47 A finite Markov chain with m states is said to have a steady state or stationary distribution if

$$\lim_{n \to \infty} p_{ij}(n) = \pi_j \qquad (2.106)$$

and

$$\sum_{j=1}^{m} \pi_j = 1 \qquad (2.107)$$

That is, the probability that the process is in state j after n transitions approaches a constant as $n \to \infty$, for all j. Given that a finite Markov chain, with transition probability matrix P, has a steady state distribution, show that

$$\Pi^T = \Pi^T P \qquad (2.108)$$

where Π is the column vector of steady state probabilities with elements π_j. That is,

$$\Pi = \begin{bmatrix} \pi_1 \\ \pi_2 \\ \vdots \\ \pi_j \\ \vdots \\ \pi_m \end{bmatrix} \qquad (2.109)$$

2.11 APPLICATIONS

From Eq. (2.99),
$$Q(n)^{\mathrm{T}} = Q(n-1)^{\mathrm{T}} P$$

By definition, Eq. (2.106) holds. Therefore,
$$\lim_{n \to \infty} Q(n)^{\mathrm{T}} = \lim_{n \to \infty} Q(n-1)^{\mathrm{T}} = \Pi^{\mathrm{T}} \qquad (2.110)$$

since the elements of $\lim_{n \to \infty} Q(n)$ are $\lim_{n \to \infty} \sum_{i=1}^{m} q_i(0) P_{ij}(n)$ and
$$\lim_{n \to \infty} \sum_{i=1}^{m} q_i(0) P_{ij}(n) = \sum_{i=1}^{m} q_i(0) \pi_j = \Pi_j$$

Hence
$$\Pi^{\mathrm{T}} = \Pi^{\mathrm{T}} P$$

Example 2.48   Given that P is the transition probability matrix of a Markov chain which has a steady state distribution, derive an expression for Π in terms of P.

From Eq. (2.108)
$$\pi_j = \sum_{I=1}^{m} \pi_i p_{ij} \qquad (2.111)$$

and
$$\sum_{j=1}^{m} \pi_j = 1$$

which leads to the following set of m equations in m unknowns.

$$\pi_1(p_{11} - 1) + \pi_2 p_{21} + \pi_3 p_{31} + \cdots + \pi_m p_{m1} = 0$$
$$\pi_1 p_{12} + \pi_2(p_{22} - 1) + \pi_3 p_{32} + \cdots + \pi_m p_{m2} = 0$$
$$\vdots$$
$$\pi_1 p_{1j} + {}_2 p_{2j} + {}_3 p_{3j} + \cdots + \pi_m p_{mj} = 0 \qquad (2.112)$$
$$\vdots$$
$$\pi_1 p_{1,m-1} + \pi_2 p_{2,m-1} + \pi_3 p_{3,m-1} + \cdots + \pi_m p_{m,m-1} = 0$$
$$\pi_1 + \pi_2 + \pi_3 + \cdots + \pi_m = 1$$

Let

$$R = \begin{bmatrix} (p_{11} - 1) & p_{21} & p_{31} & \cdots & p_{m1} \\ p_{12} & (p_{22} - 1) & p_{32} & \cdots & p_{m2} \\ \vdots & \vdots & \vdots & & \vdots \\ p_{1j} & p_{2j} & p_{3j} & \cdots & p_{mj} \\ \vdots & \vdots & \vdots & & \vdots \\ p_{1,m-1} & p_{2,m-1} & p_{3,m-1} & \cdots & p_{m,m-1} \\ 1 & 1 & 1 & & 1 \end{bmatrix} \qquad (2.113)$$

$$B = \begin{bmatrix} 0 \\ 0 \\ \vdots \\ 0 \\ \vdots \\ 0 \\ 1 \end{bmatrix} \qquad (2.114)$$

Then

$$R\Pi = B$$

and

$$\Pi = R^{-1}B \qquad (2.115)$$

Example 2.49 Find the steady state distribution for a finite Markov chain with the following transition matrix.

$$P = \begin{bmatrix} 0.0 & 0.5 & 0.5 \\ 0.2 & 0.2 & 0.6 \\ 0.5 & 0.1 & 0.4 \end{bmatrix}$$

From Eq. (2.113)

$$R = \begin{bmatrix} -1 & 0.2 & 0.5 \\ 0.5 & -0.8 & 0.1 \\ 1 & 1 & 1 \end{bmatrix}, \quad B = \begin{bmatrix} 0 \\ 0 \\ 1 \end{bmatrix}, \quad |R| = 1.47$$

$$\operatorname{adj} R = \begin{bmatrix} -0.90 & 0.30 & 0.42 \\ -0.40 & -1.50 & 0.35 \\ 1.30 & 1.20 & 0.70 \end{bmatrix}$$

$$R^{-1} = \begin{bmatrix} -0.6122 & 0.2041 & 0.2857 \\ -0.2041 & -1.0204 & 0.2381 \\ 0.8844 & 0.8163 & 0.4762 \end{bmatrix}$$

$$\Pi = \begin{bmatrix} -0.6122 & 0.2041 & 0.2857 \\ -0.2041 & -1.0204 & 0.2381 \\ .8844 & 0.8163 & 0.4762 \end{bmatrix} \begin{bmatrix} 0 \\ 0 \\ 1 \end{bmatrix} = \begin{bmatrix} 0.2857 \\ 0.2381 \\ 0.4762 \end{bmatrix}$$

Example 2.50 A manufacturing process is to be analyzed by means of a work sampling study. The system is to be checked once a day and at this time the state of the system is to be recorded. The manufacturing process may be in one of three states.

    State 1: operating
    State 2: under repair
    State 3: down and waiting for repair

2.11 APPLICATIONS

The transition probability matrix for this process is given by

$$P = \begin{bmatrix} 0.5 & 0.4 & 0.1 \\ 0.6 & 0.2 & 0.2 \\ 0.8 & 0.2 & 0.0 \end{bmatrix}$$

Determine the steady state probabilities that the process is in each of the possible states indicated above.

$$R = \begin{bmatrix} -0.5 & 0.6 & 0.8 \\ 0.4 & -0.8 & 0.2 \\ 1 & 1 & 1 \end{bmatrix}, \quad B = \begin{bmatrix} 0 \\ 0 \\ 1 \end{bmatrix}, \quad |R| = 1.34$$

$$\text{adj}(R) = \begin{bmatrix} -1.00 & 0.20 & 0.76 \\ -0.20 & -1.30 & 0.42 \\ 1.20 & 1.10 & 0.16 \end{bmatrix},$$

$$R^{-1} = \begin{bmatrix} -0.7463 & 0.1493 & 0.5672 \\ -0.1493 & -0.9701 & 0.3134 \\ 0.8955 & 0.8283 & 0.1194 \end{bmatrix}$$

$$\Pi = R^{-1}B = \begin{bmatrix} 0.5672 \\ 0.3134 \\ 0.1194 \end{bmatrix}$$

Example 2.51 A quality control system functions under one of three inspection plans. The first inspection plan, tightened inspection, is applied when product quality seems to be running substantially below standard. The second inspection plan, normal inspection, is used when product quality is running neither particularly high or low. The third inspection plan, reduced inspection, is used when product quality is running significantly above standard. For the ith plan, a sample of n_i is drawn from a lot of L units and inspected. If c_i or less defects are observed the lot is accepted. If the sample contains zero defects, the inspection level is relaxed to the next lower level; that is, from tightened inspection to normal inspection or from normal inspection to reduced inspection depending upon the current level. If $c_i + 1$ or more defects are found the lot is rejected. In addition, if four or more defects are found the level of inspection is increased to the next higher level. The inspection plans for the three levels are as follows:

reduced inspection (state 1) – $n_1 = 20$, $c_1 = 2$
normal inspection (state 2) – $n_2 = 40$, $c_2 = 2$
tightened inspection (state 3) – $n_3 = 40$, $c_3 = 1$

If incoming lots are 10 percent defective, find the steady state probabilities that the quality control system will operate under reduced, normal, and tightened inspection.

Based upon the information given, the transition probabilities can be computed as follows. Let x be the number of defects found in a given sample. Then

$$p_{11} = P[x < 4 \mid n_1, c_1], \qquad p_{12} = P[x \geq 4 \mid n_1, c_1], \qquad p_{21} = P[x = 0 \mid n_2, c_2]$$
$$p_{22} = P[1 \leq x \leq 3 \mid n_2, c_2], \qquad p_{23} = P[x \geq 4 \mid n_2, c_2]$$
$$p_{32} = P[x = 0 \mid n_3, c_3], \qquad p_{33} = P[x > 0 \mid n_3, c_3], \qquad p_{13} = p_{31} = 0$$

It will be assumed that the number of defects observed in a sample of size n is Poisson distributed with mean pn where p is the lot proportion defective. Therefore

$$P[x = k \mid n_i, c_i] = \frac{(0.1 n_i)^k}{k!} \exp[-0.1 n], \quad k = 0, 1, 2, \ldots$$

Using this equation

$$p_{11} = 0.86, \qquad p_{21} = 0.02, \qquad p_{32} = 0.02$$
$$p_{12} = 0.14, \qquad p_{22} = 0.41, \qquad p_{33} = 0.98$$
$$p_{23} = 0.57$$

The transition probability matrix is then given by

$$P = \begin{bmatrix} 0.86 & 0.14 & 0.00 \\ 0.02 & 0.41 & 0.57 \\ 0.00 & 0.02 & 0.98 \end{bmatrix}$$

$$R = \begin{bmatrix} -0.14 & 0.02 & 0.00 \\ 0.14 & -0.59 & 0.02 \\ 1 & 1 & 1 \end{bmatrix}, \quad |R| = 0.0830$$

$$\text{adj}(R) = \begin{bmatrix} -0.6100 & -0.0200 & 0.0004 \\ -0.1200 & -0.1400 & 0.0028 \\ 0.7300 & 0.1600 & 0.0798 \end{bmatrix}$$

$$R^{-1} = \begin{bmatrix} -7.3494 & -0.2410 & 0.0048 \\ -1.4458 & -1.6867 & 0.0337 \\ 8.7952 & 1.9277 & 0.9615 \end{bmatrix} \quad \text{and} \quad \Pi = \begin{bmatrix} 0.0048 \\ 0.0337 \\ 0.9615 \end{bmatrix}$$

2.11.2 Experimental Design

Example 2.52 A relationship between a dependent variable and n independent variables is to be developed from experimental data. Let y_i be the observed value of the dependent variable corresponding to the ith values of the independent variables, $x_{1i}, x_{2i}, \ldots, x_{ni}$. That is, for each observed combination of values of $x_{1i}, x_{2i}, \ldots, x_{ni}$, there is one observed value of the dependent variable, y_i. The form of the relationship is to be the following

2.11 APPLICATIONS

$$z_i = b_0 + \sum_{j=1}^{n} b_j x_{ji} \qquad (2.116)$$

where z_i is the estimate of y_i given by Eq. (2.116). The estimating equation is defined once the values of b_j, $j = 0, 1, 2, \ldots, n$, are defined. The method of least squares is to be used to define the coefficients, b_j. That is, the values of b_j are to be defined such that

$$\sum_{i=1}^{m} \left[y_i - b_0 - \sum_{j=1}^{n} b_j x_{ji} \right]^2 = \min \qquad (2.117)$$

where m is the number of sets of observations on the dependent and independent variables. Develop a method for determining the coefficients, b_j, which satisfy the criterion defined in Eq. (2.117).

Let

$$Y = \begin{bmatrix} y_1 \\ y_2 \\ \vdots \\ y_i \\ \vdots \\ y_m \end{bmatrix}, \quad B = \begin{bmatrix} b_0 \\ b_1 \\ \vdots \\ b_j \\ \vdots \\ b_n \end{bmatrix}$$

$$X = \begin{bmatrix} 1 & x_{11} & x_{21} & \cdots & x_{j1} & \cdots & x_{ni} \\ 1 & x_{12} & x_{22} & \cdots & x_{j2} & \cdots & x_{n2} \\ \vdots & \vdots & \vdots & & \vdots & & \vdots \\ 1 & x_{1i} & x_{2i} & \cdots & x_{ji} & \cdots & x_{ni} \\ \vdots & \vdots & \vdots & & \vdots & & \vdots \\ 1 & x_{1m} & x_{2m} & \cdots & x_{jm} & \cdots & x_{nm} \end{bmatrix}$$

Equation (2.117) is extremized by taking partial derivatives with respect to b_j for $j = 0, 1, 2, \ldots, n$. Let

$$E = \sum_{i=1}^{m} \left[y_i - b_0 - \sum_{j=1}^{n} b_j x_{ji} \right]^2$$

Then

$$\frac{\partial E}{\partial b_0} = \sum_{i=1}^{m} -2 \left[y_i - b_0 - \sum_{j=1}^{n} b_j x_{ji} \right] = 0$$

and

$$\frac{\partial E}{\partial b_j} = \sum_{i=1}^{m} -2 x_{ji} \left[y_i - b_0 - \sum_{j=1}^{n} b_j x_{ji} \right] = 0, \qquad j = 1, 2, \ldots, n$$

This set of equations can be represented in matrix form by

$$2X^TY - 2X^TXB = 0$$

or

$$X^TXB = X^TY \qquad (2.118)$$

The elements of X^TXB are $\sum_{k=0}^{n} b_k \sum_{i=1}^{m} x_{ji} x_{ki}$, where $x_{0i} = 1$ for all i. The elements of X^TY are $\sum_{i=1}^{m} y_i x_{ji}$. Therefore,

$$B^* = (X^TX)^{-1} X^TY \qquad (2.119)$$

The column vector B^* yields an extremum for 2.117 provided X^TX is a nonsingular matrix. Since (2.117) is a quadratic function of b_j, $j = 0, 1, 2, \ldots, n$, B^* cannot be a saddle point for (2.117) and must therefore be either a maximizing or minimizing vector. Now

$$E = (Y - XB^*)^T (Y - XB^*)$$

If E is finite, then B^* is a minimizing vector. Since E is a quadratic function it has either one minimum or one maximum. Let

$$B_1 = \begin{bmatrix} \infty \\ 0 \\ 0 \\ \vdots \\ 0 \end{bmatrix}$$

and

$$E_1 = (Y - XB_1)^T (Y - XB_1) = \infty$$

Since $E < E_1$, B^* is a minimizing vector.

Example 2.53 The following data have been collected on the independent variables x_1 and x_2 and the dependent variable y.

y	x_1	x_2
1	0	0
2	0	1
2	1	0
5	3	1
3	2	0
6	4	1
4	2	1

Find the equation of the form

$$y_i = b_0 + b_1 x_{1i} + b_2 x_{2i}$$

which best fits the above data in a least squares sense.

2.11 APPLICATIONS

Let

$$X = \begin{bmatrix} 1 & 0 & 0 \\ 1 & 0 & 1 \\ 1 & 1 & 0 \\ 1 & 3 & 1 \\ 1 & 2 & 0 \\ 1 & 4 & 1 \\ 1 & 2 & 1 \end{bmatrix}, \quad Y = \begin{bmatrix} 1 \\ 2 \\ 2 \\ 5 \\ 3 \\ 6 \\ 4 \end{bmatrix}$$

$$X^T X = \begin{bmatrix} 7 & 12 & 4 \\ 12 & 34 & 9 \\ 4 & 9 & 4 \end{bmatrix}, \quad X^T Y = \begin{bmatrix} 23 \\ 55 \\ 17 \end{bmatrix}$$

$$|X^T X| = 129, \quad \text{adj } X^T X = \begin{bmatrix} 55 & -12 & -28 \\ -12 & 12 & -15 \\ -28 & -15 & 94 \end{bmatrix}$$

$$(X^T X)^{-1} = \begin{bmatrix} \dfrac{55}{129} & -\dfrac{12}{129} & -\dfrac{28}{129} \\ -\dfrac{12}{129} & \dfrac{12}{129} & -\dfrac{15}{129} \\ -\dfrac{28}{129} & -\dfrac{15}{129} & \dfrac{94}{129} \end{bmatrix}$$

$$B^* = (X^T X)^{-1} X^T Y = \begin{bmatrix} 1 \\ 1 \\ 1 \end{bmatrix}$$

Therefore, the estimating equation is given by

$$z_i = 1 + x_{1i} + x_{2i} \quad \text{for all } i$$

Example 2.54 Show that $X(X^T X)^{-1} X^T$ is an idempotent matrix if $X^T X$ is nonsingular, where X is as defined in Example 2.52.

By Theorem 2.6, $X^T X$ is a symmetric matrix. By Properties 2 and 4

$$[(X^T X)^{-1}]^T = [(X^T X)^T]^{-1} = (X^T X)^{-1}$$

and $(X^T X)^{-1}$ is a symmetric matrix.

If $X(X^T X)^{-1} X^T$ is an idempotent matrix then

$$[X(X^T X)^{-1} X^T]^2 = X(X^T X)^{-1} X^T$$

Expanding the left side of the equality

$$[X(X^T X)^{-1} X^T]^2 = X(X^T X)^{-1} X^T X (X^T X)^{-1} X^T = X(X^T X)^{-1} X^T$$

since $X^T X (X^T X)^{-1} = I$.

2.11.3 Production Systems

Example 2.55 A manufacturer has m resources. These resources include raw materials, in process inventories, finished goods inventories, manufacturing lines, and labor. In order to produce one unit of resource j, a_{ij} units of resource i are required, $i = 1, 2, \ldots, m$. Similarly, a_{ki} units of resource k are required to obtain one unit of resource i, and so on. The process is shown schematically in Fig. 2.9. Let t_{ij} be the total number of units of resource i required to produce one unit of j taken over all possible stages of production. Determine the matrix of total requirements, T, with elements t_{ij}.

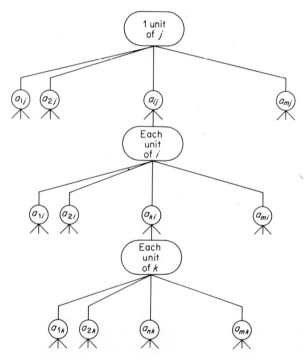

Figure 2.9 Resource Requirements for One Unit of j

Let

$$A = \begin{bmatrix} a_{11} & a_{12} & \cdots & a_{1j} & \cdots & a_{1m} \\ a_{21} & a_{22} & \cdots & a_{2j} & \cdots & a_{2m} \\ \vdots & \vdots & & \vdots & & \vdots \\ a_{i1} & a_{i2} & \cdots & a_{ij} & \cdots & a_{im} \\ \vdots & \vdots & & \vdots & & \vdots \\ a_{m1} & a_{m2} & \cdots & a_{mj} & \cdots & a_{mm} \end{bmatrix}$$

2.11 APPLICATIONS

To produce one unit of j, a_{ij} units of i are reiuired at the first stage in Fig. 2.9. At the second stage of production $\sum_{k=1}^{m} a_{ik} a_{kj}$ units of i are required. At the third stage $\sum_{k=1}^{m} a_{ik} \sum_{h=1}^{m} a_{kh} a_{hj}$ units of i are required. The matrix of requirements at the first, second, and third stages are A, A^2, and A^3 respectively. Therefore, the requirements matrix at the Nth stage is A^N. The preceding analysis applies for $i \neq j$.

For $i = j$ the analysis is similar except that one additional unit of i must be added since this is the final resource produced as well as a resource contributing to the final product. Therefore if t_{ij} is the total quantity of i required to produce one unit of j.

$$t_{ij} = \delta_{ij} + a_{ij} + \sum_{k_1=1}^{m} a_{ik_1} a_{k_1 j} + \sum_{k_1=1}^{m} a_{ik_1} \sum_{k_2=1}^{m} a_{k_1 k_2} a_{k_2 j}$$
$$+ \sum_{k_1=1}^{m} a_{ik_1} \sum_{k_2=1}^{m} a_{k_1 k_2} \sum_{k_3=1}^{m} a_{k_2 k_3} a_{k_3 j} + \cdots \qquad (2.120)$$

where

$$\delta_{ij} = \begin{cases} 1, & i = j \\ 0, & i \neq j \end{cases}$$

In matrix form,

$$T = I + \sum_{k=1}^{\infty} A^k \qquad (2.121)$$

Define A^0 as I. Then

$$T = \sum_{k=0}^{\infty} A^k = (I - A)^{-1} \qquad (2.122)$$

The validity of Eq. (2.122) is demonstrated as follows.

$$[(I - A)^{-1}]^{-1} = I - A$$

$$(I - A) \sum_{k=0}^{\infty} A^k = I \sum_{k=0}^{\infty} A^k - \sum_{k=1}^{\infty} A^k = I + \sum_{k=1}^{\infty} A^k - \sum_{k=1}^{\infty} A^k = I$$

Example 2.56 One unit of product I is to be produced. For each unit of I produced 1 unit of product II and 3 units of raw material III are required. In addition 3 units of time on production line IV are required for each unit of I manufactured. Each unit of II used requires 2 units of III and one unit of time on manufacturing line IV. Determine the total requirements matrix T.

The first stage requirements matrix is given by

$$A = \begin{bmatrix} 0 & 0 & 0 & 0 \\ 1 & 0 & 0 & 0 \\ 3 & 2 & 0 & 0 \\ 3 & 1 & 0 & 0 \end{bmatrix}, \quad I - A = \begin{bmatrix} 1 & 0 & 0 & 0 \\ -1 & 1 & 0 & 0 \\ -3 & -2 & 1 & 0 \\ -3 & -1 & 0 & 1 \end{bmatrix},$$

$$|I - A| = 1, \quad \text{adj}(I - A) = \begin{bmatrix} 1 & 0 & 0 & 0 \\ 1 & 1 & 0 & 0 \\ 5 & 2 & 1 & 0 \\ 4 & 1 & 0 & 1 \end{bmatrix}$$

Therefore

$$T = \begin{bmatrix} 1 & 0 & 0 & 0 \\ 1 & 1 & 0 & 0 \\ 5 & 2 & 1 & 0 \\ 4 & 1 & 0 & 1 \end{bmatrix}$$

Example 2.57 To produce one unit of product A, one unit of product B and 4 units of raw material C are required. Each unit of B requires 2 units of C and 20 units of raw material D. For production of one unit of A, 10 man-hours, E, are required while 3 man-hours are required for production of each unit of B. Determine the total requirements matrix, T.

$$A = \begin{bmatrix} 0 & 0 & 0 & 0 & 0 \\ 1 & 0 & 0 & 0 & 0 \\ 4 & 2 & 0 & 0 & 0 \\ 0 & 20 & 0 & 0 & 0 \\ 10 & 3 & 0 & 0 & 0 \end{bmatrix}, \quad I - A = \begin{bmatrix} 1 & 0 & 0 & 0 & 0 \\ -1 & 1 & 0 & 0 & 0 \\ -4 & -2 & 1 & 0 & 0 \\ 0 & -20 & 0 & 1 & 0 \\ -10 & -3 & 0 & 0 & 1 \end{bmatrix}$$

$$|I - A| = 1, \quad \text{adj}(I - A) = \begin{bmatrix} 1 & 0 & 0 & 0 & 0 \\ 1 & 1 & 0 & 0 & 0 \\ 6 & 2 & 1 & 0 & 0 \\ 20 & 20 & 0 & 1 & 0 \\ 13 & 3 & 0 & 0 & 1 \end{bmatrix}$$

and

$$T = \begin{bmatrix} 1 & 0 & 0 & 0 & 0 \\ 1 & 1 & 0 & 0 & 0 \\ 6 & 2 & 1 & 0 & 0 \\ 20 & 20 & 0 & 1 & 0 \\ 13 & 3 & 0 & 0 & 1 \end{bmatrix}$$

Example 2.58 In Example 2.57 suppose that 100 units of A were required. In addition, at the end of the production period the inventories of B, C, and D are to be 50, 1000, and 2000 units respectively. What are the total requirements for A, B, C, D, and E for the production period?

2.11 APPLICATIONS

Let S be the vector of final inventories for A, B, C, D, and E and let R be vector of total requirements for all manufacturing components. For the ith component,

$$r_i = \sum_{j=1}^{m} t_{ij} s_j \tag{2.123}$$

Therefore

$$R = TS \tag{2.124}$$

and

$$R = \begin{bmatrix} 1 & 0 & 0 & 0 & 0 \\ 1 & 1 & 0 & 0 & 0 \\ 6 & 2 & 1 & 0 & 0 \\ 20 & 20 & 0 & 1 & 0 \\ 13 & 3 & 0 & 0 & 1 \end{bmatrix} \begin{bmatrix} 100 \\ 50 \\ 1000 \\ 2000 \\ 0 \end{bmatrix}, \quad R = \begin{bmatrix} 100 \\ 150 \\ 1700 \\ 5000 \\ 1450 \end{bmatrix}$$

2.11.4 Linear Programming

The general linear programming problem can be defined as optimization of a linear function in several decision variables subject to a set of linear constraints. Stated mathematically, the problem is

$$\sum_{i=1}^{n} c_i x_i = \min(\max) \tag{2.125}$$

subject to

$$\sum_{i=1}^{n} a_{1i} x_i \leq b_1$$

$$\sum_{i=1}^{n} a_{2i} x_i \leq b_2 \tag{2.126}$$

$$\vdots$$

$$\sum_{i=1}^{n} a_{mi} x_i \leq b_m, \quad x_i \geq 0, \quad i = 1, 2, \ldots, n$$

where $\sum_{i=1}^{n} c_i x_i$ is to be minimized (maximized) with respect to x_i, $i = 1, 2, \ldots, n$. The relationships in (2.125) and (2.126) can be expressed in matrix form by

$$CX = \min \, (\max) \tag{2.127}$$

subject to

$$AX \leq B, \quad X \geq 0 \tag{2.128}$$

where

$$C = [c_1, c_2, \ldots, c_n] \tag{2.129}$$

$$A = \begin{bmatrix} a_{11} & a_{12} & \cdots & a_{1n} \\ a_{21} & a_{22} & \cdots & a_{2n} \\ \vdots & \vdots & & \vdots \\ a_{m1} & a_{m2} & \cdots & a_{mn} \end{bmatrix} \tag{2.130}$$

$$X = \begin{bmatrix} x_1 \\ x_2 \\ \vdots \\ x_n \end{bmatrix} \tag{2.131}$$

$$B = \begin{bmatrix} b_1 \\ b_2 \\ \vdots \\ b_m \end{bmatrix} \tag{2.132}$$

Example 2.59 A chemical plant produces three solvents, A, B, and C. The plant makes a profit of \$2.00, \$1.00, and \$4.00 on each unit of A, B, and C manufactured. The number of units of A, B, and C manufactured is x_1, x_2, and x_3 respectively. In Table 2.3, the raw material requirements for each product are given together with the total raw materials available.

TABLE 2.3

Raw Material	Units of Raw Material Required Per Unit of Final Product			Available Resources
	A	B	C	
Labor	4	5	10	100
Time (hours)	10	4	40	200

The objective of the problem is to find the values of x_1, x_2, and x_3 which maximize total profit.

This problem is expressed as a linear programming problem as follows.

$$2x_1 + x_2 + 4x_3 = \max \tag{2.133}$$

subject to

$$\begin{aligned} 4x_1 + 5x_2 + 10x_3 &\leq 100 \\ 10x_1 + 4x_2 + 40x_3 &\leq 200 \qquad x_i \geq 0, \quad i = 1, 2, 3 \end{aligned} \tag{2.134}$$

The space of feasible solutions (solutions satisfying the constraints) is shown in Fig. 2.10. The objective is to find the combination of values of x_1, x_2, and x_3 satisfying the constraints of Eqs. (2.134) which maximize Eq. (2.133).

2.11 APPLICATIONS

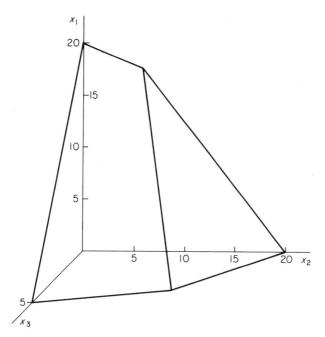

Figure 2.10 Solution Space for Example 2.59

Equation (2.133) defines a set of parallel planes. Two of these planes are shown in Fig. 2.11 and are defined by

$$2x_1 + x_2 + 4x_3 = 80 \tag{2.135}$$

and

$$2x_1 + x_2 + 4x_3 = 36 \tag{2.136}$$

As shown in Fig. 2.11, the plane defined by Eq. (2.135) does not intersect the space of feasible solutions. Therefore any solutions on this plane are infeasible. However, the plane defined by Eq. (2.136) contains feasible solutions for all points defining the shaded area of this plane as shown in Fig. 2.11.

In the above example one could attempt, by trial and error, to find the maximizing plane. The solution would then be given by all combinations of values of x_1, x_2, and x_3 lying on the plane and in the space of feasible solutions. Even for this simple problem such an approach would be tedious. Fortunately the search for a maximum (minimum) can be reduced to the investigation of a finite number of points in the solution space.

Definition 2.35 An *extreme point* of a convex set is a point in the convex set which cannot be expressed by a convex combination of any two distinct points in the convex set.

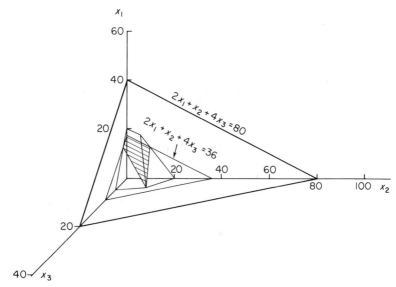

Figure 2.11 Two Hyperplanes Defined by $2x_1 + x_2 + 4x_3 = a$, $a = 80$, and $a = 36$

When the convex set of interest is defined by the intersection of several half spaces, then an extreme point is any point belonging to this intersection which lies at the intersection of n *bounding hyperplanes* where n is the dimension of the space.

Example 2.60 Define the extreme points of the convex set defined by Eqs. (2.134) in Example 2.59.

In Example 2.59, the half spaces and bounding hyperplanes for the convex set are given in Table 2.4.

TABLE 2.4

Bounding Hyperplane No.	Bounding Hyperplane Equation	Half Space
1	$4x_1 + 5x_2 + 10x_3 = 100$	$4x_1 + 5x_2 + 10x_3 \leq 100$
2	$10x_1 + 4x_2 + 40x_3 = 200$	$10x_1 + 4x_2 + 40x_3 \leq 200$
3	$x_1 = 0$	$x_1 \geq 0$
4	$x_2 = 0$	$x_2 \geq 0$
5	$x_3 = 0$	$x_3 \geq 0$

To determine the extreme points of the polyhedral convex set under study, it is necessary to determine the point of intersection of all combinations of three bounding hyperplanes. If such a point belongs to the space of feasible solutions then it is an extreme point. In this case there are five bounding hyperplanes and therefore ten combinations of three bounding hyperplanes each. The points of

2.11 APPLICATIONS

TABLE 2.5

Bounding Hyperplane Combination	Point of Intersection	Extreme Point?
1, 2, 3	$x_1 = 0$, $x_2 = 12.50$, $x_3 = 3.75$	Yes
1, 2, 4	$x_1 = 33.32$, $x_2 = 0$, $x_3 = -3.33$	No
1, 2, 5	$x_1 = 17.65$, $x_2 = 5.88$, $x_3 = 0$	Yes
1, 3, 4	$x_1 = 0$, $x_2 = 0$, $x_3 = 10$	No
1, 3, 5	$x_1 = 0$, $x_2 = 20$, $x_3 = 0$	Yes
1, 4, 5	$x_1 = 25$, $x_2 = 0$, $x_3 = 0$	No
2, 3, 4	$x_1 = 0$, $x_2 = 0$, $x_3 = 5$	Yes
2, 3, 5	$x_1 = 0$, $x_2 = 50$, $x_3 = 0$	No
2, 4, 5	$x_1 = 20$, $x_2 = 0$, $x_3 = 0$	Yes
3, 4, 5	$x_1 = 0$, $x_2 = 0$, $x_3 = 0$	Yes

intersection are defined in Table 2.5 and those which are extreme points are identified. Since each point in the above table is a point of intersection of three bounding hyperplanes, each point is an extreme point unless it does not belong to the space of feasible solutions.

Definition 2.36 A polyhedral convex set $\mathscr{X}_n$ is bounded if the elements of every vector, X_i, belonging to $\mathscr{X}_n$ are finite.

Example 2.61 Show that the polyhedral convex set, $\mathscr{X}_2$, defined by the intersection of the following half spaces is unbounded.

$$x_1 + x_2 \geq 5, \quad 3x_1 - x_2 \geq 3, \quad x_1 \geq 0, \quad x_2 \geq 0$$

The polyhedral convex set defined by the above half spaces is shown graphically in Fig. 2.12. Let X_i be the vector defined by

$$X_i = \begin{bmatrix} a + (b/3) \\ b \end{bmatrix}$$

where $a > 3$, $b > 0$. All such vectors belong to $\mathscr{X}_2$. Therefore

$$Y_i = \lim_{a \to \infty} X_i$$

also belongs to $\mathscr{X}_2$ and $\mathscr{X}_2$ is unbounded.

Example 2.62 Show that the polyhedral convex set, $\mathscr{X}_2$, defined by the intersection of the following half spaces is bounded.

$$x_1 + x_2 \leq 5, \quad 3x_1 - x_2 \leq 3, \quad x_1 \geq 0, \quad x_2 \geq 0$$

$\mathscr{X}_2$ is shown graphically in Fig. 2.13.

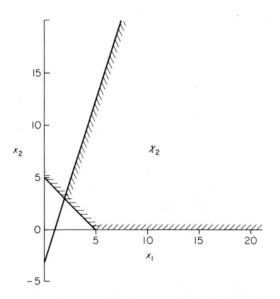

Figure 2.12 Polyhedral Convex Set, $\mathscr{X}_2$, in Example 2.61

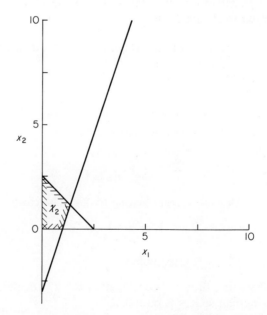

Figure 2.13 Polyhedral Convex Set, $\mathscr{X}_2$, in Example 2.62

2.11 APPLICATIONS

Let X_i be a vector of the form

$$X_i = \begin{bmatrix} a \\ b \end{bmatrix}$$

If $a < 0$ or $a > 3$, X_i cannot belong to $\mathscr{X}_2$. Similarly, if $b < 0$ or $b > 5$, X_i cannot belong to $\mathscr{X}_2$. Therefore, the elements of every vector belonging to $\mathscr{X}_2$ must be finite and $\mathscr{X}_2$ is bounded.

Theorem 2.31 If $\mathscr{X}_n$ is a bounded polyhedral convex set with extreme points E_1, $E_2, \ldots, E_m$, then every point in $\mathscr{X}_n$ can be expressed by a convex combination of the extreme points.

Theorem 2.32 The maximizing (minimizing) vector, X, for the problem

$$CX = \max(\min)$$

subject to

$$AX \leq B, \quad X \geq 0$$

where A, B, C, and X are defined in Eqs. (2.129)–(2.132), lies at an extreme point of the polyhedral convex set, $\mathscr{X}_n$, defined by the constraints $AX \leq B$, $X \geq 0$.

Proof Let the extreme points of $\mathscr{X}_n$ be $E_1, E_2, \ldots, E_m$ and let X_i be any point belonging to $\mathscr{X}_n$. By Theorem 2.31, X_i can be represented by a convex combination of the extreme points of $\mathscr{X}_n$.

$$X_i = \sum_{i=1}^{m} \lambda_i E_i$$

where $0 \leq \lambda_i \leq 1$ for all i and $\sum_{i=1}^{m} \lambda_i = 1$. Therefore

$$CX_i = C\left(\sum_{i=1}^{m} \lambda_i E_i\right) = \sum_{i=1}^{m} \lambda_i CE_i$$

Let E_n be the extreme point such that

$$CE_n = \max_i CE_i$$

Then

$$\sum_{i=1}^{m} \lambda_i CE_i \leq \sum_{i=1}^{m} \lambda_i CE_n = CE_n$$

It follows that if X_i is a point belonging to $\mathscr{X}_n$ then

$$CX_i \leq CE_n$$

Therefore, there is an extreme point, E_n for $\mathscr{X}_n$ at which CX is maximized.

The proof of the theorem for the case of minimization is left as an exercise for the reader.

Based upon the results of Theorem 2.32, the linear programming problem can be solved by evaluating the function CX at each extreme point of the solution space. For simple linear programming problems, this task is neither difficult nor time consuming. However, for most practical problems involving many variables and constraints, this approach can be useless. Fortunately an iterative technique is available which allows the analyst to carry out the search for an optimum solution in a more systematic and less time consuming manner. This technique is called the *simplex method* and can be found in any text on linear programming.

Example 2.63 Find the vector, X, which maximizes the objective function for the problem given in Example 2.59.

For the problem given in Example 2.59, there are six extreme points. To determine the maximizing vector, the objective function is evaluated at each of the extreme points. The required calculations are summarized in Table 2.6. The extreme points for this problem were defined in Example 2.60.

TABLE 2.6

Extreme Point	Value of the Objective Function
$x_1 = 0$, $x_2 = 12.5$, $x_3 = 3.75$	27.50
$x_1 = 17.65$, $x_2 = 5.88$, $x_3 = 0$	41.18
$x_1 = 0$, $x_2 = 20$, $x_3 = 0$	20.00
$x_1 = 0$, $x_2 = 0$, $x_3 = 5$	20.00
$x_1 = 20$, $x_2 = 0$, $x_3 = 0$	40.00
$x_1 = 0$, $x_2 = 0$, $x_3 = 0$	0.00

The maximizing vector is

$$X = \begin{bmatrix} 17.65 \\ 5.88 \\ 0.00 \end{bmatrix}$$

2.11.5 Engineering Applications

Example 2.64 An electrical circuit is composed of three loops as shown in Fig. 2.14 where E is the electromotive force in volts, and R_i is the value of the ith resistance in ohms. The direction of current flow in the three loops is shown by the arrows. Let I_1, I_2, and I_3 be the current flowing in amperes in the three loops. If

$$R_1 = 2, \quad R_5 = 7, \quad E = 100$$
$$R_2 = 1, \quad R_6 = 2$$
$$R_3 = 4, \quad R_7 = 2$$
$$R_4 = 6, \quad R_8 = 5$$

find the current in each loop.

2.11 APPLICATIONS

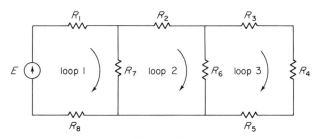

Figure 2.14

The voltage, V_i, across any resistance, R_i, is given by

$$V_i = R_i N_i$$

where N_i is the net current through R_i. Thus

$$V_1 = R_1 I_1$$

while

$$V_7 = R_7(I_1 - I_2)$$

when loop 1 is considered since the current through R_7 is the difference between I_1 and I_2, positive current flow in a given loop being that indicated by the direction of the associated arrow. Hence for loop 2

$$V_7 = R_7(I_2 - I_1)$$

Since the sum of the voltage around each loop is zero, we have

$$R_1 I_1 + R_7(I_1 - I_2) + R_8 I_1 - E = 0$$
$$R_2 I_2 + R_6(I_2 - I_3) + R_7(I_2 - I_1) = 0$$
$$R_3 I_3 + R_4 I_3 + R_5 I_3 + R_6(I_3 - I_2) = 0$$

or

$$I_1(R_1 + R_7 + R_8) - I_2 R_7 = E$$
$$-I_1 R_7 + I_2(R_2 + R_6 + R_7) - I_3 R_6 = 0$$
$$-I_2 R_6 + I_3(R_3 + R_4 + R_5 + R_6) = 0$$

Thus

$$9I_1 - 2I_2 \qquad\quad = 100$$
$$-2I_1 + 5I_2 - 2I_3 = 0$$
$$-2I_2 + 19I_3 = 0$$

and

$$\begin{bmatrix} 9 & -2 & 0 \\ -2 & 5 & -2 \\ 0 & -2 & 19 \end{bmatrix} \begin{bmatrix} I_1 \\ I_2 \\ I_3 \end{bmatrix} = \begin{bmatrix} 100 \\ 0 \\ 0 \end{bmatrix}$$

Solving for I_1, I_2, and I_3 yields

$$\begin{bmatrix} I_1 \\ I_2 \\ I_3 \end{bmatrix} = \frac{1}{743} \begin{bmatrix} 91 & 38 & 4 \\ 38 & 171 & 18 \\ 4 & 18 & 41 \end{bmatrix} \begin{bmatrix} 100 \\ 0 \\ 0 \end{bmatrix} = \begin{bmatrix} 12.25 \\ 5.11 \\ 0.54 \end{bmatrix}$$

Example 2.65 A unit of material is subjected to three mutually perpendicular normal stresses, s_x, s_y, and s_z. The corresponding strains are σ_x, σ_y, and σ_z respectively and are given by

$$\sigma_x = \frac{1}{E}[s_x - \mu s_y - \mu s_z], \qquad \sigma_y = \frac{1}{E}[-\mu s_x + s_y - \mu s_z],$$

$$\sigma_z = \frac{1}{E}[-\mu s_x - \mu s_y + s_z]$$

where E is the modulus of elasticity and μ is Poisson's ratio. Derive expressions for s_x, s_y, and s_z in terms of σ_x, σ_y, σ_z, E, and μ.

$$\Sigma = \begin{bmatrix} \sigma_x \\ \sigma_y \\ \sigma_z \end{bmatrix}, \qquad S = \begin{bmatrix} s_x \\ s_y \\ s_z \end{bmatrix}, \qquad M = \begin{bmatrix} 1 & -\mu & -\mu \\ -\mu & 1 & -\mu \\ -\mu & -\mu & 1 \end{bmatrix}$$

Then

$$\Sigma = \frac{1}{E} MS \qquad \text{and} \qquad S = EM^{-1}\Sigma$$

Now

$$|M| = 1 - 2\mu^3 - 3\mu^2 \quad \text{and} \quad \text{adj}(M) = (1 + \mu) \begin{bmatrix} (1-\mu) & \mu & \mu \\ \mu & (1-\mu) & \mu \\ \mu & \mu & (1-\mu) \end{bmatrix}$$

Hence

$$M^{-1} = \frac{1}{(1+\mu)(1-2\mu)} \begin{bmatrix} (1-\mu) & \mu & \mu \\ \mu & (1-\mu) & \mu \\ \mu & \mu & (1-\mu) \end{bmatrix}$$

and

$$S = \frac{E}{(1+\mu)(1-2\mu)} \begin{bmatrix} (1-\mu) & \mu & \mu \\ \mu & (1-\mu) & \mu \\ \mu & \mu & (1-\mu) \end{bmatrix} \begin{bmatrix} \sigma_x \\ \sigma_y \\ \sigma_z \end{bmatrix}$$

$$= \begin{bmatrix} E \dfrac{(1-\mu)\sigma_x + \mu(\sigma_y + \sigma_z)}{(1+\mu)(1-2\mu)} \\ E \dfrac{(1-\mu)\sigma_y + \mu(\sigma_x + \sigma_z)}{(1+\mu)(1-2\mu)} \\ E \dfrac{(1-\mu)\sigma_z + \mu(\sigma_x + \sigma_y)}{(1+\mu)(1-2\mu)} \end{bmatrix}$$

or

$$S_x = A + \frac{E\sigma_x}{(1+\mu)}, \quad S_y = A + \frac{E\sigma_y}{(1+\mu)}, \quad S_z = A + \frac{E\sigma_z}{(1+\mu)}$$

where

$$A = \frac{E\mu(\sigma_x + \sigma_y + \sigma_z)}{(1+\mu)(1-2\mu)}$$

Problems

1. For the matrices A, B, and C, determine which of the following products are defined
   a. AB b. BA c. AC
   d. CA e. BC f. CB

   where

   $$A = \begin{bmatrix} a_{11} \\ a_{21} \\ a_{31} \end{bmatrix} \quad B = \begin{bmatrix} b_{11} & b_{12} & b_{13} \end{bmatrix} \quad C = \begin{bmatrix} c_{11} & c_{12} & c_{13} \\ c_{21} & c_{22} & c_{23} \\ c_{31} & c_{32} & c_{33} \end{bmatrix}$$

2. Let

   $$A = \begin{bmatrix} 1 & 0 & 1 \\ 0 & 1 & 0 \\ 1 & 0 & 1 \end{bmatrix}, \quad B = \begin{bmatrix} 2 \\ 1 \\ 2 \end{bmatrix}, \quad C = \begin{bmatrix} 2 & 1 \\ 1 & 1 \\ 0 & 1 \end{bmatrix}$$

   Evaluate those products AB, BA, AC, CA, BC, CB which are defined.

3. Which of the following matrices are nonsingular?

   $$A = \begin{bmatrix} 3 & 1 & 4 \\ 1 & 1 & 1 \\ 2 & 1 & 3 \end{bmatrix}, \quad B = \begin{bmatrix} 1 & 0 & -1 \\ 0 & -1 & 0 \\ 1 & 0 & 1 \end{bmatrix}$$

   $$C = \begin{bmatrix} 3 & 0 & 6 \\ 0 & 2 & 2 \\ 1 & 2 & 4 \end{bmatrix}, \quad D = \begin{bmatrix} 1 & 1 & 2 & 0 \\ 3 & 1 & 1 & 0 \\ 2 & 1 & 2 & 0 \\ 4 & 1 & 1 & 1 \end{bmatrix}$$

4. Find the inverse of each of the matrices in Problem 3 which are nonsingular.
5. Prove Theorem 2.9 using Theorem 2.8.
6. Find the inverse of each of the following matrices

   $$A = \begin{bmatrix} 2 & 0 & 1 & 0 \\ 1 & 2 & 1 & 1 \\ 3 & 1 & 2 & 1 \\ 1 & 0 & 4 & 1 \end{bmatrix}, \quad B = \begin{bmatrix} 2 & 1 & 1 & 1 \\ 0 & 1 & 2 & 4 \\ 0 & 1 & 5 & 1 \\ 2 & 2 & 1 & 0 \end{bmatrix}$$

7. Determine the rank of the following matrices.

   $$A = \begin{bmatrix} 2 & 10 & 8 \\ 1 & 4 & 6 \\ 2 & 8 & 12 \end{bmatrix}, \quad B = \begin{bmatrix} 2 & 6 & 0 \\ 1 & 3 & 0 \\ 1 & 1 & 0 \end{bmatrix}, \quad C = \begin{bmatrix} 4 & 1 & 2 \\ 3 & 1 & 1 \\ 1 & 1 & 1 \end{bmatrix}$$

8. Prove Theorem 2.8 when two columns are interchanged.
9. By counterexample show that $|A \pm B|$ is not necessarily equal to $|A| \pm |B|$.

10. Show that the adjoint of a symmetric matrix is a symmetric matrix.
11. Show that the inverse of a nonsingular symmetric square matrix is a symmetric matrix.
12. Prove the Schwarz inequality.
$$X_1^T X_2 \leq \|X_1\| \|X_2\|$$
13. Prove the triangle inequality
$$\|X_1 + X_2\| \leq \|X_1\| + \|X_2\|$$
14. Find the vector X such that
$$AX = B, \quad \text{where} \quad A = \begin{bmatrix} 4 & 1 \\ 3 & 4 \end{bmatrix}, \quad B = \begin{bmatrix} 10 \\ 1 \end{bmatrix}$$
15. Find the vector X such that
$$AX = B, \quad \text{where} \quad A = \begin{bmatrix} 1 & 0 & 2 \\ 0 & 2 & 0 \\ 2 & 0 & 1 \end{bmatrix}, \quad B = \begin{bmatrix} 1 \\ 1 \\ 1 \end{bmatrix}$$
16. Find the vector X such that
$$AX = B, \quad \text{where} \quad A = \begin{bmatrix} 2 & 1 & 1 & 1 \\ 0 & 1 & 2 & 4 \\ 0 & 1 & 5 & 1 \\ 2 & 2 & 1 & 0 \end{bmatrix}, \quad B = \begin{bmatrix} 4 \\ 3 \\ 2 \\ 1 \end{bmatrix}$$
17. Let A, X, and B be $(n \times n)$, $(n \times 1)$, and $(n \times 1)$ matrices such that
$$AX = B$$
If A is nonsingular, then
$$X = A^{-1}B$$
Show that
$$x_k = \frac{|C|}{|A|}$$
where the matrix C is identical to A except that the kth column of A is replaced by the vector B. The expression for x_k is called Cramer's rule.
18. Let A be a square matrix of order n. Show that
$$|A^T| = |A|$$
19. Let $\mathscr{X}_n$ be a finite set of m n-dimensional vectors, where $m > 1$. Show that $\mathscr{X}_n$ is not a vector space.
20. $\mathscr{X}_n$ is the set of all n-dimensional vectors, X_i, having elements $x_{ij} > 0$. Is $\mathscr{X}_n$ a vector space?
21. Show that the zero vector is a vector space.
22. Let $\mathscr{X}_n$ be a vector space of vectors X_i. Show that
$$\sum_{i=1}^{m} X_i \in \mathscr{X}_n$$
23. Let $\mathscr{X}_n$ be a set of n-dimensional vectors. Let
$$Y_i = k_1 X_1 + k_2 X_2$$
where k_1 and k_2 are arbitrary scalars and
$$X_1 \in \mathscr{X}_n, \quad X_2 \in \mathscr{X}_n$$
Show that if Y_i belongs to $\mathscr{X}_n$ for all k_1, k_2, X_1, and X_2, then $\mathscr{X}_n$ is a vector space.

PROBLEMS

24. Show that the following vectors are mutually orthogonal.

$$X_1 = \begin{bmatrix} -1 \\ 1 \\ -1 \\ 1 \end{bmatrix}, \quad X_2 = \begin{bmatrix} 1 \\ 1 \\ -1 \\ -1 \end{bmatrix}, \quad X_3 = \begin{bmatrix} -1 \\ 1 \\ 1 \\ -1 \end{bmatrix}, \quad X_4 = \begin{bmatrix} 1 \\ 1 \\ 1 \\ 1 \end{bmatrix}$$

25. For the set of vectors given in Problem 24, show that there is no other vector of nonzero length which is orthogonal to X_1, X_2, X_3, and X_4.
26. Show that the space, $\mathscr{X}_2$, shown in Fig. 2.5 is not a vector space.
27. Show that a vector space is convex.
28. Prove Theorem 2.31 when $\mathscr{X}_n$ is a polyhedral convex set in three dimensions.
29. Prove Theorem 2.32 when CX is to be minimized.
30. Let $X_1, X_2, \ldots, X_{n-1}$ be a set of mutually orthogonal n-dimensional vectors. Let Y_1 and Y_2 be nonzero n-dimensional vectors which are both orthogonal to $X_1, X_2, \ldots, X_{n-1}$. Show that Y_2 is a scalar multiple of Y_1.
31. Let A be a symmetric idempotent matrix. Show that either

$$|A| = 0 \quad \text{or} \quad A = I$$

32. Determine the eigenvalues for the following matrices.

$$A = \begin{bmatrix} 1 & 1 & 0 \\ 1 & 2 & 0 \\ 0 & 0 & 3 \end{bmatrix}, \quad B = \begin{bmatrix} 1 & 2 \\ 2 & 6 \end{bmatrix}, \quad C = \begin{bmatrix} 4 & 0 & 2 \\ 0 & -1 & 0 \\ 2 & 0 & 3 \end{bmatrix}$$

$$D = \begin{bmatrix} 2 & 0 & 0 & 0 \\ 0 & 1 & 0 & 0 \\ 0 & 0 & 4 & 0 \\ 0 & 0 & 0 & 3 \end{bmatrix}$$

33. Find the eigenvalues for the following matrix.

$$\begin{bmatrix} 9 & 0 & 2 \\ 0 & 1 & 0 \\ 2 & 0 & 3 \end{bmatrix}$$

34. Find a set of mutually orthogonal eigenvectors for the matrices given in Problem 32. A & C only
35. Let A, C, and D in Problem 32 be the matrices of three quadratic forms. Determine whether the quadratic forms corresponding to A, C, and D are positive definite, negative definite, or indefinite.
36. Express the quadratic form q as a sum of squares where

$$q = 2x_1^2 + x_2^2 + 3x_3^2 + 4x_1 x_3$$

37. Given the following quadratic form $f(x_1, x_2, x_3) = 4x_1^2 + 2x_2^2 + x_3^2 + 6x_1 x_2 + 2x_1 x_3 + 2x_2 x_3$
    a. find the matrix, A, of the quadratic form expressed by

$$f(X) = X^T A X$$

where

$$X = \begin{bmatrix} x_1 \\ x_2 \\ x_3 \end{bmatrix}$$

   b. determine whether the quadratic form is positive definite, negative definite, positive semidefinite, negative semidefinite, or indefinite;
   c. express the quadratic form as a sum of squares.
38. Let B be a square matrix with columns a set of mutually orthogonal eigenvectors for a symmetric matrix A. By counterexample show that B is not necessarily an orthogonal matrix.

39. Let
$$q_1 = X^T A X, \qquad q_2 = X^T C X$$
where A and C are defined in Problem 32. By an orthogonal transformation, express q_1 and q_2 as a sum of squares.

40. Let A be a symmetric matrix of order n and B an orthogonal matrix such that
$$R = B^T A B$$
Show that
$$|R| = |A|$$

41. Prove Theorem 2.29.
42. Prove Theorem 2.30.
43. Let q_1 and q_2 be the quadratic forms defined in Problem 39. By means of a nonorthogonal transformation, reduce q_1 and q_2 to a sum of squares.

References

Buck, R. C., *Advanced Calculus*, New York: McGraw-Hill, 1956.
Fulks, W., *Advanced Calculus*, New York: Wiley (Interscience), 1961.
Gass, S. I., *Linear Programming*, New York: McGraw-Hill, 1964.
Graybill, F. A., *An Introduction to Linear Statistical Models*, Volume I, New York: McGraw-Hill, 1961.
Gue, R. L., and Thomas, M. E., *Mathematical Methods in Operations Research*, New York: Macmillan, 1968.
Hadley, G., *Linear Algebra*, Reading, Massachusetts: Addison-Wesley, 1961.
Hadley, G., *Linear Programming*, Reading, Massachusetts: Addison-Wesley, 1962.
Hohn, F. E., *Elementary Matrix Algebra*, New York: Macmillan, 1958.
Kemeny, J. B., Mirkil, H., Snell, J. L., and Thompson, G. L., *Finite Mathematical Structures*, Englewood Cliffs, New Jersey: Prentice-Hall, 1959.
Kemeny, J. G., and Snell, J. L., *Finite Markov Chains*, New York: Van Nostrand, 1960.
Parzen, E., *Stochastic Processes*, San Francisco: Holden-Day, 1962.
Perlis, S., *Theory of Matrices*, Reading, Massachusetts: Addison-Wesley, 1952.
Teichroew, D., *An Introduction to Management Science, Deterministic Models*, New York: Wiley, 1964.

CHAPTER 3

REAL ANALYSIS

3.1 Introduction

In this chapter the mathematical concepts forming a foundation for the material presented in subsequent chapters will be discussed. The reader with an understanding of basic calculus will undoubtedly be familiar with much of this material, although possibly only in a vague sense. However, maximum benefit may not be gained from the chapters which follow if the reader has only a casual acquaintance with the material presented here, and therefore a formal discussion of the fundamental concepts of real analysis is in order.

3.2 Sets

Definition 3.1 A set is a well-defined collection of elements.

The composition of a set is defined by the elements which are included in the set. To illustrate, the set denoted by A includes all those products manufactured by a given company. If the product a is produced by the company then it is an element of A or a is said to belong to A. Membership of element a in set A can be expressed by

$$a \in A \qquad (3.1)$$

Equation (3.1) reads "a belongs to A." If b is a product which is not manufactured by the company, then

$$b \notin A \qquad (3.2)$$

or b *does not belong to* A. As indicated in Definition 3.1 and the preceding discussion, a set is defined only when it is possible to define those elements which do and do not belong to the set. This is the implication of the phrase *well-defined collection* in Definition 3.1.

A set may be defined by listing all of those elements which belong to the set or by specifying the characteristics which an element must possess to belong to the set. For example, the set A might be defined by

$$A = \text{integer } x, \quad 0 \le x \le 10 \quad \text{or} \quad A = 0, 1, 2, 3, 4, 5, 6, 7, 8, 9, 10$$

In both cases the set A is composed of all integers between zero and ten inclusive.

Let A and B be two sets such that *all those elements not contained in A are contained in B*. If we combine the two sets to form a new set, this set is called the *universe* and is usually denoted by U. Ordinarily the universe is defined with reference to the characteristics of the set of interest. For example, suppose the set A includes all the integers between -10 and $+10$ inclusive. We might define U as the set of all integers between $-\infty$ and ∞. On the other hand, U might be the set of all real numbers, both integer and noninteger. However, it is unlikely that one would define U as, for example, everything in the world, including all numbers, people, trees, mountains, etc. Therefore, one usually defines the universe in terms of elements which possess some characteristic. The set A, which is part of the universe, contains elements having the characteristic by which the universe is defined and which have some other property or properties which distinguish the elements of A from those elements in the universe which do not belong to A.

Definition 3.2 Let A and B be two sets.

a. If every element of A is also contained in B, then A is said to be a subset of B and is expressed by

$$A \subset B \tag{3.3}$$

b. If C is a set composed of all the elements in either A or B, then C is said to be the union of A and B and is expressed by

$$C = A \cup B \tag{3.4}$$

or

$$C = A + B \tag{3.5}$$

c. If C is a set composed of all the elements belonging to both A and B, then C is said to be the intersection of A and B and is expressed by

$$C = A \cap B \tag{3.6}$$

or

$$C = AB \tag{3.7}$$

d. If A is a set which contains no elements, then A is said to be *empty* or *null* and is expressed by

$$A = \emptyset \tag{3.8}$$

e. The empty set belongs to every set.

f. Two sets, A and B, are said to be equal if each element in A also belongs to B and if each element in B also belongs to A. That is, if $A \subset B$ and $B \subset A$ then

$$A = B \tag{3.9}$$

g. Let C be the difference between two sets A and B, expressed by
$$C = A - B \tag{3.10}$$
Then C contains those elements belonging to A and not belonging to B. Therefore,
$$A - B = A - A \cap B \tag{3.11}$$

h. The complement of set A, $\bar{A}$, is defined as all those elements in the universe, U, which are not contained in A. $\bar{A}$ is often read *not A* and is expressed by
$$\bar{A} = U - A \tag{3.12}$$

Example 3.1 Let the sets A, B, C, and D be defined as follows:

A = all persons living in Baltimore;
B = all persons living in Washington;
C = all persons working in Baltimore;
D = all persons working in Washington;
$U = A \cup B \cup C \cup D$.

Define

a. $A \cup C$;
b. $A \cap C$;
c. $A \cap D$;
d. $U - A \cup C$.

Part a The union of A and C includes all persons living in Baltimore and all persons working in Baltimore. Therefore, if x is a person, then
$$x \in A \cup C$$
if x either works or lives in Baltimore.

Part b The intersection of A and C includes all persons both living and working in Baltimore. Therefore,
$$x \in A \cap C$$
if x lives and works in Baltimore.

Part c The intersection of A and D includes all persons living in Baltimore but working in Washington. Therefore
$$x \in A \cap D$$
if x lives in Baltimore but works in Washington.

Part d $A \cup C$ is the union of A and C or all persons either living or working in Baltimore. $U - A \cup C$ is then the complement of $A \cup C$ or $\overline{A \cup C}$. If
$$x \in \overline{A \cup C}$$

then x neither works nor lives in Baltimore and therefore by definition of the universe, x lives and works in Washington and

$$U - A \cup C = B \cap D$$

Example 3.2 Let the production lines in a plant, U, be classified as belonging to one or more of the following sets.

A = lines which are producing;
B = lines which have failed;
C = lines which have failed and are under repair;
D = lines which are not producing.

Define the following sets

a. $A \cup D$
b. $A \cap D$
c. $B \cap C \cap D$
d. $D - B \cup C$
e. $D \cup B \cap C$

Part a The union of A and D includes all lines which are either producing or not producing. Therefore

$$A \cup D = U$$

Part b The intersection of A and D includes all lines which are both producing and not producing and

$$A \cap D = \emptyset$$

Part c The intersection of B and C includes all lines which are down due to failure and under repair. However set C includes all such lines and

$$B \cap C = C$$

But lines which are under repair belong to the set of lines which are not producing. Therefore

$$C \subset D \quad \text{and} \quad B \cap C \cap D = C$$

Part d The union of B and C includes all lines which are down due to failure or under repair. However lines which have failed include those under repair. Therefore

$$C \subset B \quad \text{and} \quad B \cup C = B$$

Then

$$D - B \cup C = D - B$$

Further

$$B \subset D$$

Hence $D - B$ includes those lines which are down but have not failed.

Part e From Part c,
$$B \cap C = C$$
Since D includes all nonproductive lines, D includes C. Therefore
$$D \cup B \cap C = D \cup C = D \qquad \blacksquare$$

The following laws of set operation are often useful in simplifying the definition of a set.

Commutative laws
$$A \cup B = B \cup A \qquad (3.13)$$
$$A \cap B = B \cap A \qquad (3.14)$$

Associative laws
$$A \cup (B \cup C) = (A \cup B) \cup C \qquad (3.15)$$
$$A \cap (B \cap C) = (A \cap B) \cap C \qquad (3.16)$$

Distributive laws
$$A \cup (B \cap C) = (A \cup B) \cap (A \cup C) \qquad (3.17)$$
$$A \cap (B \cup C) = (A \cap B) \cup (A \cap C) \qquad (3.18)$$

Example 3.3 Show that

a. $A \cup (A \cap B) = A$;
b. $A \cap (A \cup B) = A$;
c. $\overline{A \cup B} = \bar{A} \cap \bar{B}$;
d. $A \cap \emptyset = \emptyset$.

Part a From the distributive laws
$$A \cup (A \cap B) = (A \cup A) \cap (A \cup B)$$
But
$$A \cup A = A$$
Therefore
$$A \cup (A \cap B) = A \cap (A \cup B)$$
If $x \in A \cap (A \cup B)$ then $x \in A$ and $x \in (A \cup B)$. Thus $x \in A$ and
$$A \cup (A \cap B) = A$$

Part b From Part a
$$A \cap (A \cup B) = A$$

Part c $A \cup B$ includes all elements belonging to A or B. Therefore $\overline{A \cup B}$ includes all elements which do not belong to either A or B. The elements which do not belong to A belong to $\bar{A}$ and the elements which do not belong to B belong to $\bar{B}$. Therefore, the elements belonging to $\overline{A \cup B}$ belong to both $\bar{A}$ and $\bar{B}$. Thus $\overline{A \cup B} = \bar{A} \cap \bar{B}$.

Part d The elements belonging to $A \cap \varnothing$ belong to both A and the empty set $\varnothing$. But $\varnothing \subset A$. Therefore

$$A \cap \varnothing = \varnothing$$

A convenient method of representing sets and their interrelationships is the *Venn diagram*. Let A, B, and C be three sets included in the universe, U, and such that $U \neq A \cup B \cup C$. The universe is represented by a rectangle and the sets A, B, and C by circles within the rectangle as in Fig. 3.1. In this figure the intersections $A \cap B$, $A \cap C$, $B \cap C$, and $A \cap B \cap C$ are not empty. If we wished to represent the sets A, B, and C such that $A \cap B$, $A \cap C$, and $B \cap C$ are not empty but $A \cap B \cap C$ is empty, we would be led to Fig. 3.2. In Fig. 3.2 the shaded areas represent the intersections $A \cap B$, $A \cap C$, and $B \cap C$. In constructing Venn diagrams the use of circles to represent sets is arbitrary and any convenient form might be used.

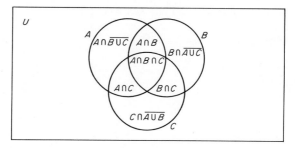

Figure 3.1 Venn Diagram

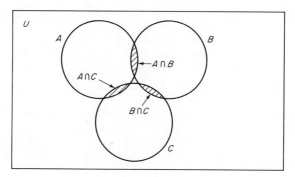

Figure 3.2 Venn Diagram Where $A \cap B \cap C = \varnothing$

3.2 SETS   213

Example 3.4 Draw Venn diagrams showing the relationships among the sets A, B, C, and D where

    a. $A \cap B \neq \emptyset$, $A \cap C = \emptyset$, $A \cap D \neq \emptyset$, $B \cap C \neq \emptyset$, $B \cap D = \emptyset$, $C \cap D \neq \emptyset$;
    b. $A \cap B \neq \emptyset$, $A \cap C \neq \emptyset$, $A \cap D \neq \emptyset$, $B \cap C = \emptyset$, $B \cap D = \emptyset$, $C \cap D = \emptyset$;
    c. $A \cup B = A$, $B \cup C = B$, $C \cup D = C$.

In all cases assume that the sets A, B, C, and D are not empty.

Part a The two-way intersections between any two sets are given in the statement of the problem. We determine the three-way intersections as follows.

$$A \cap B \cap C = B \cap (A \cap C) \quad \text{but} \quad A \cap C = \emptyset$$

Therefore

$$A \cap B \cap C = B \cap \emptyset = \emptyset$$

By a similar argument it can be shown that

$$A \cap B \cap D = \emptyset, \quad A \cap C \cap D = \emptyset, \quad B \cap C \cap D = \emptyset,$$
$$A \cap B \cap C \cap D = \emptyset$$

The relationships which must be represented in the Venn diagram are the intersections between A and B, B and C, C and D, and D and A, and are shown in Fig. 3.3.

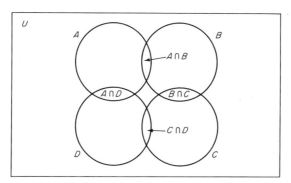

Figure 3.3   Venn Diagram for Example 3.4, Part a

Part b Since the intersections between all combinations of two sets are given, we need only consider the intersections among all combinations of three sets and the intersections among all four sets.

$$A \cap B \cap C = A \cap (B \cap C) = A \cap \emptyset = \emptyset$$
$$A \cap B \cap D = A \cap (B \cap D) = A \cap \emptyset = \emptyset$$
$$A \cap C \cap D = A \cap (C \cap D) = A \cap \emptyset = \emptyset$$
$$B \cap C \cap D = B \cap (C \cap D) = B \cap \emptyset = \emptyset$$
$$A \cap B \cap C \cap D = (A \cap B) \cap (C \cap D) = (A \cap B) \cap \emptyset = \emptyset$$

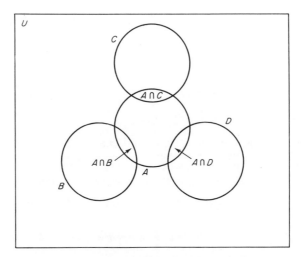

Figure 3.4 Venn Diagram for Example 3.4, Part b

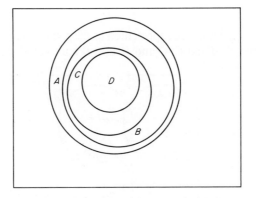

Figure 3.5 Venn Diagram for Example 3.4, Part c

The Venn diagram is given in Fig. 3.4.

Part c Since $A \cup B = A$ the set B must be a subset of A. By similar argument C is a subset of B and D is a subset of C. Therefore $D \subset C \subset B \subset A$ and the Venn diagram is as shown in Fig. 3.5.

Example 3.5 An inspection of 200 units of product led to the following results. Defect A was found in 30 units, defect B in 10 units and defect C in 20. Two units were found with both defects A and B, 5 with defects A and C, and 3 with defects B and C. All three defects were found in 2 units. How many units possessed

a. defects A and B but not C?
b. defects B and C but not A?

c. defect A only?
d. defect C only?
e. no defects?

Part a The number of units having defects A and B but not C is the number of units in the set $A \cap B - A \cap B \cap C$ or $2 - 2 = 0$.

Part b Here we are interested in the number of units in the set $B \cap C - A \cap B \cap C$. Therefore, the number of units containing defects B and C but not A is $3 - 2 = 1$.

Part c To determine the number of units containing defect A only, we must subtract from the number of units having A, the number having A and B only, the number having A and C only, and the number having A, B, and C. The number having defects A and B only is the number of units in the set $A \cap B - A \cap B \cap C$. Similarly, the number having A and C only is the number of units in the set $A \cap C - A \cap B \cap C$. Therefore, the number of units having defect A only is the number of units belonging to the set

$$A - (A \cap B - A \cap B \cap C) - (A \cap C - A \cap B \cap C) - A \cap B \cap C$$

which reduces to the set

$$A - A \cap B - A \cap C + A \cap B \cap C$$

Therefore, the number of units containing defect A only is $30 - 2 - 5 + 2 = 25$.

Part d By an argument similar to that in Part c, the number of units containing defect C is the number of units belonging to the set

$$C - A \cap C - B \cap C + A \cap B \cap C$$

and there are $10 - 5 - 3 + 2 = 4$ such units.

Part e To determine the number of units containing no defects we subtract from the number of units in the universe, 200, the number having A only, B only, C only, A and B only, A and C only, B and C only, and A, B, and C. The number of such units is the number of units belonging to the set.

$$\begin{aligned}
& U - (A - A \cap B - A \cap C + A \cap B \cap C) \\
& - (B - A \cap B - B \cap C + A \cap B \cap C) \\
& - (C - A \cap C - B \cap C + A \cap B \cap C) - (A \cap B - A \cap B \cap C) \\
& - (A \cap C - A \cap B \cap C) - (B \cap C - A \cap B \cap C) - A \cap B \cap C
\end{aligned}$$

Simplifying the expression we obtain the set

$$U - A - B - C + A \cap B + A \cap C + B \cap C - A \cap B \cap C$$

and the number of units containing zero defects is $200 - 30 - 10 - 20 + 2 + 5 + 3 - 2$ or 148 units.

3.3 Real Numerical Sets

In the previous section a set was defined as a well-defined collection of elements, and in this context an element could be any definable entity. In this section a set may include only real numbers as elements. When the set includes all the real numbers between two numbers, a and b, the set is called an interval. The numbers a and b are called end points or boundary points for the interval. We will deal with three classes of intervals in this and the remaining chapters.

Definition 3.3 An open interval is an interval which does not contain its boundary points and is denoted by (a, b), where a and b are the boundary points. Therefore

$$x \in (a, b) \quad \text{if and only if} \quad a < x < b$$

A closed interval is an interval which contains its boundary points and is denoted by $[a, b]$, where a and b are the boundary points. Therefore

$$x \in [a, b] \quad \text{if and only if} \quad a \leq x \leq b$$

As indicated in the above definition, to define an interval it is necessary to define the boundary points, a and b, and to specify whether or not these points are to be included in the interval. An interval thus defined may be represented by a straight line of length $b - a$ and is therefore sometimes referred to as a one-dimensional space or simply as one space. If the elements of a set are defined by two numbers, x and y, the set is called a two-dimensional space or two space. As in the case of intervals, a two-dimensional space is defined only when the boundary points of the space are defined. For example, if $\mathscr{X}_2$ is a two-dimensional space with elements the numbers x, y and boundaries defined by $2 \leq x \leq 10, 3 \leq y \leq 6$, then $\mathscr{X}_2$ is closed, denoted $[\mathscr{X}_2]$, and can be described by the rectangle shown in Fig. 3.6. Therefore, if x and y are numbers, the element $[x, y]$ belongs to $[\mathscr{X}_2]$ if and only if $2 \leq x \leq 10$ and $3 \leq y \leq 6$. Based upon this definition of $[\mathscr{X}_2]$, the element or point $[3, 6]$ belongs to $[\mathscr{X}_2]$ while the $[3, 8]$ does not.

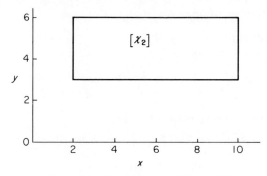

Figure 3.6 Two-Dimensional Space $[\mathscr{X}_2]$

3.3 REAL NUMERICAL SETS

A space which does not include its boundary points is called an open space. In the example just discussed, if $\mathscr{X}_2$ were defined as a space with elements the points $[x, y]$ such that $2 < x < 10, 3 < y < 6$, then the space is open and denoted by $(\mathscr{X}_2)$. In this case the point $[3, 6]$ does not belong to $(\mathscr{X}_2)$.

It is not necessary that an interval or space be either open or closed. To illustrate the interval defined by $10 < x \leq 20$ contains one but not both of its boundary points and as such, this interval does not fit either of the definitions given for open and closed intervals. Therefore, any interval or space which contains some but not all of its boundary points is neither closed nor open. In the case of an interval this condition is denoted $(a, b]$ or $[a, b)$. The notation $\mathscr{X}_n$ will be used to denote an n-dimensional space which is neither open nor closed, where $n > 1$. This notation will also be used to denote a general n-dimensional space, the boundary of which has not been defined.

Example 3.6 Determine whether the following space is open, closed, or neither.

$$[x, y] \in \mathscr{X}_2 \quad \text{for} \quad x^2 + y^2 < r$$

The boundary for this space can be expressed by

$$x = \sqrt{r - y^2} \quad \text{and} \quad y^2 \leq r$$

If $y^2 = r$, then

$$x = 0$$

and $x^2 + y^2 = r$, and $[x, y]$ does not belong to $\mathscr{X}_2$. If $y^2 = a < r$ then

$$x = \sqrt{r - a}$$

and $x^2 + y^2 = r$, and again $[x, y]$ does not belong to $\mathscr{X}_2$. Therefore, $\mathscr{X}_2$ does not contain its boundary and is open.

Let $[x_1, x_2, \ldots, x_n]$ and $[y_1, y_2, \ldots, y_n]$ be two points or vectors in n-dimensional space. In this context the brackets do not denote a closed interval or space. For simplicity we will refer to these two points as X and Y respectively. The distance between the two points X and Y, $\|X - Y\|$, is then given by

$$\|X - Y\| = \sqrt{(x_1 - y_1)^2 + (x_2 - y_2)^2 + \cdots + (x_n - y_n)^2} \quad (3.19)$$

The reader will recall that the distance between two points may be interpreted as the length of the straight line joining the points. The distance between two points is useful in defining the neighborhood of a point.

Definition 3.4 Let X be a point in n-dimensional space. If Y is also a point in n-dimensional space, Y belongs to an ϵ neighborhood about X if and only if

$$\|Y - X\| < \epsilon \quad (3.20)$$

An ϵ neighborhood about X may be considered to be an open, spherical, n-dimensional space with center X and radius ϵ. The notion of a neighborhood will be used extensively throughout the remainder of this chapter and in subsequent chapters.

Since an *n*-dimensional space, $\mathscr{X}_n$, can be considered to be a set with elements a set of numbers $[x_1, x_2, \ldots, x_n]$, we can define a universe, U_n, which includes the space $\mathscr{X}_n$. The elements of U_n are the points $[u_1, u_2, \ldots, u_n]$, at least some of which belong to $\mathscr{X}_n$. In our discussion of sets we distinguish those elements which belonged to the set and those which did not. In dealing with an *n*-dimensional space, $\mathscr{X}_n$, we will refine this distinction by categorizing elements or points as those which are interior to $\mathscr{X}_n$, those which are exterior to $\mathscr{X}_n$, and those which lie on the boundary of $\mathscr{X}_n$.

Definition 3.5 Let $\mathscr{X}_n$ be an *n*-dimensional space and X_1 a point with coordinates $[x_1, x_2, \ldots, x_n]$. The point X_1 is an *interior point* of $\mathscr{X}_n$ if there exists some neighborhood of radius $\epsilon > 0$ about X_1 such that every point in that neighborhood belongs to $\mathscr{X}_n$. The point X_1 is an *exterior point* of $\mathscr{X}_n$ if there exists some neighborhood of radius $\epsilon > 0$ about X_1 such that every point in the neighborhood does not belong to $\mathscr{X}_n$. The point X_1 is a *boundary point* of $\mathscr{X}_n$ if every neighborhood of radius $\epsilon > 0$ about X_1 contains points which belong to $\mathscr{X}_n$ and points which do not belong to $\mathscr{X}_n$.

Based upon Definition 3.5, an interior point for $\mathscr{X}_n$ always belongs to $\mathscr{X}_n$ and an exterior point for $\mathscr{X}_n$ never belongs to $\mathscr{X}_n$. However, a boundary point for $\mathscr{X}_n$ may or may not belong to $\mathscr{X}_n$ depending upon whether or not the space is open, closed, or neither. To illustrate the concepts of interior, exterior, and boundary points, consider the two-dimensional space $\mathscr{X}_2$ such that $[x, y]$ belongs to $\mathscr{X}_2$ if $2 < x \leq 4$ and $1 \leq y < 5$. Let us determine the nature of the point $[3, 4]$. If $[3, 4]$ is an interior point for $\mathscr{X}_2$ then there exists some neighborhood of radius $\epsilon > 0$ with center at $[3, 4]$ such that every point in that neighborhood belongs to $\mathscr{X}_2$. Let $[y_1, y_2]$ be any point in this neighborhood. Therefore, the required neighborhood is composed of those points $[y_1, y_2]$ such that

$$\|[y_1, y_2] - [3, 4]\| = \sqrt{(y_1 - 3)^2 + (y_2 - 4)^2}$$

If we let $\epsilon = 1$, we have

$$\sqrt{(y_1 - 3)^2 + (y_2 - 4)^2} < 1$$

Choosing $3 < y_2 < 5$ and y_1 such that

$$(y_1 - 3)^2 < 1 - (y_2 - 4)^2$$

gives us points (y_1, y_2) which lie in $\mathscr{X}_2$, as shown in Fig. 3.7. Since all points in the neighborhood of $[3, 4]$ lie in $\mathscr{X}_2$, $[3, 4]$ is an interior point for $\mathscr{X}_2$. Now let us

3.3 REAL NUMERICAL SETS

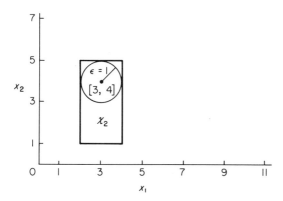

Figure 3.7 Neighborhood of [3, 4] with Radius $\epsilon = 1$

consider the point [6, 4]. If [6, 4] is an interior point for $\mathscr{X}_2$ then there exists a neighborhood of radius $\epsilon > 0$ such that all points $[y_1, y_2]$ satisfying

$$\sqrt{(y_1 - 6)^2 + (y_2 - 4)^2} < \epsilon$$

belong to $\mathscr{X}_2$. Let $\epsilon = 0.5$. Then $[y_1, y_2]$ must satisfy

$$(y_1 - 6)^2 < 0.25 - (y_2 - 4)^2$$

Since $(y_1 - 6)^2 \geq 0$, $3.5 < y_2 < 4.5$. However, every point $[y_1, y_2]$ lying in this neighborhood lies outside of the boundary of $\mathscr{X}_2$, and the point [6, 4] is an exterior point for $\mathscr{X}_2$. The neighborhood of the point [6, 4] is shown in Fig. 3.8.

As a final illustration, consider the point [3, 5] and choose any neighborhood with center [3, 5] and radius $0 < \epsilon < 4$. Let us examine the points $[3, y_2]$ where $5 - \epsilon < y_2 < 5 + \epsilon$. For $5 - \epsilon < y_2 < 5$, the points $[3, y_2]$ lie in $\mathscr{X}_2$, while for $5 < y_2 < 5 + \epsilon$ the points $[3, y_2]$ lie outside $\mathscr{X}_2$. This result holds true for all

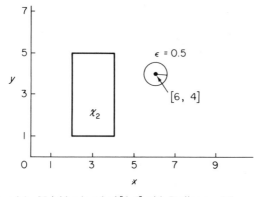

Figure 3.8 Neighborhood of [6, 4] with Radius $\epsilon = 0.5$

$0 < \epsilon < 4$. For $\epsilon > 4$, the neighborhood about [3, 5] still contains points which lie both inside and outside of $\mathscr{X}_2$. Therefore every neighborhood of [3, 5] with nonzero radius contains some points which belong to $\mathscr{X}_2$ and some points which do not belong to $\mathscr{X}_2$, and [3, 5] is a boundary point for $\mathscr{X}_2$.

The reader will notice that in attempting to determine whether a given point, X_1, is interior or exterior to a space, $\mathscr{X}_n$, we examine the nature of the points in the immediate vicinity of the point of interest. If there is any neighborhood or open space containing X_1 which is itself completely contained in the space of interest, $\mathscr{X}_n$, then X_1 is an interior point for $\mathscr{X}_n$. Similarly, if a neighborhood containing X_1 can be found such that every point in the neighborhood does not belong to $\mathscr{X}_n$, then X_1 is an exterior point for $\mathscr{X}_n$. Finally, if X_1 satisfies neither condition, then X_1 is a boundary point for $\mathscr{X}_n$.

In the examples presented thus far, finite boundaries have been specified for each of the spaces considered. When the boundary of a space is infinite in any direction then the space is said to be unbounded. For example the interval $(-\infty, 10)$ is unbounded since the lower bound is not finite. Similarly the open space ($\mathscr{X}_2$) with points $[x, y]$ defined for $0 < x < \infty$, $1 < y < 10$ is unbounded since the maximum value of x is not defined.

Definition 3.6 The n-dimensional space $\mathscr{X}_n$ with points $[x_1, x_2, \ldots, x_n]$ is bounded if and only if there exist finite numbers a_i and b_i, called *lower* and *upper bounds* respectively, such that $a_i \leq x_i \leq b_i$ for $i = 1, 2, \ldots, n$.

In Definition 3.6, we establish the principle that a space is bounded if and only if the components of the points of the space are limited to a finite range of values. Consider the interval (0, 10). The finite numbers a and b are boundaries for the interval if $a \leq 0$ and $b \geq 10$. Therefore the numbers -100, -23, -10, -5, and 0 are all lower bounds for the interval and the numbers 10, 20, 53, and 1000 are upper bounds for the interval. It should be clear that if an interval has one upper bound it has an infinite number, since if b is one upper bound then any number c such that $c > b$ is also an upper bound. A similar statement can be made with respect to lower bounds.

If an interval is bounded, then among the upper bounds for the interval there is one which is less than any other called the *least upper bound*, abbreviated l.u.b. Similarly, among the lower bounds there is one which is greater than any other called the *greatest lower bound*, abbreviated g.l.b. For the interval (0, 10), 0 is the greatest lower bound while 10 is the least upper bound.

3.4 Functions, Limits, Continuity, and the Derivative

Suppose that one square mile of land is represented by the two-dimensional closed space $[\mathscr{X}_2]$ with points $[x, y]$ where $0 \leq x \leq 1$, and $0 \leq y \leq 1$. To each point in the space we assign a number corresponding to the altitude of the

3.4 FUNCTIONS, LIMITS, CONTINUITY, AND THE DERIVATIVE

corresponding point on the land represented by the space. The rule used to establish the correspondence between the points $[x, y]$ and the numerical values assigned to $[x, y]$ representing altitude is called a function. The space $[\mathscr{X}_2]$ to which this rule applies is called the domain of the function and the number assigned by this rule to the point $[x, y]$ is called the value of the function at $[x, y]$ and is denoted by $f(x, y)$, $g(x, y)$, $F(x, y)$, $H(x, y)$, etc. Notation such as $f(x, y)$ is also used to express the fact that a function may be dependent upon the values of two variables, x and y. Usually functional relationships are expressed in the form of equations. For example, the functional relationship

$$f(x, y) = x + y^2, \qquad 0 \le x \le 10, \quad 5 \le y < 6 \qquad (3.21)$$

assigns the number $x + y^2$ to the point $[x, y]$. The domain of definition of the function is the space defined by the limits for x and y.

If a function assigns to each and every point in its domain of definition one and only one numerical value, the function is said to be single-valued. If to one or more points in the domain of definition, a function takes on more than one numerical value, the function is said to be multiple-valued. For example, the function defined in Eq. (3.21) is single-valued, while the function

$$f(x) = \sqrt{4x}, \qquad 0 < x < 100 \qquad (3.22)$$

assigns the values $\pm 2x$ to each point in the domain of definition. Since $f(x)$ assumes two values at some point in its domain, in fact at all points in this case, $f(x)$ is multiple-valued.

If there is a finite number, a, such that $f(x) \ge a$ for all x in the domain of definition of $f(x)$, then $f(x)$ is said to be bounded from below or $f(x)$ has a lower bound. Similarly if there exists a finite number b such that $f(x) \le b$ for all x in the domain of definition of $f(x)$, $f(x)$ is said to be bounded from above. If $a \le f(x) \le b$ for all x in the domain of definition of $f(x)$, $f(x)$ is bounded. Therefore if $f(x)$ is bounded, it has both a lower and an upper bound. A function may also be shown to be bounded if there exists a number c such that $|f(x)| \le c$ for all x in the domain of definition of $f(x)$.

Example 3.7 Determine whether the following functions are single- or multiple-valued, and bounded or unbounded.

a. $f(x) = x^2$, $-5 < x < 5$;
b. $g(x, y) = \sqrt{x^2 + y^2}$, $-1 < x \le 1$, $-1 \le y < 1$;
c. $h(x) = 1/(x^2 - 1)$, $1 \le x \le 4$.

Part a Since the square of any real number yields one and only one number, $f(x)$ is single-valued. If $f(x)$ has a lower bound, then there exists a number a such that $f(x) \ge a$ for all x on the interval $(-5, 5)$. Therefore, for this number

$$x^2 \ge a, \qquad -5 < x < 5$$

Since this relationship is satisfied for all $a \le 0$, any nonpositive real number is a lower bound for $f(x)$. If b is an upper bound for $f(x)$, then

$$x^2 \leq b, \qquad -5 < x < 5$$

However, this relationship is satisfied for all $b \geq 25$. Therefore, any real number greater than or equal to 25 is an upper bound for $f(x)$. Since $f(x)$ has both a lower and an upper bound it is bounded.

Part b Since the quantity $x^2 + y^2$ is a positive real number for all real x and y, and since the square root of every positive real number yields a positive and negative number, $g(x, y)$ has two values for $x \neq 0$ or $y \neq 0$. Therefore, $g(x, y)$ is a multiple-valued function. Inspection of $g(x, y)$ shows that it has a minimum value of $-\sqrt{2}$ and a maximum value of $\sqrt{2}$. Hence, any real number a such that $a \leq -\sqrt{2}$ is a lower bound for $g(x, y)$ and any real number b such that $b \geq \sqrt{2}$ is an upper bound for $g(x, y)$, and $g(x, y)$ is bounded.

Part c From part a, x^2 is a single-valued function. Therefore $1/(x^2 - 1)$ is also a single valued function on the interval $[1, 4]$. If a is a lower bound for $h(x)$, then

$$\frac{1}{x^2 - 1} \geq a, \qquad 1 \leq x \leq 4$$

Inspection indicates that $h(x)$ achieves a minimum of $\frac{1}{15}$ at $x = 4$. Therefore, a is a lower bound for $h(x)$ if $a \leq \frac{1}{15}$. If b is an upper bound for $h(x)$, then

$$\frac{1}{x^2 - 1} \leq b, \qquad 1 \leq x \leq 4 \qquad (3.23)$$

or

$$x^2 - 1 \geq \frac{1}{b}, \qquad 1 \leq x \leq 4 \qquad (3.24)$$

For $b = 1$, this inequality does not hold for $1 \leq x < \sqrt{2}$. Therefore, $b = 1$ is not an upper bound for $h(x)$ and if an upper bound exists it is greater than unity. Choose any $b > 1$ and let $x = \sqrt{1 + 1/2b}$, which is a value of x on the interval $[1, 4]$. Substituting this value of x in Eq. (3.24) leads to

$$\frac{1}{2b} \geq \frac{1}{b}$$

which is an invalid relationship for all $b > 1$. Therefore, there is no number b satisfying Eq. (3.23), and $h(x)$ does not have an upper bound. Since $h(x)$ does not have a finite upper *and* lower bound it is unbounded.

In the above example the functions $f(x)$, $g(x, y)$, and $h(x)$ have lower bounds and therefore have greatest lower bounds. These greatest lower bounds are 0, $-\sqrt{2}$, and $\frac{1}{15}$ respectively. The functions $f(x)$ and $g(x, y)$ have upper bounds and therefore have least upper bounds given by 25 and $\sqrt{2}$ respectively. Since $h(x)$ is not bounded from above, it has no least upper bound.

It is sometimes necessary to determine the behavior of a function as the

3.4 FUNCTIONS LIMITS, CONTINUITY, AND THE DERIVATIVE

variables of the function approach specific values. For example, we might ask what the function $f(x)$ does as x increases to infinity? In mathematical terms we are attempting to determine the limit of $f(x)$ as x approaches infinity. The limit of the function $f(x)$ as x approaches some number a_0 is written

$$\lim_{x \to a_0} f(x) = L \tag{3.25}$$

where L is the value $f(x)$ approaches as x approaches a_0. Unless otherwise specified, L is taken as a finite real number. We might wish to examine the limit of $f(x)$ as x approaches a_0 through values less than a_0. Such a limit is called a left-hand limit and is expressed by

$$\lim_{x \to a_0-} f(x) = L \tag{3.26}$$

If the limit is evaluated as x decreases through values greater than a_0, the limit is called a right-hand limit and is written

$$\lim_{x \to a_0+} f(x) = L \tag{3.27}$$

The validity of the general expression for a limit given in Eq. (3.25) assumes that the limits given in Eqs. (3.26) and (3.27) are identical. That is, the general limit exists if and only if the corresponding left- and right-hand limits exist and are equal. The general limit defined in Eq. (3.25) is usually referred to as a two-sided limit.

Definition 3.7 The statement

$$\lim_{x \to a_0-} f(x) = L$$

is valid if and only if corresponding to every $\epsilon > 0$ there exists a number $\delta > 0$ such that $|f(x) - L| < \epsilon$ whenever $0 < (a_0 - x) < \delta$.

In Definition 3.7 we are saying that $f(x)$ approaches L as x approaches a_0 through numbers less than a_0 if and only if the difference between $f(x)$ and L decreases as x moves closer to a_0. The use of Definition 3.7 is illustrated in the following examples.

Example 3.8 Show that

$$\lim_{x \to 5-} f(x) = 25, \quad \text{where} \quad f(x) = x^2$$

To show that this limit is valid we will attempt to define a relationship between the numbers $\epsilon > 0$ and $\delta > 0$ such that whenever $|x^2 - 25| < \epsilon$, $0 < (5 - x) < \delta$ and such that ϵ approaches zero as δ approaches zero. Since we are interested in the behavior of $f(x)$ near $x = 5$, we can restrict our attention to values of x on the interval $(0, 5)$ where

$$|x^2 - 25| = (5 - x)(5 + x)$$

For x on the interval $(0, 5)$, $(5 + x) < 10$. Therefore
$$|x^2 - 25| < 10(5 - x)$$
Let
$$\delta = \frac{\epsilon}{10}$$
Then, for $0 < (5 - x) < \delta$,
$$|x^2 - 25| < 10\delta$$
As δ approaches zero so does ϵ. Therefore as x approaches 5 from the left, x^2 approaches 25.

Example 3.9 Show that
$$\lim_{x \to 10-} \frac{1}{x} = \frac{1}{10}$$

We will attempt to demonstrate the validity of this limit in a manner similar to that used in Example 3.8. Since we are interested in the behavior of $1/x$ near $x = 10$, we may restrict our attention to values of x on the interval $(1, 10)$ for which
$$\left| \frac{1}{x} - \frac{1}{10} \right| = \frac{10 - x}{10x}$$
On this interval
$$\left| \frac{10 - x}{10x} \right| < \frac{10 - x}{10}$$
Therefore, near $x = 10$,
$$\left| \frac{1}{x} - \frac{1}{10} \right| < \frac{10 - x}{10}$$
Let
$$\delta = 10\epsilon$$
Then, for $0 < (10 - x) < \delta$
$$\left| \frac{1}{x} - \frac{1}{10} \right| < \frac{\delta}{10}$$
$$< \epsilon$$
Hence, for any $\epsilon > 0$ for which
$$\left| \frac{1}{x} - \frac{1}{10} \right| < \epsilon, \quad 0 < (10 - x) < 10\epsilon \quad \text{or} \quad 0 < (10 - x) < \delta.$$

Therefore, as x approaches 10 from the left, $1/x$ approaches $\frac{1}{10}$ since ϵ approaches zero as δ approaches zero. ∎

Definition 3.8 The statement
$$\lim_{x \to a_0+} f(x) = L$$
is valid if and only if corresponding to every $\epsilon > 0$ there exists a number $\delta > 0$ such that $|f(x) - L| < \epsilon$ whenever $0 < (x - a_0) < \delta$.

Definition 3.9 The two-sided limit, $\lim_{x \to a_0} f(x)$, exists and is equal to L if and only if
$$\lim_{x \to a_0} f(x) = \lim_{x \to a_0-} f(x) = \lim_{x \to a_0+} f(x) = L \tag{3.28}$$

In Definition 3.10, an alternative but equivalent definition of the two-sided limit is given.

Definition 3.10 The statement
$$\lim_{x \to a_0} f(x) = L \tag{3.29}$$
is valid if and only if corresponding to every $\epsilon > 0$ there exists a number $\delta > 0$ such that $|f(x) - L| < \epsilon$ whenever $0 < |x - a_0| < \delta$.

Definition 3.9 states that the two-sided limit exists if and only if the corresponding left- and right-hand limits are equal. The statement in Definition 3.10 is equivalent in that $0 < |x - a| < \delta$ implies that x lies on the interval $(a_0 - \delta, a_0 + \delta)$ and therefore x may approach a_0 from either left or right.

Example 3.10 Let
$$f(x) = \begin{cases} 1 - x^2, & -1 < x < 0 \\ x^2 - 1, & 0 \le x < 1 \end{cases}$$
Show that $\lim_{x \to 0} f(x)$ does not exist.

We first show that
$$\lim_{x \to 0-} f(x) = 1$$
In this case we are interested in the behavior of $1 - x^2$ as x approaches zero from the left. That is we must show that there exist numbers $\epsilon > 0$ and $\delta > 0$ such that $|(1 - x^2) - 1| < \epsilon$ whenever $0 < (0 - x) < \delta$. Since we are concerned with the behavior of $1 - x^2$ near but to the left of $x = 0$, we will restrict our attention to values of x on the interval $(-1, 0)$. For such values of x

$$|(1 - x^2) - 1| = x^2 \quad \text{and} \quad |(1 - x^2) - 1| < -x$$

Therefore, if we let $\epsilon = \delta$, whenever $0 < (0 - x) < \delta$, then $|(1 - x^2) - 1| < \epsilon$ and $(1 - x^2)$ approaches unity as x approaches zero. We will now show that the limit of $f(x)$ as x approaches zero from the right is -1.

For $x \geq 0$,
$$f(x) = x^2 - 1$$

If
$$\lim_{x \to 0+} f(x) = -1$$

then there exist numbers ϵ and δ such that $|(x^2 - 1) + 1| < \epsilon$ whenever $0 < (x - 0) < \delta$. Choose x on the interval $(0, 1)$. Then

$$|(x^2 - 1) + 1| = x^2 < x$$

Let $\epsilon = \delta$. For $0 < (x - 0) < \delta$, $|(x^2 - 1) + 1| < \epsilon$. Since δ approaches zero as ϵ approaches zero,
$$\lim_{x \to 0+} (x^2 - 1) = -1$$

Let us compare the left- and right-hand limits as x approaches zero
$$\lim_{x \to 0-} f(x) = 1, \quad \lim_{x \to 0+} f(x) = -1$$

Since these two limits are not equal, $\lim_{x \to 0} f(x)$ does not exist.

Example 3.11 Let
$$f(x) = \begin{cases} x^2 - 1, & -\infty < x < 1 \\ 1 - x^2, & 1 \leq x < \infty \end{cases}$$

Show that
$$\lim_{x \to 1} f(x) = 0$$

We first show that
$$\lim_{x \to 1-} f(x) = 0$$

For x on the interval $(0, 1)$
$$|(x^2 - 1) - 0| = (1 + x)(1 - x) < 2(1 - x)$$

Letting $\delta = \epsilon/2$, whenever $0 < (1 - x) < \delta$, $|(x^2 - 1) - 0| < \epsilon$ and since δ and ϵ approach zero simultaneously
$$\lim_{x \to 1-} f(x) = 0$$

Now consider the right-hand limit. We will show that
$$\lim_{x \to 1+} f(x) = 0$$

3.4 FUNCTIONS, LIMITS, CONTINUITY, AND THE DERIVATIVE

For x on the interval $(1, 2)$
$$|(1 - x^2) - 0| = (x + 1)(x - 1) < 3(x - 1)$$
Let $\delta = \epsilon/3$. Then whenever $0 < (x - 1) < \delta$, $|(1 - x^2) - 0| < \epsilon$ and
$$\lim_{x \to 1+} f(x) = 0$$
Since
$$\lim_{x \to 1-} f(x) = \lim_{x \to 1+} f(x) = 0$$
we have
$$\lim_{x \to 1} f(x) = 0$$

The definitions given for left-hand, right-hand, and two-sided limits hold provided x approaches a finite value. However, an expression of the form $\lim_{x \to \infty+} f(x)$ has no meaning, since it would imply that x approaches ∞ through values of greater than ∞. The same argument holds for $\lim_{x \to -\infty-} f(x)$. Since $\lim_{x \to -\infty-} f(x)$ and $\lim_{x \to \infty+} f(x)$ have no meaning, two-sided limits at $-\infty$ and $+\infty$ have no meaning. Thus the expressions $\lim_{x \to -\infty} f(x)$ and $\lim_{x \to \infty} f(x)$ should be interpreted as $\lim_{x \to -\infty+} f(x)$ and $\lim_{x \to \infty-} f(x)$ respectively. In addition to the interpretation of $\lim_{x \to -\infty} f(x)$ and $\lim_{x \to \infty} f(x)$, we must also change their definitions. To illustrate why these changes in definition are necessary, let us assume that

$$\lim_{x \to \infty} f(x) = L \tag{3.30}$$

where x necessarily approaches ∞ from the left. If we apply definition 3.6, for every $\epsilon > 0$ there exists a number $\delta > 0$ such that $|f(x) - L| < \epsilon$ whenever $0 < (\infty - x) < \delta$. However, since $\infty - x$ is undefined, the expression $0 < (\infty - x) < \delta$ has no meaning. That is, we cannot determine when x is within δ of ∞.

To arrive at a definition of the conditions under which the limit in Eq. (3.30) is valid, we examine what happens to $f(x)$ as x increases. If Eq. (3.30) is valid, then for $|f(x) - L| < \epsilon$, x must be greater than some finite number A. As ϵ decreases, the number A should increase. This leads to the following definition.

Definition 3.11 The limit of $f(x)$ as $x \to \infty$ is given by
$$\lim_{x \to \infty} f(x) = L \tag{3.31}$$
if and only if, corresponding to every $\epsilon > 0$, there exists a number A such that $|f(x) - L| < \epsilon$ whenever $x \geq A$.

Definition 3.12 The limit of $f(x)$ as $x \to -\infty$ is given by
$$\lim_{x \to -\infty} f(x) = L$$

if and only if, corresponding to every $\epsilon > 0$, there exists a number A such that $|f(x) - L| < \epsilon$ whenever $x \leq A$.

Example 3.12 Show that

$$\lim_{x \to \infty} f(x) = 1, \quad \text{where} \quad f(x) = \frac{x-1}{x}$$

Let

$$x = \frac{2}{\epsilon}$$

where $\epsilon > 0$. Then

$$\left|\frac{x-1}{x} - 1\right| = \frac{\epsilon}{2} < \epsilon \quad \text{whenever} \quad \left|\frac{x-1}{x} - 1\right| < \epsilon, \quad x > \frac{1}{\epsilon}$$

As $\epsilon \to 0$, $x \to \infty$. Therefore, $|(x-1)/x - 1|$ can be made arbitrarily small by making x suitably large which is accomplished in turn by making ϵ suitably small.

In stating that L is the limit of a function $f(x)$ as x approaches a_0, we are saying that $f(x)$ approaches a definable quantity as x approaches a_0. If $f(x)$ increases (or decreases) in an unbounded fashion as x approaches a_0, there is no definable quantity which $f(x)$ converges to, and in such cases we say that the limit tends to infinity and we express this tendency for either one-sided limit by

$$\lim_{x \to a_0 -} f(x) = \infty \tag{3.32}$$

or

$$\lim_{x \to a_0 +} f(x) = \infty \tag{3.33}$$

The reader should note that Eqs. (3.32) and (3.33) do not define the left- and right-hand limits as infinity, since infinity is not a definable quantity, but merely express the fact that the limits are not defined, since $f(x)$ increases without bound as x approaches a_0. If $f(x)$ decreases without bound as x approaches a_0 from the left or right, we replace ∞ by $-\infty$ in Eqs. (3.32) and (3.33).

Definition 3.13 The limit of $f(x)$ as x approaches a_0 from the left is written

$$\lim_{x \to a_0 -} f(x) = \infty \tag{3.34}$$

if and only if corresponding to every A there exists a $\delta > 0$ such that $f(x) \geq A$ whenever $0 < (a_0 - x) < \delta$.

Definition 3.14 The limit of $f(x)$ as x approaches a_0 from the right is written

$$\lim_{x \to a_0 +} f(x) = \infty \tag{3.35}$$

3.4 FUNCTIONS, LIMITS, CONTINUITY, AND THE DERIVATIVE

if and only if corresponding to every A there exists a $\delta > 0$ such that $f(x) \geq A$ whenever $0 < (x - a_0) < \delta$.

Definition 3.15 The two-sided limit of $f(x)$ as x approaches a_0 is written

$$\lim_{x \to a_0} f(x) = \infty \qquad (3.36)$$

if and only if

$$\lim_{x \to a_0-} f(x) = \infty \quad \text{and} \quad \lim_{x \to a_0+} f(x) = \infty$$

The two-sided limit of a function as x approaches a given value, a_0, may fail to exist for two reasons. First, the corresponding left- and right-handed limits may exist and be unequal. Second, one or both of the one-sided limits may not exist. Examples of such functions are $\lim_{x \to \infty} \sin(x)$ or $\lim_{x \to \infty} \cos(x)$. However, the one- and two-sided limits may exist at a_0 even though $f(a_0)$ is not defined. Therefore, in examining the limit of a function as $x \to a_0$, we are attempting to determine the behavior of the function in an arbitrarily small neighborhood of a_0, $(a_0 - \delta, a_0)$ and $(a_0, a_0 + \delta)$ for the one-sided limits and $(a_0 - \delta, a_0 + \delta)$ for the two-sided limit, but not at the point a_0 itself.

The limit theory presented for functions of a single variable can be extended to multivariate functions in a rather straightforward manner. Let

$$X = [x_1, x_2, \ldots, x_n], \qquad A_0 = [a_1, a_2, \ldots, a_n]$$

We wish to examine the limit

$$\lim_{X \to A_0} f(X) = L \qquad (3.37)$$

In dealing with functions of a single variable we examined $f(x)$ as x approached a_0 through values less than a_0 and through values greater than a_0. That is, x could approach a_0 from only two directions. However, if $f(X)$ is defined for every point in some neighborhood of A_0, X may approach A_0 over infinitely many paths. We can define such a path by an n-dimensional space which is a subspace of the domain of $f(X)$. Let this subspace be defined by $\mathscr{X}'_n$. With this definition of $\mathscr{X}'_n$ in mind we can define the multivariate equivalent of a one-sided limit.

Definition 3.16 The limit

$$\lim_{X \to A_0} f(X) = L, \qquad X \in \mathscr{X}'_n \qquad (3.38)$$

is valid if and only if corresponding to severy $\epsilon > 0$ there exists a $\delta > 0$ such that $|f(X) - L| < \epsilon$ whenever $0 < \|X - A_0\| < \delta$, where $X \in \mathscr{X}'_n$.

The multivariate analogue of the two-sided limit is the limit

$$\lim_{X \to A_0} f(X) = L$$

where the limit is valid no matter what path of approach to A_0 is taken.

Definition 3.17 The limit

$$\lim_{X \to A_0} f(X) = L \qquad (3.39)$$

is valid if and only if corresponding to every $\epsilon > 0$ there exists a $\delta > 0$ such that $|f(X) - L| < \epsilon$ whenever $0 < \|X - A_0\| < \delta$.

The reader will note that the only difference between Definitions 3.16 and 3.17 is the omission of any reference to a path of approach to a_0 in Definition 3.17.

Example 3.13 Show that

$$\lim_{X \to A_0} f(X) = 1, \quad \text{where} \quad f(X) = \frac{1}{x^2 + y^2 + 1}, \quad A_0 = [0, 0]$$

and X approaches A_0 along the path $\mathscr{X}'_2$. A point $X = [x, y]$ belongs to $\mathscr{X}'_2$ if and only if $x = y$.

To show that the limit defined conforms to the conditions given in Definition 3.16, we attempt to find a relationship between ϵ and δ such that $\|X - A_0\|$ and $|f(X) - 1|$ can be simultaneously reduced to arbitrarily small numbers. Since X must approach A_0 such that $X \in \mathscr{X}'_2$ we choose $z = x = y$. Therefore

$$0 < \|[z, z] - [0, 0]\| < \delta \quad \text{or} \quad 0 < \sqrt{2z^2} < \delta$$

from Eq. (3.19). Then

$$-\frac{\delta}{\sqrt{2}} < z < \frac{\delta}{\sqrt{2}}$$

If $z = \pm \delta/2$ this inequality is satisfied. Substituting this value of z in $|f(X) - 1|$ leads to

$$|f(X) - 1| = \left| \frac{1}{2\left(\frac{\delta}{2}\right)^2 + 1} - 1 \right| = \left| \frac{2\left(\frac{\delta}{2}\right)^2}{2\left(\frac{\delta}{2}\right)^2 + 1} \right| < 2\left(\frac{\delta}{2}\right)^2$$

Therefore, reducing δ decreases both $|f(X) - 1|$ and $\|X - A_0\|$ and the limit expressed is valid at least for the path $\mathscr{X}'_2$.

Example 3.14 Show that the limit given in Example 3.13 is valid for all paths of approach of $[x, y]$ to $[0, 0]$.

To show that

$$\lim_{X \to A_0} \frac{1}{x^2 + y^2 + 1} = 1$$

we will attempt to establish a relationship between ϵ and δ such that when either is decreased both

3.4 FUNCTIONS, LIMITS, CONTINUITY, AND THE DERIVATIVE

$$\|[x, y] - [0, 0]\| \quad \text{and} \quad \left|\frac{1}{x^2 + y^2 + 1} - 1\right|$$

are simultaneously reduced. To accomplish this we choose x and y such that

$$0 < \sqrt{x^2 + y^2} < \delta \quad \text{or} \quad 0 < x^2 + y^2 < \delta^2$$

If

$$x^2 + y^2 = \left(\frac{\delta}{2}\right)^2$$

the above inequality is satisfied. Then

$$|f(X) - 1| = \left|\frac{\left(\frac{\delta}{2}\right)^2}{\left(\frac{\delta}{2}\right)^2 + 1}\right| < \left(\frac{\delta}{2}\right)^2$$

Since the quantities $|f(X) - 1|$ and $\|X - A_0\|$ can be made arbitrarily small by choosing δ appropriately small, the limit is valid for all paths of approach to A_0.

Theorem 3.1 Let

$$\lim_{x \to a_0} f(x) = L_1 \tag{3.40}$$

and

$$\lim_{x \to a_0} g(x) = L_2 \tag{3.41}$$

where L_1 and L_2 are finite numbers. Then

a. $$\lim_{x \to a_0} [f(x) + g(x)] = L_1 + L_2 \tag{3.42}$$

b. $$\lim_{x \to a_0} [f(x)g(x)] = L_1 L_2 \tag{3.43}$$

c. and if $L_2 \neq 0$

$$\lim_{x \to a_0} \frac{f(x)}{g(x)} = \frac{L_1}{L_2} \tag{3.44}$$

Proof By Eqs. (3.40) and (3.41) there exist numbers $\epsilon_1 > 0$ and $\delta_1 > 0$ such that $|f(x) - L_1| < \epsilon_1$ whenever $0 < |x - a_0| < \delta_1$ and numbers $\epsilon_2 > 0$ and $\delta_2 > 0$ such that $|g(x) - L_2| < \epsilon_2$ whenever $0 < |x - a_0| < \delta_2$. Therefore, there exists a number $\delta > 0$ such that $|f(x) - L_1| < \epsilon_1$ and $|g(x) - L_2| < \epsilon_2$ whenever $0 < |x - a_0| < \delta$. To establish the validity of this statement assume that $\delta_1 < \delta_2$. For $0 < |x - a_0| < \delta_1$, $|f(x) - L_1| < \epsilon_1$. However, if $|g(x) -$

$L_2| < \epsilon_2$ whenever $0 < |x - a_0| < \delta_2$, then $|g(x) - L_2| < \epsilon_2$ whenever $0 < |x - a_0| < \delta_1 < \delta_2$, and δ may be taken as equal to or less than δ_1 or in general as equal to or less than the minimum of δ_1 and δ_2.

If $0 < |x - a_0| < \delta$, then

$$|f(x) - L_1| < \epsilon_1$$

To the left-hand side of the inequality add $|g(x) - L_2|$ and to the right-hand side add ϵ_2. Since $|g(x) - L_2| < \epsilon_2$ for $0 < |x - a_0| < \delta$, we are adding a smaller quantity on the left than that added on the right and the inequality is maintained. That is

$$|f(x) - L_1| + |g(x) - L_2| < \epsilon_1 + \epsilon_2 \tag{3.45}$$

Let us consider now the quantity $|f(x) + g(x) - L_1 - L_2|$ and its relationship to the sum on the left-hand side of Eq. (3.45). The square of $|f(x) + g(x) - L_1 - L_2|$ is given by

$$|f(x) + g(x) - L_1 - L_2|^2 = [f(x) + g(x) - L_1 - L_2]^2$$
$$= [f(x) - L_1]^2 + [g(x) - L_2]^2$$
$$+ 2[f(x) - L_1][g(x) - L_2]$$

The square of $|f(x) - L_1| + |g(x) - L_2|$ is given by

$$[|f(x) - L_1| + |g(x) - L_2|]^2$$
$$= [f(x) - L_1]^2 + [g(x) - L_2]^2 + 2|f(x) - L_1||g(x) - L_2|$$

Since

$$|f(x) - L_1||g(x) - L_2| \geq [f(x) - L_1][g(x) - L_2]$$

we have

$$|f(x) + g(x) - L_1 - L_2|^2 \leq [|f(x) - L_1| + |g(x) - L_2|]^2$$

or

$$|f(x) + g(x) - L_1 - L_2| \leq |f(x) - L_1| + |g(x) - L_2| \tag{3.46}$$

Therefore, whenever $0 < |x - a_0| < \delta$

$$|f(x) + g(x) - L_1 - L_2| < \epsilon \tag{3.47}$$

where $\epsilon = \epsilon_1 + \epsilon_2$ and

$$\lim_{x \to a_0} [f(x) + g(x)] = L_1 + L_2$$

For part b we first note that

$$|f(x)g(x) - L_1 L_2| = |f(x)[g(x) - L_2] + L_2[f(x) - L_1]|$$
$$\leq |f(x)||g(x) - L_2| + |L_2||f(x) - L_1|$$

3.4 FUNCTIONS, LIMITS, CONTINUITY, AND THE DERIVATIVE

Since $\lim_{x \to a_0} f(x)$ and $\lim_{x \to a_0} g(x)$ exist, there exists some neighborhood about a_0 such that for $0 < |x - a_0| < \delta$, $f(x)$ and $g(x)$ are bounded. Hence, there exists a number $M > 0$ such that $|f(x)| < M$ for $0 < |x - a_0| < \delta$ and

$$|f(x)g(x) - L_1 L_2| \leq M|g(x) - L_2| + |L_2||f(x) - L_1|$$

For $0 < |x - a_0| < \delta$, there exist $\epsilon_1 > 0$ and $\epsilon_2 > 0$ such that

$$|f(x)g(x) - L_1 L_2| \leq M\epsilon_1 + |L_2|\epsilon_2 = \epsilon \quad (3.48)$$

Hence, whenever $0 < |x - a_0| < \delta$

$$|f(x)g(x) - L_1 L_2| < \epsilon$$

and

$$\lim_{x \to a_0} [f(x)g(x)] = L_1 L_2$$

To show that Eq. (3.44) is valid, let

$$h(x) = \frac{1}{g(x)} \quad (3.49)$$

Then

$$\lim_{x \to a_0} \frac{f(x)}{g(x)} = \lim_{x \to a_0} f(x)h(x) = L_1 \lim_{x \to a_0} \frac{1}{g(x)} \quad (3.50)$$

We must show that

$$\left| \frac{1}{g(x)} - \frac{1}{L_2} \right| < \epsilon$$

whenever $0 < |x - a_0| < \delta$.

We know that

$$\left| \frac{1}{g(x)} - \frac{1}{L_2} \right| = \left| \frac{g(x) - L_2}{L_2 g(x)} \right| = \frac{|g(x) - L_2|}{|L_2||g(x)|} \quad (3.51)$$

Since

$$\lim_{x \to a_0} g(x) = L_2$$

for $\epsilon > 0$ there exists a δ_3 such that

$$|g(x) - L_2| < \tfrac{1}{2} L_2^2 \epsilon \quad (3.52)$$

whenever $0 < |x - a_0| < \delta_3$. Further, there exists a δ_4 such that

$$|g(x) - L_2| < \tfrac{1}{2}|L_2| \quad (3.53)$$

whenever $0 < |x - a_0| < \delta_4$. Write L_2 as

$$L_2 = L_2 + g(x) - g(x)$$

Then
$$|L_2| \le |g(x) - L_2| + |g(x)| < \tfrac{1}{2}|L_2| + |g(x)|$$

from Eq. (3.53). Hence
$$|L_2| < \tfrac{1}{2}|L_2| + |g(x)| \quad \text{and} \quad |g(x)| > \tfrac{1}{2}|L_2|$$

Then
$$\frac{|g(x) - L_2|}{|L_2||g(x)|} < \frac{\tfrac{1}{2}L_2^2 \epsilon}{\tfrac{1}{2}L_2^2} \tag{3.54}$$

for $0 < |x - a_0| < \delta$ where $\delta > 0$ and $\delta < \delta_3$ and $\delta < \delta_4$. Since
$$\left|\frac{1}{g(x)} - \frac{1}{L_2}\right| = \frac{|g(x) - L_2|}{|L_2||g(x)|}$$

we have
$$\left|\frac{1}{g(x)} - \frac{1}{L_2}\right| < \epsilon$$

whenever $0 < |x - a_0| < \delta$, and
$$\lim_{x \to a_0} \frac{1}{g(x)} = \frac{1}{L_2} \tag{3.55}$$

when $L_2 \ne 0$ and
$$\lim_{x \to a_0} \frac{f(x)}{g(x)} = \frac{L_1}{L_2}$$

The limiting operation has been used to describe the behavior of a function near a point but not at the point itself. The relationship between $\lim_{X \to A_0} f(X)$ and $f(A_0)$ determines whether the function $f(X)$ is continuous or discontinuous at A_0.

Definition 3.18 The function $f(X)$ is continuous at A_0 if and only if $\lim_{X \to A_0} f(X)$ exists, $f(A_0)$ is defined, and
$$\lim_{X \to A_0} f(X) = f(A_0) \tag{3.56}$$

Consideration of Definition 3.18 indicates that $f(X)$ may fail to be continuous at A_0 for several reasons. First, $f(X)$ is discontinuous at A_0 if
$$\lim_{\substack{X \to A_0 \\ X \in \mathcal{T}_{n'}}} f(X) \ne \lim_{\substack{X \to A_0 \\ X \in \mathcal{T}_{n''}}} f(X) \tag{3.57}$$

3.4 FUNCTIONS, LIMITS, CONTINUITY, AND THE DERIVATIVE

where $\mathscr{X}'_n$ is one path of approach to A_0 and $\mathscr{X}''_n$ is any other path of approach to A_0. In such a case the limit of $f(X)$ as X approaches A_0 does not exist. Second, the function is discontinuous at A_0 if $\lim_{X \to A_0} f(X) = \infty$. A third type of discontinuity exists at A_0 if $\lim_{X \to A_0} f(X)$ exists and $f(A_0)$ is defined, but

$$\lim_{X \to A_0} f(X) \neq f(A_0) \tag{3.58}$$

Finally, $f(X)$ is discontinuous at A_0 if $f(A_0)$ is undefined. The value of $f(X)$ is taken as undefined at A_0 if it is unbounded at A_0 or if it simply lacks any definition at all at A_0. Each of these discontinuities is illustrated in Figs. 3.9 through 3.12. The reader should note that in Figure 3.11 neither $\lim_{x \to 5} f(x)$ nor $f(5)$ exist since both are unbounded.

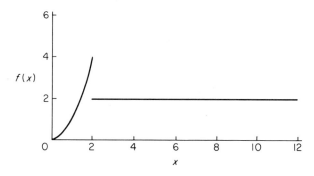

Figure 3.9 Discontinuity at $x = 2$ Where $\lim_{x \to 2-} f(x) \neq \lim_{x \to 2+} f(x)$

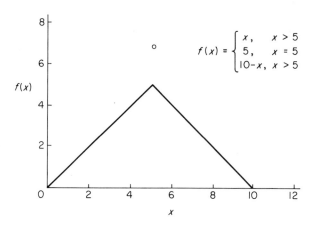

Figure 3.10 Discontinuity at $x = 5$ Where $\lim_{x \to 5} f(x)$ Exists But $\lim_{x \to 5} f(x) \neq f(5)$

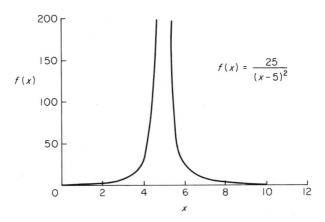

Figure 3.11   Discontinuity at $x = 5$ Where $\lim_{x \to 5} f(x) = \infty$

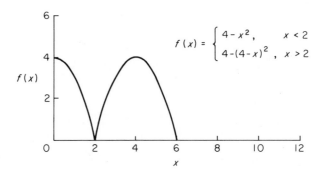

Figure 3.12   Discontinuity at $x = 2$ Where $f(x)$ Is Undefined

When $\lim_{X \to A_0} f(X)$ exists but $f(X)$ is discontinuous at A_0, the discontinuity is called removable. A removable discontinuity is one which can be eliminated by redefining the value of the function at the point at which the discontinuity occurs. Therefore, if $\lim_{X \to A_0} f(X)$ exists, defining $f(A_0)$ as $\lim_{X \to A_0} f(X)$ renders the function continuous at A_0. In Fig. 3.10, if $f(5)$ is assigned the value 5, $f(x)$ becomes continuous at $x = 5$. Similarly the discontinuity shown in Fig. 3.12 may be removed by defining $f(2)$ as 0. If $\lim_{X \to A_0} f(X)$ does not exist, the discontinuity at A_0 is essential since there is no finite value of $f(A_0)$ which will satisfy the conditions given in Definition 3.18. Examples of essential discontinuities are given in Figs. 3.9 and 3.11.

Example 3.15   Show that the following function is continuous at $[1, 1]$.

$$f(X) = (x + y)^n, \qquad n > 1$$

3.4 FUNCTIONS, LIMITS, CONTINUITY, AND THE DERIVATIVE

Let

$$A_0 = [1, 1]$$

Then

$$f(A_0) = 2^n$$

$f(X)$ is continuous at A_0 if

$$\lim_{X \to A_0} f(X) = 2^n$$

We select x and y such that

$$0 < \|[x, y] - [1, 1]\| < \delta$$

or

$$0 < \sqrt{(x - 1)^2 + (y - 1)^2} < \delta$$

For $\gamma > 1$ and $\beta > 1$ and

$$x = 1 \pm \frac{\delta}{\sqrt{2\beta}}, \qquad y = 1 \pm \frac{\delta}{\sqrt{2\gamma}}$$

the above inequality is satisfied. For these values of x and y we have

$$|(x + y)^n - 2^n| = \left| \left[\pm \delta \left(\frac{\beta + \gamma}{\sqrt{2\beta\gamma}} \right) + 2 \right]^n - 2^n \right|$$

Since we are interested in the behavior of $f(X)$ near A_0, we may choose δ arbitrarily small. Let $0 < \delta < 1$. Then

$$\left| \left[\pm \delta \left(\frac{\beta + \gamma}{\sqrt{2\beta\gamma}} \right) + 2 \right]^n - 2^n \right| < \left| \left[\pm \delta \left(\frac{\beta + \gamma}{\sqrt{2\beta\gamma}} \right) + 2 \right]^n - (2 - \delta)^n \right|$$

Each of the terms within the absolute value on the right-hand side of this inequality is expanded using the binomial expansion given by

$$(a + b)^n = \sum_{i=0}^{n} \frac{n!}{i!(n - i)!} a^i b^{n-i}$$

Therefore

$$\left[\pm \delta \left(\frac{\beta + \gamma}{\sqrt{2\beta\gamma}} \right) + 2 \right]^n = \sum_{i=0}^{n} \frac{n!}{i!(n - i)!} (\pm \delta)^i \left(\frac{\beta + \gamma}{\sqrt{2\beta\gamma}} \right)^i 2^{n-i}$$

$$(2 - \delta)^n = \sum_{i=0}^{n} \frac{n!}{i!(n - i)!} (-\delta)^i 2^{n-i}$$

and

$$\left[\pm \delta \left(\frac{\beta + \gamma}{\sqrt{2\beta\gamma}} \right) + 2 \right]^n - (2 - \delta)^n = \sum_{i=0}^{n} \frac{n!}{i!(n - i)!} \delta^i 2^{n-i} \left[\left(\frac{\pm \beta + \gamma}{\sqrt{2\beta\gamma}} \right)^i - (-1)^i \right]$$

Since $0 < \delta < 1$,

$$\left[\pm\delta\left(\frac{\beta+\gamma}{\sqrt{2\beta\gamma}}\right)+2\right]^n - (2-\delta)^n < \sum_{i=0}^{n}\frac{n!}{i!(n-i)!}\left[\left(\frac{\pm\beta\pm\gamma}{\sqrt{2\beta\gamma}}\right)^i - (-1)^i\right]2^{n-i}$$

$$< \delta\left[\left(\frac{\pm\beta\pm\gamma}{\sqrt{2\beta\gamma}}+2\right)^n - 1\right]$$

Then, for $0 < \delta < 1$,

$$|(x+y)^n - 2^n| < \delta\left|\left(\frac{\pm\beta\pm\gamma}{\sqrt{2\beta\gamma}}+2\right)^n - 1\right|$$

Letting

$$\epsilon = \delta\left|\left(\frac{\pm\beta\pm\gamma}{\sqrt{2\beta\gamma}}+2\right)^n - 1\right|$$

$|(x+y)^n - 2^n| < \epsilon$ whenever $0 < \|(x,y) - (1,1)\| < \delta$.
Therefore

$$\lim_{X \to A_0} f(X) = 2^n = f(A_0)$$

and $f(X)$ is continuous at A_0.

Example 3.16 If

$$f(x) = \begin{cases} x+1, & x \leq 1 \\ (x+1)^2, & x > 1 \end{cases}$$

show that an essential discontinuity exists at $x = 1$.
At $x = 1$,

$$f(x) = 2$$

To show that an essential discontinuity exists at $x = 1$, we show that

$$\lim_{x \to 1-} f(x) = 2 \quad \text{and} \quad \lim_{x \to 1+} f(x) = 4$$

We consider the limit from the left first. If $x = 1 - (\delta/2)$ and $\delta > 0$, x satisfies the inequality

$$0 < (1-x) < \delta \quad \text{and} \quad |f(x) - 2| = \left|2 - \frac{\delta}{2} - 2\right| = \frac{\delta}{2}$$

Letting $\epsilon = \delta/2$, $|f(x) - 2| < \epsilon$ whenever $0 < (1-x) < \delta$. Since $\epsilon \to 0$ as $\delta \to 0$

$$\lim_{x \to 1-} f(x) = 2$$

For the right-hand limit, we choose $x = 1 + (\delta/2)$ with the result that

$$0 < (x-1) < \delta$$

3.4 FUNCTIONS, LIMITS, CONTINUITY, AND THE DERIVATIVE

and
$$|f(x) - 4| = \left|\left(2 + \frac{\delta}{2}\right)^2 - 4\right| = \frac{\delta}{2}\left(4 + \frac{\delta}{2}\right)$$

If $\epsilon = (\delta/2)(4 + \delta/2)$, then $|f(x) - 4| < \epsilon$ whenever $0 < (x - 1) < \delta$. Since $\epsilon \to 0$ as $\delta \to 0$
$$\lim_{x \to 1+} f(x) = 4$$

Therefore, $\lim_{x \to 1} f(x)$ does not exist and an essential discontinuity exists at $x = 1$.

In Example 3.16
$$\lim_{x \to 1-} f(x) = f(1) \quad \text{but} \quad \lim_{x \to 1+} f(x) \neq f(1)$$
and therefore $f(x)$ is discontinuous at $x = 1$. However, $f(x)$ is said to be continuous from the left at $x = 1$, since $\lim_{x \to 1-} f(x) = f(1)$. In general, $f(x)$ is continuous from the left at $x = a_0$ if and only if
$$\lim_{x \to a_0-} f(x) = f(a_0) \tag{3.59}$$
and from the right if
$$\lim_{x \to a_0+} f(x) = f(a_0) \tag{3.60}$$
Therefore, $f(x)$ is continuous at a_0 if and only if
$$\lim_{x \to a_0-} f(x) = \lim_{x \to a_0+} f(x) = f(a_0) \tag{3.61}$$

If $f(X)$ is continuous at each X in the space $\mathcal{X}_n$ then $f(X)$ is said to be continuous on $\mathcal{X}_n$. If $\mathcal{X}_n$ is not open and A_0 is a boundary point for $\mathcal{X}_n$ such that A_0 belongs to $\mathcal{X}_n$, then $f(X)$ is continuous at A_0 if
$$\lim_{\substack{X \to A_0 \\ X \in \mathcal{X}_n}} f(X) = f(A_0) \tag{3.62}$$

That is, in the limit given in Eq. (3.62), the path of approach to A_0 is restricted to a set of points belonging to $\mathcal{X}_n$. If the limit taken over all such paths to A_0 is equal to $f(A_0)$, then $f(X)$ is continuous at this boundary point. This concept can be clarified by considering a function of a single variable. Let $f(x)$ be defined for all x on the interval $(a, b]$. If
$$\lim_{x \to b-} f(x) = f(b) \tag{3.63}$$
then $f(x)$ is continuous at $f(b)$ even though $\lim_{x \to b+} f(x)$ does not exist.

Example 3.17 Show that
$$f(x) = x^3$$
is continuous on the open interval $(0, b)$.

Let a_0 be any point on the interval $(0, b)$. Then
$$f(a_0) = a_0^3$$
If $f(x)$ is continuous at a_0, then
$$\lim_{x \to a_0} f(x) = a_0^3$$
From Theorem 3.1,
$$\lim_{x \to a_0} x^3 = \left[\lim_{x \to a_0} x\right]^3$$
Let
$$g(x) = x$$
For $0 < |x - a_0| < \delta$,
$$|g(x) - a_0| = |x - a_0| < \delta$$
Therefore, whenever $0 < |x - a_0| < \delta$, then $|g(x) - a_0| < \delta$ and
$$\lim_{x \to a_0} g(x) = a_0 \quad \text{and} \quad \lim_{x \to a_0} f(x) = a_0^3$$
Hence, $f(x)$ is continuous at all points, a_0, on the interval $(0, b)$. ▨

In Example 3.17, $f(x)$ is both continuous and bounded on the interval $(0, b)$. Suppose we have a function $f(x)$ which is unbounded on its interval of definition. Can it be continuous throughout its interval of definition? For example, consider the function $1/(x - 1)$ defined on the interval $(1, b)$. This function is unbounded on the interval $(1, b)$ since it increases without bound as x decreases towards unity. However, the function is continuous at every point on $(1, b)$. On the other hand, $1/(x - 1)$ is not continuous at every point on the interval $[1, b)$ since it is discontinuous at $x = 1$ where $\lim_{x \to 1} f(x) = \infty$. The reason that $1/(x - 1)$ is continuous throughout $(1, b)$ is that $\lim_{x \to a_0} f(x)$ and $f(a_0)$ exist and are equal for every a_0 on $(1, b)$ since this open interval does not include the point $a_0 = 1$ which is the only point for which $\lim_{x \to a_0} f(x)$ does not exist. On the other hand, the interval $[1, b)$ includes $a_0 = 1$ and therefore $f(x)$ is discontinuous at a point on $[1, b)$ and is thus not continuous throughout the interval. We prove this result in the following example.

Example 3.18 Show that
$$f(x) = \frac{1}{(x - 1)}$$
is continuous on $(1, b)$ but is not continuous on $[1, b)$.

Let a_0 be any point on the interval $(1, b)$. Then
$$f(a_0) = \frac{1}{(a_0 - 1)}$$

3.4 FUNCTIONS, LIMITS, CONTINUITY, AND THE DERIVATIVE

If
$$\lim_{x \to a_0} f(x) = \frac{1}{(a_0 - 1)}$$
then $f(x)$ is continuous at each point on $(1, b)$ and therefore throughout $(1, b)$. Choose x such that
$$0 < |x - a_0| < \delta$$
This inequality is satisfied if $x = a_0 \pm \delta/2$, where $a_0 > 1$. Therefore
$$|f(x) - f(a_0)| = \left|\frac{1}{\left(a_0 \pm \frac{\delta}{2} - 1\right)} - \frac{1}{(a_0 - 1)}\right| = \frac{|\delta|}{2}\left|\frac{1}{a_0 - 1}\right|\left|\frac{1}{a_0 \pm \frac{\delta}{2} - 1}\right|$$

Since
$$\left|a_0 \pm \frac{\delta}{2} - 1\right| \geq |a_0 - 1| - \left|\frac{\delta}{2}\right|$$
we have for $\delta < 2|a_0 - 1| > 0$
$$\left|\frac{1}{a_0 \pm \frac{\delta}{2} - 1}\right| \leq \frac{1}{|a_0 - 1| - \left|\frac{\delta}{2}\right|} < \frac{2}{|a_0 - 1| - \left|\frac{\delta}{2}\right|} > 0$$

Letting
$$\epsilon = \frac{\delta}{|a_0 - 1|\left[|a_0 - 1| - \left|\frac{\delta}{2}\right|\right]}$$

$|f(x) - f(a_0)| < \epsilon$ whenever $0 < |x - a_0| < \delta$. Since ϵ approaches 0 as δ approaches 0
$$\lim_{x \to a_0} f(x) = f(a_0)$$
and $f(x)$ is continuous throughout $(1, b)$.

To show that $f(x)$ is not continuous throughout $[1, b)$, we must show that a discontinuity exists at $x = 1$ since we have shown that a discontinuity does not exist elsewhere in the interval. We accomplish this by showing that
$$\lim_{x \to 1+} f(x) = \infty$$
We let $x = 1 + \delta/2$ and thus satisfy the inequality
$$0 < (x - 1) < \delta$$
Then
$$f(x) = \frac{2}{\delta} > \frac{1}{\delta}$$

Letting $A = 1/\delta$, we obtain $f(x) > A$ whenever $0 < (x - 1) < \delta$. Therefore, as x approaches 1 from the right, $f(x)$ increases without limit and

$$\lim_{x \to 1+} f(x) = \infty$$ ◻

Theorem 3.2 If the functions $f(x)$ and $g(x)$ are continuous at a_0, then $f(x) + g(x)$ and $f(x)g(x)$ are continuous at a_0, and $f(x)/g(x)$ is continuous at a_0 if $g(a_0) \neq 0$.

The rate of change of the function $f(x)$ at the point x_0 can be determined by evaluating the derivative of $f(x)$ at x_0. As we shall see in Chapter IV, the derivative plays an important role in classical optimization theory.

Definition 3.19 The derivative of $f(x)$ at x_0 exists and is defined by

$$f'(x_0) = \lim_{\Delta x \to 0} \frac{f(x_0 + \Delta x) - f(x_0)}{\Delta x} \qquad (3.64)$$

if the limit given exists and if $f(x)$ is defined on some neighborhood of x_0.

As indicated in the discussion of limits, the limit in Eq. (3.64) exists if and only if the left- and right-hand limits exist and are equal. The limit from the left defines the left-hand derivative given by

$$f'_-(x_0) = \lim_{\Delta x \to 0-} \frac{f(x_0 + \Delta x) - f(x_0)}{\Delta x} \qquad (3.65)$$

while the right-hand derivative is defined by

$$f'_+(x_0) = \lim_{\Delta x \to 0+} \frac{f(x_0 + \Delta x) - f(x_0)}{\Delta x} \qquad (3.66)$$

Although the derivative of $f(x)$ at the point x_0 is often expressed as $f'(x_0)$, this is by no means the only expression for the derivative. Those familiar with basic calculus are undoubtedly familiar with the equivalent expressions $df(x_0)/dx$ or $df(x)/dx|_{x_0}$. That is

$$\frac{d}{dx}f(x_0) = \frac{d}{dx}f(x)\bigg|_{x_0} = f'(x_0) \qquad (3.67)$$

Since all three forms are equivalent, they will be used interchangeably throughout the remainder of this and succeeding chapters.

Example 3.19 Find the derivative of $f(x)$ at x_0 if

$$f(x) = x^2$$

From Eq. (3.64),

$$f'(x_0) = \lim_{\Delta x \to 0} \frac{(x_0 + \Delta x)^2 - x_0^2}{\Delta x}$$

$$= \lim_{\Delta x \to 0} \frac{2\,\Delta x\, x_0 + (\Delta x)^2}{\Delta x} = \lim_{\Delta x \to 0} [2x_0 + \Delta x] = 2x_0$$

3.4 FUNCTIONS, LIMITS, CONTINUITY, AND THE DERIVATIVE

If the derivative of $f(x)$ exists at x_0, then $f(x)$ is continuous at x_0. However, the converse is not necessarily true. The validity of the first assertion is given in the proof of Theorem 3.3 and the second is demonstrated in Example 3.20.

Theorem 3.3 If $f'(x_0)$ exists, $f(x)$ is continuous at x_0.

Proof For any $\Delta x \neq 0$,

$$f(x_0 + \Delta x) - f(x_0) = \frac{f(x_0 + \Delta x) - f(x_0)}{\Delta x} \Delta x$$

Taking the limit of both sides of this equation as Δx approaches 0 and applying Eq. (3.43) yields us

$$\lim_{\Delta x \to 0} [f(x_0 + \Delta x) - f(x_0)] = \lim_{\Delta x \to 0} \frac{f(x_0 + \Delta x) - f(x_0)}{\Delta x} \lim_{\Delta x \to 0} \Delta x$$
$$= f'(x_0) \cdot 0 = 0$$

since $f'(x_0)$ exists by assumption. From Eq. (3.42),

$$\lim_{\Delta x \to 0} f(x_0 + \Delta x) = f(x_0)$$

since $f(x_0)$ is not a function of Δx. Therefore

$$\lim_{x \to x_0} f(x) = f(x_0)$$

and $f(x)$ is continuous at x_0. Note that $f(x_0)$ is defined since $f'(x_0)$ does not exist otherwise.

Example 3.20 By example, show that if a function $f(x)$ is continuous at x_0, the derivative, $f'(x_0)$, may not exist.
    Let

$$f(x) = \begin{cases} x^2, & x \leq 1 \\ x, & x > 1 \end{cases}$$

We will show that $f(x)$ is continuous at $x = 1$, but that $f'(1)$ does not exist. From the definition of $f(x)$,

$$f(1) = 1$$

To show that $f(x)$ is continuous at $x = 1$, we must show that

$$\lim_{x \to 1-} f(x) = \lim_{x \to 1+} f(x) = 1 \qquad (3.68)$$

For the left-hand limit, choose $x = 1 - \delta/2$ for $\delta > 0$. Then

$$0 < (1 - x) < \delta \quad \text{and} \quad |f(x) - 1| = \left|\left(1 - \frac{\delta}{2}\right)^2 - 1\right| = \left|\frac{\delta}{2}\right|\left|\frac{\delta}{2} - 2\right|$$

For $\delta < 1$, $|\delta/2 - 2| < 2$. Letting $\epsilon = 2(\delta/2)$, we have that $|f(x) - 1| < \epsilon$ whenever $0 < (1 - x) < \delta$. Since ϵ approaches zero as δ approaches zero, we have

$$\lim_{x \to 1-} f(x) = 1$$

For the right-hand limit, we choose $x = 1 + \delta/2$ for $\delta > 0$. Then

$$0 < (x - 1) < \delta \quad \text{and} \quad |f(x) - 1| = \left|\left(1 + \frac{\delta}{2}\right) - 1\right| = \frac{\delta}{2}$$

Therefore, $|f(x) - 1| < \delta$ whenever $(x - 1) < \delta$ and

$$\lim_{x \to 1+} f(x) = 1$$

Since Eq. (3.68) is valid, $f(x)$ is continuous at $x = 1$. The left-hand derivative at $x = 1$ is given by

$$f'_-(1) = \lim_{\Delta x \to 0-} \frac{(1 + \Delta x)^2 - 1}{\Delta x} = \lim_{\Delta x \to 0-} [2 + \Delta x] = 2$$

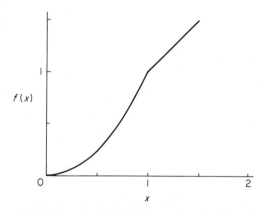

Figure 3.13 Function $f(x)$ Defined in Example 3.20

For the right-hand derivative at $x = 1$, we have

$$f'_+(1) = \lim_{\Delta x \to 0+} \frac{(1 + \Delta x) - 1}{\Delta x} = \lim_{\Delta x \to 0+} (1) = 1$$

Since

$$f'_-(1) \neq f'_+(1)$$

$f'(1)$ does not exist. The function $f(x)$ is shown graphically in Fig. 3.13.

Theorem 3.4 Let $f(x)$ and $g(x)$ be functions such that $f'(x)$ and $g'(x)$ exist on the open interval (a, b). Then, for x on (a, b)

3.4 FUNCTIONS, LIMITS, CONTINUITY, AND THE DERIVATIVE

a. $\dfrac{d}{dx}[f(x) \pm g(x)] = f'(x) \pm g'(x)$ \hfill (3.69)

b. $\dfrac{d}{dx}[f(x)g(x)] = f(x)g'(x) + g(x)f'(x)$ \hfill (3.70)

c. $\dfrac{d}{dx}\left[\dfrac{f(x)}{g(x)}\right] = \dfrac{g(x)f'(x) - f(x)g'(x)}{[g(x)]^2}, \quad g(x) \neq 0$ \hfill (3.71)

Proof The proof of part a follows directly from Theorem 3.1.

For part b we have

$$\frac{d}{dx}[f(x)g(x)] = \lim_{\Delta x \to 0}\left[\frac{f(x+\Delta x)g(x+\Delta x) - f(x)g(x)}{\Delta x}\right] \quad (3.72)$$

Adding and subtracting $f(x + \Delta x)g(x)$ in the numerator of Eq. (3.72) yields us

$$\frac{d}{dx}[f(x)g(x)]$$

$$= \lim_{\Delta x \to 0} \frac{[f(x+\Delta x) - f(x)]g(x) + f(x+\Delta x)[g(x+\Delta x) - g(x)]}{\Delta x}$$

$$= g(x)\lim_{\Delta x \to 0}\left[\frac{f(x+\Delta x) - f(x)}{\Delta x}\right] + \lim_{\Delta x \to 0} f(x+\Delta x)\left[\frac{g(x+\Delta x) - g(x)}{\Delta x}\right]$$

$$= g(x)f'(x) + f(x)g'(x)$$

from Theorem 3.1.

For part c, we have

$$\frac{d}{dx}\left[\frac{f(x)}{g(x)}\right] = \lim_{\Delta x \to 0}\left[\frac{\dfrac{f(x+\Delta x)}{g(x+\Delta x)} - \dfrac{f(x)}{g(x)}}{\Delta x}\right]$$

$$= \lim_{\Delta x \to 0}\left[\frac{\dfrac{f(x+\Delta x)g(x) - g(x+\Delta x)f(x)}{g(x+\Delta x)g(x)}}{\Delta x}\right]$$

$$= \lim_{\Delta x \to 0}\left[\frac{1}{g(x+\Delta x)g(x)}\right] \lim_{\Delta x \to 0}\left[\frac{f(x+\Delta x)g(x) - g(x+\Delta x)f(x)}{\Delta x}\right]$$

$$= \frac{1}{[g(x)]^2} \lim_{\Delta x \to 0}\left[\frac{f(x+\Delta x)g(x) - g(x+\Delta x)f(x)}{\Delta x}\right] \quad (3.73)$$

if $g(x) \neq 0$. Adding and subtracting $f(x)g(x)$ in the numerator of Eq. (3.73) gives us

$$\lim_{\Delta x \to 0}\left[\frac{f(x+\Delta x)g(x) - g(x+\Delta x)f(x)}{\Delta x}\right] = \lim_{\Delta x \to 0}\left[\frac{g(x)[f(x+\Delta x) - f(x)]}{\Delta x}\right]$$

$$- \lim_{\Delta x \to 0}\left[\frac{f(x)[g(x+\Delta x) - g(x)]}{\Delta x}\right]$$

$$= g(x)f'(x) - f(x)g'(x)$$

and

$$\frac{d}{dx}\left[\frac{f(x)}{g(x)}\right] = \frac{g(x)f'(x) - f(x)g'(x)}{g(x)^2}$$

In Theorem 3.1, we showed that the limit of the ratio of two functions is equal to the ratio of the limits of the two functions when both limits exist and when the limit of the function in the denominator is nonzero. However, the limit of a ratio may exist even when the limits of the individual functions do not exist or when the individual limits are identically zero. To illustrate, consider the functions $f(x)$ and $g(x)$ defined by

$$f(x) = ax, \qquad g(x) = bx$$

As x approaches ∞, $f(x)$ and $g(x)$ approach infinity. Although Theorem 3.1 does not apply in attempting to evaluate $\lim_{x \to \infty} f(x)/g(x)$, this limit can be easily shown to be a/b.

When the ratio of two limits is of the form $0/0$ or ∞/∞, the limit of the ratio is said to be indeterminate. In general, indeterminate forms are not resolved as simply as in the example given above. However, we now have the tools at hand to treat the problem posed by indeterminate forms under certain conditions.

Theorem 3.5 L'Hospital's Rule Let $f(x)$ and $g(x)$ be functions which are differentiable throughout the interval (b, c). If

a. $g'(x) \neq 0$ for x on (b, c)

b. $\lim_{x \to a_0} f(x) = 0$ and $\lim_{x \to a_0} g(x) = 0$ or $\lim_{x \to a_0} f(x) = \infty$

and

$$\lim_{x \to a_0} g(x) = \infty \quad \text{for} \quad a_0 \text{ on } (b, c)$$

c. $\lim_{x \to a_0} \frac{f'(x)}{g'(x)} = L$ \hfill (3.74)

then

$$\lim_{x \to a_0} \frac{f(x)}{g(x)} = L \tag{3.75}$$

where b and c may be finite or infinite.

Example 3.21 Show that

$$\lim_{x \to \infty} \frac{x}{\exp[x]} = 0$$

3.4 FUNCTIONS, LIMITS, CONTINUITY, AND THE DERIVATIVE

Since

$$\lim_{x \to \infty} x = \lim_{x \to \infty} \exp[x] = \infty$$

we have the indeterminate form ∞/∞. From Theorem 3.5, if $\lim_{x \to 0} f'(x)/g'(x)$ exists, then

$$\lim_{x \to \infty} \frac{f(x)}{g(x)} = \lim_{x \to \infty} \frac{f'(x)}{g'(x)}$$

But

$$\lim_{x \to \infty} \frac{f'(x)}{g'(x)} = \lim_{x \to \infty} \frac{1}{\exp[x]} = \lim_{x \to \infty} \exp[-x] = 0$$

Therefore

$$\lim_{x \to \infty} \frac{x}{\exp[x]} = 0$$

Example 3.22 Evaluate

$$\lim_{x \to 0} \frac{x^n}{1 - a^x}$$

Let

$$f(x) = x^n, \quad g(x) = 1 - a^x, \quad \text{and} \quad \lim_{x \to 0} \frac{x^n}{1 - a^x} = \frac{0}{0}$$

Taking derivatives of $f(x)$ and $g(x)$ gives us

$$\lim_{x \to 0} \frac{nx^{n-1}}{-a^x \ln(a)} = 0 \quad \text{and} \quad \lim_{x \to 0} \frac{x^n}{1 - a^x} = 0$$

Example 3.23 Evaluate

$$\lim_{x \to \infty} \left[\frac{a + bx}{c + dx}\right]^n$$

Let

$$f(x) = (a + bx)^n, \quad g(x) = (c + dx)^n$$

Taking first derivatives, we have

$$\lim_{x \to \infty} \frac{f'(x)}{g'(x)} = \lim_{x \to \infty} \frac{b(a + bx)^{n-1}}{d(c + dx)^{n-1}} = \frac{\infty}{\infty}$$

Thus

$$\lim_{x \to \infty} \left[\frac{a + bx}{c + dx}\right]^n \quad \text{and} \quad \lim_{x \to \infty} \frac{b(a + bx)^{n-1}}{d(c + dx)^{n-1}}$$

are both indeterminate. Reapplying Theorem 3.5 to $\lim_{x\to\infty} f'(x)/g'(x)$ yields us

$$\lim_{x\to\infty} \frac{f^2(x)}{g^2(x)} = \lim_{x\to\infty} \frac{b^2(a+bx)^{n-2}}{d^2(a+bx)^{n-2}} = \frac{\infty}{\infty}$$

Again we have an indeterminate form. Inspection of $f(x)$ and $g(x)$ indicates that application of Theorem 3.5 m times leads to

$$\lim_{x\to\infty} \frac{f^m(x)}{g^m(x)} = \lim_{x\to\infty} \frac{b^m(a+bx)^{n-m}}{d^m(c+dx)^{n-m}} = \frac{\infty}{\infty}, \qquad m < n$$

Therefore, the limit is resolved after successively applying the results of Theorem 3.5 n times, or

$$\lim_{x\to\infty} \frac{f^n(x)}{g^n(x)} = \lim_{x\to\infty} \left(\frac{b}{d}\right)^n = \left(\frac{b}{d}\right)^n \quad \text{and} \quad \lim_{x\to\infty} \left[\frac{a+bx}{c+dx}\right]^n = \left(\frac{b}{d}\right)^n \qquad \text{▨}$$

Occasionally the analyst is faced with the problem of differentiating a function with respect to a variable, x, where the function is not a simple expression in x. For example, suppose that we wish to differentiate the function $f(t)$ with respect to x, at x_0, where

$$t = g(z), \qquad z = h(x)$$

Then

$$f(t) = f[g(z)] = f\{g[h(x)]\}$$

so that f is ultimately a function of x in that t is related to x through the functions g and h. The function $f(t)$ is called a composite function. Let us consider an example.

$$f(t) = \frac{1}{t}, \qquad t = z^n, \quad n > 1, \qquad z = x + 1$$

Then

$$f(t) = f\{g[h(x)]\} = \frac{1}{(x+1)^n}$$

Obviously the derivative of $f(t)$ with respect to x at x_0 is given by

$$\left.\frac{d}{dx}f(t)\right|_{x_0} = -\frac{n}{(x_0+1)^{n+1}}$$

This approach can be used to evaluate the derivative of a composite function in general. However, the chain rule for differentiation offers an alternative which may facilitate the process. ▨

Theorem 3.6 Chain Rule for Differentiation Let $y = f(t)$ be differentiable on the interval (a, b) and let $t = g(x)$ be differentiable on an interval (c, d) such that when x belongs to (c, d), $g(x)$ belongs to (a, b). If

3.4 FUNCTIONS, LIMITS, CONTINUITY, AND THE DERIVATIVE

$$y_0 = f(t_0), \qquad t_0 = g(x_0)$$

Then
$$\left.\frac{dy}{dx}\right|_{x_0} = \left.\frac{df(t)}{dt}\right|_{t_0} \left.\frac{dg(x)}{dx}\right|_{x_0} \qquad (3.76)$$

Proof Let
$$\Delta y = f(t_0 + \Delta t) - f(t_0) \qquad (3.77)$$

and
$$\Delta t = g(x_0 + \Delta x) - g(x_0) \qquad (3.78)$$

where $\Delta t \neq 0$ and $\Delta x \neq 0$. Further let γ be defined as

$$\gamma = \frac{\Delta y}{\Delta t} - \left.\frac{dy}{dt}\right|_{t_0} \qquad (3.79)$$

where
$$\left.\frac{dy}{dt}\right|_{t_0} = \lim_{\Delta t \to 0} \frac{\Delta y}{\Delta t} \qquad (3.80)$$

since Δy is as defined in Eq. 3.77. Then Δy may be expressed by

$$\Delta y = \left.\frac{dy}{dt}\right|_{t_0} \Delta t + \gamma \Delta t \qquad (3.81)$$

Dividing Eq. (3.81) by Δx yields us

$$\frac{\Delta y}{\Delta x} = \left.\frac{dy}{dt}\right|_{t_0} \frac{\Delta t}{\Delta x} + \gamma \frac{\Delta t}{\Delta x} \qquad (3.82)$$

Since
$$\lim_{\Delta x \to 0} \frac{\Delta y}{\Delta x} = \left.\frac{dy}{dx}\right|_{x_0}$$

we have

$$\left.\frac{dy}{dx}\right|_{x_0} = \lim_{\Delta x \to 0} \left.\frac{dy}{dt}\right|_{t_0} \frac{\Delta t}{\Delta x} + \lim_{\Delta x \to 0} \gamma \frac{\Delta t}{\Delta x} = \left.\frac{dy}{dt}\right|_{t_0} \lim_{\Delta x \to 0} \frac{\Delta t}{\Delta x} + \lim_{\Delta x \to 0} \gamma \frac{\Delta t}{\Delta x} \qquad (3.83)$$

where $dy/dt|_{t_0}$ is treated as a constant and the limit of a constant is the constant itself. Now

$$\lim_{\Delta x \to 0} \frac{\Delta t}{\Delta x} = \left.\frac{dt}{dx}\right|_{x_0} \qquad (3.84)$$

and

$$\lim_{\Delta x \to 0} \gamma \frac{\Delta t}{\Delta x} = \lim_{\Delta x \to 0} \gamma \lim_{\Delta x \to 0} \frac{\Delta t}{\Delta x}$$

Examination of Eq. (3.78) indicates that as Δx approaches zero, Δt approaches zero and as Δt approaches zero, $\Delta y/\Delta t$ approaches $dy/dt\,|_{t_0}$. Therefore

$$\lim_{\Delta x \to 0} \gamma = \lim_{\Delta x \to 0} \frac{\Delta y}{\Delta t} - \frac{dy}{dt}\bigg|_{t_0} = \lim_{\Delta t \to 0} \frac{\Delta y}{\Delta t} - \frac{dy}{dt}\bigg|_{t_0} = \frac{dy}{dt}\bigg|_{t_0} - \frac{dy}{dt}\bigg|_{t_0} = 0 \quad (3.85)$$

and

$$\lim_{\Delta x \to 0} \frac{\Delta t}{\Delta x} = \frac{dt}{dx} \quad (3.86)$$

Substituting the results of Eqs. (3.84), (3.85), and (3.86) into Eq. (3.83), gives us

$$\frac{dy}{dx}\bigg|_{x_0} = \frac{dy}{dt}\bigg|_{t_0} \frac{dt}{dx}\bigg|_{x_0} \quad (3.87)$$

or

$$\frac{dy}{dx}\bigg|_{x_0} = \frac{df(t)}{dt}\bigg|_{t_0} \frac{dg(x)}{dx}\bigg|_{x_0}$$

The chain rule for differentiation of composite functions can be extended to the case where f is composed of an arbitrary number of other functions. For example, if

$$y = f(t), \quad t = g(u), \quad u = h(v), \quad v = s(x)$$

then

$$\frac{dy}{dx}\bigg|_{x_0} = \frac{df(t)}{dt}\bigg|_{t_0} \frac{dg(u)}{du}\bigg|_{u_0} \frac{dh(v)}{dv}\bigg|_{v_0} \frac{ds(x)}{dx}\bigg|_{x_0} \quad (3.88)$$

where

$$t_0 = g(u_0), \quad u_0 = h(v_0), \quad v_0 = s(x_0)$$

With this extension of the chain rule in mind, let us return to the example presented initially.

Example 3.24 Let

$$y = f(t), \quad t = g(z), \quad z = h(x)$$

where

$$f(t) = \frac{1}{t}, \quad g(z) = z^n, \quad n > 1, \quad h(x) = x + 1$$

Find $dy/dx\,|_{x_0}$.

By the chain rule we have

$$\frac{dy}{dx}\bigg|_{x_0} = \frac{df(t)}{dt}\bigg|_{t_0} \frac{dg(z)}{dz}\bigg|_{z_0} \frac{dh(x)}{dx}\bigg|_{x_0} \quad (3.89)$$

3.4 FUNCTIONS, LIMITS, CONTINUITY, AND THE DERIVATIVE

For the three derivatives on the right-hand side of Eq. (3.89) we have

$$\frac{df(t)}{dt} = -t^{-2}, \qquad \frac{dg(z)}{dz} = nz^{n-1}, \qquad \frac{dh(x)}{dx} = 1$$

By the definitions of t and z, we have

$$z_0 = (x_0 + 1), \qquad t_0 = z_0^n = (x_0 + 1)^n$$

Therefore

$$\left.\frac{df(t)}{dt}\right|_{t_0} = -[(x_0 + 1)^n]^{-2}$$

$$\left.\frac{dg(z)}{dz}\right|_{z_0} = n(x_0 + 1)^{n-1}, \qquad \left.\frac{dh(x)}{dx}\right|_{x_0} = 1$$

Multiplying these three terms together leads to

$$\left.\frac{dy}{dx}\right|_{x_0} = -\frac{n}{(x_0 + 1)^{n+1}}$$

The derivative of a function was initially defined as the rate of change of the function $f(x)$ at some point x_0. In many applications of the calculus, particularly optimization theory, the analyst is also interested in the rate of change of the derivative, $f'(x)$, at x_0. The rate of change of $f'(x)$ at x_0 is called the second derivative and is denoted by

$$f^2(x_0), \qquad \frac{d^2}{dx^2}f(x_0), \qquad \text{or} \qquad \left.\frac{d^2}{dx^2}f(x)\right|_{x_0}$$

where x_0 is the point at which the second derivative is evaluated. We therefore frequently refer to $f'(x_0)$ as the first derivative at x_0 to distinguish it from higher-order derivatives. In general, if it exists, the nth derivative of $f(x)$ at x_0, $f^n(x_0)$, can be expressed as the first derivative of $f^{n-1}(x_0)$. The superscript, n, is used to denote the order of the derivative. Therefore, if $f^n(x_0)$ exists it can be expressed by

$$f^n(x_0) = \frac{d}{dx}f^{n-1}(x_0) \tag{3.90}$$

As one might expect, $f^n(x_0)$ in Eq. (3.90) can be replaced by

$$\frac{d^n}{dx^n}f(x_0) \qquad \text{or} \qquad \left.\frac{d^n}{dx^n}f(x)\right|_{x_0}.$$

Example 3.25 Find and plot the first four derivatives of $f(x)$ on the interval $(0, 1)$ where

$$f(x) = x^4$$

The first derivative is given by

$$f'(x_0) = 4x_0^3$$

for all x_0 on (0, 1). To obtain the second derivative we take the first derivative of the first derivative.

$$f^2(x_0) = \frac{d}{dx} f'(x_0) = 12x_0^2$$

Proceeding in a similar manner to find the third derivative we have

$$f^3(x_0) = \frac{d}{dx} f^2(x_0) = 24x_0$$

Reapplying the procedure gives us

$$f^4(x_0) = 24$$

These four derivatives are shown graphically in Fig. 3.14.

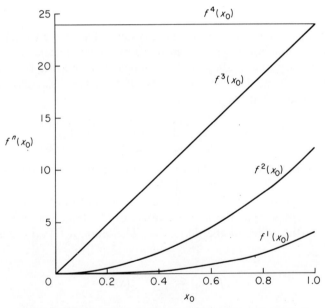

Figure 3.14 First Four Derivatives of x^4 on the Interval (0, 1)

3.4.1 Partial Differentiation

In the preceding section we defined the derivative of $f(x)$ at x_0 as the rate of change of $f(x)$ at x_0. Let

$$\Delta y = f(x_0 + \Delta x) - f(x_0)$$

3.4 FUNCTIONS, LIMITS, CONTINUITY, AND THE DERIVATIVE

Then Δy measures the incremental change in $f(x)$ corresponding to an incremental change, Δx, in x or Δy is the change in $f(x)$ corresponding to an increase in x from x_0 to $x_0 + \Delta x$. Thus $df(x)/dx$ measures the rate of change of $f(x)$ as x increases. In this sense, the derivative measures the rate of change of a function with respect to the change in x in a specified direction.

Now let us consider the problem of finding the derivative of a function of several variables. Let $f(x, y)$ be a function of the variables x and y. When we refer to the derivative of $f(x, y)$ we are again referring to the rate of change of $f(x, y)$. However, we must specify the rate of change of $f(x, y)$ with respect to a specified change in x and y. The rate of change in $f(x, y)$ in the direction x for constant y is called the partial derivative of $f(x, y)$ with respect to x.

Definition 3.20 The partial derivative of $f(x_1, x_2, \ldots, x_n)$ with respect to x_i at X_0 is defined by

$$\left. \frac{\partial}{\partial x_i} f(x_1, \ldots, x_n) \right|_{X_0}$$
$$= \lim_{\Delta x_i \to 0} \frac{f(x_{10}, \ldots, x_{i0} + \Delta x_i, \ldots, x_{n0}) - f(x_{10}, \ldots, x_{i0}, \ldots, x_{n0})}{\Delta x_i} \quad (3.91)$$

if the limit exists where

$$X_0 = [x_{10}, \ldots, x_{i0}, \ldots, x_{n0}] \quad (3.92)$$

The partial derivative expressed in Eq. (3.91) defines the rate of change of $f(x_1, \ldots, x_n)$ with respect to x_i while x_j, $j \neq i$, are held constant. That is, the change occurs in the direction specified by the vector $[\alpha_1, \alpha_2, \ldots, \alpha_i, \ldots, \alpha_n]$, where $\alpha_i = 1$ and $\alpha_j = 0, j \neq i$. Now suppose that we wish to examine the rate of change of $f(x_1, x_2, \ldots, x_n)$ in a direction other than along the coordinate x_i, $i = 1, 2, \ldots, n$. We will define a direction as a vector of unit length. Thus, $[\alpha_1, \alpha_2, \ldots, \alpha_n]$ is a direction if

$$\sqrt{\alpha_1^2 + \alpha_2^2 + \cdots + \alpha_n^2} = 1 \quad (3.93)$$

The rate of change of $f(x_1, x_2, \ldots, x_n)$ in the direction $[\alpha_1 \alpha_2, \ldots, \alpha_n]$ is called the directional derivative.

Definition 3.21 Let α be the direction defined by the vector $[\alpha_1, \alpha_2, \ldots, \alpha_n]$ and let

$$X_0 = [x_{10}, x_{20}, \ldots, x_{n0}]$$

The *directional derivative* of $f(X)$ at X_0 in the direction α is given by

$$\nabla_\alpha f(X_0) = \lim_{\lambda \to 0} \frac{f(X_0 + \lambda \alpha) - f(X_0)}{\lambda} \quad (3.94)$$

if the limit exists.

The reader will note that if $\alpha_i = 1$ and $\alpha_j = 0$ for $j \neq i$

$$\nabla_\alpha f(X_0) = \frac{\partial}{\partial x_i} f(X_0) \qquad (3.95)$$

The expression for the directional derivative can be simplified by defining the operational vector called the gradient, ∇.

$$\nabla = \begin{bmatrix} \partial/\partial x_1 \\ \partial/\partial x_2 \\ \vdots \\ \partial/\partial x_n \end{bmatrix} \qquad (3.96)$$

Then $\nabla f(x_1, x_2, \ldots, x_n)$ is an $n \times 1$ vector of partial derivatives or

$$\nabla f(x_1, x_2, \ldots, x_n) = \begin{bmatrix} \frac{\partial}{\partial x_1} f(x_1, x_2, \ldots, x_n) \\ \frac{\partial}{\partial x_2} f(x_1, x_2, \ldots, x_n) \\ \vdots \\ \frac{\partial}{\partial x_n} f(x_1, x_2, \ldots, x_n) \end{bmatrix} \qquad (3.97)$$

Multiplying the vector $\nabla f(x_1, x_2, \ldots, x_n)$ by α gives an alternative expression for the directional derivative.

$$\nabla_\alpha f(X_0) = \alpha \, \nabla f(X_0) \qquad (3.98)$$

Example 3.26 Find the directional derivative of $f(x, y, z)$ in the direction $[1/\sqrt{2}, \frac{1}{2}, \frac{1}{2}]$ at $[1, 1, 1]$ and $[0, 0, 0]$, where

$$f(x, y, z) = (x - 1)^2 + (y - 1)^2 + (z - 1)^2$$

Applying Eq. (3.94) we have

$$\alpha = \left[\frac{1}{\sqrt{2}}, \frac{1}{2}, \frac{1}{2}\right], \quad X_0 = [1, 1, 1], \quad X_0 + \lambda\alpha = \left[1 + \frac{\lambda}{\sqrt{2}}, 1 + \frac{\lambda}{2}, 1 + \frac{\lambda}{2}\right]$$

and

$$\nabla_\alpha f(X_0) = \lim_{\lambda \to 0} \frac{\left(\frac{\lambda}{\sqrt{2}}\right)^2 + \left(\frac{\lambda}{2}\right)^2 + \left(\frac{\lambda}{2}\right)^2}{\lambda} = \lim_{\lambda \to 0} (\lambda) = 0$$

At

$$X_0 = [0, 0, 0]$$

we have

$$X_0 + \lambda\alpha = \left[\frac{\lambda}{\sqrt{2}}, \frac{\lambda}{2}, \frac{\lambda}{2}\right]$$

and
$$\nabla_\alpha f(X_0) = \lim_{\lambda \to 0} \frac{\left(\frac{\lambda}{\sqrt{2}} - 1\right)^2 + \left(\frac{\lambda}{2} - 1\right)^2 + \left(\frac{\lambda}{2} - 1\right)^2 - 3}{\lambda}$$
$$= \lim_{\lambda \to 0} \frac{\lambda^2 - (2 + \sqrt{2})\lambda}{\lambda} = -(2 + \sqrt{2})$$

Example 3.27 Let
$$f(x, y, z) = \frac{x}{yz}$$

Find
$$\left.\frac{\partial}{\partial x} f(x, y, z)\right|_{X_0}, \quad \left.\frac{\partial}{\partial y} f(x, y, z)\right|_{X_0}, \quad \text{and} \quad \left.\frac{\partial}{\partial z} f(x, y, z)\right|_{X_0}$$

where $X_0 = [1, 1, 2]$.

From Eq. (3.91)
$$\frac{\partial}{\partial x} f(x, y, z) = \lim_{\Delta x \to 0} \frac{\frac{x + \Delta x}{yz} - \frac{x}{yz}}{\Delta x} = \lim_{\Delta x \to 0} \frac{1}{yz} = \frac{1}{yz}$$

and
$$\left.\frac{\partial}{\partial x} f(x, y, z)\right|_{X_0} = \frac{1}{2}$$

For $(\partial/\partial y) f(x, y, z)$, we have
$$\frac{\partial}{\partial y} f(x, y, z) = \lim_{\Delta y \to 0} \frac{\frac{x}{z(y + \Delta y)} - \frac{x}{zy}}{\Delta y}$$
$$= \frac{x}{z} \lim_{\Delta y \to 0} \frac{\frac{y - (y + \Delta y)}{y(y + \Delta y)}}{\Delta y} = \frac{x}{z} \lim_{\Delta y \to 0} \frac{-1}{y(y + \Delta y)} = -\frac{x}{y^2 z}$$

and
$$\left.\frac{\partial}{\partial y} f(x, y, z)\right|_{X_0} = -\frac{1}{2}$$

By a similar argument
$$\frac{\partial}{\partial z} f(x, y, z) = -\frac{x}{yz^2} \quad \text{and} \quad \left.\frac{\partial}{\partial z} f(x, y, z)\right|_{X_0} = -\frac{1}{4}$$

The above example illustrates the point that the partial derivative of a function with respect to x_i is derived by taking the ordinary first derivative of the function while treating the variables $x_j, j \neq i$, as constants.

As in the case of the ordinary derivative, one is frequently interested in higher-order partial derivatives. Let $f_i(x_1, x_2, \ldots, x_n)$ be the first partial derivative or simply the partial derivative of $f(x_1, x_2, \ldots, x_n)$ with respect to x_i. The second partial derivative with respect to x_i is given by

$$f_{ii}(x_1, x_2, \ldots, x_n) = \frac{\partial}{\partial x_i} f_i(x_1, x_2, \ldots, x_n) \tag{3.99}$$

In a similar manner we might wish to evaluate the second partial derivative with respect to x_i and x_j. To evaluate this second-order partial derivative we take the first partial of $f_i(x_1, x_2, \ldots, x_n)$ with respect to x_j, or equivalently the first partial of $f_j(x_1, x_2, \ldots, x_n)$ with respect to x_i. Therefore

$$f_{ij}(x_1, x_2, \ldots, x_n) = \frac{\partial}{\partial x_j} f_i(x_1, x_2, \ldots, x_n)$$

$$= \frac{\partial}{\partial x_i} f_j(x_1, x_2, \ldots, x_n) = f_{ji}(x_1, x_2, \ldots, x_n) \tag{3.100}$$

Higher-order partial derivatives will prove most useful when we treat the problem of optimization of functions of several variables.

Example 3.28 Find $f_{11}(x, y), f_{12}(x, y), f_{22}(x, y)$ where

$$f(x, y) = \frac{x}{y}$$

To find the second-order partial derivatives, we must first find $f_1(x, y)$ and $f_2(x, y)$.

$$f_1(x, y) = \frac{\partial}{\partial x} f(x, y) = \frac{1}{y}$$

$$f_2(x, y) = \frac{\partial}{\partial y} f(x, y) = -\frac{x}{y^2}$$

From Eqs. (3.96) and (3.97)

$$f_{11}(x, y) = \frac{\partial}{\partial x} f_1(x, y) = 0$$

$$f_{12}(x, y) = \frac{\partial}{\partial y} f_1(x, y) = -\frac{1}{y^2}$$

$$f_{22}(x, y) = \frac{\partial}{\partial y} f_2(x, y) = 2\frac{x}{y^3}$$

3.5 Integration

Let us consider the problem of determining the area included between the x axis and the function $f(x)$, for x on the interval $[a, b]$. To illustrate, suppose that the graph of $f(x)$ is as given in Fig. 3.15, where the shaded portion of the graph is the area of interest. We will approach the problem of determining the area enclosed by a function over a given interval by approximating the area by a series

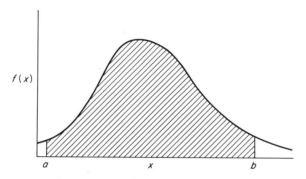

Figure 3.15 Area under $f(x)$ on the Interval $[a, b]$

of rectangles. Let $A_{a,b}[f(x)]$ be the area enclosed by $f(x)$ on the closed interval $[a, b]$. Now let us divide the interval into n_1 subintervals $[x_0, x_1]$, $[x_1, x_2]$, ..., $[x_{n_1-1}, x_{n_1}]$, where $x_0 = a$ and $x_{n_1} = b$. Let Δ_i be the width of the interval $[x_{i-1}, x_i]$. Then

$$\sum_{i=1}^{n_1} \Delta_i = b - a \qquad (3.101)$$

Let M_i and m_i be defined by

$$M_i = \max_{x_{i-1} \leq x \leq x_i} f(x) \qquad (3.102)$$

$$m_i = \min_{x_{i-1} \leq x < x_i} f(x) \qquad (3.103)$$

That is, M_i is the maximum value of $f(x)$ in the ith subinterval, while m_i is the minimum value of $f(x)$ there.

Now let us construct two sets of n_1 rectangles each. Each set will be used to approximate the area $A_{a,b}[f(x)]$. The ith rectangle in the first set has base Δ_i, height M_i, and area $M_i \Delta_i$, while the ith rectangle in the second set has base Δ_i, height m_i, and area $m_i \Delta_i$. Superimposing these rectangles on the graph of $f(x)$, we

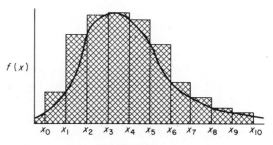

Figure 3.16

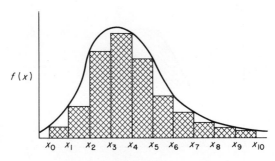

Figure 3.17

obtain two approximations to $A_{a,b}[f(x)]$ as shown in Figs. 3.16 and 3.17. These two approximations have areas given by

$$\bar{A}_{a,b,n_1}[f(x)] = \sum_{i=1}^{n_1} M_i \, \Delta_i \qquad (3.104)$$

and

$$\underline{A}_{a,b,n_1}[f(x)] = \sum_{i=1}^{n_1} m_i \, \Delta_i \qquad (3.105)$$

where $\bar{A}_{a,b,n_1}[f(x)]$ is called the *outer area* or the *upper Riemann sum* and $\underline{A}_{a,b,n_1}[f(x)]$ is called the *inner area* or the *lower Riemann sum*. Since $M_i \geq m_i$, then $\bar{A}_{a,b,n_1}[f(x)] \geq \underline{A}_{a,b,n_1}[f(x)]$.

Now suppose that the number of subintervals is increased to n_2 such that the width of each subinterval Δ_i is reduced. Then, if $f(x)$ is bounded on $[a, b]$,

$$\bar{A}_{a,b,n_2}[f(x)] \leq \bar{A}_{a,b,n_1}[f(x)] \qquad (3.106)$$

and

$$\underline{A}_{a,b,n_2}[f(x)] \geq \underline{A}_{a,b,n_1}[f(x)] \qquad (3.107)$$

Definition 3.22 If $f(x)$ is defined and bounded on $[a, b]$ and

$$\lim_{n \to \infty} \bar{A}_{a,b,n}[f(x)] = \lim_{n \to \infty} \underline{A}_{a,b,n}[f(x)] \qquad (3.108)$$

3.5 INTEGRATION

where $\Delta_i \to 0$ as $n \to \infty$ for all i, then $f(x)$ is said to be integrable on $[a, b]$ and

$$\int_a^b f(x)\,dx = \lim_{n \to \infty} \bar{A}_{a,b,n}[f(x)] = \lim_{n \to \infty} \underline{A}_{a,b,n}[f(x)] \qquad (3.109)$$

The integral defined by Eq. (3.109) is called the *Riemann* or *definite integral* of $f(x)$ on $[a, b]$.

Example 3.29 Evaluate

$$\int_a^b x\,dx$$

If x is integrable on the interval $[a, b]$ then

$$\lim_{n \to \infty} \bar{A}_{a,b,n}[x] = \lim_{n \to \infty} \underline{A}_{a,b,n}[x]$$

To evaluate the inner and outer areas, we divide the interval $[a, b]$ into n equal subintervals, each of width $(b - a)/n$ or

$$\Delta_i = \frac{b - a}{n}, \qquad i = 1, 2, \ldots, n$$

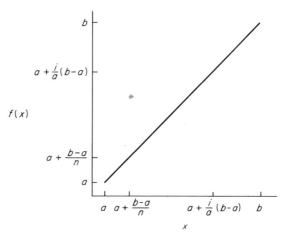

Figure 3.18 Subdivision of the Interval $[a, b]$ into n Equal Subintervals

The subdivision of this interval is shown in Fig. 3.18. The inner area for n subintervals is given by

$$\underline{A}_{a,b,n}[x] = \sum_{i=1}^{n} \left[a + \frac{(i-1)}{n}(b - a) \right] \frac{(b - a)}{n}$$

where

$$\frac{b - a}{n} = \Delta_i \qquad \text{and} \qquad a + \frac{(i-1)}{n}(b - a) = m_i$$

Then

$$\underline{A}_{a,b,n}[x] = \frac{(b-a)}{n} \sum_{i=1}^{n} \left[a + \frac{(i-1)}{n}(b-a) \right]$$

$$= \frac{(b-a)}{n} \left[na + (b-a)\frac{n(n-1)}{2n} \right] = (b-a)\left[a + \frac{(b-a)}{2}\left(1 - \frac{1}{n}\right) \right]$$

For the outer area we have

$$\Delta_i = \frac{b-a}{n}, \qquad M_i = \frac{i}{n}(b-a) + a$$

and

$$\bar{A}_{a,b,n}[x] = \sum_{i=1}^{n} \left[a + \frac{i}{n}(b-a) \right] \frac{(b-a)}{n} = (b-a)\left[a + \frac{(b-a)}{2}\left(1 + \frac{1}{n}\right) \right]$$

Taking the limit of $\underline{A}_{a,b,n}[x]$ and $\bar{A}_{a,b,n}[x]$ as $n \to \infty$ leads to

$$\lim_{n \to \infty} \underline{A}_{a,b,n}[x] = \lim_{n \to \infty} (b-a)\left[a + \frac{(b-a)}{2}\left(1 - \frac{1}{n}\right) \right] = (b-a)\frac{(b+a)}{2} = \frac{b^2 - a^2}{2}$$

and

$$\lim_{n \to \infty} \bar{A}_{a,b,n}[x] = \lim_{n \to \infty} (b-a)\left[a + \frac{(b-a)}{2}\left(1 + \frac{1}{n}\right) \right] = (b-a)\frac{(b+a)}{2} = \frac{b^2 - a^2}{2}$$

Since

$$\lim_{n \to \infty} \underline{A}_{a,b,n}[x] = \lim_{n \to \infty} \bar{A}_{a,b,n}[x]$$

x is integrable on $[a, b]$ and

$$\int_a^b x \, dx = \frac{b^2 - a^2}{2}$$

Example 3.30 Show that

$$\int_0^1 \exp[x] \, dx = e - 1$$

To evaluate the inner and outer areas we divide the interval $[0, 1]$ into n subintervals of equal width. For the inner area

$$m_i = \exp\left[\frac{i-1}{n}\right], \qquad \Delta_i = \frac{1}{n}$$

3.5 INTEGRATION

and

$$\underline{A}_{0,1,n}[\exp[x]] = \sum_{i=1}^{n} \frac{1}{n} \exp\left[\frac{i-1}{n}\right] = \frac{1}{n} \sum_{i=1}^{n} \left(\exp\left[\frac{1}{n}\right]\right)^{i-1}$$

$$= \frac{1}{n} \frac{1 - \left(\exp\left[\frac{1}{n}\right]\right)^n}{1 - \exp\left[\frac{1}{n}\right]} = \frac{1-e}{n\left(1 - \exp\left[\frac{1}{n}\right]\right)}$$

To evaluate the outer area, we have

$$M_i = \exp\left[\frac{i}{n}\right], \qquad \Delta_i = \frac{1}{n}$$

and

$$\overline{A}_{0,1,n}[\exp[x]] = \sum_{i=1}^{n} \frac{1}{n} \exp\left[\frac{i}{n}\right] = \frac{1}{n} \frac{1 - \left(\exp\left[\frac{1}{n}\right]\right)^{n+1}}{1 - \exp\left[\frac{1}{n}\right]} - 1$$

$$= \frac{\exp\left[\frac{1}{n}\right] - \exp\left[\frac{n+1}{n}\right]}{n\left(1 - \exp\left[\frac{1}{n}\right]\right)} = \frac{1-e}{n\left(\exp\left[-\frac{1}{n}\right] - 1\right)}$$

Taking the limit of $\underline{A}_{0,1,n}[\exp[x]]$ and $\overline{A}_{0,1,n}[\exp[x]]$ as $n \to \infty$ leads to

$$\lim_{n \to \infty} \underline{A}_{0,1,n}[\exp[x]] = \frac{1-e}{\lim_{n \to \infty} n\left(1 - \exp\left[\frac{1}{n}\right]\right)}$$

and

$$\lim_{n \to \infty} \overline{A}_{0,1,n}[\exp[x]] = \frac{1-e}{\lim_{n \to \infty} n\left(\exp\left[-\frac{1}{n}\right] - 1\right)}$$

Now

$$\lim_{n \to \infty} n\left(1 - \exp\left[\frac{1}{n}\right]\right) = \lim_{n \to \infty} \frac{1 - \exp\left[\frac{1}{n}\right]}{\frac{1}{n}} = \frac{0}{0}$$

Applying L'Hospital's rule to this indeterminant form yields

$$\lim_{n\to\infty} n\left(1 - \exp\left[\frac{1}{n}\right]\right) = \lim_{n\to\infty} \frac{\frac{1}{n^2}\exp\left[\frac{1}{n}\right]}{-\frac{1}{n^2}} = -1$$

Similarly, $\lim_{n\to\infty} n(\exp[-1/n] - 1)$ is indeterminant and is resolved as follows

$$\lim_{n\to\infty} n\left(\exp\left[-\frac{1}{n}\right] - 1\right) = \lim_{n\to\infty} \frac{\exp\left[-\frac{1}{n}\right] - 1}{\frac{1}{n}} = \lim_{n\to\infty} \frac{\frac{1}{n^2}\exp\left[-\frac{1}{n}\right]}{-\frac{1}{n^2}} = -1$$

Therefore

$$\lim_{n\to\infty} \underline{A}_{0,1,n}[\exp[x]] = \lim_{n\to\infty} \overline{A}_{0,1,n}[\exp[x]] = e - 1$$

and

$$\int_0^1 \exp[x]\, dx = e - 1 \qquad \boxed{}$$

The concepts underlying the definition of the definite integral of a function of a single variable can be extended to functions of several variables. For example, suppose that $f(x, y)$ is defined and bounded on the closed space $[\mathscr{X}_2]$. Let us cover $[\mathscr{X}_2]$ by a set of n_1 rectangles, as shown in Fig. 3.19. Let the area of the ith rectangle be a_i. Further let

$$M_i = \max_{[x,\, y]\,\in\, a_i\, \cap\, \mathscr{X}_2} f(x, y) \qquad (3.110)$$

$$m_i = \max_{[x,\, y]\,\in\, a_i\, \cap\, \mathscr{X}_2} f(x, y) \qquad (3.111)$$

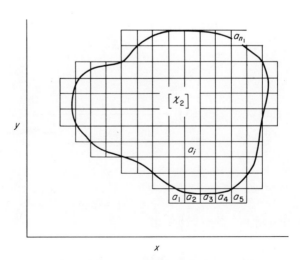

Figure 3.19 Set of n Rectangles Covering $\mathscr{X}_2$

3.5 INTEGRATION

That is, M_i is the maximum value of $f(x, y)$ for which (x, y) belongs to the intersection of a_i and $[\mathscr{X}_2]$. Similarly, m_i is the minimum value of $f(x, y)$ for which (x, y) belongs to the intersection of a_i and $[\mathscr{X}_2]$. We define the upper and lower Riemann sums as

$$\bar{A}_{\mathscr{X}_2, n_1}[f(x, y)] = \sum_{i=1}^{n_1} M_i a_i \quad (3.112)$$

$$\underline{A}_{\mathscr{X}_2, n_1}[f(x, y)] = \sum_{i=1}^{n_1} m_i a_i \quad (3.113)$$

If we increase the number of rectangles to n_2 such that the area of each rectangle is decreased, then

$$\bar{A}_{\mathscr{X}_2, n_2}[f(x, y)] \leq \bar{A}_{\mathscr{X}_2, n_1}[f(x, y)] \quad (3.114)$$

$$\underline{A}_{\mathscr{X}_2, n_2}[f(x, y)] \geq \underline{A}_{\mathscr{X}_2, n_1}[f(x, y)] \quad (3.115)$$

If we allow the number of rectangles to approach infinity such that the area of each rectangle approaches zero, we obtain a definition of the integral of a function of two variables.

Definition 3.23 If $f(x, y)$ is defined and bounded on $[\mathscr{X}_2]$ and

$$\lim_{n \to \infty} \bar{A}_{\mathscr{X}_2, n}[f(x, y)] = \lim_{n \to \infty} \underline{A}_{\mathscr{X}_2, n}[f(x, y)] \quad (3.116)$$

where $a_i \to 0$ as $n \to \infty$ for all i, then $f(x)$ is integrable on $[\mathscr{X}_2]$ and

$$\iint_{\mathscr{X}_2} f(x, y) \, dx \, dy = \lim_{n \to \infty} \bar{A}_{\mathscr{X}_2, n}[f(x, y)] = \lim_{n \to \infty} \underline{A}_{\mathscr{X}_2, n}[f(x, y)] \quad (3.117)$$

A similar approach can be used to define the integral of a function of three or more variables.

Having defined the integral of a function, let us examine the conditions under which a function is integrable and the properties of integrals.

Theorem 3.7 If $f(x)$ is continuous on $[a, b]$, then $\int_a^b f(x) \, dx$ exists; that is, $f(x)$ is integrable on $[a, b]$.

Theorem 3.8 If $f(x, y)$ is continuous on $[\mathscr{X}_2]$, then $\iint_{\mathscr{X}_2} f(x, y) \, dx \, dy$ exists.

As one might expect, the results of Theorems 3.7 and 3.8 can be extended to continuous functions of three or more variables. The remainder of the discussion of integration will be restricted to functions of a single variable. For a treatment of multiple integrals the reader should see Buck (1956) or Fulks (1961).

Theorem 3.9 If $f(x)$ is continuous on $[a, b]$ and c is an interior point of $[a, b]$, then

$$\int_a^b f(x) \, dx = \int_a^c f(x) \, dx + \int_c^b f(x) \, dx \quad (3.118)$$

Proof Since $f(x)$ is continuous throughout $[a, b]$, it is continuous on $[a, c]$ and $[c, b]$. Then, by Theorem 3.7

$$\lim_{n\to\infty} \bar{A}_{a, c, n}[f(x)] = \lim_{n\to\infty} A_{a, c, n}[f(x)] \tag{3.119}$$

$$\lim_{n\to\infty} \bar{A}_{c, b, n}[f(x)] = \lim_{n\to\infty} A_{c, b, n}[f(x)] \tag{3.120}$$

and

$$\lim_{n\to\infty} \bar{A}_{a, b, n}[f(x)] = \lim_{n\to\infty} A_{a, b, n}[f(x)] \tag{3.121}$$

Let m_1 and m_2 be the number of subintervals on $[a, c]$, and $[c, b]$ respectively, such that $n = m_1 + m_2$. Then

$$\bar{A}_{a, b, n}[f(x)] = \bar{A}_{a, c, m_1}[f(x)] + \bar{A}_{c, b, m_2}[f(x)]$$

As m_1 and m_2 approach infinity, n approaches infinity. Therefore,

$$\lim_{m_1\to\infty} \bar{A}_{a, c, m_1}[f(x)] + \lim_{m_2\to\infty} \bar{A}_{c, b, m_2}[f(x)] = \lim_{n\to\infty} \bar{A}_{a, b, n}[f(x)]$$

But by Eqs. (3.119)–(3.121)

$$\lim_{m_1\to\infty} \bar{A}_{a, c, m_1}[f(x)] = \int_a^c f(x)\,dx$$

$$\lim_{m_2\to\infty} \bar{A}_{c, b, m_2}[f(x)] = \int_c^b f(x)\,dx$$

$$\lim_{n\to\infty} \bar{A}_{a, b, n} f(x) = \int_a^b f(x)\,dx$$

and

$$\int_a^b f(x)\,dx = \int_a^c f(x)\,dx + \int_c^b f(x)\,dx$$

∎

Theorem 3.9 leads to an important result when $f(x)$ is either an even or an odd function. Let $f(x)$ be continuous on the interval $[-a, a]$. If

$$f(-x) = f(x) \tag{3.122}$$

for all x on the interval $[-a, a]$, then $f(x)$ is said to be an *even function*. If

$$f(-x) = -f(x)$$

for all x on the interval $[-a, a]$, then $f(x)$ is said to be an *odd function*. In the following example we show that

$$\int_{-a}^{a} f(x)\,dx = \begin{cases} 0, & \text{for } f(x) \text{ odd} \\ 2\int_{0}^{a} f(x)\,dx, & \text{for } f(x) \text{ even} \end{cases} \tag{3.123}$$

3.5 INTEGRATION

Example 3.31 Prove the result given in Eq. 3.123
From Theorem 3.9

$$\int_{-a}^{a} f(x)\,dx = \int_{-a}^{0} f(x)\,dx + \int_{0}^{a} f(x)\,dx$$

Divide the interval $[-a, a]$ into $2n$ subintervals such that each subinterval is of width Δx. The subintervals for which $x > 0$ will be denoted $1, 2, \ldots, n$ while those

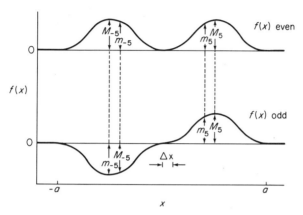

Figure 3.20 Graph of Even and Odd Functions on the Interval $[-a, a]$ Where $[-a, a]$ Is Divided into 20 Subintervals ($n = 10$)

for which $x < 0$ will be denoted $-1, -2, \ldots, -n$. Graphs of even and odd functions and the interval subdivision are shown in Fig. 3.20. Then

$$\bar{A}_{-a,0,n}[f(x)] = \sum_{i=-n}^{-1} M_i\,\Delta x, \qquad \underline{A}_{-a,0,n}[f(x)] = \sum_{i=-n}^{-1} m_i\,\Delta x$$

$$\bar{A}_{0,a,n}[f(x)] = \sum_{i=1}^{n} M_i\,\Delta x, \qquad \underline{A}_{0,a,n}[f(x)] = \sum_{i=1}^{n} m_i\,\Delta x$$

If $f(x)$ is odd on $[-a, a]$

$$M_i = -m_{-i} \qquad m_i = -M_{-i}$$

and

$$-\sum_{i=1}^{n} M_i\,\Delta x = \sum_{i=-n}^{-1} m_i\,\Delta x, \qquad -\sum_{i=1}^{n} m_i\,\Delta x = \sum_{i=-n}^{-1} M_i\,\Delta x$$

Since $f(x)$ is continuous throughout $[-a, a]$

$$\int_{-a}^{0} f(x)\,dx = \lim_{n\to\infty} \sum_{i=-n}^{-1} M_i\,\Delta x = \lim_{n\to\infty} \sum_{i=-n}^{-1} m_i\,\Delta x$$

$$= -\lim_{n\to\infty} \sum_{i=1}^{n} m_i\,\Delta x = -\lim_{n\to\infty} \sum_{i=1}^{n} M_i\,\Delta x = -\int_{0}^{a} f(x)\,dx$$

Therefore
$$\int_{-a}^{a} f(x)\,dx = -\int_{0}^{a} f(x)\,dx + \int_{0}^{a} f(x)\,dx = 0$$

If $f(x)$ is even on $[-a, a]$, then
$$M_i = M_{-i}, \qquad m_i = m_{-i}$$
and
$$\sum_{i=1}^{n} M_i\,\Delta x = \sum_{i=-n}^{-1} M_i\,\Delta x, \qquad \sum_{i=1}^{n} m_i\,\Delta x = \sum_{i=-n}^{-1} m_i\,\Delta x$$

Therefore
$$\int_{-a}^{0} f(x)\,dx = \lim_{n\to\infty} \sum_{i=-n}^{-1} M_i\,\Delta x = \lim_{n\to\infty} \sum_{i=-n}^{-1} m_i\,\Delta x$$
$$= \lim_{n\to\infty} \sum_{i=1}^{n} M_i\,\Delta x = \lim_{n\to\infty} \sum_{i=1}^{n} m_i\,\Delta x = \int_{0}^{a} f(x)\,dx$$

and
$$\int_{-a}^{a} f(x)\,dx = \int_{0}^{a} f(x)\,dx + \int_{0}^{a} f(x)\,dx = 2\int_{0}^{a} f(x)\,dx \qquad \blacksquare$$

Theorem 3.10 Mean-Value Theorem for Integrals If $f(x)$ is continuous on $[a, b]$ then there exists a point x_0 on (a, b) such that

$$\int_{a}^{b} f(x)\,dx = (b-a)f(x_0) \tag{3.124}$$

Mean value properties play an important role throughout differential and integral calculus. For example, if $f(x)$ is continuous on $[a, b]$ and differentiable on (a, b), then there exists a point x_0 on (a, b) such that

$$f'(x_0) = \frac{f(b) - f(a)}{b - a} \tag{3.125}$$

The relationship given in Eq. (3.125) is called the *mean-value theorem* or the *mean-value theorem for derivatives*. Relationships similar to Eq. (3.125) can also be developed for multivariate functions under appropriate assumptions of continuity and differentiability. A more complete discussion of mean-value properties is presented in Chapter 4 where we take up the topic of optimization of continuous functions. The proof of Theorem 3.10 can be developed based upon the results given in that chapter.

Example 3.32 Find a point x_0 on the interval $(0, 10)$ such that

$$\int_{0}^{10} f(x)\,dx = 10 f(x_0), \qquad \text{where} \quad f(x) = \exp[x], \quad 0 \le x \le 10$$

3.5 INTEGRATION

Since
$$\lim_{y \to x} f(y) = \exp[x]$$
for all x on $[0, 10]$, $f(x)$ is continuous on $[0, 10]$. Of course the limits at $x = 0$ and $x = 10$ are one-sided. Now
$$\int_0^{10} \exp[x] \, dx = \exp[10] - 1$$
This result can be shown by a simple extension of Example 3.30. Therefore
$$\exp[x_0] = \frac{\exp[10] - 1}{10} \quad \text{and} \quad x_0 = \ln\left[\frac{\exp[10] - 1}{10}\right] \quad \text{⬚}$$

Assume that $f(y)$ is continuous on $[a, b]$ and therefore integrable there. Let
$$F(x) = \int_a^x f(y) \, dy \tag{3.126}$$
where x is any point on the interval of definition of $f(y)$ and $F(a) = 0$. Then there exists a basic relationship between the derivative of $F(x)$ and the value of $f(x)$.

Definition 3.24 Let $F(x)$ be a function such that
$$F'(x) = f(x) \tag{3.127}$$
for all x on the interval $[a, b]$. Then $F(x)$ is called an *antiderivative* or *indefinite integral* of $f(x)$ on the interval $[a, b]$.

We will now show that the antiderivative of a continuous function always exists.

Theorem 3.11 If $f(x)$ is continuous on $[a, b]$, then $f(x)$ has an antiderivative on $[a, b]$.

Proof Let
$$F(x) = \int_a^x f(y) \, dy$$
where x lies on the interval $[a, b]$. The derivative of $F(x)$ is given by
$$F'(x) = \lim_{\Delta x \to 0} \frac{F(x + \Delta x) - F(x)}{\Delta x}$$
where the limit is two-sided if x is an interior point for $[a, b]$ and one-sided if $x = a$ or $x = b$. Then
$$F'(x) = \lim_{\Delta x \to 0} \frac{\int_a^{x + \Delta x} f(x) \, dx - \int_a^x f(x) \, dx}{\Delta x}$$

From Theorem 3.9,

$$\int_a^{x+\Delta x} f(x)\,dx - \int_a^x f(x)\,dx = \int_x^{x+\Delta x} f(x)\,dx$$

and

$$F'(x) = \lim_{\Delta x \to 0} \frac{\int_x^{x+\Delta x} f(x)\,dx}{\Delta x}$$

From the mean-value theorem for integrals, there exists a point x_0 such that

$$\int_x^{x+\Delta x} f(x)\,dx = \Delta x\, f(x_0)$$

where x_0 lies on the interval $(x, x + \Delta x)$. $F'(x)$ then becomes

$$F'(x) = \lim_{\Delta x \to 0} f(x_0)$$

Since x_0 lies on $(x, x + \Delta x)$, as $\Delta x \to 0$, $x_0 \to x$ and

$$F'(x) = f(x)$$

If the antiderivative of a function exists, it is not necessarily unique. For example, if $F(x)$ is an antiderivative of $f(x)$, then

$$H(x) = F(x) + c \tag{3.128}$$

is also an antiderivative of $f(x)$, where c is any constant since

$$H'(x) = F'(x) = f(x) \tag{3.129}$$

In the following theorem we show that if $f(x)$ is continuous on $[a, b]$ and its antiderivative is known there, then the integral of $f(x)$ can be evaluated on that interval.

Theorem 3.12   Fundamental Theorem of Integral Calculus   If $f(x)$ is continuous on $[a, b]$ with antiderivative $F(x)$ there, then

$$\int_a^b f(x)\,dx = F(b) - F(a) \tag{3.130}$$

Proof   Let $H(x)$ be any antiderivative of $f(x)$ on $[a, b]$. Then

$$H(x) = \int_a^x f(x)\,dx$$

By Eq. (3.218), if $F(x)$ is any other antiderivative of $f(x)$ on $[a, b]$, then

$$F(x) = H(x) + c$$

Since $H(a) = 0$, then

$$F(a) = c$$

3.5 INTEGRATION

and

$$\int_a^b f(x)\,dx = H(b) = F(b) - c = F(b) - F(a)$$

Example 3.33 Find an antiderivative of $f(x)$, $H(x)$, such that

$$H(10) = 0 \quad \text{where} \quad f(x) = \frac{1}{x^2}, \quad 10 \le x \le 100$$

One antiderivative of $f(x)$ is given by

$$F(x) = -\frac{1}{x} \quad \text{and} \quad \int_{10}^{100} f(x)\,dx = F(100) - F(10) = -\frac{1}{100} + \frac{1}{10}$$

Since $H(x)$ is defined such that

$$H(x) = \int_{10}^{x} f(x)\,dx = F(x) - F(10)$$

we have

$$H(x) = -\frac{1}{x} + \frac{1}{10}$$

If $F(x)$ is an antiderivative of $f(x)$, the evaluated integral of $f(x)$ is often expressed by

$$\int_a^b f(x)\,dx = F(x)\Big|_a^b \tag{3.131}$$

where

$$F(x)\Big|_a^b = F(b) - F(a) \tag{3.132}$$

The reader should note that Theorem 3.12 can be used to show that

$$\int_a^a f(x)\,dx = 0 \tag{3.133}$$

Theorem 3.13 If $f(x)$ and $g(x)$ are continuous on $[a, b]$, then

a. $\displaystyle\int_a^b [f(x) \pm g(x)]\,dx = \int_a^b f(x)\,dx \pm \int_a^b g(x)\,dx$ \hfill (3.134)

b. $\displaystyle\int_a^b cf(x)\,dx = c\int_a^b f(x)\,dx$ \hfill (3.135)

where c is a constant.

c. if $g(x) \le f(x)$ on $[a, b]$,

$$\int_a^b g(x)\,dx \le \int_a^b f(x)\,dx \tag{3.136}$$

d. $\int_a^b f(x)g(x)\,dx$ exists

Proof The proofs of parts a, b, and c are left as exercises for the reader. To prove part d we note the result given in Theorem 3.2. Since $f(x)$ and $g(x)$ are continuous at every point x on $[a, b]$, $f(x)g(x)$ is continuous at every point x on $[a, b]$. From Theorem 3.7, $f(x)g(x)$ is integrable on $[a, b]$. ▨

Theorem 3.12 provides the analyst with a simple method for evaluating the integral of a function given that he can identify the antiderivative of the function. However, identification of the antiderivative is not always an easy task, even when a table of integrals is available. This problem can sometimes be resolved through integration by parts.

Theorem 3.14 Integration by Parts Let $g(x)$ and $h(x)$ be continuous, differentiable functions on $[a, b]$. If $g'(x)$ and $h'(x)$ are continuous on $[a, b]$ and if

$$f(x) = g(x)h'(x) \tag{3.137}$$

then

$$\int_a^b f(x)\,dx = g(x)h(x)\Big|_a^b - \int_a^b h(x)g'(x)\,dx \tag{3.138}$$

Proof From Theorem 3.4,

$$\frac{d}{dx}[g(x)h(x)] = g(x)h'(x) + h(x)g'(x)$$

Therefore

$$f(x) = \frac{d}{dx}[g(x)h(x)] - h(x)g'(x)$$

and

$$\int_a^b f(x)\,dx = \int_a^b \left\{\frac{d}{dx}[g(x)h(x)] - h(x)g'(x)\right\}dx$$

By Theorem 3.13,

$$\int_a^b f(x)\,dx = \int_a^b \frac{d}{dx}[g(x)h(x)]\,dx - \int_a^b h(x)g'(x)\,dx$$

Since $g(x)$, $g'(x)$, $h(x)$, and $h'(x)$ are continuous, $(d/dx)[g(x)h(x)]$ is continuous, based upon the results of Theorems 3.2 and 3.4. Therefore, $(d/dx)[g(x)h(x)]$ is integrable and has as its antiderivative $g(x)h(x)$ and

$$\int_a^b f(x)\,dx = g(x)h(x)\Big|_a^b - \int_a^b h(x)g'(x)\,dx \qquad ▨$$

3.5 INTEGRATION

The use of the method of integration by parts is not always as straightforward as the proof of its validity given above. Essentially, useful application of this method of integration depends upon the ability of the analyst to identify the functions $g(x)$ and $h(x)$ such that the integral $\int_a^b h(x)g'(x)\,dx$ can be resolved, where

$$f(x) = g(x)h'(x)$$

Application of the method of integration by parts is demonstrated in the following examples.

Example 3.34 Find $F(x)$ where

$$F(x) = \int_0^x y^2 \exp[-y]\,dy$$

for finite $x > 0$.

That $y^2 \exp[-y]$ is continuous on $[0, x]$ can be shown without great difficulty. Therefore $F(x)$ exists. We wish to define the functions $g(y)$ and $h(y)$ such that

$$g(y)h'(y) = y^2 \exp[-y]$$

Let

$$g(y) = y^2, \qquad h'(y) = \exp[-y]$$

Since $g(y)$ is continuous and differentiable on $[0, x]$ and $h'(y)$ and $g'(y)$ are continuous on $[0, x]$, Theorem 3.14 can be used. The antiderivative of $h'(y)$, $h(y)$, can be expressed by

$$h(y) = -\exp[-y] \qquad \text{and} \qquad g'(y) = 2y$$

Therefore

$$\int_0^x y^2 \exp[-y]\,dy = -y^2 \exp[-y]\Big|_0^x + \int_0^x 2y \exp[-y]\,dy$$

$$= -x^2 \exp[-x] + \int_0^x 2y \exp[-y]\,dy$$

We now have the problem of resolving the integral of $2y \exp[-y]$. Since this function is continuous on $[0, x]$ it is integrable there. If we let

$$g(y) = 2y \qquad \text{and} \qquad h'(y) = \exp[-y]$$

we can reapply the method of integration by parts. Now

$$g'(y) = 2, \qquad h(y) = -\exp[-y]$$

and

$$\int_0^x 2y \exp[-y]\,dy = -2y \exp[-y]\Big|_0^x + \int_0^x 2 \exp[-y]\,dy$$

$$= -2x \exp[-x] + \int_0^x 2 \exp[-y]\,dy$$

Therefore

$$\int_0^x y^2 \exp[-y]\, dy = -x^2 \exp[-x] - 2x \exp[-x] + \int_0^x 2 \exp[-y]\, dy$$

The antiderivative of $2 \exp[-y]$ can be expressed by

$$F(x) = \int_0^x 2 \exp[-y]\, dy = -2 \exp[-x] + 2$$

and

$$\int_0^x y^2 \exp[-y]\, dy = -[x^2 + 2x + 2]\exp[-x] + 2$$

In the above example it was necessary to specify the antiderivative of $h'(y)$ and the antiderivative of $2 \exp[-y]$. Thus the method of integration by parts does not eliminate the need for finding the antiderivative of functions. However, this method can and usually does reduce the complexity of the function for which an antiderivative must be found. Usually one starts the process of integrating a function by attempting to find its antiderivative in a table of integrals. If this fails, the next step is usually an attempt at integration by parts where $g(x)$ and $h'(x)$ are defined such that

$$\int_a^b h(x)g'(x)\, dx$$

can be found in a table of integrals. If the function can be integrated by parts, but the integral of $h(x)g'(x)$ cannot be identified immediately, an attempt to integrate $h(x)g'(x)$ by parts may be made. In other words, integration of the initial function may require successive application of the method of integration by parts. This concept is illustrated in the following example.

Example 3.35 Show that

$$\int_0^t a \frac{(ax)^n}{n!} \exp[-ax]\, dx = 1 - \sum_{i=0}^n \frac{(at)^i}{i!} \exp[-at] \qquad (3.139)$$

where n is a positive integer, $a > 0$,

$$i! = i(i-1)(i-2)\cdots(2)(1) \qquad \text{and} \qquad 0! = 1$$

Let

$$g(x) = \frac{(ax)^n}{n!}, \qquad h'(x) = \exp[-ax]$$

Then

$$g'(x) = a \frac{(ax)^{n-1}}{(n-1)!}, \qquad h(x) = -\frac{1}{a} \exp[-ax]$$

3.5 INTEGRATION

and

$$\int_0^t a\frac{(ax)^n}{n!}\exp[-ax]\,dx = a\left[-\frac{(ax)^n}{an!}\exp[-ax]\Big|_0^t + \int_0^t \frac{(ax)^{n-1}}{(n-1)!}\exp[-ax]\,dx\right]$$

$$= -\frac{(at)^n}{n!}\exp[-at] + a\int_0^t \frac{(ax)^{n-1}}{(n-1)!}\exp[-ax]\,dx$$

Reapplying the method of integration by parts, let

$$g(x) = \frac{(ax)^{n-1}}{(n-1)!}, \qquad h'(x) = \exp[-ax]$$

Therefore

$$g'(x) = a\frac{(ax)^{n-2}}{(n-2)!}, \qquad h(x) = -\frac{1}{a}\exp[-ax]$$

and

$$\int_0^t a\frac{(ax)^n}{n!}\exp[-ax]\,dx = -\frac{(at)^n}{n!}\exp[-at] + a\left[-\frac{(ax)^{n-1}}{a(n-1)!}\exp[-ax]\Big|_0^t\right.$$
$$\left. + \int_0^t \frac{(ax)^{n-2}}{(n-2)!}\exp[-ax]\,dx\right]$$

$$= -\sum_{j=0}^{1}\frac{(at)^{n-j}}{(n-j)!}\exp[-at] + a\int_0^t \frac{(ax)^{n-2}}{(n-2)!}\exp[-ax]\,dx$$

Again applying the method of integration by parts, let

$$g(x) = \frac{(ax)^{n-2}}{(n-2)!}, \qquad h'(x) = \exp[-ax]$$

Then

$$g'(x) = a\frac{(ax)^{n-3}}{(n-3)!}, \qquad h(x) = -\frac{1}{a}\exp[-ax]$$

and

$$\int_0^t a\frac{(ax)^n}{n!}\exp[-ax]\,dx = -\sum_{j=0}^{2}\frac{(at)^{n-j}}{(n-j)!}\exp[-at] + a\int_0^t \frac{(ax)^{n-3}}{(n-3)!}\exp[-ax]\,dx$$

It would appear that k applications of the method of integration by parts leads to

$$\int_0^t a\frac{(ax)^n}{n!}\exp[-ax]\,dx = -\sum_{j=0}^{k-1}\frac{(at)^{n-j}}{(n-j)!}\exp[-at] + a\int_0^t \frac{(ax)^{n-k}}{(n-k)!}\exp[-ax]\,dx$$

(3.140)

where $k < n$. The validity of Eq. (3.140) can be shown by induction. Assume that Eq. (3.140) holds for the first k applications of the method of integration by parts. If Eq. (3.140) also holds for the $(k+1)$st application then it holds for all $k \leq n$. Applying integration by parts to the integral on the right-hand side of Eq. (3.140), let

$$g(x) = \frac{(ax)^{n-k}}{(n-k)!}, \qquad h'(x) = \exp[-ax]$$

As before, we have

$$g'(x) = a\frac{(ax)^{n-k-1}}{(n-k-1)!} \qquad \text{and} \qquad h(x) = -\frac{1}{a}\exp[-ax]$$

Therefore

$$\int_0^t a\frac{(ax)^n}{n!}\exp[-ax]\,dx = -\sum_{j=0}^{k-1}\frac{(at)^{n-j}}{(n-j)!}\exp[-at] + a\left[-\frac{(ax)^{n-k}}{a(n-k)!}\exp[-ax]\bigg|_0^t\right.$$

$$\left.+\int_0^t \frac{(ax)^{n-k-1}}{(n-k-1)!}\exp[-ax]\,dx\right]$$

$$= -\sum_{j=0}^{k}\frac{(at)^{n-j}}{(n-j)!}\exp[-at]$$

$$+ a\int_0^t \frac{(ax)^{n-k-1}}{(n-k-1)!}\exp[-ax]\,dx$$

This completes the proof that Eq. (3.140) holds for $k \leq n$ successive applications of the method of integration by parts. For $k = n$,

$$\int_0^t a\frac{(ax)^n}{n!}\exp[-ax]\,dx = -\sum_{j=0}^{n-1}\frac{(at)^{n-j}}{(n-j)!}\exp[-at] + a\int_0^t \exp[-ax]\,dx$$

$$= -\sum_{j=0}^{n-1}\frac{(at)^{n-j}}{(n-j)!}\exp[-at] + a\frac{\exp[-ax]}{-a}\bigg|_0^t$$

$$= -\sum_{j=0}^{n}\frac{(at)^{n-j}}{(n-j)!}\exp[-at] + 1$$

Since

$$\sum_{j=0}^{n}\frac{(at)^{n-j}}{(n-j)!}\exp[-at] = \sum_{i=0}^{n}\frac{(at)^i}{i!}\exp[-at]$$

Eq. (3.139) is valid. ▧

The reader familiar with probability theory will recognize that $a\{(ax)^n/n!\}\exp[-ax]$ is the density function of the Erlang random variable with parameters a and n. Further the function $\{(at)^i/i!\}\exp[-at]$ is the probability mass function of the Poisson random variable with parameter a for $i = 0, 1, 2, \ldots$. Now, the integral on the left-hand side of Eq. (3.139) represents the cumulative distribu-

3.5 INTEGRATION

tion function of the Erlang random variable, while the summation on the right-hand side of Eq. 3.139 is the cumulative distribution function of the Poisson random variable. Therefore, the cumulative distribution function of the Erlang random variable can be expressed in terms of that for the Poisson random variable. This relationship can be quite useful since tables of the cumulative Poisson distribution are readily available in many texts on statistical methods and quality control as well as handbooks of statistical tables and mathematical tables, while this is not the case for tables of the cumulative Erlang distribution.

Theorem 3.7 gives continuity of $f(x)$ on $[a, b]$ as a sufficient condition for the existence of the integral of $f(x)$ on the interval $[a, b]$. Since continuity is only a sufficient condition, the implication of Theorem 3.7 is that a function may be integrable even when it is not continuous on the closed interval of integration. Let us first consider the case where $f(x)$ is continuous on $(a, b]$ but discontinuous at $x = a$. The integral

$$\int_a^b f(x)\, dx$$

is then called an *improper integral*.

Definition 3.25 If $f(x)$ is continuous on $(a, b]$ and discontinuous at $x = a$, then the improper integral $\int_a^b f(x)\, dx$ exists and has the value

$$\lim_{y \to a+} \int_y^b f(x)\, dx$$

if the limit exists. If $f(x)$ is continuous on $[a, b)$ and discontinuous at $x = b$, then the improper integral $\int_a^b f(x)\, dx$ exists and has the value

$$\lim_{y \to b-} \int_a^y f(x)\, dx$$

if the limit exists.

In Definition 3.25, a and/or b may be replaced by $-\infty$ and ∞, although the interval of integration is then open. For example, consider the integral $\int_a^\infty f(x)\, dx$, where $f(x)$ is continuous on every finite closed interval $[a, x]$. Therefore $f(x)$ is properly integrable on $[a, x]$ and the improper integral exists and has value given by

$$\lim_{y \to \infty} \int_a^y f(x)\, dx$$

if the limit exists. If the limit does not exist, the improper integral is said to *diverge*.

Example 3.36 Show that

$$\int_0^\infty a \frac{(ax)^n}{n!} \exp[-ax]\, dx = 1$$

The function $a\{(ax)^n/n!\}\exp[-ax]$ is continuous and therefore properly integrable on every finite interval $[0, b]$. From Example 3.35,

$$F(x) = 1 - \sum_{i=0}^{n} \frac{(ax)^i}{i!} \exp[-ax]$$

Taking the limit of $F(x)$ as $x \to \infty$

$$\lim_{x \to \infty} F(x) = 1 - \sum_{i=0}^{n} \lim_{x \to \infty} \frac{(ax)^i}{i!} \exp[-ax]$$

But

$$\lim_{x \to \infty} \frac{(ax)^i}{i!} \exp[-ax] = \frac{1}{i!} \lim_{x \to \infty} \frac{(ax)^i}{\exp[ax]} = \frac{\infty}{\infty}$$

Applying L'Hospital's rule,

$$\lim_{x \to \infty} \frac{(ax)^i}{\exp[ax]} = \lim_{x \to \infty} \frac{\frac{d}{dx}(ax)^i}{\frac{d}{dx}\exp[ax]} = \lim_{x \to \infty} \frac{ai(ax)^{i-1}}{a \exp[ax]}$$

which is still indeterminant. Applying L'Hospital's rule i times leads to

$$\lim_{x \to \infty} \frac{(ax)^i}{\exp[ax]} = \lim_{x \to \infty} \frac{\frac{d^i}{dx^i}(ax)^i}{\frac{d^i}{dx^i}\exp[ax]} = \lim_{x \to \infty} \frac{i!}{\exp[ax]} = 0$$

Therefore

$$\lim_{x \to \infty} F(x) = 1$$

Example 3.37 Show that

$$\int_0^1 \frac{dx}{x^k}$$

diverges where $k > 1$.

Since $1/x^k$ is discontinuous at $x = 0$, the integral is improper.

$$\int_y^1 \frac{dx}{x^k} = -\frac{1}{(k-1)x^{k-1}}\bigg|_y^1 = \frac{1}{(k-1)y^{k-1}} - \frac{1}{k-1}$$

Taking the limit as $y \to 0+$,

$$\lim_{y \to 0+} \int_y^1 \frac{dx}{x^k} = \lim_{y \to 0+} \frac{1}{(k-1)y^{k-1}} - \frac{1}{k-1} = \infty$$

and the integral diverges.

Now let us consider the problem of evaluating the integral of $f(x)$ on $[a, b]$ where $f(x)$ is continuous on $[a, b]$ at all but a *finite number of points*. In particular, assume that $f(x)$ is discontinuous at the points $x_1, x_2, \ldots, x_n$ on $[a, b]$. If the integral of $f(x)$ on $[a, b]$ exists, it can be expressed by

$$\int_a^b f(x)\, dx = \int_a^{x_1} f(x)\, dx + \sum_{i=1}^{n-1} \int_{x_i}^{x_{i+1}} f(x)\, dx + \int_{x_n}^b f(x)\, dx \qquad (3.141)$$

Since $f(x)$ is discontinuous at x_i, $i = 1, 2, \ldots, n$ and continuous elsewhere on $[a, b]$, each of the integrals in Eq. (3.141) is an improper integral. If each of the improper integrals in Eq. (3.141) exists, then $\int_a^b f(x)\, dx$ exists.

Problems

1. Show that
   a. $A - B = A \cap (\overline{A \cap B})$
   b. $A \cup (\bar{A} \cap B) = A \cup B$
   c. $A \cup (\bar{A} \cup B) = B$
   d. $\overline{A \cap B} = \bar{A} \cup \bar{B}$
2. Show Venn diagrams representing the sets defined in Problem 1.
3. Let U be the set of products carried in inventory by a company. The set A represents those products which are inspected prior to being placed in inventory. The set B represents those products which are inspected after they are placed in inventory but before they are removed from the inventory. The set C includes all products inspected after they are removed from inventory. No product undergoes more than two inspections although some products are not inspected at all. Draw a Venn diagram representing the products stored by number and type of inspection.
4. For the problem given in Example 3.5, assume that a fourth defect, D, was found during the inspection. In addition to the information given in Example 3.5, 50 units possessed defect D, 4 units possessed A and D, 0 units possessed B and D, and 10 possessed C and D. Defect D was not found in combination with two or more other defects. How many units possessed
   a. defect A only?
   b. defect C only?
   c. defect D only?
5. A maintenance department has 100 repairmen. A one-week period showed 12 men were absent on Monday, 4 on Tuesday, 3 on Wednesday, 3 on Thursday, and 8 on Friday. In addition 1 man was absent on Monday and Wednesday, 5 on Monday and Friday, 2 on Tuesday and Friday, and 1 on Thursday and Friday. No one was absent on any other combination of days of the week. How many men were absent on
   a. Monday only?
   b. Wednesday only?
   c. Friday only?
6. Determine whether the following functions are single- or multiple-valued and bounded or unbounded.
   a. $f(x, y) = 2x + 3y$, $-\infty < x < \infty$, $0 < y \le 5$
   b. $g(x, y) = 1/(x + y)$, $0 < x < \infty$, $0 < y < \infty$
   c. $h(x) = x + \sqrt{x}$, $0 < x \le 6$
7. For those functions in Problem 6 which are bounded from above or below, find the corresponding least upper bounds and greatest lower bounds.
8. Show that
$$\lim_{x \to a_0} C = C$$
where C is constant.

9. Write a definition for
$$\lim_{x \to a_0 -} f(x) = -\infty, \qquad \lim_{x \to a_0 +} f(x) = -\infty$$
$$\lim_{x \to \infty} f(x) = \infty, \qquad \lim_{x \to \infty} f(x) = -\infty$$

10. Write a definition for
$$\lim_{X \to A_0} f(p) = L$$
where $X = (x, y)$ and $A_0 = (\infty, \infty)$.

11. Find

  a. $\lim_{x \to 1}(x^2 + x + 1)$
  b. $\lim_{x \to \infty} \dfrac{1}{x^2 + x}$
  c. $\lim_{x \to 1} \dfrac{1}{(x-1)}$

12. Using Theorem 3.1, show that
$$\lim_{x \to 2} \frac{x^2}{x^2 - 1} = \frac{4}{3}, \qquad \lim_{x \to 3} x(x-1)(x-2) = 6$$

13. Show that
$$\lim_{x \to a+} (1 - x) = 1 - a, \qquad \lim_{x \to 0+} \exp[x] = 1$$
$$\lim_{x \to 0-} \exp\left[-\frac{1}{x}\right] = 0, \qquad \lim_{x \to \infty} \frac{x}{x^2 - 1} = 0$$

14. Show that
$$f(x) = \begin{cases} x \sin\left(\dfrac{1}{x}\right), & x \neq 0 \\ 0, & x = 0 \end{cases}$$
is continuous at $x = 0$.

15. Show that
$$f(x) = 3x^3 - x^2 + 2x$$
is continuous on the open interval $(0, b)$.

16. Show that the derivative of the function
$$f(x) = \begin{cases} x^2, & x \leq 1 \\ 2x, & x > 1 \end{cases}$$
does not exist at $x = 1$.

17. Show that
$$\frac{d}{dx}[af(x)] = af'(x)$$
if $f'(x)$ exists.

PROBLEMS

18. Find all of the first- and second-order partial derivatives of the following functions

    a. $f(x, y) = \exp[a(x + y)]$ b. $f(x, y, z) = \dfrac{1}{xyz}$

    c. $f(x, y) = \dfrac{1}{x} \exp[-y^2]$ d. $f(x, y, z) = a^x b^y c^z$

19. Find the point X_0 such that

    $$f_{11}(x, y)\bigg|_{X_0} = 0 \quad \text{and} \quad f_{22}(x, y)\bigg|_{X_0} = 0$$

    for each of the following functions.
    a. $f(x, y) = x^3 + (y - 4)^3$
    b. $f(x, y) = 3x^3 + 2y^2 - 4x^4 - 2y + 6$

20. Find the directional derivatives of the functions given in problem 19 at the point X_0 for any direction $[\alpha_1, \alpha_2]$.

21. Find the directional derivatives of the following functions at $X_0 = [1, 5]$ in the directions indicated.

    a. $f(x, y) = x^2 + y^2$, $\alpha = \left[\dfrac{1}{\sqrt{2}}, -\dfrac{1}{\sqrt{2}}\right]$

    b. $f(x, y) = y \exp[-2x]$, $\alpha = \left[\dfrac{1}{2}, \dfrac{\sqrt{3}}{2}\right]$

22. Given the following function, find the point X_0 such that

    $$f_1(X_0) = f_2(X_0) = 0, \quad \begin{vmatrix} f_{11}(X_0) & f_{12}(X_0) \\ f_{21}(X_0) & f_{22}(X_0) \end{vmatrix} > 0$$

    $$f(x, y) = x^3 + 4y^2 - 3x - 9$$

23. Using the definition of $\int_a^b f(x)\, dx$, evaluate the following.

    a. $\displaystyle\int_0^1 (1 - x)^2\, dx$ b. $\displaystyle\int_0^1 x \exp[x]\, dx$

24. Find the antiderivatives of the following functions.
    a. x^k, $k > 0$ b. $\exp[ax]$ c. $(a + bx)^n$
25. Prove parts a, b, and c of Theorem 3.13.
26. Integrate the following by parts

    a. $\displaystyle\int_1^{10} x(a + bx)^n\, dx$ b. $\displaystyle\int_1^{10} x \log(x)\, dx$

27. Determine whether the following integrals exist

    a. $\displaystyle\int_0^\infty x\, dx$ b. $\displaystyle\int_0^1 \dfrac{dx}{a + bx}$ c. $\displaystyle\int_0^1 \dfrac{dx}{bx}$ d. $\displaystyle\int_0^\infty \dfrac{dx}{x(a + bx)}$

28. If $\int_a^b f(x)\, dx$ exists, show that

    $$\int_a^b f(x)\, dx = \lim_{n \to \infty} \sum_{i=1}^n f(\xi_i) \Delta_i$$

    where ξ_i is any point on the interval $[x_{i-1}, x_i]$.

References

Buck, R. C., *Advanced Calculus*, New York: McGraw-Hill, 1956.
Fulks, W., *Advanced Calculus: An Introduction to Analysis*, New York: Wiley, (Interscience), 1961.
Kattsoff, L. O., and Simone, A. J., *Finite Mathematics with Applications in the Social and Management Sciences*, New York: McGraw-Hill, 1965.
Kemeny, J. G., Mirkil, H., Snell, J. L., and Thompson, G. L., *Finite Mathematical Structures*, Englewood Cliffs, New Jersey: Prentice-Hall, 1960.
Miller, K. S., *Advanced Real Calculus*, New York: Harper, 1957.
Olmsted, J. M., *Real Variables*, New York: Appleton, 1956.

CHAPTER 4

CLASSICAL OPTIMIZATION THEORY

4.1 Introduction

Central to the analysis of many systems problems is the development of a model which realistically captures the essential characteristics of the system studied. However, the model itself usually will not provide the solution which is sought. It is often the case that the solution pursued requires the definition of those conditions under which optimal operation of the system will be realized. If the model developed is mathematical in form and is a reliable representation of the physical system, it may be manipulated to define the optimal operating conditions. As an illustration, suppose that a manufacturing process produces units at a constant rate of ψ per year. Manufactured units are placed in a final product inventory. Units are withdrawn from the inventory at a constant rate of λ per year, where $\lambda < \psi$. If units are manufactured continuously throughout the year, an inventory surplus of $\psi - \lambda$ units will result at the end of the year. Since only λ units are required per year, the problem to be resolved is the determination of the number of production runs to be scheduled during the year and the number of units of product to be manufactured at each run such that the sum of the costs of production and inventory are minimized. Thus, if we can develop a mathematical model representing the total annual cost of production and inventory as a function of the *number* and *length* of production runs, we can optimize the model with respect to these two *decision variables* and thus identify their values such that total annual cost is minimized.

Let us consider a second example to further illustrate the role of optimization in systems analysis. A company manufactures one of its products at a rate of ψ units per year and guarantees the performance of each product for a period of T years. Each unit which fails costs C_f. The mean failure rate of the product, u, can be controlled but at a cost which is inversely proportional to u and given by C_r/u. The problem is to determine the design value of u such that the sum of C_r/u and the expected cost of failures is minimized. Thus, if we can develop a mathematical model representing total annual cost as a function of u, we can then attempt to find the value of u which minimizes this cost.

The methods presented in this chapter are frequently useful in performing analyses of the type described in the examples just presented. However, it should be noted that the techniques presented in this chapter are not always the most effective in attempting to define the optimum values of decision variables for a given system. Where these techniques do not apply or where they are applicable but inefficient, the analyst may find it useful to resort to one of many *mathematical programming* techniques or to *search* techniques. However, many of these methods are based upon the material presented here.

This chapter is concerned with the optimization of continuous functions of one or more variables. A point will be referred to as an *extreme point* for a function if the function attains a *minimum* or a *maximum value* at that point. As the terms are used here, a function may have many extreme values on any interval of definition. For example, consider the function, $f(x)$, shown in Fig. 4.1. On the

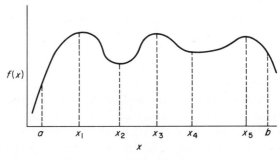

Figure 4.1

closed interval $[a, b]$, $f(x)$ has seven extreme values. $f(x)$ has minima at a, x_2, x_4, and b and maxima at x_1, x_3, and x_5. If we consider only the open interval (a, b), $f(x)$ has five extreme points. When an extreme point is defined on a finite open interval it will be referred to as an *interior extreme point*. To facilitate the discussion of extreme points and extreme values for a continuous function it is useful to introduce the concept of a convex function.

Definition 4.1 Let X be an n-dimensional vector and $f(X)$ a continuous function of X on the open space $(\mathcal{X}_n)$. Let X_1 and X_2 be any two distinct vectors in $(\mathcal{X}_n)$ for which $f(X)$ is defined and let λ be a scalar such that $0 < \lambda < 1$. If for all λ,

$$f[\lambda X_1 + (1 - \lambda)X_2] \leq \lambda f(X_1) + (1 - \lambda)f(X_2) \tag{4.1}$$

then $f(X)$ is said to be *convex from above* on $(\mathcal{X}_n)$. If

$$f[\lambda X_1 + (1 - \lambda)X_2] \geq \lambda f(X_1) + (1 - \lambda)f(X_2) \tag{4.2}$$

then $f(X)$ is said to be *convex from below* on $(\mathcal{X}_n)$. If in Eq. (4.1) (4.2) the $\leq$ ($\geq$) sign is replaced by $<$ ($>$), then $f(X)$ is *strictly convex from above (below)*.

4.1 INTRODUCTION

The implications of Definition 4.1 may require some elaboration. First, any two vectors, X_1 and X_2, for which $f(X)$ is defined are selected and the function is evaluated at those two points, $f(X_1)$ and $f(X_2)$. Now a convex combination of $f(X_1)$ and $f(X_2)$ is formed and defines a straight line joining $f(X_1)$ and $f(X_2)$. Next $f(X)$ is evaluated at $X_1 + (1 - \lambda)X_2$, $f[\lambda X_1 + (1 - \lambda)X_2]$. If each point on the line segment joining $f(X_1)$ and $f(X_2)$ lies on or above the corresponding point on the function, the function is convex from above (*convex*). On the other hand if each point on the line segment lies on or below the corresponding point on the function, the function is convex from below (*concave*). These concepts are illustrated graphically in Figs. 4.2 and 4.3 for a function of a single variable.

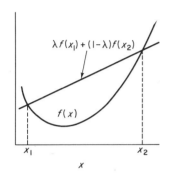
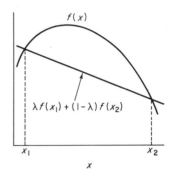

Figure 4.2 $f(x)$ Convex from Above   *Figure 4.3* $f(x)$ Convex from Below

Example 4.1 Show that the function $f(x)$ is convex from above where

$$f(x_1, x_2) = x_1^2 + x_2^2, \quad -\infty < x_1, x_2 < \infty$$

Let X_1 and X_2 be any two distinct two-dimensional vectors.

$$X_1 = \begin{bmatrix} x_{11} \\ x_{12} \end{bmatrix}, \quad X_2 = \begin{bmatrix} x_{21} \\ x_{22} \end{bmatrix}, \quad \lambda X_1 + (1 - \lambda)X_2 = \begin{bmatrix} \lambda x_{11} + (1 - \lambda)x_{21} \\ \lambda x_{12} + (1 - \lambda)x_{22} \end{bmatrix}$$

where $0 \le \lambda \le 1$. Therefore

$$f[\lambda X_1 + (1 - \lambda)X_2] = [\lambda(x_{11} - x_{21}) + x_{21}]^2 + [\lambda(x_{12} - x_{22}) + x_{22}]^2$$

and

$$\lambda f(X_1) + (1 - \lambda)f(X_2) = \lambda[x_{11}^2 + x_{12}^2] + (1 - \lambda)[x_{21}^2 + x_{22}^2]$$

Examining the difference between $f[\lambda X_1 + (1 - \lambda)X_2]$ and $\lambda f(X_1) + (1 - \lambda)f(X_2)$

$$f[\lambda X_1 + (1 - \lambda)X_2] - [\lambda f(X_1) + (1 - \lambda)f(X_2)]$$
$$= \lambda(\lambda - 1)[(x_{11} - x_{21})^2 + (x_{22} - x_{12})^2] \le 0$$

since $0 \le \lambda \le 1$ and $[(x_{11} - x_{21})^2 + (x_{22} - x_{12})^2] > 0$. By Definition 4.1, $f(x_1, x_2)$ is convex from above.

The focus of attention of this chapter will be on the analysis of continuous functions with interior extreme points, although the analysis of functions with extreme points at a boundary will also be considered. For the purposes of the remaining discussion the following definition of an extreme point is given.

Definition 4.2 Let $f(X)$ be a continuous function where X is an n-dimensional vector and let $(\mathscr{X}_n)$ be the open space on which $f(X)$ is convex (from above or below). The point X^* is an interior extreme point for $f(X)$ if X^* belongs to $(\mathscr{X}_n)$ and

$$f(X^*) \leq f(X), \qquad X \in (\mathscr{X}_n) \tag{4.3}$$

or

$$f(X^*) \geq f(X), \qquad X \in (\mathscr{X}_n) \tag{4.4}$$

Since $f(X)$ is assumed to be convex on the open space $(\mathscr{X}_n)$, if $f(X)$ has extreme points on $(\mathscr{X}_n)$, it has either one or an infinite number of extreme points on $(\mathscr{X}_n)$. In addition, if $f(X)$ has an infinite number of extreme points on $(\mathscr{X}_n)$, than for any two extreme points, X^* and X',

$$f(X^*) = f(X') \tag{4.5}$$

If $f(X)$ has only one interior extreme point on $(\mathscr{X}_n)$, then $f(X)$ is strictly convex on $(\mathscr{X}_n)$. If $f(X)$ has an infinite number of extreme points on $(\mathscr{X}_n)$, then $f(X)$ is convex there. The two cases are illustrated in Figs. 4.4 and 4.5, where $f(X)$ is a function of a single variable.

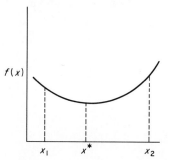

Figure 4.4 $f(x)$ Has One Extreme Point, x^*, on (x_1, x_2)

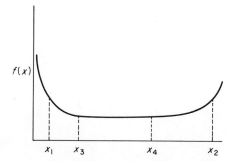

Figure 4.5 $f(x)$ Has an Infinite Number of Extreme Points, x_3 to x_4, on (x_1, x_2)

Definition 4.3 If $f(X)$ is a continuous function and X is a vector with elements $x_1, x_2, \ldots, x_n$, then $f(X)$ is called *strictly monotonic increasing* in x_i if

$$f(x_1, x_2, \ldots, x_i, \ldots, x_n) < f(x_1, x_2, \ldots, x_i + \epsilon, \ldots, x_n) \tag{4.6}$$

for $\epsilon > 0$; and is *strictly monotonic decreasing* in x_i if

$$f(x_1, x_2, \ldots, x_i, \ldots, x_n) > f(x_1, x_2, \ldots, x_i + \epsilon, \ldots, x_n) \tag{4.7}$$

4.1 INTRODUCTION

for $\epsilon > 0$. If $<$ in Eq. 4.6 is replaced by $\leq$ then $f(X)$ is monotonic increasing. Similarly if $>$ in Eq. 4.7 is replaced by $\geq$, $f(X)$ is said to be monotonic decreasing.

Examples of monotonic and strictly monotonic functions of a single variable are shown in Figs. 4.6 to 4.9.

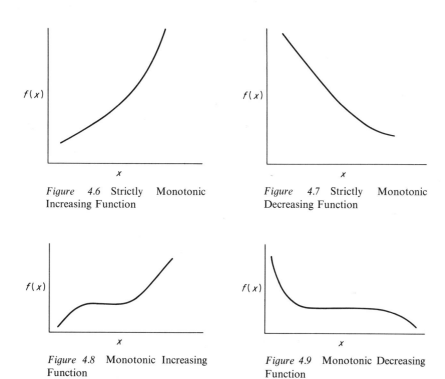

Figure 4.6 Strictly Monotonic Increasing Function

Figure 4.7 Strictly Monotonic Decreasing Function

Figure 4.8 Monotonic Increasing Function

Figure 4.9 Monotonic Decreasing Function

Based upon the definition of an extreme point, it is evident that a continuous function, $f(X)$, may have many distinct extreme points on a sufficiently large open space of definition, $(\mathscr{X}_n)$. The function shown in Fig. 4.1 is an example of such a function, where the space of definition is the closed interval $[a, b]$. In this case $f(x)$ has five interior extreme points, three maxima and two minima. When a function has more than one maximum or minimum, the maxima are referred to as *local maxima* and the minima as *local minima*. The term *local* is used since any given maximum or minimum may be a maximum or minimum only for a given, and perhaps small, region of definition of $f(X)$; a greater maximum or lesser minimum existing outside of this region. Now suppose that all of the extreme points for $f(X)$, interior or not, are identified over the entire space of definition of $f(X)$ and denoted by $X_1^*, X_2^*, \ldots, X_m^*$. Among these points there will be one, X_i^*, such that

$$f(X_i^*) \geq f(X_j^*), \qquad j = 1, 2, \ldots, m \tag{4.8}$$

That is, the maximum for $f(X)$ at X_i^* yields a value of $f(X)$ which is at least as great as that for any other local maximum. Such a point is called a *global maximum*. Similarly, there is a local minimum, X_k^*, such that

$$f(X_k^*) \leq f(X_j^*), \quad j = 1, 2, \ldots, m \tag{4.9}$$

The minimum yielding the least value of $f(X)$ among all the local minima is called a *global minimum*.

Although this chapter deals mainly with the investigation of continuous functions with interior extreme points, the operations research analyst is usually concerned with the identification of the global maximum or minimum of the function. To accomplish this it is necessary to evaluate the function at each interior extreme point, at the boundaries of the function, and at each point at which the function is discontinuous if such points exist.

Example 4.2 Identify the global maximum and minimum for the following function.

$$f(x) = \begin{cases} 10 \exp[-x], & 0 \leq x \leq 10 \\ 0.5x, & 10 < x \leq 20 \\ 30 - x, & 20 < x \leq 30 \end{cases}$$

Assume that $f(x)$ is not defined for $x < 0$ or $x > 30$.

Let us examine the function over each of the three intervals $[0, 10]$, $(10, 20]$, and $(20, 30]$. At $x = 0, f(x) = 10$. Let x be any point on the open interval $(0, 10)$ and define $\epsilon > 0$ such that $x + \epsilon$ lies on $(0, 10)$. Now

$$f(x + \epsilon) - f(x) = 10 \exp[-(x + \epsilon)] - 10 \exp[-x]$$
$$= 10 \exp[-x](\exp[-\epsilon] - 1) < 0$$

for all ϵ since $\exp[-x] > 0$ and $0 < \exp[-\epsilon] < 1$. Therefore $f(x)$ decreases uniformly on the interval $(0, 10)$ and therefore no extreme point for $f(x)$ exists on the open interval $(0, 10)$. Next we show that $f(x)$ is discontinuous at $x = 10$.

$$\lim_{x \to 10-} f(x) = 10 \exp[-10] = 0.000454 \quad \text{and} \quad \lim_{x \to 10+} f(x) = 5$$

Since $f(x) = 10 \exp[-x]$, $0 \leq x \leq 10$, $f(10) = 0.000454$. Since $f(10 - \epsilon) > f(10) < f(10 + \epsilon)$, $x = 10$ is an extreme point. On the interval $(10, 20), f(x)$ has no extreme points since

$$f(x + \epsilon) - f(x) = 0.5(x + \epsilon) - 0.5(x) = 0.5\epsilon > 0$$

for all ϵ such that $10 < x + \epsilon < 20$. Similarly $f(x)$ has no extreme points on $(20, 30)$ since

$$f(x + \epsilon) - f(x) = (30 - x - \epsilon) - (30 - x) = -\epsilon < 0$$

for all ϵ such that $20 < x + \epsilon < 30$. Thus we need to examine $f(x)$ only at $x = 20$ and $x = 30$. Note the $f(x)$ is continuous at $x = 20$ since

$$\lim_{x \to 20-} f(x) = 10, \quad \lim_{x \to 20+} f(x) = 10$$

To show that $f(x)$ has an extreme value at $x = 20$ we examine $f(20 - \epsilon)$ and $f(20 + \epsilon)$.

$$f(20 - \epsilon) = 0.5(20 - \epsilon), \qquad f(20 + \epsilon) = (30 - 20 - \epsilon)$$

Therefore

$$f(20 - \epsilon) < f(20) > f(20 + \epsilon)$$

for all $0 < \epsilon < 10$. Therefore $x = 20$ is a local maximum for $f(x)$. At $x = 30$, $f(x) = 0$. (See Table 4.1.)

TABLE 4.1

Extreme Point, x	Extreme Value, $f(x)$
0	10.000000
10	0.000454
20	10.000000
30	0.000000

Therefore, $x = 30$ is a global minimum for $f(x)$, and $x = 0$ and $x = 20$ are global maxima.

4.2 Functions of a Single Variable

Let $f(x)$ be a single-valued continuous function of x defined on the closed interval $[a, b]$ for x. Assume that $f(x)$ attains an extreme value at x^* on $[a, b]$. If

$$f(x^*) \geq f(x) \tag{4.10}$$

then x^* maximizes $f(x)$ on $[a, b]$. If

$$f(x^*) \leq f(x) \tag{4.11}$$

then x^* minimizes $f(x)$ on $[a, b]$.

Theorem 4.1 If $f(x)$ is a continuous function on the closed, interval, $[a, b]$, then $f(x)$ attains a maximum and a minimum on $[a, b]$.

Theorem 4.1 implies that a continuous function attains a maximum and minimum either at an interior point of $[a, b]$ or at the endpoints of the interval, $[a, b]$. For example

$$f(x) = x^2, \quad -5 \leq x \leq 5$$

attains a maximum at -5 and 5 and a minimum at zero. On the other hand

$$f(x) = -x^2, \quad -5 \leq x \leq 5$$

attains a maximum at zero and a minimum at -5 and 5.

In general we will be concerned with the examination of functions with finite extreme points. That is, if $f(x)$ is a continuous function defined on the open interval $(-\infty, \infty)$ and x^* is an extreme point for $f(x)$, then there exist finite numbers M_1 and M_2 such that

$$M_1 < x^* < M_2 \tag{4.12}$$

where $M_1 < M_2$. We will define conditions under which a continuous function, $f(x)$, does not have finite extreme points. However, to define these conditions it is first necessary to prove certain useful properties of continuous functions.

Theorem 4.2 Rolle's Theorem If $f(x)$ is continuous on the closed interval $[a, b]$ and differentiable on the open interval (a, b), and $f(a) = f(b)$, then there exists a point x^* on (a, b) such that

$$f'(x^*) = 0$$

Proof First, if $f(x) = f(a)$ for all x on (a, b), then $f'(x) = 0$ for all x on (a, b), and the proof is complete. Now assume that $f(x) \neq f(a)$ for at least some x on (a, b). Since the closed interval $[a, b]$ is bounded, $f(x)$ attains a minimum and a maximum on $[a, b]$. Since $f(a) = f(b)$, $f(x)$ has at least one extreme point on (a, b). Let x^* be such a point. That is, $f(x^*) \neq f(a)$. Assume

$$f(x^*) = \max_{[a, b]} f(x)$$

Then

$$\frac{f(x^* + \Delta x) - f(x^*)}{\Delta x} \leq 0 \tag{4.13}$$

for $\Delta x > 0$, and

$$\frac{f(x^* + \Delta x) - f(x^*)}{\Delta x} \geq 0 \tag{4.14}$$

for $\Delta x < 0$. Taking the right- and left-hand limits in Eqs. (4.13) and (4.14) as $\Delta x \to 0$ yields us

$$\lim_{\Delta x \to 0+} \frac{f(x^* + \Delta x) - f(x^*)}{\Delta x} = f'(x^*) \leq 0$$

and

$$\lim_{\Delta x \to 0-} \frac{f(x^* + \Delta x) - f(x^*)}{\Delta x} = f'(x^*) \geq 0$$

Since $f'(x)$ is defined for all x on (a, b)

$$f'(x^*) = 0$$

Theorem 4.3 Mean-Value Theorem If $f(x)$ is continuous on the closed interval $[a, b]$ and differentiable on (a, b), then there exists a point x_0 on (a, b) such that

$$\frac{f(b) - f(a)}{b - a} = f'(x_0)$$

Proof Let

$$F(x) = f(x) - f(a) - (x - a)\frac{f(b) - f(a)}{b - a}$$

Since $F(a) = F(b) = 0$, Rolle's Theorem can be applied to $F(x)$.

$$F'(x) = f'(x) - \frac{f(b) - f(a)}{b - a}$$

Let x_0 be a point such that $F'(x_0) = 0$. Then

$$\frac{f(b) - f(a)}{b - a} = f'(x_0)$$

Theorem 4.4 If $f(x)$ is continuous on the closed interval $[a, b]$ and differentiable on the open interval (a, b) such that $f'(x) > 0$ or $f'(x) < 0$ for $a < x < b$, then $f(x)$ does not attain an extreme value on the open interval (a, b).

Proof The proof of this theorem is given for $f'(x) > 0$ only. Let x_1 and x_2 be any two points on the open interval (a, b) such that $x_1 < x_2$. Since $f(x)$ must be continuous and differentiable on $[x_1, x_2]$, the mean-value theorem implies

$$\frac{f(x_2) - f(x_1)}{x_2 - x_1} = f'(x_0)$$

for some x_0 on the interval (x_1, x_2). Therefore

$$f(x_2) - f(x_1) = (x_2 - x_1)f'(x_0)$$

Since $f'(x) > 0$ for all x on (a, b) and since $x_2 > x_1$

$$f(x_2) > f(x_1)$$

Hence $f(x)$ is a strictly monotonic increasing function and does not attain an extreme value on (a, b).

The results of Theorem 4.4 can be extended to the case where $f(x)$ is defined for $-\infty < x < \infty$ and $f'(x) > 0$ or $f'(x) < 0$ for all x. In this case $f(x)$ attains its extreme values at $-\infty$ and $+\infty$. Even though a function may have an extreme value at $-\infty$ or $+\infty$, this does not necessarily imply that it has no finite extreme points. For example, $f(x) = x^2$ has maxima at $-\infty$ and $+\infty$ but has a minimum at zero.

Example 4.3 Which of the following functions are strictly monotonic?

a. $f(x) = 1 - \exp[x], \quad 0 < x < \infty$
b. $f(x) = \exp[-x^2], \quad -\infty < x < \infty$

c. $f(x) = \begin{cases} \exp[x], & -\infty < x \leq 0 \\ 1 + x, & 0 < x \leq 5 \\ \dfrac{(1+x)^2}{6} - (x-5), & 5 < x \leq \infty \end{cases}$

If $f(x)$ is strictly monotonic then $f'(x) < 0$ or $f'(x) > 0$ for the range of values of x for which $f(x)$ is defined. Demonstration that each of these functions is continuous is left to the reader.

a. $f'(x) = -\exp[x], \quad 0 < x < \infty$

Therefore $f(x)$ is strictly monotonic decreasing on $(0, \infty)$ since $f'(x) < 0$ on $(0, \infty)$

b. $f'(x) = -2x \exp[-x^2], \quad -\infty < x < \infty$

The sign of $f'(x)$ in this case depends upon the sign of x. That is

$$f'(x) > 0, \quad -\infty < x < 0 \qquad f'(x) < 0, \quad 0 < x < \infty$$

and

$$f'(x) = 0, \quad x = 0$$

Therefore $f(x)$ is not monotonic over its domain of definition (and therefore not strictly monotonic).

c. In this case

$$f'(x) = \begin{cases} \exp[x], & -\infty < x \leq 0 \\ 1, & 0 < x \leq 5 \\ \frac{1}{3}(1+x) - 1, & 5 < x < \infty \end{cases}$$

That $f'(x)$ is continuous on $(-\infty, 0), (0, 5), (5, \infty)$ need not be demonstrated. That $f'(x)$ is continuous at $x = 0$ and $x = 5$ is shown as follows.

$$\lim_{x \to 0-} f'(x) = \lim_{x \to 0-} \exp[x] = 1, \qquad \lim_{x \to 0+} f'(x) = \lim_{x \to 0+} (1) = 1$$

Therefore

$$\lim_{x \to 0-} f'(x) = \lim_{x \to 0+} f'(x)$$

and $f'(x)$ is continuous at $x = 0$.

$$\lim_{x \to 5-} f'(x) = \lim_{x \to 5-} (1) = 1$$

$$\lim_{x \to 5+} f'(x) = \lim_{x \to 5+} \left[\frac{1}{3}(1+x) - 1\right] = 1$$

4.2 FUNCTIONS OF A SINGLE VARIABLE

Since

$$\lim_{x \to 5-} f'(x) = \lim_{x \to 5+} f'(x)$$

$f'(x)$ is continuous at $x = 5$. Since $f'(x)$ is uniformly positive on $(-\infty, \infty)$, $f(x)$ is strictly monotonic increasing on $(-\infty, \infty)$.

Example 4.4 Show that the following continuous function is monotonic decreasing but is not strictly monotonic.

$$f(x) = \begin{cases} (3-x)^2 + 9, & 0 < x \le 3 \\ 9, & 3 < x \le 10 \\ 9 - (x-10)^2, & 10 < x < \infty \end{cases}$$

Taking the first derivative of $f(x)$ yields

$$f'(x) = \begin{cases} -2(3-x), & 0 < x \le 3 \\ 0, & 3 < x \le 10 \\ -2(x-10), & 10 < x < \infty \end{cases}$$

Therefore $f'(x) < 0$ on $(0, 3)$, $f'(x) = 0$ on $[3, 10]$, and $f'(x) < 0$ on $(10, \infty)$. Since $f'(x)$ is nonpositive throughout the open interval $(0, \infty)$, $f(x)$ is not strictly monotonic decreasing. However, since $f'(x) \le 0$ on $(0, \infty)$ it is monotonic decreasing there.

As indicated previously the global maximum or minimum for a function is usually identified by examination of each extreme point for the function. The first step, then, is to identify the extreme points for the function. The second step is to determine which of the extreme points are local minima and which are local maxima. To aid in the identification of interior extreme points for a continuous function, we shall develop conditions which must exist if a given point is an extreme point.

By the definition of an interior extreme point for a continuous function, it must lie on an open interval for which the function is convex. Further, the function must not be strictly monotonic on that interval. Therefore, if x^* is an extreme point for $f(x)$ on the interval (a, b), then $f'(x)$ must change sign on the interval. Intuitively one might guess that the extreme point, x^*, is that point at which $f'(x)$ changes sign. In other words, if x^* is an extreme point for $f(x)$ on (a, b) then

$$f'(x^*) = 0$$

if $f'(x^*)$ exists. A more rigorous proof of the *necessary condition* for the existence of an extreme point at x^* is given below.

Theorem 4.5 Let $f(x)$ be a continuous function on the closed interval $[a, b]$ and differentiable on the open interval (a, b). If x^* is an interior extreme point for $f(x)$ then

$$f'(x^*) = 0 \tag{4.15}$$

Proof Since x^* is an extreme point for $f(x)$ on (a, b) either

$$f(x^*) \geq f(x) \quad \text{or} \quad f(x^*) \leq f(x)$$

where x lies on the interval $[x^* - \Delta x, x^* + \Delta x]$ for some $\Delta x > 0$. Further, since $f(x)$ is continuous and differentiable on (a, b), the limit

$$\lim_{\Delta x \to 0} \frac{f(x^* + \Delta x) - f(x^*)}{\Delta x} = f'(x^*)$$

is defined where $\Delta x \to 0$ from $-$ or $+$. If $f(x^*) \geq f(x)$ for x on $[x^* - \Delta x, x^* + \Delta x]$, then for any $\Delta x < 0$,

$$\frac{f(x^* + \Delta x) - f(x^*)}{\Delta x} \geq 0 \tag{4.16}$$

and for any $\Delta x > 0$,

$$\frac{f(x^* + \Delta x) - f(x^*)}{\Delta x} \leq 0 \tag{4.17}$$

Therefore

$$\lim_{\Delta x \to 0} \frac{f(x^* + \Delta x) - f(x^*)}{\Delta x} = 0 \tag{4.18}$$

and $f'(x^*) = 0$ if x^* is an extreme point. A similar argument follows when $f(x^*) \leq f(x)$ for x on $[x^* - \Delta x, x^* + \Delta x]$. ∎

In brief, Theorem 4.5 states that under certain conditions for $f(x)$, the existence of an extreme point at x^* implies that $f'(x^*) = 0$. However, the converse is not necessarily true. That is, x^* may be a point for which $f'(x^*) = 0$ and yet *not* be an extreme point for $f(x)$. Such cases arise when dealing with monotonic functions. The functions shown in Figs. 4.8 and 4.9 illustrate this phenomenon. In both cases there is an open interval for which $f'(x) = 0$, and yet none of the values of x on these intervals are extreme points. Such points are called *points of inflection*. In general, a point, x^*, for which $f'(x^*) = 0$ is called a *stationary point*. Thus a stationary point may be a local minimum, maximum, or a point of inflection.

Example 4.5 Determine the interior extreme points for the function

$$f(x) = 8x^3 - 30x^2 + 36x - 12$$

on the interval $[0, 100]$

From Theorem 4.5, we know that any point, x^*, on the open interval $(0, 100)$ for which $f'(x^*) = 0$ must be considered as a possible extreme point. Therefore, to identify the extreme points for $f(x)$ we determine those values of x for which $f'(x) = 0$. In this case

$$f'(x) = 24x^2 - 60x + 36$$

FUNCTIONS OF A SINGLE VARIABLE

Solving $f'(x) = 0$ yields values of x^* of 1 and 1.5. Therefore if $f(x)$ has an extreme point on (0, 100) it is either at 1 or 1.5 or both. Let ϵ be an arbitrarily small positive number. If x^* is an extreme point then either

$$f(x^* - \epsilon) > f(x^*) < f(x^* + \epsilon) \quad \text{or} \quad f(x^* - \epsilon) < f(x^*) > f(x^* + \epsilon)$$

Now

$$f(x^* \pm \epsilon) = 8(x^* \pm \epsilon)^3 - 30(x^* \pm \epsilon)^2 + 36(x^* \pm \epsilon) - 12$$

For $x^* = 1$,

$$f(x^*) = 2, \quad f(x^* + \epsilon) = 2 + 8\epsilon^3 - 6\epsilon^2, \quad f(x^* - \epsilon) = 2 - 8\epsilon^3 - 6\epsilon^2$$

For $0 < \epsilon < .25$,

$$f(x^* + \epsilon) < f(x^*) > f(x^* - \epsilon)$$

and $x^* = 1$ is a local maximum for $f(x)$. For $x^* = 1.5$

$$f(x^*) = 1.5, \quad f(x^* + \epsilon) = 1.5 + 8\epsilon^3 + 6\epsilon^2, \quad f(x^* - \epsilon) = 1.5 - 8\epsilon^3 + 6\epsilon^2$$

For $0 < \epsilon < .25$,

$$f(x^* + \epsilon) > f(x^*) < f(x^* - \epsilon)$$

and $x = 1.5$ is a local minimum.

Example 4.6 Show that the function

$$f(x) = ax^3 + b$$

has no interior extreme points.

If $f(x)$ has interior extreme points they exist at points x^* such that $f'(x^*) = 0$. The value of x satisfying this relationship is determined by solving the equation

$$f'(x) = 3ax^2 = 0$$

Therefore

$$x^* = 0 \quad \text{and} \quad f(x^*) = b$$

Let ϵ be an arbitrarily small positive number. Then

$$f(x^* \pm \epsilon) = a(x^* \pm \epsilon)^3 + b \quad \text{and} \quad f(x^* - \epsilon) < f(x^*) < f(x^* + \epsilon)$$

for $a > 0$. For $a < 0$

$$f(x^* - \epsilon) > f(x^*) > f(x^* + \epsilon)$$

In either case x^* is not an extreme point for $f(x)$ but is a point of inflection.

In the above examples it was necessary to analyze the behavior of $f(x)$ in the neighborhood of each point, x^*, at which $f'(x^*) = 0$ to determine whether the point was a local maximum, local minimum, or point of inflection. A similar analysis can be carried out by examining the second derivatives of $f(x)$ at the point x^*, where $f'(x^*) = 0$. As already indicated if $f'(x^*) = 0$ then one of the following is true:

1. x^* is a local maximum;
2. x^* is a local minimum;
3. x^* is a point of inflection.

Let us consider the case where x^* is a local maximum. Now

$$f''(x^*) = \lim_{\Delta x \to 0^-} \frac{f'(x^* + \Delta x) - f'(x^*)}{\Delta x} \tag{4.19}$$

$$= \lim_{\Delta x \to 0^+} \frac{f'(x^* + \Delta x) - f'(x^*)}{\Delta x} \tag{4.20}$$

where both limits are assumed to exist and are equal. However, since $f'(x^*) = 0$

$$f''(x^*) = \lim_{\Delta x \to 0^-} \frac{f'(x^* + \Delta x)}{\Delta x} = \lim_{\Delta x \to 0^+} \frac{f'(x^* + \Delta x)}{\Delta x} \tag{4.21}$$

If $f''(x^*) < 0$ then

$$\lim_{\Delta x \to 0^-} \frac{f'(x^* + \Delta x)}{\Delta x} < 0 \tag{4.22}$$

and

$$\lim_{\Delta x \to 0^+} \frac{f'(x^* + \Delta x)}{\Delta x} < 0 \tag{4.23}$$

From Eq. (4.22), $f'(x^* + \Delta x) > 0$ when $\Delta x < 0$ and from Eq. (4.23), $f'(x^* + \Delta x) < 0$ when $\Delta x > 0$. But if $f'(x^* + \Delta x) > 0$ for $\Delta x < 0$ and $f'(x^* + \Delta x) < 0$ for $\Delta x > 0$ then $f(x)$ is a strictly increasing function to the left of x^* and strictly decreasing to the right of x^*. Hence for $\Delta x > 0$

$$f(x^* - \Delta x) < f(x^*) > f(x^* + \Delta x) \tag{4.24}$$

and x^* must be a local maximum.

If $f'(x^*) = 0$ the existence of the condition $f''(x^*) < 0$ is said to be sufficient for x^* to be a local maximum. This statement means that if $f'(x^*) = 0$, the analyst need only show that $f''(x^*) < 0$ to establish that x^* is a local maximum. By a similar argument it can be shown that if $f'(x^*) = 0$ and $f''(x^*) > 0$ then x^* is a local minimum for $f(x)$. The *sufficient conditions* for the existence of a local maximum and a local minimum at a point are summarized in the following theorem.

Theorem 4.6 Let $f(x)$ be continuous on the closed interval $[a, b]$ and twice differentiable at x^*, which lies on the open interval (a, b). If $f'(x^*) = 0$, then

a. x^* is a local maximum if $f''(x^*) < 0$;
b. x^* is a local minimum if $f''(x^*) > 0$;
c. x^* is a local minimum, local maximum, or a point of inflection if $f''(x^*) = 0$.

Proof The proof of part a has already been given and the proof of part b follows a similar argument.

To prove part c, the reader need only consider the functions

$$f(x) = x^3, \quad f(x) = x^4, \quad f(x) = -x^4$$

In all three cases $f'(x^*) = 0$ for $x^* = 0$ and $f''(x^*) = 0$. Examination of $f(x)$ near x^* in each of the three cases will show that x^* is a point of inflection for x^3, a minimum for x^4 and a maximum for $-x^4$. Therefore, if $f'(x^*) = 0$ and $f''(x^*) = 0$, nothing can be said about the nature of x^*, based upon the analysis presented thus far.

Example 4.7 Find the stationary points for the following continuous function and determine whether each is a local minimum, a local maximum, or a point of inflection.

$$f(x) = x^3 + x^2 - x, \quad -\infty < x < \infty$$

The interior stationary points for $f(x)$ are those values of x for which $f'(x) = 0$.

$$f'(x) = 3x^2 + 2x - 1 \quad \text{and} \quad x^* = -1, \tfrac{1}{3}$$

Taking the second derivative of $f(x)$ yields

$$f''(x) = 6x + 2$$

For $x^* = -1$

$$f''(x_0^*) = -4$$

Since $f''(-1) < 0$, $x^* = -1$ is a local maximum for $f(x)$. For $x^* = \tfrac{1}{3}$,

$$f''(x^*) = 4$$

and $x^* = \tfrac{1}{3}$ is a local minimum for $f(x)$.

Theorems 4.5 and 4.6 provide the analyst with a set of necessary and sufficient conditions for the existence of a local maximum or minimum at a point, x^*, for a function which is continuous and twice differentiable at x^*. Unfortunately the nature of every interior stationary point cannot always be determined by this set of conditions, as demonstrated in the proof of part c of Theorem 4.6. When this situation arises the analyst can examine the behavior of the function in the neighborhood of the stationary point of interest. However, the following theorem leads to a more convenient method for determining the nature of a stationary point.

Theorem 4.7 Taylor's Mean-Value Theorem Let $f(x)$ be a continuous function on $[a, b]$ with ith derivative $f^i(x)$. If $f^i(x)$, $i = 1, 2, \ldots, m$, exist and are continuous on (a, b) then for x and x_0 on (a, b) there exists a point x_1 on (a, b) such that

$$f(x) = f(x_0) + \sum_{i=1}^{n-1} \frac{(x - x_0)^i}{i!} f^i(x_0) + \frac{(x - x_0)^n}{n!} f^n(x_1) \qquad (4.25)$$

where $n \leq m$.

Proof We prove this result by induction, noting that

$$f(x) = f(x_0) + \int_{x_0}^{x} f'(y)\, dy$$

Integrating by parts, let

$$u = f'(y), \qquad du = f''(y)\, dy, \qquad v = -(x - y), \qquad dv = dy$$

where

$$\int_{y_1}^{y_2} u\, dv = uv \Big|_{y_1}^{y_2} - \int_{y_1}^{y_2} v\, du$$

Therefore

$$f(x) = f(x_0) - (x - y)f'(y)\Big|_{x_0}^{x} + \int_{x_0}^{x} (x - y)f''(y)\, dy$$

or

$$f(x) = f(x_0) + (x - x_0)f'(x_0) + \int_{x_0}^{x} (x - y)f''(y)\, dy$$

Repeating the integration by parts, let

$$u = f''(y), \qquad dv = (x - y)\, dy, \qquad du = f^3(y)\, dy, \qquad v = -\frac{(x - y)^2}{2}$$

$$f(x) = f(x_0) + (x - x_0)f'(x_0) - \frac{(x - y)^2}{2} f''(y)\Big|_{x_0}^{x} + \int_{x_0}^{x} \frac{(x - y)^2}{2} f^3(y)\, dy$$

$$= f(x_0) + (x - x_0)f'(x_0) + \frac{(x - x_0)^2}{2} f''(x_0) + \int_{x_0}^{x} \frac{(x - y)^2}{2} f^3(y)\, dy$$

We repeat the process again in an attempt to determine a generalized expression for $f(x)$. Let

$$u = f^3(y), \qquad dv = \frac{(x - y)^2}{2}\, dy, \qquad du = f^4(y)\, dy, \qquad v = -\frac{(x - y)^3}{3!}$$

4.2 FUNCTIONS OF A SINGLE VARIABLE

$$f(x) = f(x_0) + (x-x_0)f'(x_0) + \frac{(x-x_0)^2}{2}f''(x_0) - \left.\frac{(x-y)^3}{3!}f^3(y)\right|_{x_0}^{x}$$

$$+ \int_{x_0}^{x} \frac{(x-y)^3}{3!} f^4(y)\, dy$$

$$= f(x_0) + (x-x_0)f'(x_0) + \frac{(x-x_0)^2}{2}f''(x_0) + \frac{(x-x_0)^3}{3!}f^3(x_0)$$

$$+ \int_{x_0}^{x} \frac{(x-y)^3}{3!} f^4(y)\, dy$$

Based upon these three iterations, assume that for $k < n$,

$$f(x) = f(x_0) + \sum_{i=1}^{k-1} \frac{(x-x_0)^i}{i!} f^i(x_0) + \int_{x_0}^{x} \frac{(x-y)^{k-1}}{(k-1)!} f^k(y)\, dy \quad (4.26)$$

Integrating by parts, let

$$u = f^k(y), \quad du = f^{k+1}(y)\, dy, \quad v = -\frac{(x-y)^k}{k!}, \quad dv = \frac{(x-y)^{k-1}}{(k-1)!}\, dy$$

Then

$$f(x) = f(x_0) + \sum_{i=1}^{k-1} \frac{(x-x_0)^i}{i!} f^i(x_0) - \left.\frac{(x-y)^k}{k!} f^k(y)\right|_{x_0}^{x}$$

$$+ \int_{x_0}^{x} \frac{(x-y)^k}{k!} f^{k+1}(y)\, dy$$

$$= f(x_0) + \sum_{i=1}^{k} \frac{(x-x_0)^i}{i!} f^i(x_0) + \int_{x_0}^{x} \frac{(x-y)^k}{k!} f^{k+1}(y)\, dy$$

and the expansion given in (4.26) is valid. Let

$$R_{k+1}(x) = \int_{x_0}^{x} \frac{(x-y)^k}{k!} f^{k+1}(y)\, dy$$

and

$$g(y) = \frac{(x-y)^k}{k!}$$

$$h(y) = f^{k+1}(y)$$

where y is on the interval (x_0, x). Since $g(y)$ is integrable and nonnegative on (x_0, x) and since

$$\int_{x_0}^{x} g(y)\, dy \geq 0 \quad \text{on } (x_0, x)$$

then

$$\int_{x_0}^{x} g(y)h(y)\, dy = h(x_1) \int_{x_0}^{x} g(y)\, dy \qquad (4.27)$$

where x_1 is a point on the interval (x_0, x) such that

$$h(x_1) = \frac{\int_{x_0}^{x} g(y)h(y)\, dy}{\int_{x_0}^{x} g(y)\, dy}$$

Therefore

$$R_{k+1}(x) = f^{k+1}(x_1) \int_{x_0}^{x} \frac{(x-y)^k}{k!}\, dy = \frac{(x-x_0)^{k+1}}{(k+1)!} f^{k+1}(x_1) \qquad (4.28)$$

Letting $n = k + 1$, we have the desired result. ▢

Theorem 4.7 provides the analyst with a means for approximating a continuous function by a polynomial, often called a Taylor polynomial. That is, the continuous function $f(x)$ can be approximated near a point x_0 by

$$f(x) \simeq f(x_0) + \sum_{i=1}^{n-1} \frac{(x-x_0)^i}{i!} f^i(x_0) \qquad (4.29)$$

where x_0 and x lie on an interval for which $f(x)$ is continuous. The error in the approximation is given by $R_n(x)$ where

$$R_n(x) = \frac{(x-x_0)^n}{n!} f^n(x_1) \qquad (4.30)$$

where x_1 is as defined in Theorem 4.7.

Example 4.8 Approximate the function

$$f(x) = \exp[-x]$$

near $x_0 = 1$ by a second-degree Taylor polynomial and determine the maximum error in the approximation when $x = 1.5$ and 2.0.

To approximate $f(x)$ by a second-degree Taylor polynomial, we need $f(x_0)$, $f'(x_0)$, and $f''(x_0)$.

$$f(x_0) = 0.368, \qquad f'(x_0) = -0.368, \qquad f''(x_0) = 0.368$$

Therefore

$$f(x) \simeq 0.368 - 0.368(x-1) + 0.184(x-1)^2$$

The error in the approximation is given by

$$R_3(x) = -\frac{(x-1)^3}{6} \exp[-x_1]$$

4.2 FUNCTIONS OF A SINGLE VARIABLE

When $x = 1.5$,
$$R_3(x) = -0.021 \exp[-x_1]$$
since $x_0 < x_1 < x$, $1 < x_1 < 1.5$ and
$$|R_3(x)| < 0.021 \exp[-1] < 0.0077$$

For $x = 2$,
$$R_3(x) = -0.167 \exp[-x_1] \quad \text{and} \quad |R_3(x)| < 0.167 \exp[-1] < 0.061 \; \blacksquare$$

As demonstrated in the last example, the error in the approximation decreases as $x \to x_0$ since
$$\lim_{x \to x_0} R_n(x) = 0 \tag{4.31}$$

The sufficient conditions for a local minimum or maximum can be derived through the use of Taylor's mean-value theorem. Let x^* be a point such that $f'(x^*) = 0$ and assume that $f''(x^*) \neq 0$. Then, at a point, x, near x^*,

$$f(x) = f(x^*) + \frac{(x - x^*)^2}{2!} f''(x^*) + \frac{(x - x^*)^3}{3!} f^3(x_1)$$

Then
$$2 \frac{f(x) - f(x^*)}{(x - x^*)^2} = f''(x^*) + \frac{(x - x^*)}{3} f^3(x_1)$$

Since $f^i(x)$ is assumed to exist for $i = 1, 2, \ldots, n$, and x on (a, b), $f^3(x_1)$ is finite since x_1 is on (a, b). Therefore, by choosing $|x - x^*|$ sufficiently small, $f''(x^*)$ dominates $\{(x - x^*)/3\} f^3(x_1)$. That is, for all x in a sufficiently small neighborhood of x^*,

$$|f''(x^*)| > \left| \frac{(x - x^*)}{3} f^3(x_1) \right|$$

Therefore the sign of $2\{f(x) - f(x^*)/(x - x^*)^2\}$ is the same as the sign of $f''(x^*)$ in this neighborhood of x^*. If $f''(x^*) < 0$, then $f(x) - f(x^*) < 0$ for all x sufficiently close to x^* since $2/(x - x^*)^2 > 0$. However, if $f(x) - f(x^*) < 0$ for all x near x^*, then x^* is a local maximum for $f(x)$. On the other hand, if $f''(x^*) > 0$ then $f(x) - f(x^*) > 0$ for all x sufficiently near x^* and x^* is a local minimum for $f(x)$. As before, if $f''(x^*) = 0$, conclusions cannot be drawn about the nature of x^* without examining the remainder. The following theorem extends the sufficient conditions for the existence of a local maximum or minimum at a point, x^*, to the case where the first m derivatives of $f(x)$ vanish at x^*.

Theorem 4.8 Let $f(x)$ be continuous on the closed interval $[a, b]$, and let $f^i(x)$ exist and be continuous on the open interval (a, b) for $i = 1, 2, \ldots, n$. If x^* is a point on (a, b) such that

$$f^i(x^*) = 0, \quad i = 1, 2, \ldots, m \quad \text{and} \quad f^{m+1}(x^*) \neq 0, \quad m < n - 1$$

then

1. x^* is a local maximum if $m + 1$ is even and $f^{m+1}(x^*) < 0$;
2. x^* is a local minimum if $m + 1$ is even and $f^{m+1}(x^*) > 0$;
3. x^* is a point of inflection if $m + 1$ is odd.

Proof Given the above conditions and some point x_1 on (a, b)

$$f(x) = f(x^*) + \frac{(x - x^*)^{m+1}}{(m + 1)!} f^{m+1}(x^*) + \frac{(x - x^*)^{m+2}}{(m + 2)!} f^{m+2}(x_1)$$

or

$$(m + 1)! \frac{f(x) - f(x^*)}{(x - x^*)^{m+1}} = f^{m+1}(x^*) + \frac{(x - x^*)}{(m + 2)} f^{m+2}(x_1)$$

We choose x sufficiently close to x^*, and $f^{m+1}(x^*)$ dominates $\{(x - x^*)/(m + 2)\} f^{m+2}(x_1)$. Therefore, the signs of $f^{m+1}(x^*)$ and $[f(x) - f(x^*)]/(x - x^*)^{m+1}$ will be the same. If $m + 1$ is even and $f^{m+1}(x^*) < 0$, then $f(x) - f(x^*) < 0$ for all x sufficiently close to x^* since $(x - x^*)^{m+1}$ is positive if $m + 1$ is even. But if $f(x) - f(x^*) < 0$ for all x sufficiently close to x^*, then x^* must be a local maximum for $f(x)$.

If $m + 1$ is even and $f^{m+1}(x^*) > 0$, then $f(x) - f(x^*) > 0$ since $(x - x^*)^{m+1}$ is again positive. However, if $f(x) - f(x^*) > 0$ for all x sufficiently close to x^*, x^* must be a local minimum for $f(x)$.

If $m + 1$ is odd, $(x - x^*)^{m+1}$ is negative for $x < x^*$ and positive for $x > x^*$. Assume

$$f^{m+1}(x^*) < 0 \quad \text{then} \quad \frac{f(x) - f(x^*)}{(x - x^*)^{m+1}} < 0$$

But if $\{f(x) - f(x^*)/(x - x^*)^{m+1}\} < 0$ for all x sufficiently close to x^*, then for $x > x^*$, $f(x) < f(x^*)$, and for $x < x^*$, $f(x) > f(x^*)$. Therefore x^* must be a point of inflection for $f(x)$.

Next assume $m + 1$ is odd and

$$f^{m+1}(x^*) > 0$$

Then

$$\frac{f(x) - f(x^*)}{(x - x^*)^{m+1}} > 0$$

By an argument similar to that given above for $x > x^*$, $f(x) > f(x^*)$, and for $x < x^*$, $f(x) < f(x^*)$. Thus x^* is a point of inflection for $f(x)$. Hence, if the $(m + 1)$st derivative in Taylor's expansion is the first which is nonzero and if $(m + 1)$ is odd, x^* is a point of inflection for $f(x)$.

4.2 FUNCTIONS OF A SINGLE VARIABLE

Example 4.9 Identify the interior local maxima, minima, and points of inflection for the following function.

$$f(x) = 3x^4 - 4x^3$$

For any interior extreme point or point of inflection, x^*,

$$f'(x^*) = 0$$

In this case

$$f'(x) = 12x^3 - 12x^2 \quad \text{and} \quad x^* = 0, 1$$

Using the second derivative test to determine the nature of the points 0 and 1:

$$f''(x) = 36x^2 - 24x \quad \text{and} \quad f''(0) = 0, \quad f''(1) = 12$$

Therefore $x^* = 1$ is a local minimum for $f(x)$, but the nature of $x^* = 0$ is not yet determined. Taking the third derivative

$$f^3(x) = 72x - 24 \quad \text{and} \quad f^3(0) \neq 0$$

Therefore, $x^* = 0$ is a point of inflection for $f(x)$. The function analyzed here is shown in Fig. 4.10.

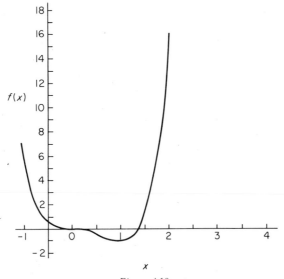

Figure 4.10

Example 4.10 Identify the interior extreme points and the points of inflection for the function

$$f(x) = \frac{1}{5}x^5 - \frac{5}{2}x^4 + \frac{35}{3}x^3 - 25x^2 + 24x + 6$$

Taking the first derivative of $f(x)$

$$f'(x) = x^4 - 10x^3 + 35x^2 - 50x + 24 = (x - 1)(x - 2)(x - 3)(x - 4)$$

Therefore
$$x^* = 1, 2, 3, 4$$
Using the second derivative test to determine the nature of each value of x^*,
$$f''(x) = 4x^3 - 30x^2 + 70x - 50, \qquad f''(1) = -6$$
$$f''(2) = 2, \qquad f''(3) = -2, \qquad f''(4) = 6$$
Therefore, local minima are located at 2 and 4 and local maxima at 1 and 3. ∎

4.3 Functions of Several Variables

Let $f(X)$ be a continuous function defined on the closed space $[\mathscr{X}_n]$. That is, $f(X)$ is defined for all X belonging to $[\mathscr{X}_n]$. For example, let
$$f(X) = ax_1^2 + bx_2^2, \qquad 0 \le x_1^2 + x_2^2 \le r^2$$
The vector X is given by
$$X = \begin{bmatrix} x_1 \\ x_2 \end{bmatrix}$$
and the space for which $f(X)$ is defined, $[\mathscr{X}_2]$, is given by the set of two-dimensional vectors, X, such that $-r \le x_1 \le r$ and $-\sqrt{r^2 - x_1^2} \le x_2 \le \sqrt{r^2 - x_1^2}$.

Let $f(X)$ be a continuous function defined on the space $[\mathscr{X}_n]$. Assume that X is an n-dimensional vector. Then $f(X)$ and $f(x_1, x_2, \ldots, x_n)$ are equivalent expressions for the functional relationship. If X^* is an interior extreme point for $f(X)$, then there exists a subspace of $[\mathscr{X}_n]$, $(\mathscr{X}_n')$, such that $f(X)$ is convex on $(\mathscr{X}_n')$. If X^* is a local maximum on $(\mathscr{X}_n')$, then
$$f(X^*) \ge f(X), \qquad X \in (\mathscr{X}_n') \tag{4.32}$$
and if X^* is a local minimum on $(\mathscr{X}_n')$, then
$$f(X^*) \le f(X), \qquad X \in (\mathscr{X}_n') \tag{4.33}$$
If X^* is a global maximum, then the relationship given in Eq. (4.32) holds for all X belonging to $[\mathscr{X}_n]$. Similarly, if X^* is a global minimum then Eq. (4.33) holds for all X belonging to $[\mathscr{X}_n]$.

To develop the necessary conditions for the existence of an extreme point on $[\mathscr{X}_n]$ and the sufficient conditions for a local maximum or minimum at an extreme point we will use an approach analogous to that used for functions of a single variable. First, the results of Theorem 4.3, the mean-value theorem, are extended to functions of several variables. Next Taylor's mean-value theorem is similarly extended to functions of several variables. For functions of several variables we will use Taylor's mean-value theorem to prove both the necessary and sufficient conditions for a local maximum or local minimum at a given extreme point.

4.3 FUNCTIONS OF SEVERAL VARIABLES

Theorem 4.9 Mean-Value Theorem for Functions of Several Variables Let X be an m-dimensional vector and let $f(X)$ be a continuous function defined on the closed convex space $[\mathscr{X}_m]$. If $(\partial/\partial x_i)f(X)$ exists and is continuous on the open space $(\mathscr{X}_m)$, then for any X and X_0 on $(\mathscr{X}_m)$ there exists a point X_1 on $(\mathscr{X}_m)$ such that

$$f(X) = f(X_0) + \sum_{j=1}^{m} (x_j - x_{0j}) \frac{\partial}{\partial x_j} f(X) \bigg|_{X_1} \quad (4.34)$$

where x_j and x_{0j} are elements of X and X_0 respectively.

Proof Let the vector Y be defined by a convex combination of X and X_0.

$$Y = \lambda X + (1 - \lambda) X_0 \quad (4.35)$$

where λ lies on the closed interval $[0, 1]$. Therefore, since X and X_0 belong to $(\mathscr{X}_m)$ and since $(\mathscr{X}_m)$ is convex, Y belongs to $(\mathscr{X}_m)$. Now define the function $g(\lambda)$ as

$$g(\lambda) = f[\lambda X + (1 - \lambda) X_0] \quad (4.36)$$

Since $g(\lambda)$ is defined on the closed interval $[0, 1]$,

$$\frac{g(1) - g(0)}{1 - 0} = g'(\lambda_1) \quad (4.37)$$

from the mean-value theorem for functions of a single variable where λ_1 lies on the open interval $(0, 1)$. Differentiating $g(\lambda)$ with respect to λ yields

$$\frac{dg(\lambda)}{d\lambda} = \sum_{j=1}^{m} \frac{\partial g(\lambda)}{\partial y_j} \frac{\partial y_j}{\partial \lambda} \quad (4.38)$$

From Eq. (4.35)

$$\frac{\partial y_j}{\partial \lambda} = (x_j - x_{0j})$$

Therefore

$$\frac{dg(\lambda)}{d\lambda} = \sum_{j=1}^{m} (x_j - x_{0j}) \frac{\partial g(\lambda)}{\partial y_j} \quad (4.39)$$

and from Eq. (4.37)

$$g(1) = g(0) + \sum_{j=1}^{m} (x_j - x_{0j}) \frac{\partial g(\lambda)}{\partial y_j} \bigg|_{\lambda_1} \quad (4.40)$$

From the definition of $g(\lambda)$

$$g(1) = f(X) \quad \text{and} \quad g(0) = f(X_0)$$

Let

$$X_1 = \lambda_1 X + (1 - \lambda_1) X_0 \quad (4.41)$$

Then
$$g(\lambda_1) = f(X_1) \quad \text{and} \quad \frac{\partial g(\lambda)}{\partial y_j}\bigg|_{\lambda=\lambda_1} = \frac{\partial}{\partial x_j}f(X)\bigg|_{X_1}$$

Equation (4.40) then reduces to
$$f(X) = f(X_0) + \sum_{j=1}^{m}(x_j - x_{0j})\frac{\partial}{\partial x_j}f(X)\bigg|_{X_1}$$

Since X_1 is defined by a convex combination of points on $(\mathcal{X}_m)$, X_1 must lie on $(\mathcal{X}_m)$. Further, the fact that X_1 is defined by a convex combination of X and X_0, implies that X_1 lies on the line segment joining X and X_0. ◪

Example 4.11 Let X be a three-dimensional vector and define $f(X)$ as
$$f(X) = x_1^2 + x_2^2 + x_3^2$$

Find the vector X_1 such that
$$f(X) = f(X_0) + \sum_{j=1}^{3}(x_j - x_{0j})\frac{\partial}{\partial x_j}f(X)\bigg|_{X_1}$$

Assume that $f(X)$ is defined for all positive vectors X. That is
$$X \geq 0$$

From the proof of the mean-value theorem for functions of several variables, X_1 can be defined as a point on the line segment joining X and X_0. Therefore
$$X_1 = \lambda X + (1 - \lambda)X_0 \quad \text{or} \quad X_1 = X_0 + \lambda(X - X_0)$$

Now
$$\frac{\partial}{\partial x_j}f(X) = 2x_j, \quad j = 1, 2, 3$$

Therefore
$$f(X) = f(X_0) + 2(X - X_0)^T X_1$$
$$= f(X_0) + 2(X - X_0)^T X_0 + 2\lambda(X - X_0)^T(X - X_0)$$

and
$$\lambda = \frac{f(X) - f(X_0) - 2(X - X_0)^T X_0}{2(X - X_0)^T(X - X_0)}$$

To illustrate, let
$$X = \begin{bmatrix} 2 \\ 2 \\ 2 \end{bmatrix}, \quad X_0 = \begin{bmatrix} 1 \\ 1 \\ 1 \end{bmatrix}$$

4.3 FUNCTIONS OF SEVERAL VARIABLES

Then

$$f(X) = 12 \quad f(X_0) = 3 \quad \text{and} \quad \lambda = \frac{12 - 3 - 2(3)}{2(3)} = \frac{1}{2}$$

Therefore X_1 is given by

$$X_1 = \begin{bmatrix} \frac{3}{2} \\ \frac{3}{2} \\ \frac{3}{2} \\ \frac{3}{2} \end{bmatrix}$$

Although the necessary conditions for the existence of an interior extreme point for a continuous function $f(X)$ can be developed using the mean-value theorem for functions of several variables, this development is left as an exercise for the reader. Instead the necessary conditions for the existence of an interior extreme point are developed through Taylor's theorem for functions of several variables.

Theorem 4.10 Taylor's Theorem for Functions of Several Variables Let $f(X)$ be a continuous function defined on the closed convex space $[\mathscr{X}_m]$, where X is an m-dimensional vector with elements x_k. If the first n partial derivatives of $f(X)$ with respect to x_k, $k = 1, 2, \ldots, m$, exist and are continuous on the open space $(\mathscr{X}_m)$, then for X and X_0 on $(\mathscr{X}_m)$ there exists a point X_1 on the line segment joining X and X_0 such that

$$f(X) = f(X_0) + \sum_{i=1}^{n-1} \frac{1}{i!} D^i f(X) \bigg|_{X_0} + R_n(X) \bigg|_{X_1} \quad (4.42)$$

where X_0 has elements x_{0k}, X_1 has elements x_{1k}, and D is the operator defined by

$$D = \sum_{k=1}^{m} (x_k - x_{0k}) \frac{\partial}{\partial x_k} \quad (4.43)$$

and

$$R_n(X) \bigg|_{X_1} = \frac{1}{n!} D^n f(X) \bigg|_{X_1} \quad (4.44)$$

Proof To prove this theorem, let X, X_0, and X_1 be m-dimensional vectors on $(\mathscr{X}_m)$ with elements x_j, x_{0j}, and x_{1j} respectively. Let

$$Y = \lambda X + (1 - \lambda) X_0, \quad 0 \le \lambda \le 1 \quad (4.45)$$

Since $(\mathscr{X}_m)$ is convex, Y belongs to $(\mathscr{X}_m)$. Let

$$g(\lambda) = f[X_0 + \lambda(X - X_0)] \quad (4.46)$$

Therefore

$$\frac{dg}{d\lambda} = \frac{\partial g}{\partial y_1} \frac{\partial y_1}{\partial \lambda} + \frac{\partial g}{\partial y_2} \frac{\partial y_2}{\partial \lambda} + \cdots + \frac{\partial g}{\partial y_m} \frac{\partial y_m}{\partial \lambda}$$

$$= (x_1 - x_{01}) \frac{\partial g}{\partial y_1} + (x_2 - x_{02}) \frac{\partial g}{\partial y_2} + \cdots + (x_m - x_{0m}) \frac{\partial g}{\partial y_m} \quad (4.47)$$

From Taylor's mean-value theorem for continuous functions of a single variable,

$$g(\lambda) = g(\lambda_0) + \sum_{i=1}^{n-1} \frac{(\lambda - \lambda_0)^i}{i!} \frac{d^i g(\lambda)}{d\lambda^i}\bigg|_{\lambda=\lambda_0} + \frac{(\lambda - \lambda_0)^n}{n!} \frac{d^n g(\lambda)}{d\lambda^n}\bigg|_{\lambda=\lambda_1}$$

where λ_0, λ, and λ_1 lie on the interval $[0, 1]$. From Eq. (4.43)

$$\frac{d^i}{d\lambda^i} = \left[\sum_{j=1}^{m} (x_j - x_{0j}) \frac{\partial}{\partial y_j}\right]^i$$

Therefore

$$g(\lambda) = g(\lambda_0) + \sum_{i=1}^{n-1} \frac{(\lambda - \lambda_0)^i}{i!} \left[\sum_{j=1}^{m} (x_j - x_{0j}) \frac{\partial}{\partial y_j}\right]^i g(\lambda)\bigg|_{\lambda=\lambda_0}$$
$$+ \frac{(\lambda - \lambda_0)^n}{n!} \left[\sum_{j=1}^{m} (x_j - x_{0j}) \frac{\partial}{\partial y_j}\right]^n g(\lambda)\bigg|_{\lambda=\lambda_1} \quad (4.48)$$

Let $\lambda = 1$, $\lambda_0 = 0$. Then

$$g(\lambda) = f(X) \quad \text{and} \quad g(\lambda_0) = f(X_0)$$

For λ_1, define

$$X_1 = X_0 + \lambda_1 (X - X_0)$$

Then

$$f(X) = f(X_0) + \sum_{i=1}^{n-1} \frac{1}{i!} D^i f(X)\bigg|_{X_0} + R_n(X)\bigg|_{X_1}$$

where

$$R_n(X)\bigg|_{X_1} = \frac{(\lambda - \lambda_0)^n}{n!} \left[\sum_{j=1}^{m} (x_j - x_{0j}) \frac{\partial}{\partial y_j}\right]^n g(\lambda)\bigg|_{\lambda=\lambda_1}$$
$$= \frac{1}{n!} \left[\sum_{j=1}^{m} (x_j - x_{0j}) \frac{\partial}{\partial x_j}\right]^n f(X)\bigg|_{X=X_1} \quad (4.49)$$

Since X_1 is defined by a convex combination of X_0 and X, X_1 lies on a line segment joining X_0 and X. Further, since $f(X)$ is defined on the convex space $(\mathscr{X}_m)$, $f(X_1)$ is defined since X_1 belongs to $(\mathscr{X}_m)$. ◘

Example 4.12 Let

$$f(X) = 2x_1^3 - x_2^3 + x_1 x_2$$

and let

$$X = \begin{bmatrix} 1 \\ 1 \end{bmatrix}, \quad X_0 = \begin{bmatrix} 2 \\ 3 \end{bmatrix}$$

4.3 FUNCTIONS OF SEVERAL VARIABLES

Find a vector, X_1, on the line segment joining X_0 and X such that

$$f(X) = f(X_0) + \sum_{k=1}^{2} (x_k - x_{0k}) \frac{\partial}{\partial x_k} f(X) \bigg|_{X_0} + R_2(X) \bigg|_{X_1}$$

Taking partial derivatives of $f(X)$ with respect to x_1 and x_2 gives us

$$\frac{\partial}{\partial x_1} f(X) = 6x_1^2 + x_2, \qquad \frac{\partial}{\partial x_2} f(X) = -3x_2^2 + x_1$$

and

$$\frac{\partial}{\partial x_1} f(X) \bigg|_{X_0} = 27, \qquad \frac{\partial}{\partial x_2} f(X) \bigg|_{X_0} = -25$$

From Taylor's theorem for functions of several variables,

$$R_2(X) \bigg|_{X_1} = \frac{(x_1 - x_{01})^2}{2} \frac{\partial^2}{\partial x_1^2} f(X) \bigg|_{X_1}$$

$$+ (x_1 - x_{01})(x_2 - x_{02}) \frac{\partial^2}{\partial x_1 \partial x_2} f(X) \bigg|_{X_1} + \frac{(x_2 - x_{02})^2}{2} \frac{\partial^2}{\partial x_2^2} f(X) \bigg|_{X_1}$$

$$\frac{\partial^2}{\partial x_1^2} f(X) = 12x_1, \qquad \frac{\partial^2}{\partial x_1 \partial x_2} f(X) = 1, \qquad \frac{\partial^2}{\partial x_2^2} f(X) = -6x_2$$

Let

$$X_1 = \begin{bmatrix} x_{11} \\ x_{12} \end{bmatrix}$$

Therefore

$$f(X) = 2 = -5 + 27(-1) - 25(-2) + \tfrac{1}{2}(12x_{11}) + (-1)(-2)(1) + 2(-6x_{12})$$

and

$$6x_{11} - 12x_{12} = -18 \qquad \text{or} \qquad x_{11} = 2x_{12} - 3$$

Since X_1 is to lie on the line segment joining X_0 and X

$$\begin{bmatrix} 2x_{12} - 3 \\ x_{12} \end{bmatrix} = \lambda \begin{bmatrix} 1 \\ 1 \end{bmatrix} + (1 - \lambda) \begin{bmatrix} 2 \\ 3 \end{bmatrix}$$

Solving for λ and x_{12}

$$\lambda = \tfrac{1}{3}, \qquad x_{12} = \tfrac{7}{3}$$

Therefore

$$X_1 = \begin{bmatrix} \tfrac{5}{3} \\ \tfrac{7}{3} \end{bmatrix}$$

Theorem 4.11 If $f(X)$ is continuous on the closed space $[\mathscr{X}_m]$ and $\partial f(X)/\partial x_k$ is defined on the open space $(\mathscr{X}_m)$ for $k = 1, 2, \ldots, m$, then if X^* is an extreme point for $f(X)$

$$\left.\frac{\partial}{\partial x_k} f(X)\right|_{X*} = 0, \quad k = 1, 2, \ldots, m \tag{4.50}$$

where X is an m-dimensional vector.

Proof From Taylor's theorem for multivariate functions,

$$f(X) = f(X^*) + \sum_{k=1}^{m} (x_k - x_k^*) \left.\frac{\partial}{\partial x_k} f(X)\right|_{X*} + \left.R_2(X)\right|_{X_1}$$

or

$$f(X) - f(X^*) = \sum_{k=1}^{m} (x_k - x_k^*) \left.\frac{\partial}{\partial x_k} f(X)\right|_{X*} + \left.R_2(X)\right|_{X_1}$$

We prove this theorem by contradiction. Assume that

$$\left.\frac{\partial}{\partial x_k} f(X)\right|_{X*} = \begin{cases} 0, & k \neq i \\ c, & k = i \end{cases}$$

where c is nonzero. Let $(\mathscr{X}'_m)$ be a subspace of $(\mathscr{X}_m)$ containing X^*, chosen such that for all X ($\neq X^*$) belonging to $(\mathscr{X}'_m)$

$$\left|(x_i - x_i^*) \left.\frac{\partial}{\partial x_i} f(X)\right|_{X*}\right| > \left|\left.R_2(X)\right|_{X_1}\right| \tag{4.51}$$

Therefore in the neighborhood of X^* defined by $(\mathscr{X}'_m)$, $f(X) - f(X^*)$ and $(x_i - x_i^*)\{\partial f(X)/\partial x_i\}|_{X*}$ have the same sign. Assume that X^* is a local maximum for $f(X)$. Then

$$f(X) - f(X^*) \leq 0 \tag{4.52}$$

for all X in $(\mathscr{X}'_m)$. Since $(\mathscr{X}'_m)$ is an open space there exist vectors belonging to $(\mathscr{X}'_m)$ such that

$$(x_i - x_i^*) \left.\frac{\partial}{\partial x_i} f(X)\right|_{X*} < 0$$

and vectors such that

$$(x_i - x_i^*) \left.\frac{\partial}{\partial x_i} f(X)\right|_{X*} > 0$$

But since $f(X) - f(X^*)$ and $(x_i - x_i^*)\{\partial f(X)/\partial x_i\}|_{X*}$ have the same sign, the inequality in 4.52 cannot hold and X^* cannot be a local maximum if $\partial f(X)/\partial x_i|_{X*} \neq 0$. By a similar argument it can be shown that X^* is not a local minimum. Therefore if X^* is an extreme point for $f(X)$ $\partial f(X)/\partial x_i|_{X*} = 0$. It

4.3 FUNCTIONS OF SEVERAL VARIABLES

follows that if one partial derivative must vanish at X^*, all must vanish (see problems). Therefore a necessary condition for the existence of an interior extreme point at X^* is that

$$\left.\frac{\partial}{\partial x_k}f(X)\right|_{X*} = 0, \quad k = 1, 2, \ldots, m$$

Example 4.13 Find the extreme points of the following function.

$$f(X) = x_1^3 + x_2^3 + 2x_1^2 + 4x_2^2$$

From Theorem 4.11, the necessary conditions for the existence of an extreme point at X^* are that

$$\left.\frac{\partial}{\partial x_k}f(X)\right|_{X*} = 0, \quad k = 1, 2 \qquad (4.53)$$

Therefore it is first necessary to identify the points, X^*, satisfying Eq. (4.53). Taking first partial derivatives with respect to x_1 and x_2

$$\frac{\partial}{\partial x_1}f(X) = 3x_1^2 + 4x_1 \qquad \frac{\partial}{\partial x_2}f(X) = 3x_2^2 + 8x_2$$

The vectors for which the first partial derivatives vanish are

$$\begin{bmatrix} 0 \\ 0 \end{bmatrix}, \quad \begin{bmatrix} 0 \\ -\frac{8}{3} \end{bmatrix}, \quad \begin{bmatrix} -\frac{4}{3} \\ 0 \end{bmatrix}, \quad \begin{bmatrix} -\frac{4}{3} \\ -\frac{8}{3} \end{bmatrix}$$

Therefore, if the function has interior extreme points, they lie among these points. Let X^* be any of these points and let

$$E = \begin{bmatrix} \epsilon_1 \\ \epsilon_2 \end{bmatrix}$$

To determine whether X^* is an extreme point for $f(X)$ we examine the difference $f(X^*) - f(X^* + E)$ for all $X^* + E$ in a sufficiently small neighborhood of X^*. If

$$f(X^*) - f(X^* + E) \geq 0$$

for all $X^* + E$ in a sufficiently small neighborhood of X^*, then X^* is a local maximum. If

$$f(X^*) - f(X^* + E) \leq 0$$

for all $X^* + E$ in a small neighborhood of X^*, then X^* is a local minimum. If neither case holds then X^* is not an extreme point for $f(X)$. Let

$$X^* = \begin{bmatrix} 0 \\ 0 \end{bmatrix}$$

Then

$$f(X^*) = 0 \quad \text{and} \quad f(X^*) - f(X^* + E) = -\epsilon_1^2(\epsilon_1 + 2) - \epsilon_2^2(\epsilon_2 + 4)$$

For $\epsilon_1 > -2$ and $\epsilon_2 > -4$, $f(X^*) - f(X^* + E) < 0$ and X^* is a local minimum for $f(X)$. Now let

$$X^* = \begin{bmatrix} 0 \\ -\frac{8}{3} \end{bmatrix}$$

from which

$$f(X^*) = \frac{256}{27} \quad \text{and} \quad f(X^*) - f(X^* + E) = -\epsilon_1^2(\epsilon_1 + 2) + \epsilon_2^2(4 - \epsilon_2)$$

Every neighborhood of X^* must include a set of points for which $\epsilon_1 = 0$ and a set of points for which $\epsilon_2 = 0$. For $\epsilon_1 = 0$, choose $\epsilon_2 > -4$. Then $f(X^*) - f(X^* + E) > 0$. For $\epsilon_2 = 0$, choose $\epsilon_1 > -2$. Then $f(X^*) - f(X^* + E) < 0$. Therefore every neighborhood of X^* contains points for which $f(X^*) - f(X^* + E)$ is positive and points for which this difference is negative. Hence X^* is not an extreme point for $f(X)$. By a similar argument, it can be shown that

$$X^* = \begin{bmatrix} -\frac{4}{3} \\ 0 \end{bmatrix}$$

is not an extreme point for $f(X)$.

Let

$$X^* = \begin{bmatrix} -\frac{4}{3} \\ -\frac{8}{3} \end{bmatrix}$$

for which

$$f(X^*) = \frac{288}{27} \quad \text{and} \quad f(X^*) - f(X^* + E) = \epsilon_1^2(2 - \epsilon_1) + \epsilon_2^2(4 - \epsilon_2)$$

Choosing a neighborhood of points about X^* such that $\epsilon_1 < 2$ and $\epsilon_2 < 4$ shows that X^* is a maximum for $f(X)$ since $f(X^*) - f(X^* + E) > 0$ for all such points.

As in the case of functions of a single variable, a point may satisfy the necessary conditions for an extreme point and yet not be an extreme point. For functions of a single variable such a point was called a point of inflection. In the case of functions of several variables such a point is called a *saddle point*. An example of such a point is shown graphically in Fig. 4.11 for the two-dimensional case. As in the case of functions of a single variable, any point, X^*, satisfying the necessary conditions for $f(X)$ to take on an extreme value is called a stationary point whether it is in fact an extreme point or not.

To develop sufficient conditions for the existence of a local maximum or minimum at a point X^*, it will be necessary to examine the matrix of second partial derivatives of $f(X)$ evaluated at X^*. For this purpose let

$$f_{ij}(X^*) = \frac{\partial^2}{\partial x_i \, \partial x_j} f(X) \bigg|_{X^*} \quad (4.54)$$

4.3 FUNCTIONS OF SEVERAL VARIABLES

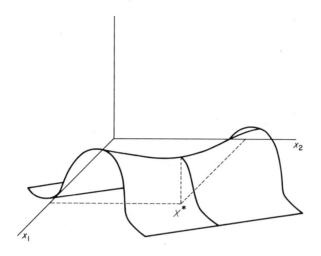

Figure 4.11 Saddle Point at X^*

and define the matrix H as

$$H = \begin{bmatrix} f_{11}(X^*) & f_{12}(X^*) & \cdots & f_{1m}(X^*) \\ f_{21}(X^*) & f_{22}(X^*) & \cdots & f_{2m}(X^*) \\ \vdots & \vdots & & \vdots \\ f_{m1}(X^*) & f_{m2}(X^*) & \cdots & f_{mm}(X^*) \end{bmatrix} \quad (4.55)$$

The matrix H is sometimes referred to as the *Hessian* of the function $f(X)$. In particular, we will be interested in determining whether H is positive definite, negative definite, or indefinite. The reader will recall that the criteria used to determine the definiteness of a matrix were based upon the assumption that the matrix was symmetric. Fortunately the matrix H possesses this property since

$$f_{ij}(X^*) = f_{ji}(X^*) \quad (4.56)$$

Theorem 4.12 Let $f(X)$ be continuous on the closed space $[\mathscr{X}_m]$ and let $\partial^2 f(X)/\partial x_i \, \partial x_j$ exist and be continuous on the open space $(\mathscr{X}_m)$ for $i, j = 1, 2, \ldots, m$. If X^* is a point on $(\mathscr{X}_m)$ such that

$$\left. \frac{\partial}{\partial x_k} f(X) \right|_{X*} = 0, \quad k = 1, 2, \ldots, m \quad (4.57)$$

and H is the matrix of second partial derivatives evaluated at $X_\emptyset^*$, then

1. X^* is a local minimum if H is positive definite
2. X^* is a local maximum if H is negative definite
3. X^* is a saddle point if H is indefinite.

Proof From Taylor's theorem for functions of several variables

$$f(X) = f(X^*) + Df(X)|_{X*} + \tfrac{1}{2}D^2f(X)|_{X*} + R_3(X)|_{X_1} \tag{4.58}$$

where X and X_1 belong to $(\mathcal{X}_m)$ and D is given by

$$D = \sum_{k=1}^{m} (x_k - x_k^*) \frac{\partial}{\partial x_k}$$

If

$$\left.\frac{\partial}{\partial x_k} f(X)\right|_{X*} = 0, \quad k = 1, 2, \ldots, m$$

then

$$f(X) - f(X^*) = \tfrac{1}{2}D^2f(X)|_{X*} + R_3(X)|_{X_1} \tag{4.59}$$

where

$$\tfrac{1}{2}D^2f(X)|_{X*} = \tfrac{1}{2}(X - X^*)^T H(X - X^*) \tag{4.60}$$

where H is the Hessian matrix defined in Eq. (4.55). If H is positive definite, then the quadratic form $(X - X^*)^T H(X - X^*)$ is also positive definite. Then for a sufficiently small neighborhood about X^*, $f(X) - f(X^*) > 0$ for all X belonging to that neighborhood since $|(X - X^*)^T H(X - X^*)| > |R_3(X_1)|$ for such a neighborhood. Hence X^* is a local minimum for $f(X)$ if H is positive definite.

If H is negative definite, $(X - X^*)^T H(X - X^*)$ is also negative definite. Choosing a sufficiently small neighborhood about X^*, $f(X) - f(X^*)$ has the same sign as $\tfrac{1}{2}(X - X^*)^T H(X - X^*)$. That is, $f(X) - f(X^*) < 0$ for all X in the specified neighborhood about X^*. Thus, if H is negative definite, X^* is a local maximum for $f(X)$.

The proof that X^* is a saddle point if H is indefinite is left as an exercise for the reader. ◻

Example 4.14 Identify the interior extreme points for the following function and determine whether each is a local minimum or a local maximum.

$$f(X) = \exp[-(x_1^2 + x_2^2 + x_3^2)]$$

Taking first partial derivatives leads us to

$$\frac{\partial}{\partial x_i} f(X) = -2x_i \exp[-(x_1^2 + x_2^2 + x_3^2)], \quad i = 1, 2, 3$$

The only finite point for which $\partial f(X)/\partial x_i$ vanishes for $i = 1, 2, 3$ is the origin. Therefore, the only point, X^*, which may be an interior extreme point is

$$X^* = \begin{bmatrix} 0 \\ 0 \\ 0 \end{bmatrix}$$

4.3 FUNCTIONS OF SEVERAL VARIABLES

To determine the nature of X^*, we first take second partial derivatives of $f(X)$.

$$\frac{\partial^2}{\partial x_i^2} f(X) = 2(2x_i^2 - 1)\exp[-(x_1^2 + x_2^2 + x_3^2)], \qquad i = 1, 2, 3$$

$$\frac{\partial^2}{\partial x_i \, \partial x_j} f(X) = 4x_i x_j \exp[-(x_1^2 + x_2^2 + x_3^2)], \qquad i \neq j$$

The matrix of second partial derivatives evaluated at X^* is given by

$$H = \begin{bmatrix} -2 & 0 & 0 \\ 0 & -2 & 0 \\ 0 & 0 & -2 \end{bmatrix}$$

The determinants of the principal minors of H, $|H_1|$, $|H_2|$, and $|H_3|$, are given by

$$|H_1| = -2, \qquad |H_2| = 4, \qquad |H_3| = -8$$

Since $|H_1| < 0$, $|H_2| > 0$, $|H_3| < 0$, H is negative definite and X^* is a local maximum for $f(X)$.

Example 4.15 Identify the interior maxima, minima, and saddle points for the function given in Example 4.13.

$$f(X) = x_1^3 + x_2^3 + 2x_1^2 + 4x_2^2$$

The second partial derivatives of $f(X)$ are given by

$$\frac{\partial^2}{\partial x_1^2} f(X) = 6x_1 + 4, \qquad \frac{\partial^2}{\partial x_1 \, \partial x_2} f(X) = 0, \qquad \frac{\partial^2}{\partial x_2^2} f(X) = 6x_2 + 8$$

From Example 4.13, the points satisfying the necessary conditions for an extreme point are

$$\begin{bmatrix} 0 \\ 0 \end{bmatrix}, \quad \begin{bmatrix} 0 \\ -\frac{8}{3} \end{bmatrix}, \quad \begin{bmatrix} -\frac{4}{3} \\ 0 \end{bmatrix}, \quad \begin{bmatrix} -\frac{4}{3} \\ -\frac{8}{3} \end{bmatrix}$$

hereafter denoted X_1, X_2, X_3, and X_4 respectively.
Let

$$|H_1| = \frac{\partial^2}{\partial x_1^2} f(X_i), \qquad |H_2| = \frac{\partial^2}{\partial x_1^2} f(X_i) \frac{\partial^2}{\partial x_2^2} f(X_i) - \left[\frac{\partial^2}{\partial x_1 \, \partial x_2} f(X_i) \right]^2$$

Therefore X_i is a local minimum if $|H_1| > 0$ and $|H_2| > 0$, a local maximum if $|H_1| < 0$ and $|H_2| > 0$, and a saddle point if $|H_2| < 0$. The calculations required to identify the nature of the points X_1, X_2, X_3, and X_4 are summarized in Table 4.2.

TABLE 4.2

i	X_i^T	$\lvert H_1 \rvert$	$\lvert H_2 \rvert$	Nature of X_i^T
1	$[0, 0]$	4	32	Minimum
2	$[0, -\frac{8}{3}]$	4	-32	Saddle point
3	$[-\frac{4}{3}, 0]$	-4	-32	Saddle point
4	$[-\frac{4}{3}, -\frac{8}{3}]$	-4	32	Maximum

In using the Hessian, we are attempting to determine whether a function $f(X)$ is strictly convex from below or above in some neighborhood of the stationary point X^*. Specifically if $f(X)$ is strictly convex from below in some neighborhood of X^*, then X^* must be a local maximum. On the other hand, if $f(X)$ is strictly convex from above in some neighborhood of X^*, then X^* is a local minimum. Therefore if H is positive definite, then $f(X)$ is strictly convex from above in some neighborhood of X^* and strictly convex from below if H is negative definite. The reader can see that the matrix of second partial derivatives can be used to determine whether a function $f(X)$ is convex in the neighborhood of any point X. Let $H(X)$ be the matrix of second partial derivatives evaluated at X. If $H(X)$ is positive definite, then $f(X)$ is strictly convex from above near X. Similarly, if $H(X)$ is negative definite, then $f(X)$ is strictly convex from below near X. If it happens that $H(X)$ is positive definite (negative definite) for all X, then $f(X)$ is strictly convex from above (below) throughout its space of definition.

Example 4.16 Find and identify the extreme points for the following function.

$$f(X) = 4 + 3x_1 + 2x_2 - x_1^2 - 5x_2^2$$

From Theorem 4.11,

$$\frac{\partial}{\partial x_1} f(X) \bigg|_{X_0} = 3 - 2x_1 = 0, \qquad \frac{\partial}{\partial x_2} f(X) \bigg|_{X_0} = 2 - 10x_2 = 0$$

and

$$X^* = \begin{bmatrix} \frac{3}{2} \\ \frac{1}{5} \end{bmatrix}$$

The matrix of second partial derivatives at X^* is given by

$$H(X) = \begin{bmatrix} -2 & 0 \\ 0 & -10 \end{bmatrix}$$

Since $H(X)$ is negative definite for all X, it is negative definite at X^*, and X^* is a maximum for $f(X)$. Further since $H(X)$ is negative definite everywhere, X^* is a global maximum for $f(X)$.

From Example 4.16, one can deduce that the necessary conditions for the existence of an interior extreme point at X^* are also sufficient if the function $f(X)$

4.4 Optimization Subject to Constraints

is strictly convex from above or below. That is, if $f(X)$ is everywhere strictly convex from above, then any point X^* satisfying the necessary conditions for an interior extreme point is a global minimum, and is a global maximum if $f(X)$ is everywhere strictly convex from below.

4.4 Optimization Subject to Constraints

In the preceding sections of this chapter, we dealt with continuous functions of one or more variables in which no restrictions were placed upon the values which the variables might assume. Usually, however, the assumption that these variables may assume a limitless range of values is impractical and an analysis based upon such an assumption often leads to an infeasible solution to the problem under study. To illustrate this problem consider the following example.

Example 4.17 In the production of a certain product, a solvent is required. The production process operates continuously and uses the solvent at a uniform rate of p units per year. To maintain the required supply of solvent, an inventory of solvent is carried and is replenished every t years. The quantity of solvent placed in inventory every t years is pt. Every time the inventory is replenished a fixed cost C_0 is incurred. In addition, it costs C_I to carry one unit of solvent for one year. Let a cycle be the time between successive inventory replenishments, t. The inventory level varies as shown in Fig. 4.12.

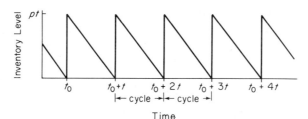

Figure 4.12 Variation of Inventory Level

The cost per cycle, $C_c(t)$, is given by

$$C_c(t) = C_0 + C_I \frac{pt^2}{2}$$

The number of cycles per year is $1/t$ and the annual cost of maintaining the solvent inventory, $C_T(t)$ is

$$C_T(t) = \frac{C_0}{t} + C_I \frac{pt}{2}$$

If

$$C_0 = \$200.00, \quad p = 100,000, \quad C_I = \$.40$$

find the value of t which minimizes the annual cost of the inventory system if the inventory will carry a maximum of 5,000 units of solvent.

Taking first and second derivatives of $C_T(t)$ with respect to t gives us

$$\frac{dC_T(t)}{dt} = -C_0 t^{-2} + C_I \frac{p}{2}, \quad \frac{d^2 C_T(t)}{dt^2} = 2C_0 t^{-3}$$

The interior extreme points for $C_T(t)$ are given by

$$t = \pm \sqrt{\frac{2C_0}{pC_I}}$$

and

$$\left. \frac{d^2 C_T(t)}{dt^2} \right|_{+\sqrt{2C_0/pC_I}} = C_I p \sqrt{\frac{pC_I}{2C_0}} > 0, \quad \left. \frac{d^2 C_T(t)}{dt^2} \right|_{-\sqrt{2C_0/pC_I}} = -C_I p \sqrt{\frac{pC_I}{2C_0}} < 0$$

Therefore, $t = \sqrt{2C_0/pC_I}$ is the minimum point for $C_T(t)$, and

$$t = 0.1 \text{ years}$$

which implies that 10,000 units of solvent will be placed in inventory at each replenishment, which is more than the holding capacity of the inventory and is thus an infeasible solution to the problem.

In the above example, a constraint exists on the amount of solvent which may be placed in inventory at one point in time. That is, the quantity of solvent placed in inventory upon replenishment must be 5000 units or less. In the discussion which follows, two types of constraints will be discussed. The first class of constraints discussed will be equality constraints. In general an *equality constraint* involving the vector X takes the form

$$g(X) = 0 \tag{4.61}$$

That is, an equality constraint involves an equality relationship containing the variables constrained. The second class of constraints discussed are *inequality constraints* and take the general form

$$g(X) \leq 0 \tag{4.62}$$

The constraint given in Example 4.17 is of the inequality type.

4.4.1 Equality Constraints

The problem treated here can be stated as that of determining the vector X such that

$$f(X) = \min(\max) \tag{4.63}$$

4.4 OPTIMIZATION SUBJECT TO CONSTRAINTS

subject to

$$g_1(X) = 0$$
$$\vdots \qquad (4.64)$$
$$g_m(X) = 0$$

where X is an n-dimensional vector and $m \leq n$. That is, the continuous function $f(X)$ is to be optimized subject to a set of m restrictions or constraints, each in the form of equality relationships involving some or all of the elements of X.

Let us consider first a function of two variables restricted by a single constraint. That is

$$f(X) = \min(\max) \qquad (4.65)$$

subject to

$$g(X) = 0 \qquad (4.66)$$

where $X = [x_1, x_2]$. One method of solving the optimization problem posed in Eqs. (4.65) and (4.66) is to express x_1 as a function of x_2 through Eq. (4.66). Thus $f(X)$ is reduced to a function of the single variable x_2 and the methods discussed under optimization of functions of a single variable may be applied. To illustrate this approach to the problem consider the following example.

Example 4.18

$$2x_1^2 + x_2^2 = \min$$

subject to

$$x_1 + x_2 - 1 = 0$$

Let

$$x_1 = 1 - x_2$$

Then

$$f(X) = 2(1 - x_2)^2 + x_2^2$$

and the function to be minimized is reduced to a function of a single variable. Taking the first derivative of $f(X)$

$$\frac{d}{dx_2}f(X) = 6x_2 - 4 \quad \text{and} \quad x_2^* = \tfrac{2}{3}$$

From the constraint placed upon the problem, $x_1^* = \tfrac{1}{3}$. Since

$$\frac{d^2}{dx_2^2}f(X) = 6 > 0$$

the point $[\tfrac{1}{3}, \tfrac{2}{3}]$ is a minimum.

Now let us generalize the above problem to the optimization of a function of n variables subject to m equality constraints where $m < n$.

$$f(X) = \min(\max) \tag{4.67}$$

subject to

$$\begin{aligned} g_1(X) &= 0 \\ &\vdots \\ g_m(X) &= 0 \end{aligned} \tag{4.68}$$

We will attempt to develop conditions which are necessary for the existence of an interior extreme point at X^*. To develop these conditions we introduce the *total differential* defined by

$$df(X) = \sum_{i=1}^{n} \frac{\partial}{\partial x_i} f(X) \, dx_i \tag{4.69}$$

If X^* is an interior extreme point, then

$$\left. \frac{\partial}{\partial x_i} f(X) \right|_{X^*} = 0, \quad i = 1, 2, \ldots, n \tag{4.70}$$

from Theorem 4.11 and therefore $df(X^*)$ must vanish at X^*. However, since the variables $x_1, x_2, \ldots, x_n$ are related through the constraints $g_k(X) = 0$, $k = 1, 2, \ldots, m$, and are thus not independent variables, we cannot assume that Eq. (4.70) is valid although we may state that $df(X^*) = 0$. We will attempt to resolve this problem by expressing the change in x_i, dx_i for $i = 1, 2, \ldots, m$ as a function of the change in x_j, dx_j, $j = m + 1, m + 2, \ldots, n$. Thus we reduce $df(X)$ to a function of the independent variations, dx_j, $j = m + 1, m + 2, \ldots, n$, with the result that the coefficients of dx_j vanish at $X = X^*$.

Since $g_k(X) = 0$ from Eq. (4.68), any variation in $x_1, x_2, \ldots, x_n$ must be such that each of the constraints is maintained. Therefore

$$dg_k(X) = \sum_{i=1}^{n} \frac{\partial}{\partial x_i} g_k(X) \, dx_i = 0 \tag{4.71}$$

where the total differential, $dg_k(X)$, is taken only for variations in $x_1, x_2, \ldots, x_n$ such that $g_k(X) = 0$. Let

$$\Delta_k = \sum_{i=1}^{m} \frac{\partial}{\partial x_i} g_k(X) \, dx_i = - \sum_{i=m+1}^{n} \frac{\partial}{\partial x_i} g_k(X) \, dx_i \tag{4.72}$$

and

$$G' = \begin{bmatrix} g_{1_1}(X) & g_{1_2}(X) & \cdots & g_{1_m}(X) \\ g_{2_1}(X) & g_{2_2}(X) & \cdots & g_{2_m}(X) \\ \vdots & \vdots & & \vdots \\ g_{m_1}(X) & g_{m_2}(X) & \cdots & g_{m_m}(X) \end{bmatrix} \tag{4.73}$$

4.4 OPTIMIZATION SUBJECT TO CONSTRAINTS

$$D = \begin{bmatrix} dx_1 \\ dx_2 \\ \vdots \\ dx_m \end{bmatrix} \tag{4.74}$$

$$\Delta = \begin{bmatrix} \Delta_1 \\ \Delta_2 \\ \vdots \\ \Delta_m \end{bmatrix} \tag{4.75}$$

where

$$g_{k_i}(X) = \frac{\partial}{\partial x_i} g_k(X) \tag{4.76}$$

Then

$$G'D = \Delta \tag{4.77}$$

By *Cramer's rule*,

$$dx_k = \frac{|G'_{\Delta_k}|}{|G'|}, \quad k = 1, 2, \ldots, m \tag{4.78}$$

if $|G'| \neq 0$ and where G'_{Δ_k} is identical to G' except that the kth column of G' is replaced by the column vector Δ. Since

$$\Delta_k = - \sum_{i=m+1}^{n} g_{k_i}(X) \, dx_i$$

we have

$$|G'_{\Delta_k}| = \begin{vmatrix} g_{1_1}(X) & \cdots & g_{1_{k-1}}(X) & -\sum_{i=m+1}^{n} g_{1_i}(X) \, dx_i & g_{1_{k+1}}(X) & \cdots & g_{1_m}(X) \\ g_{2_1}(X) & \cdots & g_{2_{k-1}}(X) & -\sum_{i=m+1}^{n} g_{2_i}(X) \, dx_i & g_{2_{k+1}}(X) & \cdots & g_{2_m}(X) \\ \vdots & & \vdots & \vdots & \vdots & & \vdots \\ g_{m_1}(X) & \cdots & g_{m_{k-1}}(X) & -\sum_{i=m+1}^{n} g_{m_i}(X) \, dx_i & g_{m_{k+1}}(X) & \cdots & g_{m_m}(X) \end{vmatrix} \tag{4.79}$$

or

$$|G'_{\Delta_k}| = - \sum_{i=m+1}^{n} \begin{vmatrix} g_{1_1}(X) & \cdots & g_{1_{k-1}}(X) & g_{1_i}(X) & g_{1_{k+1}}(X) & \cdots & g_{1_m}(X) \\ g_{2_1}(X) & \cdots & g_{2_{k-1}}(X) & g_{2_i}(X) & g_{2_{k+1}}(X) & \cdots & g_{2_m}(X) \\ \vdots & & \vdots & \vdots & \vdots & & \vdots \\ g_{m_1}(X) & \cdots & g_{m_{k-1}}(X) & g_{m_i}(X) & g_{2_{k+1}}(X) & \cdots & g_{m_m}(X) \end{vmatrix} dx_i \tag{4.80}$$

The determinant inside the summation in Eq. (4.80) is called the *Jacobian* and will be denoted by $|J_{k_i}|$. The determinant defined by $|J_{k_i}|$ is identical to $|G'|$ except that the kth column of $|G'|$ is replaced by the vector $[g_1(X), g_2(X), \ldots, g_{m_i}(X)]^T$. Then

$$dx_k = \frac{-\sum_{i=m+1}^{n} |J_{k_i}|\, dx_i}{|G'|}, \qquad k = 1, 2, \ldots, m \qquad (4.81)$$

Now

$$df(X) = \sum_{i=1}^{n} \frac{\partial}{\partial x_i} f(X)\, dx_i = \sum_{i=1}^{m} \frac{\partial}{\partial x_i} f(X)\, dx_i + \sum_{i=m+1}^{n} \frac{\partial}{\partial x_i} f(X)\, dx_i \qquad (4.82)$$

From Eq. (4.81)

$$df(X) = -\frac{1}{|G'|} \sum_{k=1}^{m} \frac{\partial}{\partial x_k} f(X) \sum_{i=m+1}^{n} |J_{k_i}|\, dx_i + \sum_{i=m+1}^{n} \frac{\partial}{\partial x_i} f(X)\, dx_i$$

$$= \frac{1}{|G'|} \sum_{i=m+1}^{n} \left[|G'| \frac{\partial}{\partial x_i} f(X) - \sum_{k=1}^{m} \frac{\partial}{\partial x_k} f(X) |J_{k_i}| \right] dx_i \qquad (4.83)$$

Now let

$$|J'_{k_i}| = \begin{vmatrix} g_{1_i}(X) & g_{1_1}(X) & \cdots & g_{1_{k-1}}(X) & g_{1_{k+1}}(X) & \cdots & g_{1_m}(X) \\ g_{2_i}(X) & g_{2_1}(X) & \cdots & g_{2_{k-1}}(X) & g_{2_{k+1}}(X) & \cdots & g_{2_m}(X) \\ \vdots & \vdots & & \vdots & \vdots & & \vdots \\ g_{m_i}(X) & g_{m_1}(X) & \cdots & g_{m_{k-1}}(X) & g_{m_{k+1}}(X) & \cdots & g_{m_m}(X) \end{vmatrix}$$

$$= (-1)^{k-1} |J_{k_i}| \qquad (4.84)$$

Then

$$df(X) = \frac{1}{|G'|} \sum_{i=m+1}^{n} \left[|G'| \frac{\partial}{\partial x_i} f(X) + \sum_{k=1}^{m} \frac{\partial}{\partial x_k} f(X)(-1)^k |J'_{k_i}| \right] dx_i \qquad (4.85)$$

But

$$|G'| f_i(X) + \sum_{k=1}^{m} (-1)^k f_k(X) |J'_{k_i}|$$

$$= \begin{vmatrix} f_i(X) & f_1(X) & f_2(X) & \cdots & f_k(X) & \cdots & f_m(X) \\ g_{1_i}(X) & g_{1_1}(X) & g_{1_2}(X) & \cdots & g_{1_k}(X) & \cdots & g_{1_m}(X) \\ \vdots & \vdots & \vdots & & \vdots & & \vdots \\ g_{m_i}(X) & g_{m_1}(X) & g_{m_2}(X) & \cdots & g_{m_k}(X) & \cdots & g_{m_m}(X) \end{vmatrix} \qquad (4.86)$$

Denoting the determinant on the right by $|J_i|$, we have

$$df(X) = \frac{1}{|G'|} \sum_{i=m+1}^{n} |J_i|\, dx_i \qquad (4.87)$$

4.4 OPTIMIZATION SUBJECT TO CONSTRAINTS

Therefore if X^* is an extreme point for $f(X)$ subject to the constraints given in Eq. (4.68), and since $df(X^*) = 0$, it is necessary that

$$|J_i| = 0, \quad i = m+1, m+2, \ldots, n \quad (4.88)$$

if $|G'| \neq 0$. The results of the last derivation are summarized in the following theorem.

Theorem 4.13 Let $f(X)$ and $g_k(X)$, $k = 1, 2, \ldots, m$ be continuous and differentiable functions. If X^* is an extreme point for $f(X)$ subject to the constraints

$$g_k(X) = 0, \quad k = 1, 2, \ldots, m \quad (4.89)$$

then it is necessary that

$$\begin{vmatrix} f_i(X^*) & f_1(X^*) & f_2(X^*) & \cdots & f_m(X^*) \\ g_{1_i}(X^*) & g_{1_1}(X^*) & g_{1_2}(X^*) & \cdots & g_{1_m}(X^*) \\ \vdots & \vdots & \vdots & & \vdots \\ g_{m_i}(X^*) & g_{m_1}(X^*) & g_{m_2}(X^*) & \cdots & g_{m_m}(X^*) \end{vmatrix} = 0,$$

$$i = m+1, m+2, \ldots, n \quad (4.90)$$

where

$$\begin{vmatrix} g_{1_1}(X^*) & g_{1_2}(X^*) & \cdots & g_{1_m}(X^*) \\ g_{2_1}(X^*) & g_{2_2}(X^*) & \cdots & g_{2_m}(X^*) \\ \vdots & \vdots & & \vdots \\ g_{m_1}(X^*) & g_{m_2}(X^*) & \cdots & g_{m_m}(X^*) \end{vmatrix} \neq 0 \quad (4.91)$$

Example 4.19 Find the stationary points of

$$f(X) = x_1^2 + x_2^2 + x_3^2$$

subject to

$$g(X) = x_1 + x_2 + x_3 - 2 = 0$$

From Eq. (4.91),

$$|G'| = g_{1_1}(X) = 1$$

Since $|G'| \neq 0$, the results of Theorem 4.13 may be applied to the problem given here. From Eq. (4.90)

$$|J_2| = \begin{vmatrix} 2x_2 & 2x_1 \\ 1 & 1 \end{vmatrix} = 2(x_2 - x_1) \quad \text{and} \quad |J_3| = \begin{vmatrix} 2x_3 & 2x_1 \\ 1 & 1 \end{vmatrix} = 2(x_3 - x_1)$$

If X^* is an extreme point for $f(X)$ under the restriction given by $g(X) = 0$, then

$$2(x_2^* - x_1^*) = 0, \quad 2(x_3^* - x_1^*) = 0$$

or

$$x_2^* = x_1^*, \quad x_3^* = x_1^*$$

From the constraint $g(X) = 0$, we have

$$x_1^* = \tfrac{2}{3}$$

Therefore, the only stationary point for $f(X)$ under the constraint given is

$$X^* = [\tfrac{2}{3}, \tfrac{2}{3}, \tfrac{2}{3}]$$ ◪

Example 4.20 Let

$$f(X) = (x_1 - 3) + (x_2 + 2)^2 + (x_3 + 1)^2 + x_4^2$$

Find the stationary points for $f(X)$ given the following constraints.

$$g_1(X) = x_1 + x_2 + x_3 + x_4 = 0, \qquad g_2(X) = x_2 - 2x_3 = 0$$

$|G'|$ is given by

$$|G'| = \begin{vmatrix} 1 & 1 \\ 0 & 1 \end{vmatrix} = 1$$

From Eq. (4.90),

$$|J_3| = \begin{vmatrix} 2(x_3 + 1) & 1 & 2(x_2 + 2) \\ 1 & 1 & 1 \\ -2 & 0 & 1 \end{vmatrix} = 4x_2 + 2x_3 + 7$$

and

$$|J_4| = \begin{vmatrix} 2x_4 & 1 & 2(x_2 + 2) \\ 1 & 1 & 1 \\ 0 & 0 & 1 \end{vmatrix} = 2x_4 - 1$$

From $g_2(X)$ and $|J_3|$,

$$x_3^* = -0.7, \qquad x_2^* = -1.4$$

From $|J_4|$,

$$x_4^* = 0.5$$

Finally from $g_1(X)$,

$$x_1^* = 1.6$$

Therefore, the stationary point for $f(X)$ subject to the constraints $g_1(X) = 0$ and $g_2(X) = 0$ is given by

$$X^* = [1.6, -1.4, -0.7, 0.5]$$ ◪

The reader should note that Theorem 4.13 is valid *only if the matrix G' is nonsingular*, where the elements of G' are g_{k_i}, the partial derivative of the kth constraint with respect to the ith variable for $k = 1, 2, \ldots, m$ and $i = 1, 2, \ldots, m$. However, G' may be composed of partial derivatives with respect to any m of the n variables. Therefore, if $|G'|$ is as defined in Eq. (4.91) but $|G'| = 0$, we may replace any column of G' by g_{k_j}, $k = 1, 2, \ldots, m$, where $j > m$. By repeated substitutions of this type, one may be able to find a matrix G' which is nonsingular.

When the matrix G' is altered yielding a matrix different from that given in Eq. (4.91), the matrices J_i also change. Examination of Eq. (4.90) indicates that the matrix defined by the last $m - 1$ rows and columns of J_i is the matrix G'. Thus any redefinition of G' must be reflected in the matrices J_i. Specifically, if the hth column of G' in Eq. (4.91) is replaced by the column vector

$$[g_{1_j}(X), g_{2_j}(X), \ldots, g_{m_j}(X)]^T$$

then J_i becomes

$$J_i = \begin{bmatrix} f_i(X) & f_1(X) & \cdots & f_j(X) & \cdots & f_m(X) \\ g_{1_i}(X) & g_{1_1}(X) & \cdots & g_{1_j}(X) & \cdots & g_{1_m}(X) \\ \vdots & \vdots & & \vdots & & \vdots \\ g_{m_i}(X) & g_{m_1}(X) & \cdots & g_{m_j}(X) & \cdots & g_{m_m}(X) \end{bmatrix}$$

where $j > m$ and $i = h, m + 1, m + 2, \ldots, n$ but $i \neq j$.

Example 4.21 Find the stationary points of

$$f(X) = 3x_1^3 + 2x_2^2 + x_3$$

subject to

$$g(X) = x_2 + x_3^2 = 0$$

From Eq. (4.91), $|G'|$ is given by

$$|G'| = g_1(X) = 0$$

Since G' is singular, Theorem 4.13 does not apply. Let

$$|G'| = g_2(X) = 1$$

Based upon this definition of G', $|J_1|$ and $|J_3|$ become

$$|J_1| = \begin{vmatrix} f_1(X) & f_2(X) \\ g_1(X) & g_2(X) \end{vmatrix} = \begin{vmatrix} 9x_1^2 & 4x_2 \\ 0 & 1 \end{vmatrix} = 9x_1^2$$

and

$$|J_3| = \begin{vmatrix} f_3(X) & f_2(X) \\ g_3(X) & g_2(X) \end{vmatrix} = \begin{vmatrix} 1 & 4x_2 \\ 2x_3 & 1 \end{vmatrix} = 1 - 8x_2 x_3$$

From $|J_1|$ and Theorem 4.13

$$x_1^* = 0$$

From $g(X)$ and $|J_3|$

$$x_2^* = -0.25, \quad x_3^* = -0.50$$

and X^* is given by

$$X^* = [0, -0.25, -0.50]$$

Thus far in the treatment of optimization subject to equality constraints, only the necessary conditions for the existence of an interior extreme point have been presented. This analysis could be extended to the development of sufficient conditions for the existence of an interior extreme point. Instead, we will develop a new and more conventional approach to the problem of optimization subject to equality constraints, called the Lagrange multiplier technique. In the treatment of the Lagrange multiplier technique, both the necessary and sufficient conditions for the existence of an interior extreme point will be discussed.

4.4.2 Lagrange Multipliers

The problem for which we are seeking a solution is that given in Eqs. (4.67) and (4.68). Let us define the following function called the *Lagrange function*.

$$L(X, \lambda) = f(X) - \sum_{k=1}^{m} \lambda_k g_k(X) \tag{4.92}$$

where

$$\lambda = [\lambda_1, \lambda_2, \ldots, \lambda_m] \tag{4.93}$$

and $\lambda_1, \lambda_2, \ldots, \lambda_m$ are constants called *Lagrange multipliers*. The total differential of $L(X, \lambda)$ is given by

$$\begin{aligned} dL(X, \lambda) &= df(X) - \sum_{k=1}^{m} \lambda_k \, dg_k(X) \\ &= \sum_{j=1}^{n} f_j(X) \, dx_j - \sum_{k=1}^{m} \lambda_k \sum_{j=1}^{n} g_{k_j}(X) \, dx_j \end{aligned} \tag{4.94}$$

Consider the Jacobian

$$|J_j| = \begin{vmatrix} f_j(X) & f_1(X) & f_2(X) & \cdots & f_m(X) \\ g_{1_j}(X) & g_{1_1}(X) & g_{1_2}(X) & \cdots & g_{1_m}(X) \\ \vdots & \vdots & \vdots & & \vdots \\ g_{m_j}(X) & g_{m_1}(X) & g_{m_2}(X) & \cdots & g_{m_m}(X) \end{vmatrix} \tag{4.95}$$

where $j = 1, 2, \ldots, n$. Now let J_{jk} be the matrix obtained by deleting the first column and $(k+1)$st row of J_j. Then

$$|J_{jk}| = (-1)^k \begin{vmatrix} f_1(X) & f_2(X) & \cdots & f_m(X) \\ g_{1_1}(X) & g_{1_2}(X) & \cdots & g_{1_m}(X) \\ \vdots & \vdots & & \vdots \\ g_{k-1_1}(X) & g_{k-1_2}(X) & \cdots & g_{k-1_m}(X) \\ g_{k+1_1}(X) & g_{k+1_2}(X) & \cdots & g_{k+1_m}(X) \\ \vdots & \vdots & & \vdots \\ g_{m_1}(X) & g_{m_2}(X) & \cdots & g_{m_m}(X) \end{vmatrix}, \quad k = 1, 2, \ldots, m$$

$$\tag{4.96}$$

4.4 OPTIMIZATION SUBJECT TO CONSTRAINTS

where $j = 1, 2, \ldots, m$. Then

$$|J_j| = |G'|f_j(X) + \sum_{k=1}^{m} |J_{jk}|g_{k_j}(X) \qquad (4.97)$$

and

$$\frac{|J_j|}{|G'|} = f_j(X) + \sum_{k=1}^{m} \frac{|J_{jk}|}{|G'|} g_{k_j}(X) \qquad (4.98)$$

But

$$df(X) = \sum_{j=1}^{m} \frac{|J_j|}{|G'|} dx_j + \sum_{j=m+1}^{n} \frac{|J_j|}{|G'|} dx_j \qquad (4.99)$$

where $|J_j| = 0$ for $j = 1, 2, \ldots, m$ since the first and $(j+1)$st columns of J_j are identical for such values of j. Let X^* be a stationary point for $f(X)$ subject to the constraints given and let

$$\lambda_k = -\frac{|J_{jk}|}{|G'|} \qquad (4.100)$$

where $|J_{jk}|$ and $|G'|$ are evaluated at X^*. Then

$$df(X^*) = \sum_{j=1}^{n} \left[f_j(X^*) - \sum_{k=1}^{m} \lambda_k g_{k_j}(X^*) \right] dx_j \qquad (4.101)$$

or

$$df(X^*) = dL(X^*, \lambda) \qquad (4.102)$$

Therefore $df(X)$ and $dL(X, \lambda)$ are equal at every stationary point for $f(X)$. However, since

$$|J_j| = 0, \quad j = 1, 2, \ldots, m \qquad (4.103)$$

and

$$|J_j| = 0, \quad j = m+1, m+2, \ldots, n \qquad (4.104)$$

by Theorem 4.13, we have

$$f_j(X^*) - \sum_{k=1}^{m} \lambda_k g_{k_j}(X^*) = 0, \quad j = 1, 2, \ldots, n \qquad (4.105)$$

for every extreme point of $f(X)$ such that

$$g_k(X^*) = 0, \quad k = 1, 2, \ldots, m \qquad (4.106)$$

But Eqs. (4.105) and (4.106) are given by

$$\left. \frac{\partial L(X, \lambda)}{\partial x_j} \right|_{X^*} = 0, \quad j = 1, 2, \ldots, n \qquad (4.107)$$

$$\left. \frac{\partial L(X, \lambda)}{\partial \lambda_k} \right|_{X^*} = 0, \quad k = 1, 2, \ldots, m \qquad (4.108)$$

The above derivation has shown that the Lagrange function can be used to identify the stationary points of a continuous function, $f(X)$, restricted by equality constraints. The results of this derivation are summarized in the following theorem.

Theorem 4.14 Let $f(X)$ and $g_k(X)$, $k = 1, 2, \ldots, m$, be continuous and differentiable functions. If X^* is an interior extreme point for $f(X)$ subject to the constraints

$$g_k(X) = 0, \quad k = 1, 2, \ldots, m$$

then it is necessary that

$$\left.\frac{\partial L(X, \lambda)}{\partial x_j}\right|_{X^*} = \left.\left[f_j(X) - \sum_{k=1}^{m} \lambda_k g_{k_j}(X)\right]\right|_{X^*} = 0, \quad j = 1, 2, \ldots, n \quad (4.109)$$

$$\left.\frac{\partial L(X, \lambda)}{\partial \lambda_k}\right|_{X^*} = \left.g_k(X)\right|_{X^*} = 0, \quad k = 1, 2, \ldots, m \quad (4.110)$$

where

$$\lambda_k = \frac{1}{|G'|}(-1)^{k+1}$$

$$\times \begin{vmatrix} f_1(X^*) & f_2(X^*) & \cdots & f_m(X^*) \\ g_{1_1}(X^*) & g_{1_2}(X^*) & \cdots & g_{1_m}(X^*) \\ \vdots & & & \vdots \\ g_{k-1_1}(X^*) & g_{k-1_2}(X^*) & \cdots & g_{k-1_m}(X^*) \\ g_{k+1_1}(X^*) & g_{k+1_2}(X^*) & \cdots & g_{k+1_m}(X^*) \\ \vdots & & & \vdots \\ g_{m_2}(X^*) & g_{m_2}(X^*) & \cdots & g_{m_m}(X^*) \end{vmatrix} \quad k = 1, 2, \ldots, m \quad (4.111)$$

and $|G'| \neq 0$.

Example 4.22 Solve the problem given in Example 4.19 using Theorem 4.14.
The Lagrange function is given by

$$L(X, \lambda) = x_1^2 + x_2^2 + x_3^2 - \lambda_1(x_1 + x_2 + x_3 - 2)$$

From Eqs. (4.109) and (4.110)

$$\frac{\partial L(X, \lambda)}{\partial x_1} = 2x_1 - \lambda_1 = 0, \qquad \frac{\partial L(X, \lambda)}{\partial x_2} = 2x_2 - \lambda_1 = 0$$

$$\frac{\partial L(X, \lambda)}{\partial x_3} = 2x_3 - \lambda_1 = 0, \qquad \frac{\partial L(X, \lambda)}{\partial \lambda_1} = -(x_1 + x_2 + x_3 - 2) = 0$$

Solving for x_1, x_2, and x_3 in terms of λ_1 yields

$$x_j = \frac{\lambda_1}{2}, \quad j = 1, 2, 3 \quad \text{and} \quad \lambda_1 = \frac{4}{3}$$

4.4 OPTIMIZATION SUBJECT TO CONSTRAINTS

Therefore

$$x_j^* = \tfrac{2}{3}, \quad j = 1, 2, 3 \quad \text{and} \quad X^* = [\tfrac{2}{3}, \tfrac{2}{3}, \tfrac{2}{3}]$$

is a stationary point for $f(X)$ subject to the constraint given.

Example 4.23 Solve the problem given in Example 4.20 by the Lagrange multiplier technique.

From Example 4.20, the Lagrange function is given by

$$L(X, \lambda) = (x_1 - 3) + (x_2 + 2)^2 + (x_3 + 1)^2 \\ + x_4^2 - \lambda_1(x_1 + x_2 + x_3 + x_4) - \lambda_2(x_2 - 2x_3)$$

From Theorem 4.14,

$$\frac{\partial L(X, \lambda)}{\partial x_1} = 1 - \lambda_1 = 0, \qquad \frac{\partial L(X, \lambda)}{\partial x_2} = 2(x_2 + 2) - \lambda_1 - \lambda_2$$

$$\frac{\partial L(X, \lambda)}{\partial x_3} = 2(x_3 + 1) - \lambda_1 + 2\lambda_2 = 0, \qquad \frac{\partial L(X, \lambda)}{\partial x_4} = 2x_4 - \lambda_1 = 0$$

$$\frac{\partial L(X, \lambda)}{\partial \lambda_1} = -(x_1 + x_2 + x_3 + x_4) = 0, \qquad \frac{\partial L(X, \lambda)}{\partial \lambda_2} = -(x_2 - 2x_3) = 0$$

Solving for x_1, x_2, x_3, and x_4 in terms of λ_1 and λ_2 leads to

$$x_1 = -\frac{3\lambda_1 - \lambda_2 - 6}{2}, \quad x_2 = \frac{\lambda_1 + \lambda_2 - 4}{2}, \quad x_3 = \frac{\lambda_1 - 2\lambda_2 - 2}{2}, \quad x_4 = \frac{\lambda_1}{2}$$

From

$$\frac{\partial L(X, \lambda)}{\partial x_1} = 0 \quad \text{we have} \quad \lambda_1 = 1$$

and from

$$\frac{\partial L(X, \lambda)}{\partial \lambda_2} = 0 \quad \text{we have} \quad \lambda_2 = \frac{1}{5}$$

Therefore

$$X^* = [1.6, -1.4, -0.7, 0.5]$$

To develop the *sufficient conditions* for a local extreme point we shall adopt an approach similar to that taken in the case of unconstrained optimization. However, when dealing with a constrained function we must recognize the interdependence of the variables $x_1, x_2, \ldots, x_n$. From Eq. (4.58) of Theorem 4.12, if X^* is a stationary point for $f(X)$, then

$$f(X) - f(X^*) = \frac{1}{2} D^2 f(X) \bigg|_{X_1} \qquad (4.112)$$

where

$$D = \sum_{j=1}^{n} (x_j - x_j^*) \frac{\partial}{\partial x_j}$$

and where X_1 is a point on the line segment joining X and X^*. Let

$$dx_j = x_j - x_j^*$$

Then

$$f(X) - f(X^*) = \frac{1}{2} \left[\sum_{j=1}^{n} dx_j \frac{\partial}{\partial x_j} \right]^2 f(X) \bigg|_{X_1} \qquad (4.113)$$

The expression $[\sum_{j=1}^{n} dx_j \, \partial/\partial x_j]^2 f(X)$ is the second total differential of $f(X)$ when the components of X are independent. Expressing the second total differential as $d^2 f(X)$, we have

$$f(X) - f(X^*) = \frac{1}{2} d^2 f(X) \bigg|_{X_1} \qquad (4.114)$$

If $d^2 f(X)|_{X*} > 0$, then $d^2 f(X)|_{X_1} > 0$ for all X_1 in some neighborhood of X^*, and X^* is a local minimum for $f(X)$. If $d^2 f(X)|_{X*} < 0$, then $d^2 f(X)|_{X_1} < 0$ for all X_1 in some neighborhood of X^*, and X^* is a local maximum for $f(X)$. To develop the sufficient conditions for the existence of a local minimum or maximum at a stationary point X^*, we shall express $d^2 f(X)|_{X*}$ as a sum of squares just as we did in the case of optimization of unconstrained functions.

Theorem 4.15 Let $f(X)$ and $g_k(X)$, $k = 1, 2, \ldots, m$, be continuous and differentiable functions. Let X^* be a stationary point for $f(X)$ which is constrained by

$$g_k(X) = 0, \qquad k = 1, 2, \ldots, m$$

and let G, V, O, and A be the matrices defined by

$G = m \times n$ matrix with elements $g_{i_j}(X^*)$ where $i = 1, 2, \ldots, m$ and $j = 1, 2, \ldots, n$;

$O = m \times m$ matrix of zeroes;

$V = n \times n$ matrix with elements $f_{ij}(X^*) - \sum_{k=1}^{m} \lambda_k g_{k_{ij}}(X^*)$ where $i = 1, 2, \ldots, n$ and $j = 1, 2, \ldots, n$

and

$$A = \begin{bmatrix} O & G \\ G^T & V \end{bmatrix} \qquad (4.115)$$

Let A_k be the kth leading principal minor of A. If

1. m is even, then
   a. X^* is a local minimum for $f(X)$ if

   $$|A_{2m+k}| > 0, \qquad k = 1, 2, \ldots, n - m$$

4.4 OPTIMIZATION SUBJECT TO CONSTRAINTS

    b. X^* is a local maximum for $f(X)$ if
$$(-1)^k |A_{2m+k}| > 0, \qquad k = 1, 2, \ldots, n - m$$

2. m is odd, then
    a. X^* is a local minimum for $f(X)$ if
$$|A_{2m+k}| < 0 \qquad \text{for} \qquad k = 1, 2, \ldots, n - m$$
    b. X^* is a local maximum for $f(X)$ if
$$(-1)^{k+1} |A_{2m+k}| > 0 \qquad \text{for} \qquad k = 1, 2, \ldots, n - m$$

Proof The proof of this theorem will be given for $n = 3$, $m = 1$.
From Eq. 4.114

$$f(X) - f(X^*) = \frac{1}{2} d^2 f(X) \Big|_{X_1}$$

In the case considered here, $df(X)$ is given by

$$df(X) = f_1(X)\, dx_1 + f_2(X)\, dx_2 + f_3(X)\, dx_3 \tag{4.116}$$

The second total differential is the first total differential of $df(X)$. Therefore

$$\begin{aligned}
d^2 f(X) &= d[f_1(X)\, dx_1 + f_2(X)\, dx_2 + f_3(X)\, dx_3] \\
&= df_1(X)\, dx_1 + f_1(X)\, d(dx_1) + df_2(X)\, dx_2 \\
&\quad + f_2(X)\, d(dx_2) + df_3(X)\, dx_3 + f_3(X)\, d(dx_3) \\
&= [f_{11}(X)\, dx_1 + f_{12}(X)\, dx_2 + f_{13}(X)\, dx_3]\, dx_1 \\
&\quad + [f_{21}(X)\, dx_1 + f_{22}(X)\, dx_2 + f_{23}(X)\, dx_3]\, dx_2 \\
&\quad + [f_{31}(X)\, dx_1 + f_{32}(X)\, dx_2 + f_{33}(X)\, dx_3]\, dx_3 \\
&\quad + f_1(X)\, d^2 x_1 + f_2(X)\, d^2 x_2 + f_3(X)\, d^2 x_3 \\
&= f_{11}(X)(dx_1)^2 + f_{22}(X)(dx_2)^2 + f_{33}(X)(dx_3)^2 + 2f_{12}(X)\, dx_1\, dx_2 \\
&\quad + 2f_{13}(X)\, dx_1\, dx_3 + 2f_{23}(X)\, dx_2\, dx_3 + f_1(X)\, d^2 x_1 \\
&\quad + f_2(X)\, d^2 x_2 + f_3(X)\, d^2 x_3
\end{aligned} \tag{4.117}$$

where

$$d^2 x_j = d(dx_j), \qquad j = 1, 2, 3 \tag{4.118}$$

when x_1, x_2, and x_3 are independent, $d^2 x_j = 0, j = 1, 2, 3$. However, this is not the situation we have here, since these variables are related through the constraint given. We will use the constraint $g_1(X) = 0$ to solve for $d^2 x_1$. By a development similar to that given for $d^2 f(X)$ we have

$$\begin{aligned}
d^2 g_1(X) &= g_{1_{11}}(X)(dx_1)^2 + g_{1_{22}}(X)(dx_2)^2 + g_{1_{33}}(X)(dx_3)^2 \\
&\quad + 2g_{1_{12}}(X)\, dx_1\, dx_2 + 2g_{1_{13}}(X)\, dx_1\, dx_3 + 2g_{1_{23}}(X)\, dx_2\, dx_3 \\
&\quad + g_{1_1}(X)\, d^2 x_1 + g_{1_2}(X)\, d^2 x_2 + g_{1_3}(X)\, d^2 x_3 = 0
\end{aligned} \tag{4.119}$$

Solving Eq. (4.119) for d^2x_1 yields

$$d^2x_1 = -\frac{1}{g_{1_1}(X)}[g_{1_{11}}(X)(dx_1)^2 + g_{1_{22}}(X)(dx_2)^2 + g_{1_{33}}(X)(dx_3)^2$$
$$+ 2g_{1_{12}}(X)\, dx_1\, dx_2 + 2g_{1_{13}}(X)\, dx_1\, dx_3 + 2g_{1_{23}}(X)\, dx_2\, dx_3$$
$$+ g_{1_2}(X)\, d^2x_2 + g_{1_3}(X)\, d^2x_3] \qquad (4.120)$$

Substituting Eq. (4.120) for d^2x_1 in Eq. (4.117) and evaluating $d^2f(X)$ at X^*, gives us

$$d^2f(X^*) = \sum_{i=1}^{3}\sum_{j=1}^{3}\left[f_{ij}(X^*) - \frac{f_1(X^*)}{g_{1_1}(X^*)}g_{1_{ij}}(X^*)\right] dx_i\, dx_j$$
$$+ \sum_{i=2}^{3}\left[f_i(X^*) - \frac{f_1(X^*)}{g_{1_1}(X^*)}g_{1_i}(X^*)\right] d^2x_i \qquad (4.121)$$

From Eq. (4.105)

$$\frac{f_1(X^*)}{g_{1_1}(X^*)} = \lambda \qquad (4.122)$$

Then

$$d^2f(X^*) = \sum_{i=1}^{3}\sum_{j=1}^{3}[f_{ij}(X^*) - \lambda g_{1_{ij}}(X^*)]\, dx_i\, dx_j + \sum_{i=2}^{3}[f_i(X^*) - \lambda g_{1_i}(X^*)]\, d^2x_i$$

Since X^* is a stationary point,

$$f_i(X^*) - \lambda g_{1_i}(X^*) = 0, \qquad i = 2, 3 \qquad (4.123)$$

by Theorem 4.14, and

$$d^2f(X^*) = \sum_{i=1}^{3}\sum_{j=1}^{3}[f_{ij}(X^*) - \lambda g_{1_{ij}}(X^*)]\, dx_i\, dx_j \qquad (4.124)$$

Since

$$dg_1(X^*) = g_{1_1}(X^*)\, dx_1 + g_{1_2}(X^*)\, dx_2 + g_{1_3}(X^*)\, dx_3 = 0 \qquad (4.125)$$

we have

$$dx_1 = -\frac{g_{1_2}(X^*)}{g_{1_1}(X^*)}dx_2 - \frac{g_{1_3}(X^*)}{g_{1_2}(X^*)}dx_3 \qquad (4.126)$$

Let

$$\Delta_{ij} = f_{ij}(X^*) - \lambda g_{1_{ij}}(X^*) \qquad (4.127)$$

and

$$g_{1_i} = g_{1_i}(X^*) \qquad (4.128)$$

$$g_{1_{ij}} = g_{1_{ij}}(X^*) \qquad (4.129)$$

4.4 OPTIMIZATION SUBJECT TO CONSTRAINTS

From Eqs. (4.124) and (4.126) we have

$$d^2f(X^*) = -\frac{1}{g_{1_1}^2}\{[-g_{1_2}^2\,\Delta_{11} - g_{1_1}^2\,\Delta_{22} + 2g_{1_1}g_{1_2}\,\Delta_{12}](dx_2)^2$$
$$+ [-g_{1_3}^2\,\Delta_{11} - g_{1_1}^2\,\Delta_{33} + 2g_{1_1}g_{1_3}\,\Delta_{13}](dx_3)^2$$
$$+ 2[-g_{1_2}g_{1_3}\,\Delta_{11} + g_{1_1}g_{1_3}\,\Delta_{12} + g_{1_1}g_{1_2}\,\Delta_{13}$$
$$- g_{1_1}^2\,\Delta_{23}]\,dx_2\,dx_3\} \tag{4.130}$$

Let

$$a_{11} = -g_{1_2}^2\,\Delta_{11} - g_{1_1}^2\,\Delta_{22} + 2g_{1_1}g_{1_2}\,\Delta_{12}$$
$$a_{22} = -g_{1_3}^2\,\Delta_{11} - g_{1_1}^2\,\Delta_{33} + 2g_{1_1}g_{1_3}\,\Delta_{13}$$
$$a_{12} = -g_{1_2}g_{1_3}\,\Delta_{11} + g_{1_1}g_{1_3}\,\Delta_{12} + g_{1_1}g_{1_2}\,\Delta_{13} - g_{1_1}^2\,\Delta_{33}$$

Then

$$d^2f(X^*) = -\frac{1}{g_{1_1}^2}[a_{11}(dx_2)^2 + 2a_{12}\,dx_2\,dx_3 + a_{22}(dx_3)^2] \tag{4.131}$$

Defining the matrix A as

$$A = \begin{bmatrix} 0 & g_{1_1} & g_{1_2} & g_{1_3} \\ g_{1_1} & \Delta_{11} & \Delta_{12} & \Delta_{13} \\ g_{1_2} & \Delta_{21} & \Delta_{22} & \Delta_{23} \\ g_{1_3} & \Delta_{31} & \Delta_{32} & \Delta_{33} \end{bmatrix} \tag{4.132}$$

where

$$\Delta_{ij} = \Delta_{ji} \quad \text{for all } i \text{ and } j$$

we have

$$d^2f(X^*) = -\frac{1}{g_{1_1^2}}|A_3|\left(dx_2 + \frac{a_{12}}{a_{11}}dx_3\right)^2 + \frac{|A_4|}{|A_3|}(dx_3)^2$$
$$= \frac{|A_3|}{|A_2|}\left(dx_2 + \frac{a_{12}}{a_{11}}dx_3\right)^2 + \frac{|A_4|}{|A_3|}(dx_3)^2 \tag{4.133}$$

If $|A_3| < 0$ and $|A_4| < 0$, then $d^2f(X^*) > 0$ and X^* is a local minimum for $f(X)$. If $|A_3| > 0$ and $|A_4| < 0$, then $d^2f(X^*) < 0$ and X^* is a local maximum for $f(X)$.

Example 4.24 Determine whether the stationary point in Example 4.23 is a local maxima or minima.
From Example 4.23,

$$X^* = [1.6, -1.4, -0.7, 0.5]$$

Applying the results of Theorem 4.15, we have
$$g_{1_i}(X^*) = 1, \quad i = 1, 2, 3, 4, \qquad g_{2_1}(X^*) = 0$$
$$g_{2_2}(X^*) = 1, \qquad g_{2_3}(X^*) = -2, \qquad g_{2_4}(X^*) = 0$$
and
$$G = \begin{bmatrix} 1 & 1 & 1 & 1 \\ 0 & 1 & -2 & 0 \end{bmatrix}$$

Further,
$$L_{1i}(X^*, \lambda) = 0, \quad i = 1, 2, 3, 4,$$
$$L_{22}(X^*, \lambda) = 2, \qquad L_{2i}(X^*, \lambda) = 0, \quad i \neq 2,$$
$$L_{33}(X^*, \lambda) = 2, \qquad L_{3i}(X^*, \lambda) = 0, \quad i \neq 3,$$
$$L_{44}(X^*, \lambda) = 2, \qquad L_{4i}(X^*, \lambda) = 0, \quad i \neq 4$$

and
$$V = \begin{bmatrix} 0 & 0 & 0 & 0 \\ 0 & 2 & 0 & 0 \\ 0 & 0 & 2 & 0 \\ 0 & 0 & 0 & 2 \end{bmatrix}$$

Therefore
$$A = \begin{bmatrix} 0 & 0 & 1 & 1 & 1 & 1 \\ 0 & 0 & 0 & 1 & -2 & 0 \\ 1 & 0 & 0 & 0 & 0 & 0 \\ 1 & 1 & 0 & 2 & 0 & 0 \\ 1 & -2 & 0 & 0 & 2 & 0 \\ 1 & 0 & 0 & 0 & 0 & 2 \end{bmatrix}$$

$$|A_5| = 10 \quad \text{and} \quad |A_6| = 20$$

Since m is even and $|A_5| > 0$ and $|A_6| > 0$, X^* is a local minimum.

Example 4.25 Find and identify the extreme points for the following function.
$$f(X) = 2x_1 + x_2$$
subject to
$$x_1 + x_2 = 5, \qquad x_1 - x_3^2 = 0, \qquad x_2 - x_4^2 = 0$$

The Lagrange function is given by
$$L(X, \lambda) = 2x_1 + x_2 - \lambda_1(x_1 + x_2 - 5) - \lambda_2(x_1 - x_3^2) - \lambda_3(x_2 - x_4^2)$$

Taking first partial derivatives leads to
$$\frac{\partial L(X, \lambda)}{\partial x_1} = 2 - \lambda_1 - \lambda_2 = 0 \tag{4.134}$$

4.4 OPTIMIZATION SUBJECT TO CONSTRAINTS

$$\frac{\partial L(X, \lambda)}{\partial x_2} = 1 - \lambda_1 - \lambda_3 = 0 \tag{4.135}$$

$$\frac{\partial L(X, \lambda)}{\partial x_3} = 2\lambda_2 x_3 = 0 \tag{4.136}$$

$$\frac{\partial L(X, \lambda)}{\partial x_4} = 2\lambda_3 x_4 = 0 \tag{4.137}$$

$$\frac{\partial L(X, \lambda)}{\partial \lambda_1} = -(x_1 + x_2 - 5) = 0 \tag{4.138}$$

$$\frac{\partial L(X, \lambda)}{\partial \lambda_2} = -(x_1 - x_3^2) = 0 \tag{4.139}$$

$$\frac{\partial L(X, \lambda)}{\partial \lambda_3} = -(x_2 - x_4^2) = 0 \tag{4.140}$$

Equations (4.136) and (4.137) indicate that one of the following conditions must hold.

1. $\lambda_2 = 0$, $\lambda_3 = 0$, $x_3 = 0$, $x_4 = 0$
2. $\lambda_2 = 0$, $\lambda_3 = 0$, $x_3 \neq 0$, $x_4 \neq 0$
3. $\lambda_2 \neq 0$, $\lambda_3 \neq 0$, $x_3 = 0$, $x_4 = 0$
4. $\lambda_2 = 0$, $\lambda_3 \neq 0$, $x_3 \neq 0$, $x_4 = 0$
5. $\lambda_2 \neq 0$, $\lambda_3 = 0$, $x_3 = 0$, $x_4 \neq 0$

Conditions 1 and 2 imply that $\lambda_1 = 2$ and $\lambda_1 = 1$ by Eqs. (4.134) and (4.135), and this solution is rejected. Condition 3 implies that $x_1 = 0$ and $x_2 = 0$ by Eqs. (4.139) and (4.140). Since this solution violates Eq. (4.138), this solution is also rejected. If $\lambda_2 = 0$ and $x_4 = 0$, then

$$\lambda_1 = 2, \qquad \lambda_3 = -1, \qquad x_2^* = 0, \qquad x_1^* = 5$$

and $X^* = [5, 0]$ may be an extreme point. If $\lambda_3 = 0$ and $x_3 = 0$, then

$$\lambda_1 = 1, \qquad \lambda_2 = 1, \qquad x_1^* = 0, \qquad x_2^* = 5$$

and $X^* = [0, 5]$ may be an extreme point. To determine whether these points are in fact extreme points, and if so their nature, we use Theorem 4.15. The matrices G, O, and V are given by

$$G = \begin{bmatrix} 1 & 1 & 0 & 0 \\ 1 & 0 & -2x_3 & 0 \\ 0 & 1 & 0 & -2x_4 \end{bmatrix}, \qquad O = \begin{bmatrix} 0 & 0 & 0 \\ 0 & 0 & 0 \\ 0 & 0 & 0 \end{bmatrix}$$

$$V = \begin{bmatrix} 0 & 0 & 0 & 0 \\ 0 & 0 & 0 & 0 \\ 0 & 0 & 2\lambda_2 & 0 \\ 0 & 0 & 0 & 2\lambda_3 \end{bmatrix}$$

Then

$$A = \begin{bmatrix} 0 & 0 & 0 & 1 & 1 & 0 & 0 \\ 0 & 0 & 0 & 1 & 0 & -2x_3 & 0 \\ 0 & 0 & 0 & 0 & 1 & 0 & -2x_4 \\ 1 & 1 & 0 & 0 & 0 & 0 & 0 \\ 1 & 0 & 1 & 0 & 0 & 0 & 0 \\ 0 & -2x_3 & 0 & 0 & 0 & 2\lambda_2 & 0 \\ 0 & 0 & -2x_4 & 0 & 0 & 0 & 2\lambda_3 \end{bmatrix}$$

Since $n = 4$ and $m = 3$, we must calculate $|A_7|$. By the cofactor method,

$$|A_7| = -8(\lambda_2 x_4^2 + \lambda_3 x_3^2)$$

For $\lambda_2 = 0$ and $x_4 = 0$, $x_3^2 = 5$ and

$$|A_7| = 40 > 0$$

Since m is odd, the point $X^* = [5, 0]$ is a maximum for $f(X)$. For $\lambda_3 = 0$ and $x_3 = 0$, $x_4^2 = 5$ and

$$|A_7| = -40 < 0$$

and the point $X^* = [0, 5]$ is a minimum for $f(X)$.

In Theorem 4.14, the necessary conditions for an extreme point are based upon the assumption that the matrix G' is nonsingular at X^*. If the elements of G' are $g_{k_j}(X^*)$, $k = 1, 2, \ldots, m$ and $j = 1, 2, \ldots, m$, and if G' is singular, then any column of G' may be replaced by the vector $[g_{1_i}(X^*), g_{2_i}(X^*), \ldots, g_{m_i}(X^*)]^T$, where $i > m$, in an attempt to define a nonsingular form of G'. If this attempt fails, then the conditions given in Theorem 4.14 are not necessary for the existence of an extreme point. That is, an extreme point may exist at a point where the conditions given in Theorem 4.14 are not met. To illustrate this problem consider the following example.

Example 4.26 Find the stationary points of

$$f(X) = x_1^2 + x_2^2$$

subject to

$$(x_1 - 1)^2 + (x_2 - 1)^2 = 0$$

The Lagrange function is given by

$$L(X, \lambda) = x_1^2 + x_2^2 - \lambda_1[(x_1 - 1)^2 + (x_2 - 1)^2]$$

Applying the conditions for an extreme point given in Theorem 4.14

$$\frac{\partial L(X, \lambda)}{\partial x_1} = 2x_1 - 2\lambda_1(x_1 - 1) = 0, \quad \frac{\partial L(X, \lambda)}{\partial x_2} = 2x_2 - 2\lambda_1(x_2 - 1) = 0$$

$$\frac{\partial L(X, \lambda)}{\partial \lambda_1} = -[(x_1 - 1)^2 + (x_2 - 1)^2] = 0$$

4.4 OPTIMIZATION SUBJECT TO CONSTRAINTS

Therefore
$$x_1^* = x_2^* = 1, \quad \lambda_1 = 2, \quad \text{and} \quad X^* = [1, 1]$$
is the indicated stationary point. However
$$\frac{\partial L(X^*, \lambda)}{\partial x_1} = 2 \quad \frac{\partial L(X^*, \lambda)}{\partial x_2} = 2$$
Thus we have identified a point which does not satisfy the necessary conditions for an extreme point. ▨

In Example 4.26 the reader will note that $|G'| = 0$ for every choice of G'. Problems of the type posed in Example 4.26 may be resolved by redefining the Lagrange function. Let

$$L(X, \lambda) = \lambda_0 f(X) - \sum_{k=1}^{m} \lambda_k g_k(X) \tag{4.141}$$

When $\lambda_0 > 0$, the solution generated, X^*, is that for which $|G'| \neq 0$. For $\lambda_0 = 0$, the solution generated is that for which $|G'| = 0$. In other words, we first apply Theorem 4.14 for $\lambda_0 > 0$, and then for $\lambda_0 = 0$ to generate all of the stationary points for the problem studied. For convenience, when $\lambda_0 > 0$, we usually set λ_0 equal to unity, although any positive value of λ_0 will do. As an illustration of this approach consider the following reformulation of the problem given in Example 4.26.

Example 4.27 Solve the problem given in Example 4.26 when the Lagrange function is defined by Eq. (4.141).

From Example 4.26 and Eq. (4.141),
$$L(X, \lambda) = \lambda_0(x_1^2 + x_2^2) - \lambda_1[(x_1 - 1)^2 + (x_2 - 1)^2]$$
Applying the results of Theorem 4.14

$$\frac{\partial L(X, \lambda)}{\partial x_1} = 2\lambda_0 x_1 - 2\lambda_1(x_1 - 1) = 0 \tag{4.142}$$

$$\frac{\partial L(X, \lambda)}{\partial x_2} = 2\lambda_0 x_2 - 2\lambda_1(x_2 - 1) = 0 \tag{4.143}$$

$$\frac{\partial L(X, \lambda)}{\partial \lambda_1} = -[(x_1 - 1)^2 + (x_2 - 1)^2] = 0 \tag{4.144}$$

For $\lambda_0 = 1$, we obtain the solution given in Example 4.26. For $\lambda_0 = 0$, we have
$$x_1 = x_2 = 1$$
and the conditions given in Eqs. (4.142)–(4.144) are satisfied. ▨

4.4.3 Inequality Constraints

Although equality constrained problems are not unusual, more often than not practical problems involve constraints in the form of inequalities. The problem discussed in Example 4.17 illustrates the manner in which such constraints arise. Specifically we are interested in solving a problem of the form

$$f(X) = \min(\max) \tag{4.145}$$

subject to

$$g_1(X) \le 0$$
$$\vdots$$
$$g_m(X) \le 0 \tag{4.146}$$

The problem formulated in Eqs. (4.145) and (4.146) includes those optimization problems where one or more of the constraints are of the form $h(X) \ge 0$, since $-h(X) \le 0$ is an equivalent expression for this constraint.

To resolve the problem given above we will change the inequality constraints to equality constraints. Let X be an n-dimensional vector. To the kth constraint we add the variable y_k^2, where y_k^2 is defined such that

$$g_k(X) + y_k^2 = 0 \tag{4.147}$$

That is, y_k^2 is defined such that it takes up the slack between $g_k(X)$ and zero, and is thus referred to as a slack variable. We square y_k since a positive quantity must be added to $g_k(X)$ to change the inequality to an equality. Adding a slack variable to each of the constraints in Eq. (4.146) we have the problem

$$f(X) = \min(\max) \tag{4.148}$$

subject to

$$g_1(X) + y_1^2 = 0$$
$$\vdots$$
$$g_m(X) + y_m^2 = 0 \tag{4.149}$$

We may solve this problem using the Lagrange multiplier technique where

$$L(X, Y, \lambda) = f(X) - \sum_{k=1}^{m} \lambda_k [g_k(X) + y_k^2] \tag{4.150}$$

and where the vector Y is defined by $[y_1, y_2, \ldots, y_m]$.

Taking partial derivatives with respect of $x_j, j = 1, 2, \ldots, n$, y_k, and λ_k, $k = 1, 2, \ldots, m$ we have

$$\left.\frac{\partial L(X, Y, \lambda)}{\partial x_j}\right|_{X*} = \left.\left[f_j(X) - \sum_{k=1}^{m} \lambda_k g_{k_j}(X)\right]\right|_{X*} = 0, \quad j = 1, 2, \ldots, n \tag{4.151}$$

4.4 OPTIMIZATION SUBJECT TO CONSTRAINTS

$$\left.\frac{\partial L(X, Y, \lambda)}{\partial y_k}\right|_{X*} = -2\lambda_k y_k \bigg|_{X*} = 0, \qquad k = 1, 2, \ldots, m \qquad (4.152)$$

$$\left.\frac{\partial L(X, Y, \lambda)}{\partial \lambda_k}\right|_{X*} = -[g_k(X) + y_k^2]\bigg|_{X*} = 0, \qquad k = 1, 2, \ldots, m \qquad (4.153)$$

where X^* is an extreme point for $f(X)$ subject to the constraints given in Eq. (4.146). From Eq. (4.153)

$$y_k^2 = -g_k(X), \qquad k = 1, 2, \ldots, m \qquad (4.154)$$

Therefore, Eqs. (4.152) and (4.154) imply that

$$\lambda_k g_k(X) = 0, \qquad k = 1, 2, \ldots, m \qquad (4.155)$$

From this development we have the following necessary conditions for the existence of an extreme point for $f(X)$ subject to inequality constraints.

Theorem 4.16 Kuhn–Tucker Conditions Let X^* be an extreme point for $f(X)$ subject to

$$g_k(X) \leq 0, \qquad k = 1, 2, \ldots, m$$

Then it is necessary that

$$\left[f_j(X) - \sum_{k=1}^{m} \lambda_k g_{k_j}(X)\right]\bigg|_{X*} = 0, \qquad j = 1, 2, \ldots, n \qquad (4.156)$$

$$\lambda_k y_k \bigg|_{X*} = 0, \qquad k = 1, 2, \ldots, m \qquad (4.157)$$

$$g_k(X)\bigg|_{X*} \leq 0, \qquad k = 1, 2, \ldots, m \qquad (4.158)$$

if $|G'| \neq 0$.

Example 4.28 Solve the following linear programming problem

$$4x_1 + x_2 = \max$$

subject to

$$x_1 + 3x_2 \leq 100, \qquad x_1 \geq 0, \qquad x_2 \geq 0$$

The Lagrange function for this problem is given by

$L(X, Y, \lambda)$
$= 4x_1 + x_2 - \lambda_1(x_1 + 3x_2 + y_1^2 - 100) + \lambda_2(x_1 - y_2^2) + \lambda_3(x_2 - y_3^2)$

From Theorem 4.16,

$$\frac{\partial L(X, Y, \lambda)}{\partial x_1} = 4 - \lambda_1 + \lambda_2 = 0 \qquad (4.159)$$

$$\frac{\partial L(X, Y, \lambda)}{\partial x_2} = 1 - 3\lambda_1 + \lambda_3 = 0 \tag{4.160}$$

$$\lambda_1 y_1 = 0 \tag{4.161}$$

$$\lambda_2 y_2 = 0 \tag{4.162}$$

$$\lambda_3 y_3 = 0 \tag{4.163}$$

Then

$$\lambda_1 = \frac{1 + \lambda_3}{3} \tag{4.164}$$

$$\lambda_2 = -\frac{11 - \lambda_3}{3} \tag{4.165}$$

Equations (4.159) and (4.160) imply that all of the Lagrange multipliers may not be zero. Specifically

$$\lambda_1 = \lambda_2 = \lambda_3 \neq 0, \quad \lambda_1 = \lambda_2 \neq 0, \quad \lambda_1 = \lambda_3 \neq 0$$

lead to infeasible solutions; that is, solutions which violate the necessary conditions given in Eqs. (4.159) and (4.160) since $|G'| \neq 0$. Similarly, if $\lambda_1 \neq 0$, $\lambda_2 \neq 0$, $\lambda_3 \neq 0$, then $x_1 = 0$, $x_2 = 0$, and $x_1 + x_2 = 100$ which is clearly not a solution. Therefore, at least one λ_k, $k = 1, 2, 3$ is zero. At this point we then have the following possible solutions

a. $\lambda_1 \neq 0, \quad \lambda_2 = 0, \quad \lambda_3 = 0$
b. $\lambda_1 = 0, \quad \lambda_2 \neq 0, \quad \lambda_3 \neq 0$
c. $\lambda_1 \neq 0, \quad \lambda_2 = 0, \quad \lambda_3 \neq 0$
d. $\lambda_1 \neq 0, \quad \lambda_2 \neq 0, \quad \lambda_3 = 0$

Solution a is infeasible by Eqs. (4.164) and (4.165).

For solution b we have

$$\lambda_2 = -4, \quad \lambda_3 = -1$$

Therefore

$$y_2 = 0, \quad x_1 = 0, \quad y_3 = 0, \quad x_2 = 0, \quad y_1 = 10$$

For solution c,

$$\lambda_1 = 4, \quad \lambda_3 = 11$$

and

$$y_1 = 0, \quad y_2 = 10, \quad x_1 = 100, \quad y_3 = 0, \quad x_2 = 0$$

Finally for solution d,

$$\lambda_1 = \frac{1}{3}, \quad \lambda_2 = -\frac{11}{3}, \quad y_1 = 0,$$

$$y_2 = 0, \quad x_1 = 0, \quad y_3 = \sqrt{\frac{100}{3}}, \quad x_1 = \frac{100}{3}$$

4.4 OPTIMIZATION SUBJECT TO CONSTRAINTS

As illustrated in the preceding examples, we could use the results of Theorem 4.15 to determine the nature of the three stationary points just derived. However, a simpler approach, at least in this problem, is to substitute the three solutions into the objective function and determine by inspection which yields the maximum.

$$f(0, 0) = 0, \qquad f(100, 0) = 400, \qquad f(0, \tfrac{100}{3}) = \tfrac{100}{3}$$

Therefore, the solution $x_1 = 100$, $x_2 = 0$ maximizes the function $4x_1 + x_2$ subject to the constraints imposed on the problem.

Example 4.29 Find the vector X^* which minimizes

$$x_1^2 + x_2^2$$

subject to

$$x_1 \le 100, \qquad x_2 \le 50$$

The Lagrange function for this problem is given by

$$L(X, Y, \lambda) = x_1^2 + x_2^2 - \lambda_1(x_1 + y_1^2 - 100) - \lambda_2(x_2 + y_2^2 - 50)$$

and

$$\frac{\partial L(X, Y, \lambda)}{\partial x_1} = 2x_1 - \lambda_1 = 0, \qquad \frac{\partial L(X, Y, \lambda)}{\partial x_2} = 2x_2 - \lambda_2 = 0,$$

$$\lambda_1 y_1 = 0, \qquad \lambda_2 y_2 = 0$$

Therefore

$$x_1 = \frac{\lambda_1}{2}, \qquad x_2 = \frac{\lambda_2}{2}$$

For $\lambda_1 = \lambda_2 = 0$, we have $x_1 = x_2 = 0$. For $\lambda_1 = 0$, $\lambda_2 \ne 0$,

$$x_1 = 0, \qquad y_2 = 0, \qquad x_2 = 50$$

For $\lambda_1 \ne 0$, $\lambda_2 = 0$,

$$y_1 = 0, \qquad x_1 = 100, \qquad x_2 = 0$$

Finally, for $\lambda_1 \ne 0$, $\lambda_2 \ne 0$,

$$y_1 = 0, \qquad x_1 = 100, \qquad y_2 = 0, \qquad x_2 = 50$$

The solution which minimizes $x_1^2 + x_2^2$ is obviously $x_1 = 0$, $x_2 = 0$.

The reader will notice that the two constraints, $x_1 \le 100$, $x_2 \le 50$, in Example 4.29 did not affect the solution of the problem. That is, the same minimum would have been found if the constraints had been ignored. In such cases the constraints are said to be inactive. A similar situation arose in Example 4.28 with respect to the second constraint, $x_1 \ge 0$ since the maximizing solution lead to $x_1 = 100$. As demonstrated by these examples, whenever a given constraint is *inactive* with respect to a given solution, the corresponding Lagrange multiplier is zero. The converse is also true. That is, whenever $\lambda_k = 0$, the kth constraint is inactive.

Throughout the discussion of optimization subject to inequality constraints, we have expressed constraints in the form $g_k(X) \leq 0$. Let us express this constraint as $g_k(X) \leq b_k$ and examine the behavior of $f(X)$ as b_k changes. Since x_k is related to b_k through the constraint $g_k(X) \leq b_k$, we have

$$\left.\frac{\partial}{\partial b_k} f(X)\right|_{X*} = \sum_{j=1}^{n} \frac{\partial}{\partial x_j} f(X) \left.\frac{\partial x_j}{\partial b_k}\right|_{X*} \tag{4.166}$$

and

$$\left.\frac{\partial}{\partial b_i}[g_k(X) + y_k^2]\right|_{X*} = \sum_{j=1}^{n} \frac{\partial}{\partial x_j}[g_k(X) + y_k^2] \left.\frac{\partial x_j}{\partial b_i}\right|_{X*} \tag{4.167}$$

Since

$$g_k(X) + y_k^2 = b_k \tag{4.168}$$

we have

$$\left.\frac{\partial}{\partial b_i}[g_k(X) + y_k^2]\right|_{X*} = \begin{cases} 1, & i = k \\ 0, & i \neq k \end{cases} \tag{4.169}$$

Multiplying Eq. (4.169) by $-\lambda_k$, summing over k, and adding the result to Eq. (4.166) leads to

$$\left.\frac{\partial}{\partial b_k} f(X)\right|_{X*} - \sum_{i=1}^{m} \lambda_i \frac{\partial}{\partial b_i}[g_k(X) + y_k^2]\bigg|_{X*} = \left.\frac{\partial}{\partial b_k} f(X)\right|_{X*} - \lambda_k \tag{4.170}$$

But

$$\left.\frac{\partial}{\partial b_k} f(X)\right|_{X*} - \sum_{i=1}^{m} \lambda_i \frac{\partial}{\partial b_i}[g_k(X) + y_k^2]\bigg|_{X*}$$

$$= \sum_{j=1}^{n} \left\{ \frac{\partial}{\partial x_j} f(X) - \sum_{i=1}^{m} \lambda_i \frac{\partial}{\partial x_j}[g_k(X) + y_k^2] \right\} \left.\frac{\partial x_j}{\partial b_k}\right|_{X*} \tag{4.171}$$

and from the necessary conditions for an interior extreme point at X^*,

$$\left.\frac{\partial}{\partial x_j} f(X) - \sum_{i=1}^{m} \lambda_i \frac{\partial}{\partial x_j}[g_k(X) + y_k^2]\right|_{X*} = 0$$

Therefore

$$\left.\frac{\partial}{\partial b_k} f(X)\right|_{X*} = \lambda_k \tag{4.172}$$

If X^* is a local maximum for $f(X)$, then

$$\left.\frac{\partial}{\partial b_k}f(X)\right|_{X^*} \geq 0$$

Alternatively if X^* is a local minimum, then

$$\left.\frac{\partial}{\partial b_k}f(X)\right|_{X^*} \leq 0$$

Therefore, if X^* is a local maximum for $f(X)$ then $\lambda_k \geq 0$ for $k = 1, 2, \ldots, m$, and $\lambda_k \leq 0$ for $k = 1, 2, \ldots, m$ if X^* is a local minimum. These conditions are then necessary for a local maximum and minimum respectively.

Although the conditions just presented are not sufficient for the identification of the nature of an extreme point, they are useful in eliminating certain stationary points as candidates for a minimum or maximum. For example, suppose that $f(X)$ is to be maximized subject to m inequality constraints. Let X_1^*, X_2^*, X_3^* be three stationary points with associated Lagrange multipliers $\lambda_{1k}, \lambda_{2k}, \lambda_{3k}, k = 1, 2, \ldots, m$. If $\lambda_{1k} < 0$ for $k = 1, 2, \ldots, m$, X_1^* is not an interior maximum since the necessary conditions for a local maximum given above are not satisfied. If $\lambda_{2k} < 0$ for some values of k, while $\lambda_{2k} > 0$ for the remaining values of k, X_2^* is not a local maximum (nor in fact a local minimum) by the same argument. If $\lambda_{3k} \geq 0$ for $k = 1, 2, \ldots, m$, X_3^* may be a local maximum.

Although the conditions on λ_k specified above are in general not sufficient for a local minimum or maximum, if the function $f(X)$ and $g_k(X)$, $k = 1, 2, \ldots, m$ behave properly these conditions on λ_k become sufficient for a local maximum or minimum. For example suppose that $f(X)$ is strictly convex from below and the constraints $g_k(X) = 0$, $k = 1, 2, \ldots, m$ are convex from above. If $\lambda_k \geq 0$, $k = 1, 2, \ldots, m$, then $L(X, \lambda)$ is convex from below since $-\lambda_k g_k(X)$ is convex from below for $k = 1, 2, \ldots, m$. Therefore if X^* is a stationary point for $L(X, \lambda)$ it must be a local maximum if $\lambda_k \geq 0$, $k = 1, 2, \ldots, m$ since $L(X, \lambda)$ is convex from below for such values of the λ_k's. On the other hand if $f(X)$ is strictly convex from above, $g_k(X)$ is convex from above, and $\lambda_k \leq 0$ for all k, then the stationary point X^* must be a local minimum since $L(X, \lambda)$ is convex from above in this case.

4.5 Applications

4.5.1 Hypothesis Testing

The null hypothesis that the mean of a given population is less than or equal to u_0 is to be tested against the alternative that the mean is greater than u_1. If u is the population mean then the null and alternative hypotheses may be expressed as

$$H_0: u \leq u_0, \qquad H_1: u > u_0$$

Previous research indicates that the population mean may assume one of two values, u_0 and u_1 where $u_0 < u_1$, and that the probability that the mean equals u_0 is p $(0 < p < 1)$.

To test the null hypothesis a sample of size N will be drawn from the population and the sample mean, $\bar{x}$, will be computed, where

$$\bar{x} = \frac{1}{N} \sum_{i=1}^{N} x_i \qquad (4.173)$$

The variable x_i is the value of the ith unit drawn from the population. If $\bar{x} \leq U$ then the null hypothesis is accepted. Otherwise the null hypothesis is rejected. The sample size, N, is fixed but U is to be determined.

If the null hypothesis is rejected when $u = u_0$ then a cost C_R is incurred. On the other hand if the null hypothesis is accepted when $u = u_1$ then a cost C_A results. If the null hypothesis is accepted when $u = u_0$ or rejected when $u = u_1$ no cost is incurred. Therefore, if $f(\bar{x} \mid u_i)$, $i = 0, 1$, is the density function of $\bar{x}$ given $u = u_i$, then the cost of testing the hypothesis, $C_T(U)$, is given by

$$C_T(U) = \begin{cases} C_I N + C_R \int_U^{\infty} f(\bar{x} \mid u_0) \, d\bar{x}, & u = u_0 \\ C_I N + C_A \int_{-\infty}^{U} f(\bar{x} \mid u_1) \, d\bar{x}, & u = u_1 \end{cases} \qquad (4.174)$$

where C_I is the cost of drawing each unit from the population. The expected value of $C_T(U)$ is given by

$$E[C_T(U)] = C_I N + p C_R \int_U^{\infty} f(\bar{x} \mid u_0) \, d\bar{x} + (1 - p) C_A \int_{-\infty}^{U} f(\bar{x} \mid u_1) \, d\bar{x} \qquad (4.175)$$

We wish to find the value of U which minimizes $E[C_T(U)]$. Taking the first derivative of $E[C_T(U)]$ with respect to U leads to

$$\frac{d}{dU} E[C_T(U)] = -p C_R f(U \mid u_0) + (1 - p) C_A f(U \mid u_1) \qquad (4.176)$$

Assume that $f(\bar{x} \mid u_i)$ is given by

$$f(\bar{x} \mid u_i) = \frac{N}{\sigma \sqrt{2\pi}} \exp\left[-\frac{N(\bar{x} - u_i)^2}{2\sigma^2}\right], \qquad -\infty < \bar{x} < \infty \qquad (4.177)$$

Then

$$\frac{d}{dU} E[C_T(U)] = -p C_R \frac{\sqrt{N}}{\sigma \sqrt{2\pi}} \exp\left[-\frac{N(U - u_0)^2}{2\sigma^2}\right]$$

$$+ (1 - p) C_A \frac{\sqrt{N}}{\sigma \sqrt{2\pi}} \exp\left[-\frac{N(U - u_1)^2}{2\sigma^2}\right] \qquad (4.178)$$

4.5 APPLICATIONS

Setting the derivative equal to zero and solving for U leads us to

$$-pC_R \exp\left[-\frac{N(U-u_0)^2}{2\sigma^2}\right] + (1-p)C_A \exp\left[-\frac{N(U-u_1)^2}{2\sigma^2}\right] = 0 \quad (4.179)$$

or

$$\exp\left[-\frac{N}{2\sigma^2}[(U-u_1)^2 - (U-u_0)^2]\right] = \frac{pC_R}{(1-p)C_A} \quad (4.180)$$

Taking the natural log of both sides, we have

$$(U-u_1)^2 - (U-u_0)^2 = -\frac{2\sigma^2}{N} \ln\left[\frac{pC_R}{(1-p)C_A}\right]$$

and

$$U^* = \frac{u_0 + u_1}{2} - \frac{\sigma^2}{N(u_0 - u_1)} \ln\left[\frac{pC_R}{(1-p)C_A}\right] \quad (4.181)$$

To determine whether U^* is an interior minimum, maximum, or point of inflection we take the second partial derivative of $E[C_T(U)]$ with respect to U.

$$\frac{d^2}{dU^2} E[C_T(U)] = \frac{pC_R N^{3/2}}{\sigma^3 \sqrt{2\pi}} (U-u_0)\exp\left[-\frac{N(U-u_0)^2}{2\sigma^2}\right]$$
$$- \frac{(1-p)C_A N^{3/2}}{\sigma^3 \sqrt{2\pi}} (U-u_1)\exp\left[-\frac{N(U-u_1)^2}{2\sigma^2}\right] \quad (4.182)$$

If $(d^2/dU^2)E[C_T(U^*)] > 0$, then U^* is an interior minimum. Therefore, we will determine the conditions under which this relationship holds. If $(d^2/dU^2)E[C_T(U^*)] > 0$ then

$$pC_R(U^* - u_0) > (1-p)C_A(U^* - u_1)\exp\left[-\frac{N}{2\sigma^2}[(U^* - u_1)^2 - (U^* - u_0)^2]\right]$$

or

$$pC_R(U^* - u_0) > (1-p)C_A(U^* - u_1)\exp\left[-\frac{N}{2\sigma^2}(u_0 - u_1)[2U^* + u_0 + u_1]\right]$$

and

$$\frac{pC_R}{(1-p)C_A}(U^* - u_0) > (U^* - u_1)\exp\left[-\frac{N}{2\sigma^2}(u_0 - u_1)[2U^* + u_0 + u_1]\right] \quad (4.183)$$

Since the power of e is positive if $u_0 < U^* < u_1$, Eq. (4.183) holds and U^* is a minimum for $E[C_T(U)]$.

Example 4.30 Let
$$u_0 = 10, \quad u_1 = 20, \quad N = 100, \quad \sigma^2 = 9$$
$$p = 0.9, \quad C_R = \$100.00, \quad C_A = \$500.00$$

Find the value of U which minimizes $E[C_T(U)]$.

From Eq. (4.181)
$$U^* = \frac{10 + 20}{2} - \frac{9}{100(10 - 20)} \ln\left[\frac{0.9(100)}{0.1(500)}\right] = 15.0044$$

Since $u_0 < U^* < u_1$, U^* is a minimum for $E[C_T(U)]$. ◪

4.5.2 Production and Inventory Systems

Consider a product which is manufactured at a continuous and constant rate of ψ units per year. Each unit manufactured requires 6 raw materials. Specifically, m_j units of the jth raw material are required for each unit of final product, $j = 1, 2, \ldots, 6$. To provide the required supply of raw materials on hand, the inventories of raw materials are replenished periodically. That is, every t_j years q_j units of raw material j are received and placed in inventory. Thus the total quantity of raw material j placed in inventory during a one year period is $m_j \psi$ and the number of orders placed for replenishment of the jth raw material is $m_j \psi / q_j$.

Let C_{0j} be the cost of each order placed for replenishment of the jth inventory and let C_{Ij} be the cost of carrying one unit of the jth raw material in inventory for one year. If t_j is the time between successive replenishments of the jth inventory, the cycle length, then the variation of the inventory level over time is as shown in Fig. 4.13.

The cost of maintaining the jth inventory for one cycle is $C_{0j} + \frac{1}{2}(q_j^2/m_j\psi)C_{Ij}$ since
$$t_j = \frac{q_j}{m_j \psi}$$

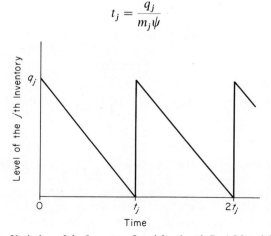

Figure 4.13 Variation of the Inventory Level for the jth Raw Material with Time

4.5 APPLICATIONS

Therefore, the annual cost of maintaining the jth inventory, $C_{Tj}(q_j)$, is given by

$$C_{Tj}(q_j) = \frac{m_j \psi}{q_j} C_{0j} + \frac{q_j}{2} C_{Ij}, \quad j = 1, 2, \ldots, 6 \tag{4.184}$$

and the total cost of maintaining all 6 raw materials inventories, $C_T(Q)$, is

$$C_T(Q) = \sum_{j=1}^{6} \left[\frac{m_j \psi}{q_j} C_{0j} + \frac{q_j}{2} C_{Ij} \right] \tag{4.185}$$

where

$$Q = [q_1, q_2, \ldots, q_6]$$

We will now find the order quantities, q_j^*, $j = 1, 2, \ldots, 6$, which minimize $C_T(Q)$. From the necessary conditions for an interior extreme point, Q^*, we have

$$\left. \frac{\partial C_T(Q)}{\partial q_j} \right|_{Q^*} = \frac{C_{Ij}}{2} - \left. \frac{m_j \psi C_{0j}}{q_j^2} \right|_{Q^*} = 0, \quad j = 1, 2, \ldots, 6 \tag{4.186}$$

or

$$q_j^* = \sqrt{\frac{2 m_j \psi C_{0j}}{C_{Ij}}} \tag{4.187}$$

To determine whether this solution for q_j^* is a local minimum we note that

$$\frac{\partial^2 C_T(Q)}{\partial q_j \, \partial q_i} = \begin{cases} 0, & i \neq j \\ \dfrac{2 m_j \psi C_{0j}}{q_j^3}, & i = j \end{cases} \tag{4.188}$$

and

$$\left. \frac{\partial^2 C_T(Q)}{\partial q_j \, \partial q_i} \right|_{Q^*} = \begin{cases} 0, & i \neq j \\ \dfrac{C_{Ij}^{3/2}}{\sqrt{2 m_j \psi C_{0j}}}, & i = j \end{cases} \tag{4.189}$$

The Hessian is given by

$$H = \begin{bmatrix} \dfrac{C_{I1}^{3/2}}{\sqrt{2 m_1 \psi C_{01}}} & 0 & \cdots & 0 \\ 0 & \dfrac{C_{I2}^{3/2}}{\sqrt{2 m_2 \psi C_{02}}} & \cdots & 0 \\ \vdots & \vdots & & \vdots \\ 0 & 0 & \cdots & \dfrac{C_{I6}^{3/2}}{\sqrt{2 m_6 \psi C_{0n}}} \end{bmatrix} \tag{4.190}$$

Since H is positive definite, Q^* is a local minimum. Further if $q_j > 0, j = 1, 2, \ldots, 6$, the matrix of second partial derivatives, $H(Q)$, is also positive definite. However, the only feasible solutions for $C_T(Q)$ are those for which the elements of Q are positive, and $C_T(Q)$ is positive definite for all feasible vectors Q. Therefore, $C_T(Q)$ is strictly convex from above for all feasible Q, and Q^* is a global minimum.

4.5.3 The Gradient Method

Consider the problem of attempting to find an extreme point for a function where the methods presented in this chapter do not apply. In particular, let $f(X)$ be an unconstrained function such that the set of equations

$$\frac{\partial f(X)}{\partial x_j} = 0, \quad j = 1, 2, \ldots, n \tag{4.191}$$

cannot be solved for the extreme point or points, X_0. We shall attempt to develop an iterative procedure through which extreme points may be located or at least approximated.

Suppose we start the search for an extreme point at X_1, which may be any point in the space for which $f(X)$ is defined. If we are attempting to minimize $f(X)$, the problem is to find another point, X_2, such that

$$f(X_2) < f(X_1) \tag{4.192}$$

If the problem is one of maximization, then X_2 should have the property that

$$f(X_2) > f(X_1) \tag{4.193}$$

At X_2 we will then find another point X_3 such that

$$f(X_3) < f(X_2) \quad \text{or} \quad f(X_3) > f(X_2)$$

depending upon the criterion of optimality. We then repeat the process iteratively until the search for the extreme point is terminated.

The first problem encountered at each iteration of the search is to determine the direction, from the present point, along which a better solution will lie. Let X_i be the ith point examined in the search for an extreme point. Let α be the direction defined by

$$\alpha = [\alpha_1, \alpha_2, \ldots, \alpha_n] \tag{4.194}$$

From Chapter 3, the rate of change of $f(X)$ at X_i in the direction α was defined as

$$\nabla_\alpha f(X_i) = \alpha \, \nabla f(X_i) \tag{4.195}$$

4.5 APPLICATIONS

where

$$\nabla f(X) = \begin{bmatrix} \dfrac{\partial f(X)}{\partial x_1} \\ \dfrac{\partial f(X)}{\partial x_2} \\ \vdots \\ \dfrac{\partial f(X)}{\partial x_n} \end{bmatrix} \qquad (4.196)$$

or

$$\nabla_\alpha f(X_i) = \sum_{j=1}^{n} \alpha_j \dfrac{\partial f(X)}{\partial x_j}\bigg|_{X_i} \qquad (4.197)$$

If we wish to maximize $f(X)$, then we must determine α such that Eq. (4.197) is maximized. Otherwise we choose α such that Eq. (4.197) is minimized.

Let the distance from X_i to X_{i+1} be r. Then

$$\|X_{i+1} - X_i\| = r \qquad (4.198)$$

Since

$$X_{i+1} = [x_{1i} + \alpha_1, x_{2i} + \alpha_2, \ldots, x_{ni} + \alpha_n] \qquad (4.199)$$

we have

$$\sum_{j=1}^{n} \alpha_j^2 = r^2 \qquad (4.200)$$

Therefore, we must solve the problem given by

$$\nabla_\alpha f(X_i) = \max(\min) \qquad (4.201)$$

subject to

$$\sum_{j=1}^{n} \alpha_j^2 = r^2 \qquad (4.202)$$

Solving this problem by the method of Lagrange multipliers, we have

$$L(\alpha, \lambda) = \sum_{j=1}^{n} \alpha_j \dfrac{\partial f(X)}{\partial x_j}\bigg|_{X_i} - \lambda_1 \left[\sum_{j=1}^{n} \alpha_j^2 - r^2 \right] \qquad (4.203)$$

$$\dfrac{\partial L(\alpha, \lambda)}{\partial \alpha_j} = \dfrac{\partial f(X)}{\partial x_j}\bigg|_{X_i} - 2\lambda_1 \alpha_j = 0, \qquad j = 1, 2, \ldots, n \qquad (4.204)$$

and

$$\dfrac{\partial L(\alpha, \lambda)}{\partial \lambda_1} = -\sum_{j=1}^{n} \alpha_j^2 + r^2 = 0 \qquad (4.205)$$

From Eq. (4.204),

$$\alpha_j = \frac{1}{2\lambda_1} \left.\frac{\partial f(X)}{\partial x_j}\right|_{X_i} \tag{4.206}$$

and from Eq. (4.205),

$$\sum_{j=1}^{n} \alpha_j^2 = r^2$$

Therefore

$$\sum_{j=1}^{n} \alpha_j^2 = \frac{1}{4\lambda_1^2} \sum_{j=1}^{n} \left[\left.\frac{\partial f(X)}{\partial x_j}\right|_{X_i}\right]^2 \tag{4.207}$$

or

$$\lambda_1 = \pm \frac{\sqrt{\sum_{j=1}^{n} \left[\left.\frac{\partial f(X)}{\partial x_j}\right|_{X_i}\right]^2}}{2r} \tag{4.208}$$

and

$$\alpha_j = \pm \frac{r \left.\frac{\partial f(X)}{\partial x_j}\right|_{X_i}}{\sqrt{\sum_{j=1}^{n} \left[\left.\frac{\partial f(X)}{\partial x_j}\right|_{X_i}\right]^2}} \tag{4.209}$$

Finally

$$\alpha = \pm \frac{r}{\|\nabla f(X_i)\|} \nabla f(X_i) \tag{4.210}$$

Thus we have two solutions for the direction α. We shall apply Theorem 4.15 to determine the nature of these solutions.

$$g_{1,j}(\alpha) = 2\alpha_j, \qquad L_{hj}(\alpha, \lambda) = -2\lambda_1, \quad h = j, \qquad L_{hj}(\alpha, \lambda) = 0, \quad h \neq j$$

Therefore

$$A = \begin{bmatrix} 0 & 2\alpha_1 & 2\alpha_2 & \cdots & 2\alpha_n \\ 2\alpha_1 & -2\lambda_1 & 0 & \cdots & 0 \\ 2\alpha_2 & 0 & -2\lambda_1 & \cdots & 0 \\ \vdots & \vdots & \vdots & & \vdots \\ 2\alpha_n & 0 & 0 & \cdots & -2\lambda_1 \end{bmatrix} \tag{4.211}$$

and

$$|A_i| = 2^i(-1)^{i-1}\lambda_1^{i-2} \sum_{j=1}^{i} \alpha_j^2, \qquad i = 3, 4, \ldots, n+1 \tag{4.212}$$

4.5 APPLICATIONS

If $\lambda_1 < 0$, $|A_i| < 0$, $i = 3, 4, \ldots, n + 1$ and the direction

$$\alpha = \frac{r}{\|\nabla f(X_i)\|} \nabla f(X_i) \qquad (4.213)$$

leads to a maximum. If $\lambda_1 > 0$, $|A_i|$ alternates in sign starting with $|A_3| > 0$, and the direction

$$\alpha = -\frac{r}{\|\nabla f(X_i)\|} \nabla f(X_i) \qquad (4.214)$$

leads to a minimum. The point X_{i+1} is thus given by

$$X_{i+1} = X_i + \alpha$$

Example 4.31 Let

$$f(X) = x_1^2(\exp[-x_1] + 1) + x_2^2 \exp[-0.1x_2]$$

If

$$X_1 = [4, 4]$$

carry out five iterations of the gradient search for a minimum where $r = 1$.

From Eq. (4.196),

$$\nabla f(X) = \begin{bmatrix} x_1 \exp[-x_1](2 - x_1) + 2x_1 \\ x_2 \exp[-0.1x_2](2 - 0.1x_2) \end{bmatrix}$$

and

$$\|\nabla f(X)\| = \sqrt{[x_1 \exp[-x_1](2 - x_1) + 2x_1]^2 + [x_2 \exp[-0.1x_2](2 - 0.1x_2)]^2}$$

Then

$$\nabla f(X_1) = \begin{bmatrix} 8(1 - \exp[-4]) \\ 6.4 \exp[-0.4] \end{bmatrix} = \begin{bmatrix} 7.86 \\ 4.29 \end{bmatrix}, \qquad \|\nabla f(X_1)\| = 8.95$$

Therefore

$$\alpha = -\frac{1}{8.95} \begin{bmatrix} 7.86 \\ 4.29 \end{bmatrix} = -\begin{bmatrix} 0.88 \\ 0.48 \end{bmatrix} \quad \text{and} \quad X_2 = \begin{bmatrix} 3.12 \\ 3.52 \end{bmatrix}$$

At X_2 we have

$$\nabla f(X_2) = \begin{bmatrix} 6.09 \\ 4.07 \end{bmatrix}, \qquad \|\nabla f(X_2)\| = 7.32$$

Then

$$\alpha = -\begin{bmatrix} 0.83 \\ 0.56 \end{bmatrix} \quad \text{and} \quad X_3 = \begin{bmatrix} 2.29 \\ 2.96 \end{bmatrix}$$

At X_3,
$$\nabla f(X_3) = \begin{bmatrix} 4.51 \\ 3.66 \end{bmatrix}, \quad \|\nabla f(X_3)\| = 5.81$$

$$\alpha = -\begin{bmatrix} 0.78 \\ 0.63 \end{bmatrix}, \quad \text{and} \quad X_4 = \begin{bmatrix} 1.51 \\ 2.33 \end{bmatrix}$$

At X_4,
$$\nabla f(X_4) = \begin{bmatrix} 2.86 \\ 3.25 \end{bmatrix}, \quad \|\nabla f(X_4)\| = 4.33, \quad \alpha = -\begin{bmatrix} 0.66 \\ 0.75 \end{bmatrix}$$

Therefore
$$X_5 = \begin{bmatrix} 0.85 \\ 1.58 \end{bmatrix}$$

At X_5,
$$\nabla f(X_5) = \begin{bmatrix} 1.28 \\ 2.57 \end{bmatrix}, \quad \|\nabla f(X_5)\| = 2.87, \quad \alpha = -\begin{bmatrix} 0.45 \\ 0.90 \end{bmatrix}$$

and
$$X_6 = \begin{bmatrix} 0.40 \\ 0.68 \end{bmatrix}$$

The progress of the search through the first five iterations is shown graphically in Fig. 4.14. Inspection of $f(X)$ indicates that the minimum lies at [0, 0]. As shown in Fig. 4.14, the search is progressing toward this point.

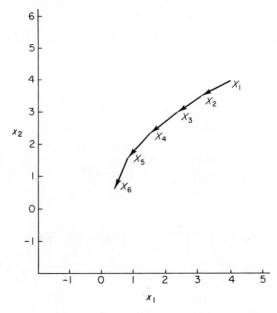

Figure 4.14 First Five Iterations of the Gradient Search for the Problem Given in Example 4.31

4.5.4 Engineering Applications

Example 4.32 A quantum of x-ray, called a photon, strikes a free electron. The photon is scattered by the electron, imparting energy to the recoiling electron as shown in Fig. 4.15. The kinetic energy, y, imparted to the electron is given by

$$y = \frac{\dfrac{h^2\lambda^2}{mc^2}[1 - \cos(\alpha)]}{1 + \dfrac{h\lambda}{mc^2}[1 - \cos(\alpha)]}$$

where h = Planck's constant, λ = frequency of the impacting photon, c = velocity of light, and m = mass of the electron at rest. Find the angle, α, such that the kinetic energy of the recoiling electron is maximized.

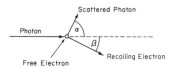

Figure 4.15 Photon Scattering due to the Collision of a Photon with a Free Electron

Taking the first derivative of y with respect to α yields us

$$\frac{dy}{d\alpha} = \frac{\dfrac{h^2\lambda^2}{mc^2}\sin(\alpha)}{1 + \dfrac{h\lambda}{mc^2}[1 - \cos(\alpha)]} - \frac{\dfrac{h^3\lambda^3}{m^2c^4}[1 - \cos(\alpha)]\sin(\alpha)}{\left\{1 + \dfrac{h\lambda}{mc^2}[1 - \cos(\alpha)]\right\}^2}$$

$$= \frac{\dfrac{h^2\lambda^2}{mc^2}\sin(\alpha)}{\left\{1 + \dfrac{h\lambda}{mc^2}[1 - \cos(\alpha)]\right\}^2} = 0$$

Hence

$$\sin(\alpha) = 0 \quad \text{and} \quad \cos(\alpha) = 1, -1 \quad \text{or} \quad \alpha^* = 0°, 180°$$

For the sufficient conditions, we must determine $d^2y/d\alpha^2$.

$$\frac{d^2y}{d\alpha^2} = \frac{\dfrac{h^2\lambda^2}{mc^2}\cos(\alpha)}{\left\{1 + \dfrac{h\lambda}{mc^2}[1 - \cos(\alpha)]\right\}^2} - \frac{2\dfrac{h^3\lambda^3}{m^2c^4}\cos(\alpha)\sin(\alpha)}{\left\{1 + \dfrac{h\lambda}{mc^2}[1 - \cos(\alpha)]\right\}^3}$$

Hence for $\alpha = 0°$, we have $\sin(\alpha) = 0$, $\cos(\alpha) = 1$, and

$$\left.\frac{d^2y}{d\alpha^2}\right|_{\alpha^* = 0°} = \frac{h^2\lambda^2}{mc^2} > 0$$

For $\alpha = 180°$, we have $\sin(\alpha) = 0$, $\cos(\alpha) = -1$, and

$$\left.\frac{d^2 y}{d\alpha^2}\right|_{\alpha* = 180°} = -\frac{\frac{h^2\lambda^2}{mc^2}}{\left[1 + 2\frac{h\lambda}{mc^2}\right]^2} < 0$$

Thus, $\alpha* = 180°$ maximizes the kinetic energy of the recoiling electron.

Example 4.33 A generator supplies an electromotive force E and is connected in series with a fixed impedance Z_f and a load impedance Z_l where

$$Z_f = R_f + iX_f, \quad Z_l = R_l + iX_l$$

and R_f and R_l are the fixed and load resistances respectively, X_f and X_l are the fixed and load reactances respectively, and $i = \sqrt{-1}$. The power to the load, P, is given by

$$P = \frac{E^2 R_l}{(R_f + R_l)^2 + (X_f + X_l)^2}$$

The values of R_l and X_l are to be determined such that the power to the load is maximized.

The necessary conditions for maximization of P are given by

$$\frac{\partial P}{\partial R_l} = \frac{E^2}{(R_f + R_l)^2 + (X_f + X_l)^2} - \frac{2(R_f + R_l)R_l E^2}{[(R_f + R_l)^2 + (X_f + X_l)^2]^2}$$

$$= \frac{E^2[(R_f^2 - R_l^2) - (X_f + X_l)^2]}{[(R_f + R_l)^2 + (X_f + X_l)^2]^2} = 0$$

$$\frac{\partial P}{\partial X_l} = \frac{-2(X_f + X_l)E^2 R_l}{[(R_f + R_l)^2 + (X_f + X_l)^2]^2} = 0$$

Hence

$$X_l^* = -X_f, \quad R_l^* = R_f$$

To determine the nature of the point (X_l^*, R_l^*) we have

$$\frac{\partial^2 P}{\partial R_l^2} = -\frac{2R_l E^2}{[(R_f + R_l)^2 + (X_f + X_l)^2]^2}$$

$$- \frac{4E^2(R_f + R_l)[(R_f^2 - R_l^2) + (X_f + X_l)^2]}{[(R_f + R_l)^2 + (X_f + X_l)^2]^3}$$

and

$$\left.\frac{\partial^2 P}{\partial R_l^2}\right|_{(X_l^*, R_l^*)} = -\frac{E^2(3R_f + 1)}{32 R_f^4}$$

4.5 APPLICATIONS

In addition

$$\frac{\partial^2 P}{\partial X_l^2} = -\frac{2E^2 R_l}{[(R_f + R_l)^2 + (X_f + X_l)^2]^2} + \frac{8E^2 R_l (X_f + X_l)^2}{[(R_f + R_l)^2 + (X_f + X_l)^2]^3}$$

and

$$\left.\frac{\partial^2 P}{\partial X_l^2}\right|_{(X_l^*, R_l^*)} = -\frac{E^2}{32 R_f^3}$$

Finally

$$\frac{\partial^2 P}{\partial R_l \, \partial X_l} = \frac{2E^2(X_f + X_l)}{[(R_f + R_l)^2 + (X_f + X_l)^2]^2}$$
$$- \frac{4(X_f + X_l)E^2[(R_f^2 - R_l^2) + (X_f + X_l)^2]}{[(R_f + R_l)^2 + (X_f + X_l)^2]^3}$$

and

$$\left.\frac{\partial^2 P}{\partial R_l \, \partial X_l}\right|_{(X_l^*, R_l^*)} = 0$$

Hence the Hessian is given by

$$H = \begin{bmatrix} -\dfrac{E^2(3R_f + 1)}{32 R_f^4} & 0 \\ 0 & -\dfrac{E^2}{32 R_f^3} \end{bmatrix}$$

Since H is negative definite for $R_f > 0$, (X_l^*, R_l^*) maximizes the power to the load. ◻

Example 4.34 The mean rate of flow of vehicles, q, in vehicles per hour, has been related to mean traffic concentration, C, jam concentration, C_j, measured in vehicles per mile, and vehicle free speed, S, measured in miles per hour by

$$q = CS\left[1 - \left(\frac{C}{C_j}\right)\right]$$

where $C \leq C_j$. Find the mean traffic concentration, C, which maximizes the mean traffic flow rate.

The Lagrange function to be maximized is given by

$$L(C, y, \lambda) = CS\left[1 - \left(\frac{C}{C_j}\right)\right] - \lambda(C - C_j + y^2)$$

The necessary conditions for a maximum are given by

$$\frac{\partial L}{\partial C} = S - 2S\frac{C}{C_j} - \lambda = 0, \qquad \frac{\partial L}{\partial y} = -2\lambda y = 0$$

$$\frac{\partial L}{\partial \lambda} = -(C - C_j + y^2) = 0$$

For $\lambda \neq 0$, $y = 0$, and
$$C = C_j$$
For $\lambda = 0$, $y \neq 0$, and
$$C = \tfrac{1}{2}C_j$$
If $C = C_j$, we have
$$q = 0$$
while $C = \tfrac{1}{2}C_j$ yields
$$q = \tfrac{1}{4}SC_j$$
Hence
$$C^* = \tfrac{1}{2}C_j$$

Example 4.35 At time $t = 0$, ship A is m miles due north of ship B. At this time ship B was sailing due north at 12 mph and ship A was sailing due east at 18 mph. If they continue on their current courses, when will they be nearest one another? (See Fig. 4.16.)

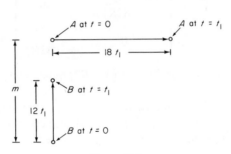

Figure 4.16

Let $A(t)$ = vector denoting the position of A at time t relative to the origin, $A(0)$; $B(t)$ = vector denoting the position of B at time t relative to the origin, $A(0)$.

$$A(0) = \begin{bmatrix} 0 \\ 0 \end{bmatrix}, \quad A(t) = \begin{bmatrix} 18t \\ 0 \end{bmatrix}, \quad B(t) = \begin{bmatrix} 0 \\ m - 12t \end{bmatrix}$$

Then, the distance between $A(t)$ and $B(t)$ is given by $D(t)$, where

$$D(t) = \|A(t) - B(t)\| = \sqrt{324t^2 + (m - 12t)^2} = \sqrt{m^2 - 24mt + 468t^2}$$

Now

$$\frac{dD(t)}{dt} = \frac{1}{2}(936t - 24m)(m^2 - 24mt + 468t^2)^{-1/2} = 0 \quad \text{and} \quad t^* = \frac{m}{39}$$

4.5 APPLICATIONS

t^* is shown to minimize $D(t)$ as follows.

$$\frac{d^2D(t)}{dt^2} = 468(m^2 - 24mt + 468t^2)^{-1/2}$$

$$- \frac{1}{4}(936t - 24m)^2(m^2 - 24mt + 468t^2)^{-3/2}$$

and

$$\frac{d^2D(t)}{dt^2}\bigg|_{t^*} = \frac{324m^2}{(m^2 - 24mt + 468t^2)^{3/2}}\bigg|_{t^*} = \frac{324}{(\frac{9}{13})^{3/2}m^3} > 0$$

Example 4.36 The ABC company has three customers located at the points (x_1, y_1), (x_2, y_2), and (x_3, y_3). A new facility is to be located to serve these three customers. The annual demand from the three customers is estimated to be d_1, d_2, and d_3 units respectively. If the new facility is located at (x_4, y_4) then the annual cost of shipping to customer i is given by

$$i = C_i d_i [(x_i - x_4)^2 + (y_i - y_4)^2], \quad i = 1, 2, 3$$

Hence, the total annual shipping cost, C_T, is given by

$$C_T = \sum_{i=1}^{3} C_i d_i [(x_i - x_4)^2 + (y_i - y_4)^2]$$

Find the location, (x_4, y_4), of the new facility such that the total annual shipping cost, C_T, is minimized.

The necessary conditions for a minimum are given by

$$\frac{\partial C_T}{\partial x_4} = -2 \sum_{i=1}^{3} C_i d_i x_i + 2x_4 \sum_{i=1}^{3} C_i d_i = 0$$

$$\frac{\partial C_T}{\partial y_4} = -2 \sum_{i=1}^{3} C_i d_i y_i + 2y_4 \sum_{i=1}^{3} C_i d_i = 0$$

and

$$x_4^* = \frac{\sum_{i=1}^{3} C_i d_i x_i}{\sum_{i=1}^{3} C_i d_i}, \qquad y_4^* = \frac{\sum_{i=1}^{3} C_i d_i x_i}{\sum_{i=1}^{3} C_i d_i}$$

For the sufficient conditions we have

$$\frac{\partial^2 C_T}{\partial x_4^2} = 2 \sum_{i=1}^{3} C_i d_i, \qquad \frac{\partial^2 C_T}{\partial y_4^2} = 2 \sum_{i=1}^{3} C_i d_i, \qquad \frac{\partial^2 C_T}{\partial x_4 \partial y_4} = 0$$

and the Hessian is given by

$$H = \begin{bmatrix} 2\sum_{i=1}^{3} C_i d_i & 0 \\ 0 & 2\sum_{i=1}^{3} C_i d_i \end{bmatrix}$$

Since, H is positive definite, (x_4^*, y_4^*) is a minimum for C_T.

Problems

1. Determine the values of x for which the following functions are strictly monotonic increasing and decreasing and strictly convex from above and below.
   a. $f(x) = \exp[-|x|]$, $\quad -\infty < x < \infty$
   b. $f(x) = 3x^3 - 9x^2 + 9x + 1$, $\quad -\infty < x < \infty$
2. By example, show that
   a. a function, $f(x)$, may be strictly monotonic without being strictly convex.
   b. a function, $f(x)$, may be strictly convex without being strictly monotonic.
3. Let
$$f(x) = a\exp[-x^2], \quad -1 \le x \le 1$$
Find a point x_0 such that
$$\frac{f(1) - f(-1)}{2} = f'(x_0)$$
4. Let $x = 3$ and $x_0 = 0$. Find the value of x_1 such that
$$f(x) = f(x_0) + (x - x_0)\frac{d}{dx}f(x_0) + \frac{(x-x_0)^2}{2}\frac{d^2}{dx^2}f(x_0) + \frac{(x-x_0)^3}{3!}\frac{d^3}{dx^3}f(x_1)$$
where $\quad f(x) = x^4 - 4x^3 + 3x^2 - x$.

5. Let $f(x)$ be a continuous differentiable function on the interval (a, b) and let x^* be an interior extreme point for $f(x)$. Using Taylor's mean-value theorem, show that $f'(x^*) = 0$.
6. Prove Theorem 4.4 when $f'(x) < 0$ for $a < x < b$.
7. Given Taylor's mean-value theorem and assuming validity of all attendant conditions, show that x can be chosen sufficiently close to x_0 to insure that
$$|f^m(x_0)| > \left|\frac{(x-x_0)}{(m+1)}f^{m+1}(x_1)\right|$$
where
$$f(x) = f(x_0) + \sum_{i=1}^{m} \frac{(x-x_0)^i}{i!}f^i(x_0) + \frac{(x-x_0)^{m+1}}{(m+1)!}f^{m+1}(x_1) \quad \text{and} \quad f^m(x_0) \ne 0$$

8. Find and identify the stationary points for
$$f(x) = \frac{a}{4}x^4 - \frac{a}{3}x^3 + b$$
   a. for $a < 0$
   b. for $a > 0$
9. Examine the following functions for interior minima and maxima.
   a. $f(X) = x_1^2 - 4x_1 + 4x_2^2$
   b. $f(X) = x_1 x_2 + x_1^2$
   c. $f(X) = x_1^3 - 3x_1 - x_2^2$
10. Given the equation
$$f(x) = ax^3 + bx^2 + cx + d$$
find conditions for a, b, c, and d such that any value, x^*, satisfying
$$f'(x^*) = 0$$
is a point of inflection for $f(x)$.

PROBLEMS

11. Let the demand for a certain product be x units and $f(x)$ its probability density function; then $f(x)\,dx$ is the probability that demand is between x and $x + dx$ units. Let C_1 and C_2 be the costs per unit of overestimation and underestimation of demand respectively. Then the total expected cost, if s units are produced, is given by

$$E(C) = C_1 \int_0^s (s - x) f(x)\, dx + C_2 \int_s^\infty (x - s) f(x)\, dx$$

Show that if $E(C)$ is minimized, s must be chosen to satisfy the equation

$$\int_0^s f(x)\, dx = C_2/(C_1 + C_2)$$

noting that

$$\int_0^s f(x)\, dx = 1 - \int_s^\infty f(x)\, dx$$

12. Find the value of β which minimizes the function

$$f(\beta) = \int_{-\infty}^\infty (x - \beta)^2 \, \frac{1}{\sqrt{2\pi}} \exp\left[\frac{-(x - u)^2}{2}\right] dx$$

Note that

$$\int_{-\infty}^\infty \frac{1}{\sqrt{2\pi}} \exp\left[\frac{-(x - u)^2}{2}\right] dx = 1$$

and

$$\int_{-\infty}^\infty x\, \frac{1}{\sqrt{2\pi}} \exp\left[\frac{-(x - u)^2}{2}\right] dx = u$$

13. The density function of the random variable x is given by $f(x; \theta)$ where θ is a parameter of the density function. A random sample of n observations of x, $x_1, x_2, \ldots, x_n$ are taken. The joint density function of these n random variables is given by

$$f(x_1; \theta) f(x_2; \theta) \cdots f(x_n; \theta)$$

and is called the likelihood function of the random sample and is denoted by $L(\theta, x_1, x_2, \ldots, x_n)$. The likelihood function is often used to estimate the value of the parameter θ from the n sampled values. Specifically, the estimate of θ, denoted $\hat{\theta}$, is that value of θ which maximizes $L(\theta, x_1, x_2, \ldots, x_n)$, and is therefore referred to as the maximum likelihood estimate of θ. Given the following likelihood functions, find the maximum likelihood estimators of θ.

a. $L(\theta, x_1, x_2, \ldots, x_n) = \theta^n \exp\left[-\theta \sum_{i=1}^n x_i\right] \quad 0 < x_i < \infty, \quad i = 1, 2, \ldots, n$

b. $L(\theta, x_1, x_2, \ldots, x_n) = \prod_{i=1}^n \frac{m!}{x_i!(m - x_i)!} \theta^{x_i}(1 - \theta)^{m - x_i} \quad x_i = 0, 1, \ldots, m, \quad i = 1, 2, \ldots, n$

Hint: The value of θ which maximizes $L(\theta, x_1, x_2, \ldots, x_n)$ also maximizes $\ln[L(\theta, x_1, x_2, \ldots, x_n)]$

14. The manufacturer of a particular product guarantees its operation for T years. If a unit fails within the guarantee period, the unit is replaced at a cost C_f. L units are produced each year. The cost of producing units with a mean life of t years is proportional to t and is given by $C_t t$ per unit. Time until failure, x, is a random variable which is approximately normally distributed with mean t and variance σ^2. That is

$$f(x \mid t) = \frac{1}{\sigma \sqrt{2\pi}} \exp\left[-\frac{(x - t)^2}{2\sigma^2}\right], \quad -\infty < x < \infty$$

Therefore, the average annual cost of producing and replacing units is given by

$$C_T(t) = C_t tL + C_f L \int_{-\infty}^{T} \frac{1}{\sigma\sqrt{2\pi}} \exp\left[-\frac{(x-t)^2}{2\sigma^2}\right] dx$$

Find the mean design life, t, which will minimize $C_T(t)$.

15. A particular product contains a critical dimension, x. The specification limits for this dimension are C_L and C_U. If $C_L < x < C_U$, the unit is acceptable. Otherwise it is rejected. Units are produced in lots of size L. The dimension x is a random variable with density function given by

$$f(x\,|\,u) = \frac{1}{\sigma\sqrt{2\pi}} \exp\left[-\frac{(x-u)^2}{2\sigma^2}\right], \quad -\infty < x < \infty$$

where u is the mean dimension and σ is the standard deviation of this dimension. If N is the number of rejectable units per lot of size L, the expected number of rejectable units per lot is given by

$$E(N\,|\,u) = L\left[\int_{-\infty}^{C_L} f(x\,|\,u)\,dx + \int_{C_U}^{\infty} f(x\,|\,u)\,dx\right]$$

Find the value of u, the design mean dimension, which minimizes $E(N\,|\,u)$. Note that $\int_{-\infty}^{\infty} f(x\,|\,u)\,dx = 1$

16. Let

$$F(x) = \int_{a(x)}^{b(x)} f(x, y)\,dy$$

Show that

$$F'(x) = \int_{a(x)}^{b(x)} \frac{\partial}{\partial x} f(x, y)\,dy + b'(x)f[x, b(x)] - a'(x)f[x, a(x)]$$

Hint: Note that

$$F'(x) = \lim_{\Delta x \to 0} \frac{\int_{a(x)}^{b(x)} f(x+\Delta x, y)\,dy - \int_{a(x)}^{b(x)} f(x, y)\,dy}{\Delta x}$$

$$+ \lim_{\Delta x \to 0} \frac{\int_{b(x)}^{b(x+\Delta x)} f(x+\Delta x, y)\,dy - \int_{a(x)}^{a(x+\Delta x)} f(x+\Delta x, y)\,dy}{\Delta x}$$

17. Let

$$F(x) = \int_{v(x)}^{w(x)} \int_{t(x,\,y)}^{u(x,\,y)} f(x, y, z)\,dz\,dy$$

Show that

$$F'(x) = \int_{v(x)}^{w(x)} \int_{t(x,\,y)}^{u(x,\,y)} \frac{\partial}{\partial x} f(x, y, z)\,dz\,dy + \int_{v(x)}^{w(x)} \frac{\partial u(x, y)}{\partial x} f[x, y, u(x, y)]\,dy$$

$$- \int_{v(x)}^{w(x)} \frac{\partial t(x, y)}{\partial x} f[x, y, t(x, y)]\,dy + w'(x) \int_{t[x,\,w(x)]}^{u[x,\,w(x)]} f[x, w(x), z]\,dz$$

$$- v'(x) \int_{t[x,\,v(x)]}^{u[x,\,v(x)]} f[x, v(x), z]\,dz$$

Hint: Let

$$H(x, y) = \int_{t(x,\,y)}^{u(x,\,y)} f(x, y, z)\,dz$$

PROBLEMS

18. Find and identify the stationary points for the Rosenbrock function given by

$$f(X) = 100(x_2 - x_1^2)^2 + (1 - x_1)^2$$

Is this function strictly convex from below or above?

19. Identify the stationary points of the following function and classify each as a maxima, minima, or saddle point.

$$f(x_1, x_2) = x_1^3 + \tfrac{1}{2}x_2^2$$

20. Find and identify the stationary points of

$$f(X) = x_1^3 + 2x_1^2 - 3x_1 + 4x_2^2 - 3x_1 x_2$$

21. A company manufactures a product, X, at a continuous and constant rate of ψ units per year. Each unit of X produced requires m_1 units of product A and m_2 units of product B. Products A and B are manufactured internally and are used only in the production of X. The production rate for A is $\phi_1 > \psi$ units per year and the rate for B is $\phi_2 > \psi$ units per year. Since A and B are used only in the production of X, production of A and B is not continuous throughout the year. The production quantities for each run of A and B are q_1 and q_2 respectively. The cost of each setup for production of A is C_{01} and C_{02} for B. The inventory carrying costs per unit-year for A and B are C_{11} and C_{12}. The costs of production for one unit of A and B are C_{p1} and C_{p2} respectively. The total annual cost of producing A and B in quantitites q_1 and q_2, $C_T(Q)$, is

$$C_T(Q) = \frac{m_1 \psi}{q_1} C_{01} + \frac{m_2 \psi}{q_2} C_{02} + \frac{q_1}{2}\left(1 - \frac{m_1 \psi}{\phi_1}\right)C_{11} + \frac{q_2}{2}\left(1 - \frac{m_2 \psi}{\phi_2}\right)C_{12} + m_1 \psi C_{p1} + m_2 \psi C_{p2}$$

where $Q = [q_1, q_2]$.

If

$$\psi = 100{,}000, \quad \phi_1 = 900{,}000, \quad \phi_2 = 300{,}000, \quad m_1 = 3, \quad m_2 = 1$$

$$C_{01} = \$100.00, \quad C_{02} = \$400.00, \quad C_{11} = \$1.00, \quad C_{12} = \$3.00, \quad C_{p1} = \$0.60, \quad C_{p2} = \$0.80$$

find the production quantities q_1 and q_2, which minimize $C_T(Q)$.

22. A subcontractor undertakes to supply two types of Diesel engines, A and B, to a truck manufacturer. The following information is available.

	A	B
Cost/day of holding an engine in stock	$C_1 = \$0.50$	$C_3 = \$0.40$
Cost/day for failing to deliver an engine on schedule	$C_2 = \$10.00$	$C_4 = \$20.00$
Contracted number of engines/day	$R_1 = 25$	$R_2 = 35$

The subcontractor owns a warehouse which will hold up to 600 engines, and the warehouse is to be stocked at the beginning of each month. Given the following total cost equation, find:
a. The optimal initial inventory levels for A and B, z_1 and z_2 respectively.
b. Solve (a) if the warehouse capacity is unlimited.

$$C_T(z_1, z_2) = \frac{C_1 z_1^2}{2R_1} + \frac{C_2(30R_1 - z_1)^2}{2R_1} + \frac{C_3 z_2^2}{2R_2} + \frac{C_4(30R_2 - z_2)^2}{2R_2}$$

23. Let X_0 be a point such that

$$\frac{\partial}{\partial x_i} f(X) \bigg|_{X_0} = -\frac{\partial}{\partial x_j} f(X) \bigg|_{X_0} \neq 0, \quad i \neq j$$

and
$$\frac{\partial}{\partial x_k} f(X)\big|_{X_0} = 0, \quad k = 1, 2, \ldots, n, \quad k \neq i, \quad k \neq j$$

Then
$$\sum_{k=1}^{n} \frac{\partial}{\partial x_k} f(X)\bigg|_{X_0} = 0$$

Show that X_0 is not an interior extreme point.

24. Let X be an n-dimensional vector. Show that if $(\partial/\partial x_j) f(X) = 0$, $j = 1, 2, \ldots, n$, then the directional derivative also vanishes at X for all directions α.
25. Prove part 3 of Theorem 4.12.
26. Show that the function $g(\lambda)$ defined in Eq. (4.46) is continuous with respect to λ.
27. Given the continuous function of four variables
$$f(X) = x_1^2 - x_2^2 + x_3^2 - x_4^2$$

define the vector X_1 such that

$$f(X) = f(X_0) + \sum_{j=1}^{4} (x_j - x_{0j}) \frac{\partial}{\partial x_j} f(x)\bigg|_{X_1} \quad \text{if} \quad X = \begin{bmatrix} 1 \\ 4 \\ 2 \\ 1 \end{bmatrix} \quad \text{and} \quad X_0 = \begin{bmatrix} 2 \\ 2 \\ 2 \\ 2 \end{bmatrix}$$

28. Construct a function of two variables, $f(X)$, which has a unique maximum at X^* such that
$$\frac{\partial}{\partial x_1} f(x)\big|_{X^*} \neq 0, \quad \frac{\partial}{\partial x_2} f(x)\big|_{X^*} \neq 0$$

29. Let $f(X)$ be a convex function. Show that
$$f(X_1) \geq f(X_2) + (X_1 - X_2)^T \nabla f(X_2)$$

for all X_1 and X_2 where $\nabla f(X_2)$ is the gradient vector evaluated at X_2 and is assumed to exist for all X.

30. Let
$$\nabla^2 f(X) = \nabla[\nabla f(X)]$$

Show that $f(X)$ is convex if and only if $\nabla^2 f(X)$ is positive semidefinite. Assume that $\nabla^2 f(X)$ exists for all X.

31. Let $f(X)$ and $g(X)$ be strictly convex functions from above. Show that $f(X)g(X)$ is not necessarily a convex function.

32. Let $f_1(X)$ and $f_2(X)$ be strictly convex functions of X, each having an interior minimum. Then
$$g(X) = f_1(X) + f_2(X)$$

is also strictly convex. Let X_1, X_2, and X_3 be interior minima for $f_1(X), f_2(X)$, and $g(X)$ respectively. Show that X_3 is not necessarily given by a convex combination of X_1 and X_2.

33. A research and development group wishes to conduct a certain experiment in which two chemical compounds must be used. The group has received permission to obtain a total of ten units. That is, if x_1 is the number of units of compound A and x_2 is the number of units of compound B, then $x_1 + x_2 \leq 10$. The cost of the experiment is given by
$$C(x_1, x_2) = C_1 x_1^2 + C_2 x_2$$

where C_1 and C_2 are known positive constants. The success of the experiment is known to be proportional to the total number of chemical units used. How many units of each compound should be requisitioned if $C_2/C_1 = 10$.

34. Examine the function $2x_1^2 + x_2 + 2$ for an interior minimum or maximum subject to the constraint

$$\int_0^{x_2} x_1 \, dz = 16$$

35. The number A is to be divided into three parts, x, y, z, such that

$$A = x + y + z$$

Find the values of x, y, and z such that

$$C = xyz$$

is maximized subject to

$$x > 0, \quad y > 0, \quad z > 0$$

36. Using the Lagrange function, show that if $\lambda_k = 0$, the kth constraint is inactive.
37. Using the Lagrange function, show that λ_k is given by Eq. (4.111).
38. Prove Theorem 4.15 when $n = 3$, $m = 2$.
39. Upper and lower tolerance limits, U and L, are to be developed for a certain dimension, x. The density function of this dimension is given by

$$f(x) = \begin{cases} 0, & x \leq a \\ \dfrac{2(x-a)}{(b-a)(u-a)}, & a < x < u \\ \dfrac{2(b-x)}{(b-a)(b-u)}, & u < x < b \\ 0, & x \geq b \end{cases}$$

Each unit of product is inspected and any unit for which $x < L$ or $x > U$ is rejected at a cost C_R. If $L < x < U$, the unit is acceptable. However, the quality of the unit is dependent upon the deviation of the dimension, x, from the desired dimension u. Specifically, the cost of a unit having dimension x is given by $C_A(u-x)/(x-a)$ for $x < u$ and $C_A(x-u)/(b-x)$ for $x > u$. Therefore, the total cost of tolerance limits L and U is given by

$$C(L, U) = C_A \left[\int_L^u \frac{u-x}{x-a} f(x) \, dx + \int_u^U \frac{x-u}{b-x} f(x) \, dx \right] + C_R \left[\int_a^L f(x) \, dx + \int_U^b f(x) \, dx \right]$$

Find the values of L and U which minimize $C(L, U)$.

40. Solve the following problem using Lagrange multipliers.

$$f(X) = 10x_1 + x_2 = \max$$

subject to

$$x_1 + x_2 = 100, \quad x_1 \geq 0, \quad x_2 \geq 0$$

References

Bellman, R. E., and Dreyfus, S. E., *Applied Dynamic Programming*. Princeton, New Jersey: Princeton Univ. Press, 1962.

Bernholtz, B., A New Derivation of the Kuhn-Tucker Conditions, *Operations Res.* **12**, No. 2, 1964.

Beveridge, G. S. G., and Schechter, R. S., *Optimization: Theory and Practice*, New York: McGraw-Hill, 1970.

Denn, M. M., *Optimization by Variational Methods*, New York: McGraw-Hill, 1969.

Fulks, W., *Advanced Calculus*, New York: Wiley (Interscience), 1961.
Gue, R. L. and Thomas, M. E., *Mathematical Methods in Operation Research*, London: Macmillan, 1968.
Hadley, G., *Nonlinear and Dynamic Programming*, Reading, Massachusetts: Addison-Wesley, 1964.
Hancock, H., *Theory of Maxima and Minima*. New York: Dover, 1960.
Kuhn, H. W., and Tucker, A. W., Nonlinear Programming, *Proceedings of the Second Berkeley Symposium on Mathematical Statistics and Probability* (J. Neyman, editor), pp. 481–492. Berkeley, California: Univ. California Press, 1951.
Kunzi, H. P., Krelle, W., and Oettli, W., *Nonlinear Programming*, London: Blaisdell, 1966.
Saaty, T. L., and Bram, J., *Nonlinear Mathematics*, New York: McGraw-Hill, 1964.
Teichroew, D., *An Introduction to Management Science*, New York: Wiley (Interscience), 1964.
Wilde, D. J. and Beightler, C. S., *Foundations of Optimization*. Englewood Cliffs, New Jersey: Prentice-Hall, 1967.
Zangwill, W. I., *Nonlinear Programming*, Englewood Cliffs, New Jersey: Prentice-Hall, 1969.

CHAPTER 5

CALCULUS
OF FINITE DIFFERENCES

5.1 Introduction

Until now all functions and variables have been assumed to be continuous. However, many problems faced by the operations research analyst involve variables which are inherently *discrete*. For example, consider finding the optimal capacity of a parking lot. The number of parking spaces provided is a discrete variable. Other examples include the determination of the optimal sample size and acceptance number for an acceptance sampling plan, the optimal size of a repair crew, and the reorder point and reorder quantity for a spare parts inventory. In each case the decision variable or variables are discrete. Frequently, reliable approximate mathematical models can be developed based upon the assumption that these variables are continuous. However, when discrete variables are to be treated as such, continuous operations such as integration and differentiation must be replaced by their discrete equivalent. Even when the operations of differentiation and integration are appropriate, it is sometimes necessary to carry out these operations through approximate methods. The calculus of finite differences plays a significant role in the development of approximations for these operations.

Before pursuing the idea of a *finite difference* let us examine the concept of a discrete variable and its distinction from a continuous variable. If x is a continuous variable with values on the open interval (a, b), then on any finite subinterval, (a', b'), x may assume an infinite number of values. If x is discrete, then on any finite interval (a', b'), x may assume a finite number of values. Another way of looking at a discrete variable is that between any two successive values of a discrete variable there is a finite open interval which does not contain a permissible value of the variable. For example, suppose the variable x may assume only integer values on the interval $(-\infty, \infty)$. Let a and b be finite numbers. On the interval (a, b), x may assume no more than $b - a$ values and between any two successive values of x there is an open interval of unit length. Although most of the discrete variables encountered in operations research problems are restricted in value to the integers, this restriction does not apply to discrete variables in general and any variable satisfying the conditions given above is classified as discrete. For example, let $x_0 = 1$ and let $y_i = (x_{i-1} + 1)/x_{i-1}, i = 1, 2, \ldots$. Since y_i satisfies the required conditions for a discrete variable, it is discrete although it is not restricted in value to the integers.

Let $f(x)$ be a function of the discrete variable x and assume that $f(x)$ is defined for $x = x_0, x_1, x_2, \ldots, x_n$. In this chapter $f(x)$ will sometimes be presumed to have a value of zero for any $x \neq x_0, x_1, x_2, \ldots, x_n$. Therefore a complete definition of $f(x)$ would be given by

$$f(x) = \begin{cases} f(x), & x = x_0, x_1, x_2, \ldots, x_n \\ 0, & \text{otherwise} \end{cases}$$

An example of such a function is the discrete probability mass function. However, cases will also be encountered where $f(x)$ is not defined for $x \neq x_0, x_1, x_2, \ldots, x_n$. In this case a complete definition of $f(x)$ would be given by

$$f(x) = \begin{cases} f(x), & x = x_0, x_1, x_2, \ldots, x_n \\ \text{undefined}, & \text{otherwise} \end{cases}$$

An illustration of the latter cases is given in the following example.

Example 5.1 A company receives a supply of a particular raw material at the beginning of each week. The raw material is received in cases of 1000 units each and a partial case cannot be ordered. The problem is to determine the number of cases, x, received each week. The number of cases used during production each week is a random variable. Any raw material left over at the end of the week is disposed of at a loss. In addition if the supply of raw material is not sufficient to meet the weeks requirements a cost of lost production is incurred. Let $f(x)$ be the total cost incurred if x cases are received at the beginning of each week. The problem is then to find the value of x which minimizes $f(x)$, where $f(x)$ is given by

$$f(x) = \begin{cases} f(x), & x = 0, 1, 2, \ldots \\ \text{undefined}, & \text{otherwise} \end{cases}$$

In this example a cost cannot be assigned when x is negative valued or when x assumes a noninteger positive value.

When dealing with functions of discrete variables, the analyst usually does not consider values of the variables which are not permitted. Thus, by careful treatment of a function of discrete variables, the analyst can avoid specification of the value of the function for values of the variables which are not permitted.

5.2 The Divided Difference

The discrete analog of the derivative is called the *divided difference*. From basic calculus the derivative is defined by

$$\frac{df(x)}{dx} = \lim_{\Delta x \to 0} \frac{f(x + \Delta x) - f(x)}{\Delta x} \tag{5.1}$$

5.2 THE DIVIDED DIFFERENCE

where $f(x)$ is assumed to be continuous on an interval (a, b) which includes x. In Eq. (5.1), eliminate the limiting operation and let $x_i = x$ and $x_{i+1} = x + \Delta x$, where it is assumed that $x_i < x_{i+1}$ for all i. Then $\Delta x = x_{i+1} - x_i > 0$ and the resulting ratio is called the divided difference of $f(x)$ at x_i and is denoted by $\Delta f(x_i)$, or

$$\Delta f(x_i) = \frac{f(x_{i+1}) - f(x_i)}{x_{i+1} - x_i} \tag{5.2}$$

or equivalently

$$\Delta f(x_i) = \frac{f(x_i) - f(x_{i+1})}{x_i - x_{i+1}} \tag{5.3}$$

If the difference $x_{i+1} - x_i = \Delta x$ for all i, the divided difference is given by

$$\Delta f(x_i) = \frac{f(x_i + \Delta x) - f(x_i)}{\Delta x} \tag{5.4}$$

In Eqs. (5.2) and (5.3) the subscript i was used to define the difference $x_{i+1} - x_i$. Since this difference is a constant in Eq. (5.4), there is no need to retain the subscript. More often than not, $\Delta x = 1$ in Eq. (5.4) and $\Delta f(x)$ becomes

$$\Delta f(x) = f(x + 1) - f(x) \tag{5.5}$$

When Eq. (5.5) applies, the divided difference is referred to as the *simple difference* or just the *difference*.

Example 5.2 Let $f(x)$ be defined by

$$f(x) = \frac{x}{x - 1}, \quad x = 2, \frac{5}{2}, 3, \frac{7}{2}, \ldots$$

Find an expression for $\Delta f(x)$ and evaluate $\Delta f(x)$ at $x = 2, \frac{7}{2}, 7$.

Since $x_{i+1} - x_i = 0.5$ for all i, Eq. (5.4) may be used to determine the divided difference of $f(x)$

$$\Delta f(x) = \frac{f(x + 0.5) - f(x)}{0.5} = 2\left[\frac{x + 0.5}{x - 0.5} - \frac{x}{x - 1}\right] = -\frac{1}{x^2 - 1.5x + 0.5}$$

Therefore

$$\Delta f(2) = -\frac{1}{4 - 3 + 0.5} = -\frac{2}{3}$$

$$\Delta f\left(\frac{7}{2}\right) = -\frac{1}{12.25 - 5.25 + 0.5} = -\frac{2}{15}$$

$$\Delta f(7) = -\frac{1}{49 - 10.5 + 0.5} = -\frac{1}{39}$$

Example 5.3 Let
$$f(x) = \frac{1 - a^x}{1 - a}, \quad x = 1, 2, \ldots$$
where a is a positive constant. Show that $\Delta f(x) = a^x$.
$$\Delta f(x) = \frac{1 - a^{x+1}}{1 - a} - \frac{1 - a^x}{1 - a} = \frac{a^x(1 - a)}{1 - a} = a^x \quad \square$$

Example 5.4 Let
$$f(x) = \sum_{y=0}^{x} \frac{\lambda^y}{y!} \exp[-\lambda], \quad x = 0, 1, 2, \ldots$$
Show that
$$\Delta f(x) = \frac{\lambda^{x+1}}{(x + 1)!} \exp[-\lambda]$$

$$\Delta f(x) = \sum_{y=0}^{x+1} \frac{\lambda^y}{y!} \exp[-\lambda] - \sum_{y=0}^{x} \frac{\lambda^y}{y!} \exp[-\lambda]$$

$$= \frac{\lambda^{x+1}}{(x + 1)!} \exp[-\lambda] + \sum_{y=0}^{x} \frac{\lambda^y}{y!} \exp[-\lambda] - \sum_{y=0}^{x} \frac{\lambda^y}{y!} \exp[-\lambda]$$

$$= \frac{\lambda^{x+1}}{(x + 1)!} \exp[-\lambda] \quad \square$$

Example 5.5 Find $\Delta f(x)$ where

a. $f(x) = x!, \quad x = 0, 1, 2, \ldots$
b. $f(x) = x^{(n)}, \quad x = 1, 2, 3, \ldots$
c. $f(x) = \dfrac{1}{x^{[n]}}, \quad x = 1, 2, 3, \ldots$

In cases b and c, $x^{(n)}$ and $x^{[n]}$ are called factorial polynomials and are defined as
$$x^{(n)} = x(x - 1)(x - 2) \cdots (x - n + 1),$$
$$x^{[n]} = x(x + 1)(x + 2) \cdots (x + n - 1)$$
In each case $x_{i+1} - x_i = 1$. For part a,
$$\Delta f(x) = (x + 1)! - x! = x![(x + 1) - 1] = xx!$$
For part b,
$$\Delta f(x) = (x + 1)^{(n)} - x^{(n)}$$
$$= (x + 1)x(x - 1) \cdots (x - n + 2) - x(x - 1)(x - 2) \cdots (x - n + 1)$$
$$= (x + 1)x^{(n-1)} - (x - n + 1)x^{(n-1)} = nx^{(n-1)}$$

5.2 THE DIVIDED DIFFERENCE

For part c,

$$\Delta f(x) = \frac{1}{(x+1)^{[n]}} - \frac{1}{x^{[n]}}$$

$$= \frac{1}{(x+1)(x+2)\cdots(x+n)} - \frac{1}{x(x+1)\cdots(x+n-1)}$$

$$= \frac{x}{x^{[n+1]}} - \frac{x+n}{x^{[n+1]}} = -\frac{n}{x^{[n+1]}}$$ ▨

Based upon the similarity between Eqs. (5.1) and (5.2), it should not be surprising to find other similarities between the derivative of a function and its discrete analog. From the calculus we have the relationships

$$\frac{d}{dx}[f(x) \pm g(x)] = \frac{df(x)}{dx} \pm \frac{dg(x)}{dx} \tag{5.6}$$

$$\frac{d}{dx}[f(x)g(x)] = f(x)\frac{dg(x)}{dx} + g(x)\frac{df(x)}{dx} \tag{5.7}$$

$$\frac{d}{dx}\left[\frac{f(x)}{g(x)}\right] = \frac{1}{[g(x)]^2}\left[g(x)\frac{df(x)}{dx} - f(x)\frac{dg(x)}{dx}\right] \tag{5.8}$$

In the following theorem similar relationships are given when $f(x)$ and $g(x)$ are functions of discrete variables.

Theorem 5.1 Let $h(x)$ and $g(x)$ be functions of the discrete variable x and let x assume the values $x_1, x_2, \ldots$. Then.

a. $\Delta[h(x_i) \pm g(x_i)] = \Delta h(x_i) \pm \Delta g(x_i)$ (5.9)
b. $\Delta[h(x_i)g(x_i)] = h(x_{i+1})\Delta g(x_i) + g(x_i)\Delta h(x_i)$ (5.10)
c. $\Delta\left[\dfrac{h(x_i)}{g(x_i)}\right] = \dfrac{1}{g(x_i)g(x_{i+1})}[g(x_i)\Delta h(x_i) - h(x_i)\Delta g(x_i)]$ (5.11)

Proof To prove this theorem we start with the definition of the divided difference.

$$\Delta f(x_i) = \frac{f(x_{i+1}) - f(x_i)}{x_{i+1} - x_i}$$

To establish part a let

$$f(x) = h(x) \pm g(x)$$

Then

$$\Delta f(x_i) = \Delta[h(x_i) \pm g(x_i)] = \frac{[h(x_{i+1}) \pm g(x_{i+1})] - [h(x_i) \pm g(x_i)]}{x_{i+1} - x_i}$$

$$= \frac{[h(x_{i+1}) - h(x_i)] \pm [g(x_{i+1}) - g(x_i)]}{x_{i+1} - x_i} = \Delta h(x_i) \pm \Delta g(x_i)$$

For part b, let
$$f(x_i) = h(x_i)g(x_i)$$
Then
$$\Delta f(x_i) = \Delta[h(x_i)g(x_i)] = \frac{h(x_{i+1})g(x_{i+1}) - h(x_i)g(x_i)}{x_{i+1} - x_i}$$

Adding and subtracting $h(x_{i+1})g(x_i)$ in the numerator yields

$$\Delta[h(x_i)g(x_i)] = \frac{h(x_{i+1})[g(x_{i+1}) - g(x_i)] + g(x_i)[h(x_{i+1}) - h(x_i)]}{x_{i+1} - x_i}$$
$$= h(x_{i+1})\,\Delta g(x_i) + g(x_i)\,\Delta h(x_i)$$

The proof of part c is developed using a similar approach, but is left as an exercise for the reader. ▨

Example 5.6 Let
$$f(x) = \frac{(x-1)(x)(2x-1)}{6}, \qquad x = 1, 2, \ldots$$

Show that
$$\Delta f(x) = x^2$$

Let
$$h(x) = \frac{(x-1)}{6}, \qquad g(x) = x, \qquad s(x) = (2x-1)$$

Then by Theorem 5.1, part b,
$$\Delta f(x) = h(x+1)\,\Delta[g(x)s(x)] + g(x)s(x)\,\Delta h(x)$$
and
$$\Delta[g(x)s(x)] = g(x+1)\,\Delta s(x) + s(x)\,\Delta g(x)$$
which leads to
$$\Delta f(x) = h(x+1)[g(x+1)\,\Delta s(x) + s(x)\,\Delta g(x)] + g(x)s(x)\,\Delta h(x)$$
now
$$\Delta h(x) = \tfrac{1}{6}, \qquad \Delta g(x) = 1, \qquad \Delta s(x) = 2$$

Therefore
$$\Delta f(x) = \frac{x[2(x+1) + (2x-1)] + x(2x-1)}{6} = x^2 \qquad ▨$$

5.2 THE DIVIDED DIFFERENCE

Example 5.7 Let

$$f(x) = \frac{x}{x+1}, \qquad x = 1, 2, 3, \ldots$$

Show that

$$\Delta f(x) = \frac{1}{(x+1)(x+2)}$$

Let

$$h(x) = x, \qquad g(x) = x + 1$$

Then

$$f(x) = \frac{h(x)}{g(x)}$$

and

$$\Delta h(x) = 1, \qquad \Delta g(x) = 1$$

From part c of Theorem 5.1,

$$\Delta f(x) = \frac{1}{g(x)g(x+1)} [g(x) \Delta h(x) - h(x) \Delta g(x)]$$

$$= \frac{1}{(x+1)(x+2)} [(x+1) - x] = \frac{1}{(x+1)(x+2)} \quad \boxed{}$$

Although the relationships given in Theorem 5.1 are useful in developing the divided difference, it is sometimes just as convenient to use the basic definition of the divided difference instead of these expressions. As an illustration consider the following example.

Example 5.8 Find $\Delta f(x)$ for all values of x where

$$f(x) = \begin{cases} 0, & x = -(n+1), -(n+2), \ldots \\ -x^2, & x = -n, -(n-1), \ldots, 0 \\ x^2, & x = 1, 2, \ldots, n \\ 0, & x = n+1, n+2, \ldots \end{cases}$$

For $x \leq -(n+2)$,

$$\Delta f(x) = 0$$

For $x = -(n+1)$,

$$\Delta f(x) = f(x+1) - f(x) = -n^2 - 0 = -n^2$$

For $x = -n, -(n-1), \ldots, -1$,

$$\Delta f(x) = f(x+1) - f(x) = -(x+1)^2 + x^2 = -2x - 1$$

For $x = 0$,
$$\Delta f(x) = f(x+1) - f(x) = (1)^2 + (0)^2 = 1$$
For $x = 1, 2, \ldots, n-1$
$$\Delta f(x) = f(x+1) - f(x) = (x+1)^2 - x^2 = 2x + 1$$
For $x = n$,
$$\Delta f(x) = f(x+1) - f(x) = 0 - n^2 = -n^2$$
For $x \geq n + 1$,
$$\Delta f(x) = 0$$
Therefore,
$$\Delta f(x) = \begin{cases} 0, & x = -(n+2), -(n+3), \ldots \\ -n^2, & x = -(n+1) \\ -2x - 1, & x = -n, -(n-1), \ldots, -1 \\ 1, & x = 0 \\ 2x + 1, & x = 1, 2, \ldots, n-1 \\ -n^2, & x = n \\ 0, & x = n+1, n+2, \ldots \end{cases}$$

Example 5.9 Given the function
$$f(x) = \frac{2^x}{x+2} - \frac{1}{2}, \quad x = 1, 2, \ldots$$
find $\Delta f(x)$ using part c of Theorem 5.1 and Eq. (5.5).

First, from Theorem 5.1,
$$f(x) = \frac{h(x)}{g(x)}$$
where
$$h(x) = 2^{x+1} - (x+2), \quad g(x) = 2(x+2)$$
Taking the difference of $h(x)$ and $g(x)$ gives us
$$\Delta h(x) = 2^{x+1} - 1, \quad \Delta g(x) = 2$$
Therefore,
$$\Delta f(x) = \frac{1}{4(x+2)(x+3)}[2(x+2)(2^{x+1} - 1) - 2(2^{x+1} - x - 2)]$$
$$= \frac{2^x(x+1)}{(x+2)(x+3)}$$

5.2 THE DIVIDED DIFFERENCE

Using Eq. (5.5) we have

$$\Delta f(x) = \frac{2^{x+1}}{(x+3)} - \frac{2^x}{(x+2)}$$

$$= \frac{(x+2)2^{x+1} - (x+3)2^x}{(x+2)(x+3)} = \frac{2^x(x+1)}{(x+2)(x+3)}$$

Thus far the discussion has been restricted to finding the divided difference of simple functions. Let us now turn our attention to the problem of finding the divided difference of a function which involves a *summation*. Let

$$C(x) = \sum_{y=a(x)}^{b(x)} f(x, y)$$

$$= f[x, a(x)] + f[x, a(x) + \Delta y] + f[x, a(x) + 2\Delta y] + \cdots$$
$$+ f[x, b(x) - \Delta y] + f[x, b(x)] \tag{5.12}$$

where x and y are discrete variables such that $y_{i+1} - y_i = \Delta y$ and $x_{i+1} - x_i = \Delta x$ for all i. In Eq. (5.12) it is further assumed that $a(x)$ and $b(x)$ are permissible values of y for all values of x, that

$$a(x) < b(x)$$

and that $a(x)$ and $b(x)$ are increasing functions of x. That is,

$$a(x) < a(x + \Delta x), \qquad b(x) < b(x + \Delta x)$$

To find $\Delta C(x)$ for the expression given in Eq. (5.12), Eq. (5.4) may be used and this is sometimes the simplest approach to the problem. However, the expression for $\Delta C(x)$ given in the following theorem is also quite useful in this determination.

Theorem 5.2 If $C(x)$ is defined by Eq. (5.12), then

$$\Delta C(x) = \sum_{y=a(x)}^{b(x)} \Delta_x f(x, y) + \sum_{y=b(x)+\Delta y}^{b(x+\Delta x)} \frac{f(x + \Delta x, y)}{\Delta x}$$
$$- \sum_{y=a(x)}^{a(x+\Delta x) - \Delta y} \frac{f(x + \Delta x, y)}{\Delta x} \tag{5.13}$$

where $\Delta_x f(x, y)$ is the divided difference of $f(x, y)$ with respect to x.

Proof From Eq. (5.4),

$$\Delta C(x) = \frac{C(x + \Delta x) - C(x)}{\Delta x} = \frac{\sum_{y=a(x+\Delta x)}^{b(x+\Delta x)} f(x + \Delta x, y) - \sum_{y=a(x)}^{b(x)} f(x, y)}{\Delta x}$$

The summation in the expression for $C(x + \Delta x)$ can be divided into three parts as follows.

$$\sum_{y=a(x+\Delta x)}^{b(x+\Delta x)} f(x + \Delta x, y) = \sum_{y=a(x)}^{b(x)} f(x + \Delta x, y) + \sum_{y=b(x)+\Delta y}^{b(x+\Delta x)} f(x + \Delta x, y)$$
$$- \sum_{y=a(x)}^{a(x+\Delta x) - \Delta y} f(x + \Delta x, y)$$

The reader will note that the first summation on the right-hand side of the above equation differs from that on left-hand side in that the terms $b(x) + \Delta y$ to $b(x + \Delta x)$ have been left out while the terms $a(x)$ to $a(x + \Delta x) - \Delta y$ have been added. Therefore the sum over the first series of terms must be added and the sum over the second series must be subtracted to maintain the equality. The reader should remember that the sequence of values of y are

$$a(x), a(x) + \Delta y, \ldots, a(x + \Delta x) - \Delta y, a(x + \Delta x), a(x + \Delta x) + \Delta y,$$
$$\ldots, b(x), b(x) + \Delta y, \ldots, b(x + \Delta x) - \Delta y, b(x + \Delta x)$$

since y increases in increments Δy.

From the above development, $\Delta C(x)$ is given by

$$\Delta C(x) = \frac{\sum_{y=a(x)}^{b(x)} [f(x + \Delta x, y) - f(x, y)]}{\Delta x} + \frac{\sum_{y=b(x)+\Delta y}^{b(x+\Delta x)} f(x + \Delta x, y)}{\Delta x}$$

$$- \frac{\sum_{y=a(x)}^{a(x+\Delta x)-\Delta y} f(x + \Delta x, y)}{\Delta x}$$

$$= \sum_{y=a(x)}^{b(x)} \Delta_x f(x, y) + \frac{\sum_{y=b(x)+\Delta y}^{b(x+\Delta x)} f(x + \Delta x, y) - \sum_{y=a(x)}^{a(x+\Delta x)-\Delta y} f(x + \Delta x, y)}{\Delta x}$$

In Theorem 5.2, if $\Delta y = \Delta x = 1$, Eq. (5.13) becomes

$$\Delta C(x) = \sum_{y=a(x)}^{b(x)} \Delta_x f(x, y) + \sum_{y=b(x)+1}^{b(x+1)} f(x + 1, y) - \sum_{y=a(x)}^{a(x+1)-1} f(x + 1, y) \quad (5.14)$$

Example 5.10 If

$$C(x) = \sum_{y=c}^{d} f(x, y)$$

Show that

$$\Delta C(x) = \sum_{y=c}^{d} \Delta_x f(x, y) \quad (5.15)$$

if c and d are constants and the increments for x and y are Δx and Δy respectively.

Since the increments for x and y are constant, then

$$\Delta C(x) = \frac{C(x + \Delta x) - C(x)}{\Delta x} = \frac{\sum_{y=c}^{d} f(x + \Delta x, y) - \sum_{y=c}^{d} f(x, y)}{\Delta x}$$

$$= \sum_{y=c}^{d} \frac{f(x + \Delta x, y) - f(x, y)}{\Delta x} = \sum_{y=c}^{d} \Delta_x f(x, y)$$

Example 5.11 Let

$$C(x) = \sum_{y=1}^{n} (y^2 + x)$$

Find $\Delta C(x)$ where x and y are integer valued.

5.2 THE DIVIDED DIFFERENCE

Using the result of Example 5.10, we have

$$\Delta C(x) = \sum_{y=1}^{n} \Delta_x(y^2 + x), \qquad \Delta_x(y^2 + x) = 1$$

Therefore,

$$\Delta C(x) = \sum_{y=1}^{n} (1) = n$$

Example 5.12 If $\Delta y = 1$, find $\Delta C(x)$ where

$$C(x) = \sum_{y=x}^{x^2} (x + 2), \qquad x = 2, 3, 4, \ldots$$

From Eq. (5.14),

$$\Delta C(x) = \sum_{y=a(x)}^{b(x)} \Delta_x f(x, y) + \sum_{y=b(x)+1}^{b(x+1)} f(x+1, y) - \sum_{y=a(x)}^{a(x+1)-1} f(x+1, y)$$

In this case

$$a(x) = x, \qquad a(x+1) = x+1, \qquad b(x) = x^2$$
$$b(x+1) = (x+1)^2, \qquad \Delta_x f(x, y) = 1, \qquad f(x+1, y) = x+3$$

Therefore

$$\Delta C(x) = \sum_{y=x}^{x^2} (1) + \sum_{y=x^2+1}^{(x+1)^2} (x+3) - \sum_{y=x}^{x} (x+3)$$
$$= (x^2 - x + 1) + [(x+1)^2 - (x^2+1) + 1](x+3) - (x+3)$$
$$= 3x^2 + 5x + 1$$

Example 5.13 Find $\Delta C(x)$ if

$$C(x) = \sum_{y=x}^{x+n} f(x, y), \qquad x = 0, \pm 1, \pm 2, \ldots$$

where n is a positive integer and where

$$f(x, y) = (y - x), \qquad y = x, x+1, x+2, \ldots$$

Since Δx and Δy are unity, Eq. (5.14) is used to determine $\Delta C(x)$. Therefore

$$a(x) = x, \qquad a(x+1) = x+1, \qquad b(x) = x+n, \qquad b(x+1) = x+n+1$$
$$f(x+1, y) = (y - x - 1), \qquad y = x, x+1, x+2, \ldots,$$
$$\Delta_x f(x, y) = -1, \qquad y = x, x+1, \ldots$$

$$\Delta C(x) = \sum_{y=x}^{x+n} (-1) + \sum_{y=x+n+1}^{x+n+1} (y - x - 1) - \sum_{y=x}^{x} (-1) = -(n+1) + n + 1 = 0$$

Example 5.14 Find $\Delta C(x)$ where
$$C(x) = \sum_{y=x}^{x+n} (x + y + a)$$
Assume that x and y have increments $\Delta x = \Delta y = \frac{1}{2}$ and that n is a positive integer.
From Eq. (5.13),
$$\Delta C(x) = \sum_{y=x}^{x+n} \Delta_x(x + y + a) + \sum_{y=x+n+\frac{1}{2}}^{x+n+\frac{1}{2}} 2(x + y + a + \tfrac{1}{2}) - \sum_{y=x}^{x} 2(x + y + a + \tfrac{1}{2})$$
where
$$\Delta_x(x + y + a) = 1$$
Therefore
$$\Delta C(x) = (2n + 1) + 2(2x + n + a + 1) - 2(2x + a + \tfrac{1}{2}) = 4n + 2 \quad \blacksquare$$

Example 5.15 Let $C(x)$ be defined by
$$C(x) = \sum_{y=x}^{x+a} f(x, y)$$
where x and y may assume the values
$$x = 0, \pm\tfrac{1}{2}, \pm 1, \pm\tfrac{3}{2}, \pm 2, \ldots, \qquad y = 0, \pm 1, \pm 2, \ldots$$
and a is a positive integer. Develop an expression for $\Delta C(x)$.
$$\Delta C(x) = 2\left[\sum_{y=x+\frac{1}{2}}^{x+a+\frac{1}{2}} f(x + \tfrac{1}{2}, y) - \sum_{y=x}^{x+a} f(x, y)\right]$$
By the definition of the values of x and y
$$f(x, y) = \begin{cases} f(x, y), & x = 0, \pm\tfrac{1}{2}, \pm 1, \ldots, \text{ and } y = 0, \pm 1, \pm 2, \ldots \\ 0, & \text{otherwise} \end{cases}$$
Two cases must be considered; where x is an integer and where x is not an integer. For integer x,
$$f(x + \tfrac{1}{2}, x + \tfrac{1}{2}) = 0, \qquad f(x + \tfrac{1}{2}, x + a + \tfrac{1}{2}) = 0$$
Therefore
$$\sum_{y=x+\frac{1}{2}}^{x+a+\frac{1}{2}} f(x + \tfrac{1}{2}, y) = \sum_{y=x+1}^{x+a} f(x + \tfrac{1}{2}, y)$$
and
$$\Delta C(x) = 2\left[\sum_{y=x+1}^{x+a} f(x + \tfrac{1}{2}, y) - \sum_{y=x}^{x+a} f(x, y)\right]$$
or
$$\Delta C(x) = \sum_{y=x+1}^{x+a} \Delta_x f(x, y) - 2f(x, x)$$

5.2 THE DIVIDED DIFFERENCE

For noninteger x,

$$f(x, x) = 0, \quad f(x, x + a) = 0 \quad \text{and} \quad \sum_{y=x}^{x+a} f(x, y) = \sum_{y=x+\frac{1}{2}}^{x+a-\frac{1}{2}} f(x, y)$$

Therefore

$$\Delta C(x) = 2\left[\sum_{y=x+\frac{1}{2}}^{x+a+\frac{1}{2}} f(x + \tfrac{1}{2}, y) - \sum_{y=x+\frac{1}{2}}^{x+a-\frac{1}{2}} f(x, y)\right]$$

$$= \sum_{y=x+\frac{1}{2}}^{x+a-\frac{1}{2}} \Delta_x f(x, y) + 2f(x + \tfrac{1}{2}, x + a + \tfrac{1}{2}) \qquad \text{⑤}$$

The reader may wonder why Theorem 5.2 was not used to develop the expression for $\Delta C(x)$ in Example 5.15. Theorem 5.2 is based upon the conditions given for Eq. 5.12. Specifically, in Eq. (5.12) it was assumed that $a(x)$ and $b(x)$ were permissible values of y for all x. However, this was not the case in Example 5.15. It should also be noted that $a(x)$ and $b(x)$ were assumed to be *increasing functions of* x. When this assumption is violated Eq. (5.13) must be modified. Example 5.16 illustrates the problem involved when $a(x)$ and $b(x)$ are not increasing functions of x.

Example 5.16 Show that Theorem 5.2 does not apply when $C(x)$ is as defined in Example 5.10.

From Example 5.10,

$$C(x) = \sum_{y=c}^{d} f(x, y) \quad \text{and} \quad \Delta C(x) = \sum_{y=c}^{d} \Delta_x f(x, y)$$

Clearly $a(x)$ and $b(x)$ are not increasing functions of x since

$$a(x + \Delta x) = c = a(x), \quad b(x + \Delta x) = d = b(x)$$

Let us apply the result of Theorem 5.2 to see what happens when this type of misapplication occurs. From Theorem 5.2,

$$\Delta C(x) = \sum_{y=c}^{d} \Delta_x f(x, y) + \sum_{y=d+\Delta y}^{d} \frac{f(x + \Delta x, y)}{\Delta x} - \sum_{y=c}^{c-\Delta y} \frac{f(x + \Delta x, y)}{\Delta x}$$

This expression for $\Delta C(x)$ is valid only if the last two summations are defined as zero. ⑤

Example 5.16 provides a key which the analyst may use to detect improper application of the result of Theorem 5.2. If application of Eq. (5.12) yields summations over limits such that the lower limit is greater than the upper limit, then the analyst is advised to check the assumptions underlying Theorem 5.2 against the properties of the summations he is attempting to difference.

The expression for $\Delta C(x)$ is extended to the case where $C(x)$ includes a *double summation* in Theorem 5.3. The result given in Theorem 5.3 is not as important as the method used for its development since the approach used can be applied iteratively to develop the divided difference for functions of multiple summations in general.

Theorem 5.3 Let $C(x)$ be defined by

$$C(x) = \sum_{y=a(x)}^{b(x)} \sum_{z=g(x,y)}^{h(x,y)} f(x, y, z) \tag{5.16}$$

where x, y, and z have constant increments Δx, Δy, and Δz respectively such that $\Delta x = \Delta y = \Delta z$, where $a(x)$ and $b(x)$ are permissible values of y for all x, where $g(x, y)$ and $h(x, y)$ are permissible values of z for all x and y, and where

$$a(x) < b(x), \quad a(x) < a(x + \Delta x), \quad b(x) < b(x + \Delta x), \quad g(x, y) < h(x, y)$$
$$g(x, y) < g(x + \Delta x, y), \qquad g(x, y) < g(x, y + \Delta y),$$
$$g(x, y) < g(x + \Delta x, y + \Delta y), \quad h(x, y) < h(x + \Delta x, y),$$
$$h(x, y) < h(x, y + \Delta y), \qquad h(x, y) < h(x + \Delta x, y + \Delta y)$$

Then

$$\Delta C(x) = \sum_{y=a(x)}^{b(x)} \sum_{z=g(x,y)}^{h(x,y)} \Delta_x f(x, y, z)$$

$$+ \sum_{y=a(x)}^{b(x)} \left[\sum_{z=h(x,y)+\Delta z}^{h(x+\Delta x, y)} \frac{f(x + \Delta x, y, z)}{\Delta x} - \sum_{z=g(x,y)}^{g(x+\Delta x, y)-\Delta z} \frac{f(x + \Delta x, y, z)}{\Delta x} \right]$$

$$+ \sum_{y=b(x)+\Delta y}^{b(x+\Delta x)} \sum_{z=g(x+\Delta x, y)}^{h(x+\Delta x, y)} \frac{f(x + \Delta x, y, z)}{\Delta x}$$

$$- \sum_{y=a(x)}^{a(x+\Delta x)-\Delta y} \sum_{z=g(x+\Delta x, y)}^{h(x+\Delta x, y)} \frac{f(x + \Delta x, y, z)}{\Delta x} \tag{5.17}$$

Proof In Eq. (5.16), let

$$F(x, y) = \sum_{z=g(x,y)}^{h(x,y)} f(x, y, z)$$

Then

$$C(x) = \sum_{y=a(x)}^{b(x)} F(x, y)$$

and

$$\Delta C(x) = \sum_{y=a(x)}^{b(x)} \Delta_x F(x, y) + \sum_{y=b(x)+\Delta y}^{b(x+\Delta x)} \frac{F(x + \Delta x, y)}{\Delta x}$$

$$- \sum_{y=a(x)}^{a(x+\Delta x)-\Delta y} \frac{F(x + \Delta x, y)}{\Delta x} \tag{5.18}$$

5.2 THE DIVIDED DIFFERENCE

Applying Theorem 5.2 to $\Delta_x F(x, y)$ yields

$$\Delta_x F(x, y) = \sum_{z=g(x, y)}^{h(x, y)} \Delta_x f(x, y, z) + \sum_{z=h(x, y)+\Delta z}^{h(x+\Delta x, y)} \frac{f(x + \Delta x, y, z)}{\Delta x}$$

$$- \sum_{z=g(x, y)}^{g(x+\Delta x, y)-\Delta z} \frac{f(x + \Delta x, y, z)}{\Delta x}$$

Therefore,

$$\sum_{y=a(x)}^{b(x)} \Delta_x F(x, y)$$

$$= \sum_{y=a(x)}^{b(x)} \sum_{z=g(x, y)}^{h(x, y)} \Delta_x f(x, y, z)$$

$$+ \sum_{y=a(x)}^{b(x)} \left[\sum_{z=h(x, y)+\Delta z}^{h(x+\Delta x, y)} \frac{f(x + \Delta x, y, z)}{\Delta x} - \sum_{z=g(x, y)}^{g(x+\Delta x, y)-\Delta z} \frac{f(x + \Delta x, y, z)}{\Delta x} \right] \quad (5.19)$$

Replacing x by $x + \Delta x$ in $F(x, y)$ leads to

$$\sum_{y=b(x)+\Delta y}^{b(x+\Delta x)} \frac{F(x + \Delta x, y)}{\Delta x} = \sum_{y=b(x)+\Delta y}^{b(x+\Delta x)} \sum_{z=g(x+\Delta x, y)}^{h(x+\Delta x, y)} \frac{f(x + \Delta x, y, z)}{\Delta x} \quad (5.20)$$

and

$$\sum_{y=a(x)}^{a(x+\Delta x)-y} \frac{F(x + \Delta x, y)}{\Delta x} = \sum_{y=a(x)}^{a(x+\Delta x)-\Delta y} \sum_{z=g(x+\Delta x, y)}^{h(x+\Delta x, y)} \frac{f(x + \Delta x, y, z)}{\Delta x} \quad (5.21)$$

Replacing each term in Eq. (5.18) by its equivalent in Eqs. (5.19)–(5.21) gives us the desired result. ∎

In Theorem 5.3, if $\Delta x = \Delta y = \Delta z = 1$, then $\Delta C(x)$ is given by

$$\Delta C(x) = \sum_{y=a(x)}^{b(x)} \sum_{z=g(x, y)}^{h(x, y)} \Delta_x f(x, y, z)$$

$$+ \sum_{y=a(x)}^{b(x)} \left[\sum_{z=h(x, y)+1}^{h(x+1, y)} f(x + 1, y, z) - \sum_{z=g(x, y)}^{g(x+1, y)-1} f(x + 1, y, z) \right]$$

$$+ \sum_{y=b(x)+1}^{b(x+1)} \sum_{z=g(x+1, y)}^{h(x+1, y)} f(x + 1, y, z)$$

$$- \sum_{y=a(x)}^{a(x+1)-1} \sum_{z=g(x+1, y)}^{h(x+1, y)} f(x + 1, y, z) \quad (5.22)$$

Example 5.17 Find $\Delta C(x)$ where

$$C(x) = \sum_{y=x}^{x^2} \sum_{z=x+y}^{x+y+n} z$$

where $\Delta x = \Delta y = \Delta z = 1$ and n is a positive integer.
From the equation for $C(x)$

$$a(x) = x, \quad a(x + \Delta x) = x + 1, \quad b(x) = x^2, \quad b(x + \Delta x) = (x + 1)^2$$

$$g(x, y) = x + y, \qquad g(x + \Delta x, y) = x + y + 1$$

$$h(x, y) = x + y + n, \quad h(x + \Delta x, y) = x + y + n + 1$$

Since the conditions of Theorem 5.3 are satisfied

$$\Delta C(x) = \sum_{y=x}^{x^2} \sum_{z=x+y}^{x+y+n} \Delta_x f(x, y, z)$$

$$+ \sum_{y=x}^{x^2} \left[\sum_{z=x+y+n+1}^{x+y+n+1} f(x + 1, y, z) - \sum_{z=x+y}^{x+y} f(x + 1, y, z) \right]$$

$$+ \sum_{y=x^2+1}^{(x+1)^2} \sum_{z=x+y+1}^{x+y+n+1} f(x + 1, y, z) - \sum_{y=x}^{x} \sum_{z=x+y+1}^{x+y+n+1} f(x + 1, y, z)$$

where

$$\Delta_x f(x, y, z) = 0, \quad f(x + 1, y, z) = z$$

Therefore

$$\Delta C(x) = \sum_{y=x}^{x^2} [(x + y + n + 1) - (x + y)]$$

$$+ \sum_{y=x^2+1}^{(x+1)^2} \left[\frac{(x + y + n + 1)(x + y + n + 2)}{2} - \frac{(x + y)(x + y + 1)}{2} \right]$$

$$- \sum_{y=x}^{x} \left[\frac{(x + y + n + 1)(x + y + n + 2)}{2} - \frac{(x + y)(x + y + 1)}{2} \right]$$

where

$$\sum_{w=s}^{t} w = \frac{t(t + 1)}{2} - \frac{(s - 1)s}{2}$$

5.2 THE DIVIDED DIFFERENCE

Evaluating each of these summations yields us

$$\sum_{y=x}^{x^2}[(x+y+n+1)-(x+y)] = \sum_{y=x}^{x^2}(n+1) = (n+1)(x^2-x+1)$$

$$\sum_{y=x^2+1}^{(x+1)^2}\left[\frac{(x+y+n+1)(x+y+n+2)}{2} - \frac{(x+y)(x+y+1)}{2}\right]$$

$$= \sum_{y=x^2+1}^{(x+1)^2}\frac{(n+1)}{2}[2x+2y+n+2]$$

$$= \frac{(n+1)}{2}(2x+1)[2(x+1)^2+(n+2)]$$

$$\sum_{y=x}^{x}\left[\frac{(x+y+n+1)(x+y+n+2)}{2} - \frac{(x+y)(x+y+1)}{2}\right]$$

$$= \sum_{y=x}^{x}\frac{(n+1)}{2}[2x+2y+n+2] = \frac{(n+1)}{2}[4x+n+2]$$

and

$$\Delta C(x) = (n+1)(x^2-x+1) + \frac{(n+1)}{2}(2x+1)[2(x+1)^2+(n+2)]$$

$$- \frac{(n+1)}{2}[4x+n+2]$$

$$= (n+1)[2(x+1)^3 + x(n-3)] \quad \blacksquare$$

Example 5.18 Find $\Delta C(x)$ where

$$C(x) = \sum_{y=a}^{b(x)} \sum_{z=c}^{h(x, y)} f(x, y, z) \tag{5.23}$$

and

$$a < b(x) < b(x+\Delta x), \quad c < h(x, y) < h(x+\Delta x, y)$$

and $\Delta x = \Delta y = \Delta z$.

$$\Delta C(x) = \frac{C(x+\Delta x) - C(x)}{\Delta x}$$

$$= \frac{\sum_{y=a}^{b(x+\Delta x)} \sum_{z=c}^{h(x+\Delta x,\, y)} f(x+\Delta x, y, z) - \sum_{y=a}^{b(x)} \sum_{z=c}^{h(x,\, y)} f(x, y, z)}{\Delta x}$$

The first double summation can be divided into parts as follows.

$$\sum_{y=a}^{b(x+\Delta x)} \sum_{z=c}^{h(x+\Delta x,\, y)} f(x+\Delta x, y, z) = \sum_{y=a}^{b(x)} \sum_{z=c}^{h(x+\Delta x,\, y)} f(x+\Delta x, y, z)$$

$$+ \sum_{y=b(x)+\Delta y}^{b(x+\Delta x)} \sum_{z=c}^{h(x+\Delta x,\, y)} f(x+\Delta x, y, z)$$

$$= \sum_{y=a}^{b(x)} \sum_{z=c}^{h(x,\, y)} f(x+\Delta x, y, z)$$

$$+ \sum_{y=a}^{b(x)} \sum_{z=h(x,\, y)+\Delta z}^{h(x+\Delta x,\, y)} f(x+\Delta x, y, z)$$

$$+ \sum_{y=b(x)+\Delta y}^{b(x+\Delta x)} \sum_{z=c}^{h(x+\Delta x,\, y)} f(x+\Delta x, y, z)$$

Substituting this expression in the relationship for $\Delta C(x)$ leads to

$$\Delta C(x) = \sum_{y=a}^{b(x)} \sum_{z=c}^{h(x,\, y)} \Delta_x f(x, y, z) + \sum_{y=a}^{b(x)} \sum_{z=h(x,\, y)+\Delta z}^{h(x+\Delta x,\, y)} \frac{f(x+\Delta x, y, z)}{\Delta x}$$

$$+ \sum_{y=b(x)+\Delta y}^{b(x+\Delta x)} \sum_{z=c}^{h(x+\Delta x,\, y)} \frac{f(x+\Delta x, y, z)}{\Delta x} \qquad (5.24)$$

Example 5.19 Using the result of Example 5.18 find $\Delta C(x)$ where

$$C(x) = \sum_{y=0}^{x} \sum_{z=0}^{x+y} (x+z)$$

and $\Delta x = \Delta y = \Delta z = 1$.
  From Eq. (5.24)

$$\Delta C(x) = \sum_{y=0}^{x} \sum_{z=0}^{x+y} (1) + \sum_{y=0}^{x} \sum_{z=x+y+1}^{x+y+1} (x+z+1) + \sum_{y=x+1}^{x+1} \sum_{z=0}^{x+y+1} (x+z+1)$$

Treating each double summation separately

$$\sum_{y=0}^{x} \sum_{z=0}^{x+y} (1) = \sum_{y=0}^{x} (x+y+1)$$

$$= (x+1)(x+1) + \frac{x(x+1)}{2} = \frac{x+1}{2}(3x+2)$$

$$\sum_{y=0}^{x} \sum_{z=x+y+1}^{x+y+1} (x+z+1) = \sum_{y=0}^{x} (2x+y+2) = 2(x+1)^2 + \sum_{y=0}^{x} y$$

$$= 2(x+1)^2 + \frac{x(x+1)}{2} = \frac{(x+1)(5x+4)}{2}$$

$$\sum_{y=x+1}^{x+1}\sum_{z=0}^{x+y+1}(x+z+1) = \sum_{y=x+1}^{x+1}(x+1)(x+y+2)$$
$$+ \sum_{y=x+1}^{x+1}\frac{(x+y+1)(x+y+2)}{2}$$
$$= \sum_{y=x+1}^{x+1}(x+y+2)\frac{(3x+y+3)}{2}$$
$$= \frac{(2x+3)(4x+4)}{2} = 2(x+1)(2x+3)$$

and $\Delta C(x)$ is given by

$$\Delta C(x) = \frac{(x+1)(3x+2)}{2} + \frac{(x+1)(5x+4)}{2} + 2(x+1)(2x+3)$$
$$= (x+1)(8x+9)$$

5.3 Optimization of Discrete Functions

Since the operations research analyst must frequently deal with functions of discrete variables, he is faced with the resulting problem of having to find the optimum for such functions. In the case of continuous functions the derivative was used to determine the interior extreme points for the function. When dealing with discrete functions the divided difference can be used for this purpose. To illustrate the problem of concern here, consider the function shown in Fig. 5.1 where $\Delta x = 1$. If we restrict our search for *interior extreme points* for $f(x)$ to permissible values of x on the open interval $(1, 33)$, we find minima at $x = 2, 6, 7, 18, 29$ and maxima at $x = 3, 11, 12, 13, 22, 32$.

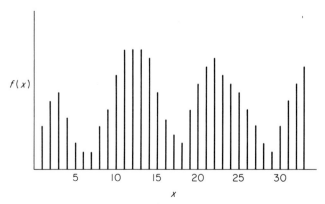

Figure 5.1 Discrete Function with Multiple Extreme Points

In treating continuous functions, a point x^* is considered an interior extreme point for $f(x)$ on an open interval (a, b) if $f(x)$ is convex on (a, b), if x^* belongs to (a, b) and if

$$f(x^*) \leq f(x), \qquad x \in (a, b) \tag{5.25}$$

or

$$f(x^*) \geq f(x), \qquad x \in (a, b) \tag{5.26}$$

By an appropriate modification, a similar definition can be developed for a function of a discrete variable. However, it is first necessary to define a *convex function of a discrete variable*. The reader will recall that a continuous function, $f(x)$, is convex on the open interval (a, b) if

$$f[\lambda x_1 + (1 - \lambda)x_2] \leq \lambda f(x_1) + (1 - \lambda)f(x_2) \tag{5.27}$$

or

$$f[\lambda x_1 + (1 - \lambda)x_2] \geq \lambda f(x_1) + (1 - \lambda)f(x_2) \tag{5.28}$$

where x_1 and x_2 are any two points on (a, b) and $0 \leq \lambda \leq 1$. For continuous functions Eqs. (5.27) and (5.28) must hold for all λ on the interval $[0, 1]$.

When dealing with functions of a discrete variable the definition of a convex function must be slightly altered, since the point $\lambda x_1 + (1 - \lambda)x_2$, on the interval (a, b), is not necessarily a permissible value of x. In fact, by the above definition, x_1 and x_2 may not be permissible values of x. Thus for functions of discrete variables, the definition of a convex function must restrict the points considered to be permissible values of x.

Definition 5.1 Let the open interval (a, b) contain the permissible values of the discrete variable x, $x_0, x_1, \ldots, x_n$, and let λ be a scalar on the interval $[0, 1]$ such that $\lambda x_i + (1 - \lambda)x_j$ is a permissible value of x for $i, j = 0, 1, \ldots, n$. Then $f(x)$ is *convex from above* on (a, b) if

$$f[\lambda x_i + (1 - \lambda)x_j] \leq \lambda f(x_i) + (1 - \lambda)f(x_j) \tag{5.29}$$

and *convex from below* on (a, b) if

$$f[\lambda x_i + (1 - \lambda)x_j] \geq \lambda f(x_i) + (1 - \lambda)f(x_j) \tag{5.30}$$

If $\leq$ is replaced by $<$ and $\geq$ is replaced by $>$, then $f(x)$ is *strictly convex* on (a, b).

In general, to make meaningful statements about the convexity of a function of a discrete variable the interval (a, b) should contain four or more values of x, since every such function is convex on an open interval containing only three points.

5.3 OPTIMIZATION OF DISCRETE FUNCTIONS

Definition 5.2 If $f(x)$ is convex on the open interval (a, b) containing the permissible values of x, $x_0, x_1, \ldots, x_n$, and x_i is an *interior extreme point* for $f(x)$ on (a, b), then either

$$f(x_{i-1}) \geq f(x_i) \leq f(x_{i+1}) \tag{5.31}$$

or

$$f(x_{i-1}) \leq f(x_i) \geq f(x_{i+1}) \tag{5.32}$$

where $i = 1, 2, \ldots, n-1$.

The reader should notice that the points x_0 and x_n cannot be interior extreme points according to Definition 5.2, although either or both will be extreme points since $f(x)$ is assumed to be convex on (a, b). Definition 5.2 leads to the necessary conditions for an interior extreme point for a function of a discrete variable given in the following theorem.

Theorem 5.4 Let $f(x)$ be a function of the discrete variable x on the open interval (a, b). If x_i is an interior extreme point for $f(x)$, then either

$$\Delta f(x_{i-1}) \leq 0 \leq \Delta f(x_i) \tag{5.33}$$

or

$$\Delta f(x_{i-1}) \geq 0 \geq \Delta f(x_i) \tag{5.34}$$

Proof If x_i is a local minimum for $f(x)$ on (a, b), then there is an open interval (a', b') about x_i for which $f(x)$ is convex from above. Let $x_{i-k}, x_{i-k+1}, \ldots, x_{i+m}$ be the values of x on the interval (a', b'). If x_i is a unique minimum on (a', b') then

$$f(x_{i-1}) > f(x_i) < f(x_{i+1})$$

and

$$f(x_i) - f(x_{i-1}) < 0 \tag{5.35}$$
$$f(x_{i+1}) - f(x_i) > 0 \tag{5.36}$$

Dividing Eq. (5.35) by $x_i - x_{i-1}$ and Eq. (5.36) by $x_{i+1} - x_i$ yields

$$\frac{f(x_i) - f(x_{i-1})}{x_i - x_{i-1}} < 0 < \frac{f(x_{i+1}) - f(x_i)}{x_{i+1} - x_i}$$

or

$$\Delta f(x_{i-1}) < 0 < \Delta f(x_i) \tag{5.37}$$

If x_i is not a unique minimum on (a', b') then one of three cases arise.

$$f(x_{i-1}) = f(x_i) < f(x_{i+1}) \tag{5.38}$$
$$f(x_{i-1}) > f(x_i) = f(x_{i+1}) \tag{5.39}$$
$$f(x_{i-1}) = f(x_i) = f(x_{i+1}) \tag{5.40}$$

From Eq. (5.38)

$$\frac{f(x_i) - f(x_{i-1})}{x_i - x_{i-1}} = 0 < \frac{f(x_{i+1}) - f(x_i)}{x_{i+1} - x_i} \qquad (5.41)$$

From Eq. (5.39)

$$\frac{f(x_i) - f(x_{i-1})}{x_i - x_{i-1}} < 0 = \frac{f(x_{i+1}) - f(x_i)}{x_{i+1} - x_i} \qquad (5.42)$$

From Eq. (5.40)

$$\frac{f(x_i) - f(x_{i-1})}{x_i - x_{i-1}} = 0 = \frac{f(x_{i+1}) - f(x_i)}{x_{i+1} - x_i} \qquad (5.43)$$

Combining the conditions given in Eqs. (5.37), (5.41), (5.42), and (5.43), if x_i is a local minimum for $f(x)$ on (a, b) then

$$\Delta f(x_{i-1}) \leq 0 \leq \Delta f(x_i)$$

The argument when x_i is a local maximum is identical to that just presented but is left as an exercise for the reader.

Since the conditions given in Eqs. (5.33) and (5.34) are necessary for the existence of an extreme point at x_i, a point satisfying one of these conditions may or may not be an extreme point and therefore is considered only a possible extreme point. For continuous functions of one variable a point satisfying the necessary conditions for an extreme point but which was not an extreme point was called a point of inflection. Examples of points of inflection for functions of discrete variables are shown in Figs. 5.2–5.4. In each case consider the point x_i on the interval (x_{i-6}, x_{i+6}). In all three cases x_i satisfies the conditions for an extreme point. However, with reference to the interval (x_{i-6}, x_{i+6}), x_i is neither a maximum nor a minimum for $f(x)$, $g(x)$, or $h(x)$. If we consider the interval

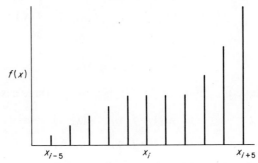

Figure 5.2 Point of Inflection, x_i, for a Discrete Function

5.3 OPTIMIZATION OF DISCRETE FUNCTIONS

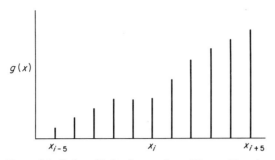

Figure 5.3 Point of Inflection, x_i, for a Discrete Function

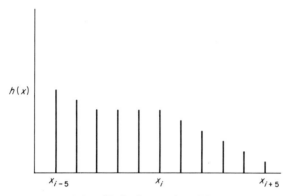

Figure 5.4 Point of Inflection, x_i, for a Discrete Function

(x_{i-2}, x_{i+2}), x_i is both a minimum and maximum for $f(x)$, a minimum for $g(x)$, and a maximum for $h(x)$. Thus, whether the point x_i is a *point of inflection depends upon the interval considered.*

Definition 5.3 Let $f(x)$ be a function of the discrete variable x, defined on the open interval (a, b) which includes the values of x, $x_0, x_1, \ldots, x_n$. If $x_j, \ldots, x_k$ are points satisfying the necessary conditions for an extreme point, then $x_j, \ldots, x_k$, $0 < j < k < n$ are points of inflection if either

$$f(x_{j-1}) < f(x_j) = f(x_{j+1}) = \cdots = f(x_i) = f(x_{i+1}) = \cdots$$
$$= f(x_k) < f(x_{k+1}) \quad (5.44)$$

or

$$f(x_{j-1}) > f(x_j) = f(x_{j+1}) = \cdots = f(x_i) = f(x_{i+1}) = \cdots$$
$$= f(x_k) > f(x_{k+1}) \quad (5.45)$$

Example 5.20 Find the possible interior extreme points for $f(x)$ on $[0, \infty)$, where

$$f(x) = \frac{a^x}{x!} \exp[-a], \quad x = 0, 1, 2, \ldots$$

and $a > 0$.

If x^* is an interior extreme point for $f(x)$ then either
$$\Delta f(x^* - 1) \leq 0 \leq \Delta f(x^*) \quad \text{or} \quad \Delta f(x^* - 1) \geq 0 \geq \Delta f(x^*)$$
where $x_i - x_{i-1} = 1$ for all $i > 0$.
$$\Delta f(x) = \frac{a^{x+1}}{(x+1)!} \exp[-a] - \frac{a^x}{x!} \exp[-a] = \frac{a^x}{x!} \exp[-a] \left[\frac{a}{x+1} - 1\right]$$
First we will attempt to find interior points x^* such that
$$\Delta f(x^* - 1) \leq 0 \leq \Delta f(x^*)$$
where
$$\Delta f(x^* - 1) = \frac{a^{x^*-1}}{(x^* - 1)!} \exp[-a] \left[\frac{a}{x^*} - 1\right]$$
$$\Delta f(x^*) = \frac{a^{x^*}}{x^*!} \exp[-a] \left[\frac{a}{(x^* + 1)} - 1\right]$$
Therefore x^* must satisfy the relationship
$$\frac{a^{x^*-1}}{(x^* - 1)!} \exp[-a] \left[\frac{a}{x^*} - 1\right] \leq 0 \leq \frac{a^{x^*}}{x^*!} \exp[-a] \left[\frac{a}{(x^* + 1)} - 1\right]$$
or
$$\left[\frac{a}{x^*} - 1\right] \leq 0 \leq \frac{a}{x^*} \left[\frac{a}{(x^* + 1)} - 1\right]$$
The inequality
$$\frac{a}{x^*} - 1 \leq 0$$
implies $x^* \geq a$ while the inequality
$$\frac{a}{x^*} \left[\frac{a}{x^* + 1} - 1\right] \geq 0$$
implies $x^* + 1 \leq a$, and $x^* + 1 \leq a \leq x^*$ which is a contradiction. Therefore, there is no interior extreme point satisfying the required relationship for a minimum for $f(x)$ on $[0, \infty)$. We now examine $f(x)$ to determine the interior extreme points for $f(x)$ on $[0, \infty)$ satisfying the relationship
$$\Delta f(x^* - 1) \geq 0 \geq \Delta f(x^*)$$
The following relationship must hold
$$\frac{a^{x^*-1}}{(x^* - 1)!} \exp[-a] \left[\frac{a}{x^*} - 1\right] \geq 0 \geq \frac{a^{x^*}}{x^*!} \exp[-a] \left[\frac{a}{(x^* + 1)} - 1\right]$$
or
$$\left[\frac{a}{x^*} - 1\right] \geq 0 \geq \frac{a}{x^*} \left[\frac{a}{(x^* + 1)} - 1\right]$$

5.3 OPTIMIZATION OF DISCRETE FUNCTIONS

The inequality

$$\left[\frac{a}{x^*} - 1\right] \geq 0$$

implies $x^* \leq a$ and

$$\frac{a}{x^*}\left[\frac{a}{(x^*+1)} - 1\right] \leq 0$$

implies $x^* + 1 \geq a$. Therefore a possible interior extreme point for $f(x)$ is the integer x^* such that

$$x^* \leq a \leq x^* + 1$$

Example 5.21 Find the possible interior extreme points for $f(x)$ on the open interval $(-\infty, \infty)$ where

$$f(x) = 2x^3 - 6x^2 + 1, \quad x = 0, \pm 1, \pm 2, \ldots$$

We first examine $f(x)$ for interior extreme points, x^*, such that

$$\Delta f(x^* - 1) \leq 0 \leq \Delta f(x^*), \quad \Delta f(x) = 6x^2 - 6x - 4$$

Therefore x^* must satisfy

$$6(x^* - 1)^2 - 6(x^* - 1) - 4 \leq 0 \leq 6(x^*)^2 - 6x^* - 4$$

The only integer value of x^* satisfying this relationship is $x^* = 2$. Thus $x^* = 2$ may be an interior extreme point for $f(x)$. If x^* is an interior extreme point satisfying

$$\Delta f(x^* - 1) \geq 0 \geq \Delta f(x^*)$$

then

$$6(x^* - 1)^2 - 6(x^* - 1) - 4 \geq 0 \geq 6(x^*)^2 - 6x^* - 4$$

Simple trial and error will show that $x^* = 0$ is the only integer value of x^* satisfying this relationship and therefore may be an interior extreme point for $f(x)$ on $(-\infty, \infty)$.

Example 5.22 Show that $x = 2$ and $x = 3$ are possible extreme points for $f(x)$ where

$$f(x) = \frac{5!}{x!(5-x)!}(0.5)^5, \quad x = 0, 1, 2, 3, 4, 5$$

$$\Delta f(x) = \frac{5!}{(x+1)!(5-x-1)!}(0.5)^5 - \frac{5!}{x!(5-x)!}(0.5)^5$$

$$= \frac{5!}{x!(5-x)!}(0.5)^5 \left[\frac{5-x}{x+1} - 1\right]$$

If x^* is an extreme point for $f(x)$ then either

$$\Delta f(x^* - 1) \leq 0 \leq \Delta f(x^*) \quad \text{or} \quad \Delta f(x^* - 1) \geq 0 \geq \Delta f(x^*)$$

where

$$\Delta f(x^* - 1) = \frac{5!}{(x^* - 1)!(5 - x^* + 1)!}(0.5)^5\left[\frac{5 - x^* + 1}{x^*} - 1\right]$$

If x^* satisfies the first condition

$$\frac{5!}{(x^* - 1)!(5 - x^* + 1)!}(0.5)^5\left[\frac{5 - x^* + 1}{x^*} - 1\right] \leq 0 \leq \frac{5!}{x^*!(5 - x^*)!}(0.5)^5\left[\frac{5 - x^*}{x^* + 1} - 1\right]$$

or

$$\frac{x^*}{6 - x^*}\left[\frac{6 - x^*}{x^*} - 1\right] \leq 0 \leq \left[\frac{5 - x^*}{x^* + 1} - 1\right]$$

The left-hand side of the inequality implies $x^* \geq 3$ while the right side implies $x^* \leq 2$. Therefore there is no value of x satisfying the first of the necessary conditions for an extreme point. For the second condition

$$\frac{x^*}{6 - x^*}\left[\frac{6 - x^*}{x^*} - 1\right] \geq 0 \geq \left[\frac{5 - x^*}{x^* + 1} - 1\right]$$

and $2 \leq x^* \leq 3$. Therefore $x^* = 2, 3$ may be extreme points for $f(x)$.

Example 5.23 Show that $x = 0$ and $x = 1$ are points of inflection for the following function.

$$f(x) = 3x^3 - 4x^2 + x, \quad x = 0, \pm 1, \pm 2, \ldots$$

Since $\Delta x = 1$,

$$\Delta f(x) = 3(x + 1)^3 - 4(x + 1)^2 + (x + 1) - 3x^3 + 4x^2 - x = x(9x + 1)$$
$$\Delta f(-1) = 8, \quad \Delta f(0) = 0, \quad \Delta f(1) = 10$$

Since

$$\Delta f(-1) > 0 = \Delta f(0) \quad \text{and} \quad \Delta f(0) = 0 < \Delta f(1)$$

$x = 0$ and $x = 1$ are possible interior extreme points for $f(x)$. To show that these points are points of inflection note that

$$f(-1) = -8, \quad f(0) = 0, \quad f(1) = 0, \quad f(2) = 10$$

Therefore, the relationship given in Eq. (5.44) is satisfied and both points are points of inflection.

Thus far we have discussed only the necessary conditions for the existence of an interior extreme point. To ascertain the nature of an extreme point, *sufficient*

5.3 OPTIMIZATION OF DISCRETE FUNCTIONS

conditions for the existence of a local minimum and maximum are needed. Suppose that x_i is an extreme point such that

$$f(x_{i-1}) > f(x_i) < f(x_{i+1}) \tag{5.46}$$

Obviously x_i is a *local minimum* and

$$\Delta f(x_{i-1}) < 0, \qquad \Delta f(x_i) > 0$$

Therefore,

$$\Delta f(x_{i-1}) < 0 < \Delta f(x_i) \tag{5.47}$$

is a *sufficient condition for the existence of a local minimum* at x_i. Similarly if

$$f(x_{i-1}) < f(x_i) > f(x_{i+1})$$

x_i is a *local maximum* for $f(x)$ and

$$\Delta f(x_{i-1}) > 0 > \Delta f(x_i) \tag{5.48}$$

is a *sufficient condition for the existence of a local maximum* at x_i.

To illustrate the inadequacy of Eqs. (5.46) and (5.48) in determining the nature of a discrete extreme point, consider the function shown in Fig. 5.5. On the interval (x_{i-5}, x_{i+5}), x_i is a local minimum for $f(x)$. But

$$\Delta f(x_{i-1}) = 0, \qquad \Delta f(x_i) = 0$$

and the sufficient conditions given in Eq. (5.47) would not indicate that x_i is a local minimum. The following theorem presents a generalization of the sufficient conditions for the existence of a local maximum and minimum.

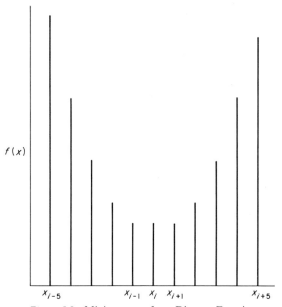

Figure 5.5 Minimum, x_i, for a Discrete Function

Theorem 5.5 Let $f(x)$ be a function of the discrete variable x on the open interval (a, b). If x_i is an interior point for the interval (a, b) then

1. if $\Delta f(x_{i-1}) < 0 < \Delta f(x_i)$, x_i is a local minimum.
2. if $\Delta f(x_{i-1}) > 0 > \Delta f(x_i)$, x_i is a local maximum.
3. if $x_{i-j}, x_{i-j+1}, \ldots, x_{i+k-1}, x_{i+k}$ belong to the interval (a, b) and

$$\Delta f(x_m) = 0, \quad m = i - j + 1, i - j + 2, \ldots, i + k - 2$$

and

$$\Delta f(x_{i-j}) < 0 < \Delta f(x_{i+k-1})$$

then x_i is a local minimum.

4. if $x_{i-j}, x_{i-j+1}, \ldots, x_{i+k-1}, x_{i+k}$ belong to the interval (a, b) and

$$\Delta f(x_m) = 0, \quad m = i - j + 1, i - j + 2, \ldots, i + k - 2$$

and

$$\Delta f(x_{i-j}) > 0 > \Delta f(x_{i+k-1})$$

then x_i is a local maximum.

Proof To prove this theorem it is first noted that in each of the four cases the necessary conditions for the existence of an extreme point at x_i are met.
For case 1,

$$\Delta f(x_{i-1}) < 0 < \Delta f(x_i) \quad \text{or} \quad \frac{f(x_i) - f(x_{i-1})}{x_i - x_{i-1}} < 0 < \frac{f(x_{i+1}) - f(x_i)}{x_{i+1} - x_i}$$

Since $x_{i-1} < x_i < x_{i+1}$,

$$f(x_i) - f(x_{i-1}) < 0 < f(x_{i+1}) - f(x_i)$$

Therefore

$$f(x_i) < f(x_{i-1}), \quad f(x_{i+1}) > f(x_i)$$

or

$$f(x_{i-1}) > f(x_i) < f(x_{i+1})$$

and x_i is a local minimum. The same argument may be used to prove case 2, but is left as an exercise for the reader.

For case 3,

$$\Delta f(x_{i-j}) < 0 = \Delta f(x_{i-j+1}) = \cdots = \Delta f(x_{i+k-2}) = 0 < \Delta f(x_{i+k-1})$$

Therefore

$$\frac{f(x_{m+1}) - f(x_m)}{x_{m+1} - x_m} = 0, \quad m = i - j + 1, i - j + 2, \ldots, i + k - 2$$

and

$$f(x_m) = f(x_{m+1}), \quad m = i - j + 1, i - j + 2, \ldots, i + k - 2$$

5.3 OPTIMIZATION OF DISCRETE FUNCTIONS

Further

$$\frac{f(x_{i-j+1}) - f(x_{i-j})}{x_{i-j+1} - x_{i-j}} < 0 \quad \text{or} \quad f(x_{i-j+1}) < f(x_{i-j})$$

Finally

$$\frac{f(x_{i+k}) - f(x_{i+k-1})}{x_{i+k} - x_{i+k-1}} > 0 \quad \text{or} \quad f(x_{i+k}) > f(x_{i+k-1})$$

Combining the above relationships yields

$$f(x_{i-j}) > f(x_{i-j+1}) = f(x_{i-j+2}) = \cdots = f(x_i) = \cdots = f(x_{i+k-1}) < f(x_{i+k})$$

and x_i is a local minimum for $f(x)$. The proof of case 4 is left as an exercise for the reader.

Example 5.24 Identify the nature of the possible extreme points for the function given in Example 5.21.

From Example 5.21,

$$f(x) = 2x^3 - 6x^2 + 1, \quad x = 0, \pm 1, \pm 2, \ldots$$

Then $x^* = 2$ satisfies

$$\Delta f(x^* - 1) < 0 < \Delta f(x^*)$$

since

$$\Delta f(1) = -4, \quad \Delta f(2) = 8$$

Therefore $x^* = 2$ is a local minimum for $f(x)$. The value of x satisfying

$$\Delta f(x^* - 1) \geq 0 \geq \Delta f(x^*)$$

is $x^* = 0$ for which

$$\Delta f(-1) = 8, \quad \Delta f(0) = -4$$

Thus

$$\Delta f(x^* - 1) > 0 > \Delta f(x^*)$$

and $x^* = 0$ is a local maximum for $f(x)$.

Example 5.25 Find and identify the interior extreme points for the following function on the interval $(-\infty, \infty)$.

$$f(x) = \tfrac{1}{4}x^4 - 2x^2, \quad x = 0, \pm 1, \pm 2, \ldots$$

Since $\Delta x = 1$,

$$\Delta f(x) = \frac{1}{4}(x+1)^4 - 2(x+1)^2 - \frac{1}{4}x^4 + 2x^2 = \frac{2x+1}{4}[2x(x+1) - 7]$$

and
$$\Delta f(x-1) = \frac{2x-1}{4}[2(x-1)x - 7]$$

We first attempt to find values of x^* such that
$$\Delta f(x^* - 1) \leq 0 \leq \Delta f(x^*)$$
or
$$\frac{2x^* - 1}{4}[2(x^* - 1)x^* - 7] \leq 0 \leq \frac{2x^* + 1}{4}[2x^*(x^* + 1) - 7]$$

The only integer values of x^* satisfying this relationship are $x^* = \pm 2$.
$$\Delta f(2-1) = -\tfrac{9}{4}, \qquad \Delta f(2) = \tfrac{25}{4}, \qquad \text{and} \qquad \Delta f(1) < 0 < \Delta f(2),$$

Therefore $x^* = 2$ is a local minimum for $f(x)$. For $x^* = -2$,
$$\Delta f(-2-1) = -\tfrac{25}{4}, \qquad \Delta f(-2) = \tfrac{9}{4}, \qquad \text{and} \qquad \Delta f(-3) < 0 < \Delta f(-2)$$

and $x^* = -2$ is a local minimum for $f(x)$. We now attempt to find values of x^* such that
$$\Delta f(x^* - 1) \geq 0 \geq \Delta f(x^*)$$
or
$$\frac{2x^* - 1}{4}[2(x^* - 1)x^* - 7] \geq 0 \geq \frac{2x^* + 1}{4}[2x^*(x^* + 1) - 7]$$

The only such value is $x^* = 0$.
$$\Delta f(-1) = \tfrac{9}{4}, \qquad \Delta f(0) = -\tfrac{7}{4}, \qquad \text{and} \qquad \Delta f(-1) > 0 > \Delta f(0)$$

Therefore $x^* = 0$ is a local maximum for $f(x)$.

5.4 The Antidifference and Summation of Series

The operational inverse of the difference is called the *antidifference* and is the discrete analog of the *antiderivative*. If $Sf(x)$ is the antidifference of $f(x)$ then
$$\Delta[Sf(x)] = f(x) \tag{5.49}$$
where S is taken as the antidifference operator. Determination of the antidifference is not as straightforward as finding the difference. However, determination of $Sf(x)$ can sometimes be facilitated if $\int f(x)\,dx$ is known where x is treated as a continuous variable. The following example illustrates this approach.

Example 5.26 Find the antidifference of $f(x)$ where
$$f(x) = x, \qquad x = 0, 1, 2, \ldots$$

5.4 THE ANTIDIFFERENCE AND SUMMATION OF SERIES

Since x is integer valued, $\Delta x = 1$. To determine $Sf(x)$ it is necessary to find a function $F(x)$ such that

$$\Delta F(x) = f(x)$$

From calculus,

$$\int x\, dx = \frac{x^2}{2} + C_1$$

where C_1 is a constant.
Therefore, let us try

$$F(x) = \frac{x^2}{2} + C_1$$

Since

$$\Delta F(x) = \frac{(x+1)^2 - x^2}{2} = \frac{2x+1}{2} = x + \frac{1}{2}$$

$F(x)$ is not the desired function. As given above

$$\Delta F(x) = f(x) + \tfrac{1}{2}$$

Let $G(x)$ be such that

$$\Delta G(x) = \tfrac{1}{2}$$

Then

$$\Delta[F(x) - G(x)] = f(x)$$

If

$$G(x) = \frac{x}{2} + C_2 \quad \text{then} \quad \Delta G(x) = \frac{(x-1) - x}{2} = \frac{1}{2}$$

Therefore, let us redefine $F(x)$ as

$$F(x) = \frac{x^2 - x}{2} + C_3 = \frac{x(x-1)}{2} + C_3 \quad \text{and} \quad \Delta F(x) = x$$

The antidifference is then

$$Sf(x) = \frac{x(x-1)}{2} + C_3$$

Example 5.27 If

$$f(x) = x^2, \quad x = 0, 1, 2, \ldots$$

find $Sf(x)$.

From calculus,
$$\int x^2\, dx = \frac{x^3}{3} + C_1$$

Therefore, let
$$F(x) = \frac{x^3}{3} + C_1, \qquad \Delta F(x) = \frac{1}{3}[(x+1)^3 - x^3] = x^2 + x + \frac{1}{3}$$

and $F(x)$ is not the proper function. Define $G(x)$ and $H(x)$ such that
$$\Delta G(x) = x, \qquad \Delta H(x) = \tfrac{1}{3}$$

From Example 5.26,
$$G(x) = \frac{x(x-1)}{2} + C_2 \quad \text{and} \quad H(x) = \frac{x}{3} + C_3$$

Then $F(x) - G(x) - H(x)$ should be the antidifference of $f(x)$.
$$F(x) - G(x) - H(x) = \frac{x^3}{3} - \frac{x(x-1)}{2} - \frac{x}{3} + C_4$$

Let $F(x)$ be redefined as this function.
$$F(x) = \frac{2x^3 - 3x(x-1) - 2x}{6} + C_4 = \frac{x(x-1)(2x-1)}{6} + C_4$$

Checking this result gives us
$$\Delta F(x) = \frac{(x+1)x(2x+1) - x(x-1)(2x-1)}{6} = x^2$$

and
$$Sf(x) = \frac{x(x-1)(2x-1)}{6} + C_4$$

Example 5.28 Find the antidifference of $f(x)$ where
$$f(x) = \exp[x], \qquad x = 0, 1, 2, \ldots$$

From calculus,
$$\int \exp[x]\, dx = \exp[x] + C_1$$

Let
$$F(x) = \exp[x] + C_1, \qquad \Delta F(x) = \exp[x](e - 1)$$

Therefore
$$\Delta F(x) = (e-1)f(x)$$

5.4 THE ANTIDIFFERENCE AND SUMMATION OF SERIES

Redefine $F(x)$ as

$$F(x) = \frac{\exp[x]}{e-1} + C_2$$

Then

$$\Delta F(x) = \exp[x] \quad \text{and} \quad Sf(x) = \frac{\exp[x]}{e-1} + C_2 \quad \text{◻}$$

The approach illustrated in the above examples will sometimes meet with failure. For this reason an antidifference table is provided in Table 2, Appendix.

The antidifference has an important application in the summation of series. Consider the sum

$$\sum_{x=a}^{b} f(x) \tag{5.50}$$

If $Sf(x)$ is the antidifference of $f(x)$ then

$$\sum_{x=a}^{b} f(x) = \sum_{x=a}^{b} \Delta Sf(x) \tag{5.51}$$

In attempting to evaluate this sum and in the discussion of summation of series in general, it will be assumed that x is integer valued. ◻

Theorem 5.6 Summation of Series If $f(x)$ is a function of the integer valued variable x, then

$$\sum_{x=a}^{b} f(x) = Sf(b+1) - Sf(a) \tag{5.52}$$

where $a \leq b$.

Proof To prove this theorem we note that

$$\Delta Sf(x) = Sf(x+1) - Sf(x) \tag{5.53}$$

Then by Eq. (5.51)

$$\sum_{x=a}^{b} f(x) = \sum_{x=a}^{b} \Delta Sf(x) = \sum_{x=a}^{b} [Sf(x+1) - Sf(x)]$$

$$= \sum_{x=a+1}^{b+1} Sf(x) - \sum_{x=a}^{b} Sf(x)$$

$$= Sf(b+1) + \sum_{x=a+1}^{b} Sf(x) - \sum_{x=a+1}^{b} Sf(x) - Sf(a) \tag{5.54}$$

Therefore

$$\sum_{x=a}^{b} f(x) = Sf(b+\cdot 1) - Sf(a)$$

The result of Theorem 5.6 implies that the sum of a series may be evaluated if the antidifference of function summed is known. A similar problem arises in calculus. If $f(x)$ is continuous on the interval (a, b), then

$$\int_{a}^{b} f(x)\, dx = F(b) - F(a) \tag{5.55}$$

where $F(x)$ is the antiderivative of $f(x)$. Therefore, evaluation of the integral in Eq. (5.55) can be achieved if the antiderivative of $f(x)$ is known. If the antidifference in Eq. (5.52) or the antiderivative in Eq. (5.55) are not known, then it is necessary to resort to other means in carrying out the required operations.

When the antiderivative is unknown, the integral in Eq. (5.55) may sometimes be resolved through integration by parts. Specifically let

$$u\, dv = f(x)\, dx \tag{5.56}$$

Then

$$\int_{a}^{b} f(x)\, dx = \int_{a}^{b} u\, dv = uv \bigg|_{a}^{b} - \int_{a}^{b} v\, du \tag{5.57}$$

A similar result is given in the following theorem where the operation to be performed is summation instead of integration.

Theorem 5.7 Summation By Parts   Let $f(x)$, $h(x)$, and $g(x)$ be functions of the integer valued variable x such that

$$f(x) = h(x)\, \Delta g(x) \tag{5.58}$$

Then

$$\sum_{x=a}^{b} f(x) = h(b+1)g(b+1) - h(a)g(a) - \sum_{x=a}^{b} g(x+1)\, \Delta h(x) \tag{5.59}$$

where $a \leq b$.

Proof   From Eq. (5.58)

$$\sum_{x=a}^{b} f(x) = \sum_{x=a}^{b} h(x)\, \Delta g(x) = \sum_{x=a}^{b} h(x)[g(x+1) - g(x)]$$

5.4 THE ANTIDIFFERENCE AND SUMMATION OF SERIES

To the right-hand side of this equation add and subtract $h(b + 1)g(b + 1)$.

$$\sum_{x=a}^{b} f(x) = h(b + 1)g(b + 1) - h(a)g(a) - h(b + 1)g(b + 1)$$

$$+ \sum_{x=a}^{b} h(x)g(x + 1) - \sum_{x=a+1}^{b} h(x)g(x)$$

$$= h(b + 1)g(b + 1) - h(a)g(a) + \sum_{x=a}^{b} h(x)g(x + 1)$$

$$- \sum_{x=a}^{b} h(x + 1)g(x + 1)$$

$$= h(b + 1)g(b + 1) - h(a)g(a) - \sum_{x=a}^{b} g(x + 1)[h(x + 1) - h(x)]$$

Therefore

$$\sum_{x=a}^{b} f(x) = h(b + 1)g(b + 1) - h(a)g(a) - \sum_{x=a}^{b} g(x + 1)\, \Delta h(x) \quad \text{⑤}$$

Example 5.29 Evaluate the following sums.

a. $\sum_{x=0}^{b} x^2$

b. $\sum_{x=a}^{b} \exp[x]$

where x is integer valued.

For part a,

$$S(x^2) = \tfrac{1}{6}x(x - 1)(2x - 1) + C_1$$

from Example 5.27. Using Eq. (5.52), we find

$$\sum_{x=0}^{b} x^2 = \tfrac{1}{6}[(b + 1)b(2b + 1) - (0)(-1)(-1)] = \tfrac{1}{6}(b + 1)b(2b + 1)$$

Using the result of Example 5.28 for part b, we have

$$S(\exp[x]) = \frac{\exp[x]}{e - 1} + C_1$$

Therefore,

$$\sum_{x=a}^{b} \exp[x] = \frac{1}{e - 1}[\exp[b + 1] - \exp[a]] \quad \text{⑤}$$

Example 5.30 Evaluate

$$\sum_{x=a}^{b} f(x), \quad \text{where} \quad 0 < a \le b$$

a. $f(x) = \dfrac{1}{x(x+1)}$

b. $f(x) = xx!$

c. $f(x) = \dfrac{x}{(x+1)!}$

For all three parts of this problem, the antidifference table in the Appendix will be used. For part a,

$$S\left[\frac{1}{x(x+1)}\right] = \frac{x-1}{x}$$

Therefore

$$\sum_{x=a}^{b} \frac{1}{x(x+1)} = \frac{b}{b+1} - \frac{a-1}{a} = \frac{b-a+1}{(b+1)a}$$

For part b,

$$S(xx!) = x! \quad \text{and} \quad \sum_{x=a}^{b} xx! = (b+1)! - a!$$

For part c,

$$S\left[\frac{x}{(x+1)!}\right] = -\frac{1}{x!}$$

and

$$\sum_{x=a}^{b} \frac{x}{(x+1)!} = -\frac{1}{(b+1)!} + \frac{1}{a!} = \frac{(b+1)! - a!}{a!(b+1)!}$$

Example 5.31 Evaluate the following summations by parts.

a. $\sum_{x=0}^{n} xa^x$

b. $\sum_{x=0}^{n} \dfrac{x(x-1)^2(x-1)!}{2}$

In part a, let

$$h(x) = x \qquad \Delta g(x) = a^x$$

To evaluate $g(x)$ we note that

$$g(x) = S[\Delta g(x)] = S[a^x] = \frac{a^x - 1}{a - 1}$$

5.4 THE ANTIDIFFERENCE AND SUMMATION OF SERIES

from the antidifference table. Using Eq. (5.59)

$$\sum_{x=0}^{n} xa^x = (n+1)\frac{a^{n+1}-1}{a-1} - (0)\frac{a^0-1}{a-1} - \sum_{x=0}^{n}\frac{a^{x+1}-1}{a-1}\Delta(x)$$

$$= (n+1)\frac{a^{n+1}-1}{a-1} - \sum_{x=0}^{n}\frac{a^{x+1}-1}{a-1}$$

since $\Delta(x) = 1$,

$$\sum_{x=0}^{n}\frac{a^{x+1}-1}{a-1} = \sum_{x=0}^{n}\frac{a^{x+1}}{a-1} - \sum_{x=0}^{n}\frac{1}{a-1} = \frac{a}{a-1}\sum_{x=0}^{n}a^x - \frac{1}{a-1}\sum_{x=0}^{n}(1)$$

$$S(a^x) = \frac{a^x - 1}{a - 1} \qquad S(1) = x$$

Therefore,

$$\sum_{x=0}^{n}\frac{a^{x+1}-1}{a-1} = \frac{1}{a-1}\left[\frac{a^{n+2}-a}{a-1} - n - 1\right]$$

and

$$\sum_{x=0}^{n} xa^x = \frac{1}{(a-1)^2}[(n+1)(a^{n+2} - a - a^{n+1} + 1) - a^{n+2} + a + (n+1)(a-1)]$$

$$= \frac{a}{(a-1)^2}[(n+1)(a-1)a^n - a^{n+1} + 1]$$

In part b, let

$$h(x) = \frac{x(x-1)}{2} \qquad \Delta g(x) = (x-1)(x-1)!$$

From the antidifference table,

$$g(x) = (x-1)! \qquad \text{and} \qquad \Delta h(x) = x$$

Therefore

$$\sum_{x=0}^{n}\frac{x(x-1)^2(x-1)!}{2} = \frac{n(n+1)}{2}n! - (0) - \sum_{x=0}^{n}xx! = \frac{n(n+1)!}{2} - \sum_{x=0}^{n}xx!$$

From the antidifference table,

$$S(xx!) = x!$$

and

$$\sum_{x=0}^{n}\frac{x(x-1)^2(x-1)!}{2} = \frac{n(n+1)!}{2} - (n+1)! + (1)$$

$$= \frac{(n-2)(n+1)!}{2} + 1$$

5.5 Higher-Order Differences

In calculus the second derivative can be expressed as the first derivative of the first derivative. That is

$$\frac{d^2}{dx^2} f(x) = \frac{d}{dx}\left[\frac{d}{dx} f(x)\right] \qquad (5.60)$$

For functions of discrete variables the *second divided difference* at x_i is defined as

$$\Delta^2 f(x_i) = \frac{1}{x_{i+2} - x_i} [\Delta f(x_{i+1}) - \Delta f(x_i)] \qquad (5.61)$$

In general the *n*th *divided difference*[†] at x_i may be expressed by

$$\Delta^n f(x_i) = \frac{1}{x_{i+n} - x_i} [\Delta^{n-1} f(x_{i+1}) - \Delta^{n-1} f(x_i)] \qquad (5.62)$$

Expanding the second and third divided differences at x_i using Eq. (5.62) leads to

$$\Delta^2 f(x_i) = \frac{1}{x_{i+2} - x_i} [\Delta f(x_{i+1}) - \Delta f(x_i)]$$

$$= \frac{1}{x_{i+2} - x_i} \left[\frac{f(x_{i+2}) - f(x_{i+1})}{x_{i+2} - x_{i+1}} - \frac{f(x_{i+1}) - f(x_i)}{x_{i+1} - x_i} \right] \qquad (5.63)$$

and

$$\Delta^3 f(x_i) = \frac{1}{x_{i+3} - x_i} [\Delta^2 f(x_{i+1}) - \Delta^2 f(x_i)]$$

$$= \frac{1}{x_{i+3} - x_i} \left[\frac{\Delta f(x_{i+2}) - \Delta f(x_{i+1})}{x_{i+3} - x_{i+1}} - \frac{\Delta f(x_{i+1}) - \Delta f(x_i)}{x_{i+2} - x_i} \right]$$

$$= \frac{1}{x_{i+3} - x_i} \left\{ \frac{1}{x_{i+3} - x_{i+1}} \left[\frac{f(x_{i+3}) - f(x_{i+2})}{x_{i+3} - x_{i+2}} \right. \right.$$

$$\left. - \frac{f(x_{i+2}) - f(x_{i+1})}{x_{i+2} - x_{i+1}} \right] - \frac{1}{x_{i+2} - x_i} \left[\frac{f(x_{i+2}) - f(x_{i+1})}{x_{i+2} - x_{i+1}} \right.$$

$$\left. \left. - \frac{f(x_{i+1}) - f(x_i)}{x_{i+1} - x_i} \right] \right\} \qquad (5.64)$$

If $x_{i+1} - x_i = \Delta x$ for all i, then Eqs. (5.63) and (5.64) become

$$\Delta^2 f(x) = \frac{1}{2(\Delta x)^2} [f(x + 2\Delta x) - 2f(x + \Delta x) + f(x)] \qquad (5.65)$$

[†] When $(x_{i+1} - x_i) = \Delta x$ for all i it is sometimes convenient to define $\Delta^n f(x_i)$ as $\Delta f^n(x_i) = \Delta[\Delta^{n-1} f(x_i)]$.

5.5 HIGHER-ORDER DIFFERENCES

and

$$\Delta^3 f(x) = \frac{1}{6(\Delta x)^3}[f(x + 3\Delta x) - 3f(x + 2\Delta x) + 3f(x + \Delta x) - f(x)] \quad (5.66)$$

Example 5.32 If $x_{i+1} - x_i = \Delta x$ for all i, show that for $n > 0$,

$$\Delta^n f(x) = \frac{1}{n!(\Delta x)^n} \sum_{i=0}^{n} (-1)^{n-i} \frac{n!}{i!(n-i)!} f(x + i\Delta x) \quad (5.67)$$

We demonstrate the validity of Eq. (5.67) by induction. By the basic definition of the difference

$$\Delta f(x) = \frac{f(x + \Delta x) - f(x)}{\Delta x}$$

$$= \frac{1}{1!\Delta x} \sum_{i=0}^{1} (-1)^{1-i} \frac{1!}{i!(1-i)!} f(x + i\Delta x) \quad (5.68)$$

From Eqs. (5.62) and (5.65)

$$\Delta^2 f(x) = \frac{1}{2\Delta x}[\Delta f(x + \Delta x) - \Delta f(x)]$$

$$= \frac{1}{2\Delta x} \left\{ \frac{1}{1!\Delta x} \sum_{i=0}^{1} (-1)^{1-i} \frac{1!}{i!(1-i)!} \right.$$

$$\left. \times \{f[x + (i+1)\Delta x] - f[x + i\Delta x]\} \right\}$$

$$= \frac{1}{2!(\Delta x)^2}[-f(x + \Delta x) + f(x) + f(x + 2\Delta x) - f(x + \Delta x)]$$

$$= \frac{1}{2!(\Delta x)^2} \sum_{i=0}^{2} (-1)^{2-i} \frac{2!}{i!(2-i)!} f(x + i\Delta x) \quad (5.69)$$

and

$$\Delta^3 f(x) = \frac{1}{3\Delta x}[\Delta^2 f(x + \Delta x) - \Delta^2 f(x)]$$

$$= \frac{1}{3\Delta x} \left\{ \frac{1}{2!(\Delta x)^2} \sum_{i=0}^{2} (-1)^{2-i} \frac{2!}{i!(2-i)!} \right.$$

$$\left. \times \{f[x + (i+1)\Delta x] - f[x + i\Delta x]\} \right\}$$

Expanding the summation yields

$$\Delta^3 f(x) = \frac{1}{3!(\Delta x)^3}[f(x + \Delta x) - f(x) - 2f(x + 2\Delta x) + 2f(x + \Delta x)$$

$$+ f(x + 3\Delta x) - f(x + 2\Delta x)]$$

$$= \frac{1}{3!(\Delta x)^3} \sum_{i=0}^{3} (-1)^{3-i} \frac{3!}{i!(3-i)!} f(x + i\Delta x) \quad (5.70)$$

Equations (5.68)–(5.70) imply that

$$\Delta^m f(x) = \frac{1}{m!(\Delta x)^m} \sum_{i=0}^{m} (-1)^{m-i} \frac{m!}{i!(m-i)!} f(x + i\Delta x) \qquad (5.71)$$

If we can show that the method used to develop $\Delta^2 f(x)$ and $\Delta^3 f(x)$ leads to Δ^{m+1} given by

$$\Delta^{m+1} f(x) = \frac{1}{(m+1)!(\Delta x)^{m+1}} \sum_{i=0}^{m+1} (-1)^{m+1-i} \frac{(m+1)!}{i!(m+1-i)!} f(x + i\Delta x)$$

using Eq. (5.71), then the proof by induction is complete.

$$\Delta^{m+1} f(x) = \frac{1}{(m+1)\Delta x} [\Delta^m f(x + \Delta x) - \Delta^m f(x)]$$

$$= \frac{1}{(m+1)\Delta x} \left\{ \frac{1}{m!(\Delta x)^m} \sum_{i=0}^{m} (-1)^{m-i} \frac{m!}{i!(m-i)!} \right.$$

$$\left. \times \{ f[x + (i+1)\Delta x] - f[x + i\Delta x] \} \right\}$$

Expanding the summation yields

$$\sum_{i=0}^{m} (-1)^{m-i} \frac{m!}{i!(m-i)!} \{ f[x + (i+1)\Delta x] - f[x + i\Delta x] \}$$

$$= f[x + (m+1)\Delta x] - (-1)^m f(x)$$

$$+ \sum_{j=1}^{m} (-1)^{m-j+1} \frac{m!}{(j-1)!(m-j+1)!} f[x + j\Delta x]$$

$$- \sum_{j=1}^{m} (-1)^{m-j} \frac{m!}{j!(m-j)!} f[x + j\Delta x]$$

Since $(-1)^{m-j+1} = -(-1)^{m-j}$

$$\sum_{i=0}^{m} (-1)^{m-i} \frac{m!}{i!(m-i)!} \{ f[x + (i+1)\Delta x] - f[x + i\Delta x] \}$$

$$= f[x + (m+1)\Delta x] - (-1)^m f(x)$$

$$- \sum_{j=1}^{m} (-1)^{m-j} \left[\frac{m!}{(j-1)!(m-j+1)!} + \frac{m!}{j!(m-j)!} \right] f[x + j\Delta x]$$

$$= f[x + (m+1)\Delta x] + (-1)^{m+1} f(x)$$

$$+ \sum_{j=1}^{m} (-1)^{m+1-j} \frac{(m+1)!}{j!(m+1-j)!} f(x + j\Delta x)$$

5.5 HIGHER-ORDER DIFFERENCES

where

$$\left[\frac{m!}{(j-1)!(m-j+1)!} + \frac{m!}{j!(m-j)!}\right] = \frac{(m+1)!}{j!(m+1-j)!}$$

Therefore

$$\Delta^{m+1}f(x) = \frac{1}{(m+1)!(\Delta x)^{m+1}} \sum_{i=0}^{m+1} (-1)^{m+1-i} \frac{(m+1)!}{i!(m+1-i)!} f(x + i\Delta x)$$

and the proof is complete.

Example 5.33 Evaluate the first three differences of $f(x)$ at $x = 1, 2, 3, 4$ where

$$f(x) = (x+1)^2, \quad x = 0, 1, 2, \ldots$$

The expressions for the required differences are given by

$$\Delta f(x) = 2x + 3$$
$$\Delta^2 f(x) = \tfrac{1}{2}[(x+1)^2 - 2(x+2)^2 + (x+3)^2] = 1$$
$$\Delta^3 f(x) = \tfrac{1}{6}[-(x+1)^2 + 3(x+2)^2 - 3(x+3)^2 + (x+4)^2] = 0$$

Evaluating these differences at the specified values of x yields Table 5.1.

TABLE 5.1

x	$f(x)$	$\Delta f(x)$	$\Delta^2 f(x)$	$\Delta^3 f(x)$
1	4	5	1	0
2	9	7	1	0
3	16	9	1	0
4	25	11	1	0

A slight rearrangement and expansion of Table 5.1 yields a table which is useful in evaluating higher order differences. The basis for this table is that the nth divided difference can be developed if the $(n-1)$st differences at x_i and x_{i+1} are known. That is

$$\Delta^n f(x_i) = \frac{1}{x_{i+n} - x_i} [\Delta^{n-1} f(x_{i+1}) - \Delta^{n-1} f(x_i)] \tag{5.72}$$

The form of the *difference table* is shown in Table 5.2. On the left-hand side of the table the values of x_i and $f(x_i)$ are entered columnwise. Since $\Delta f(x_i)$ is based upon $f(x_{i+1})$ and $f(x_i)$, $\Delta f(x_i)$ is entered between and to the right of the two values of $f(x)$ it is derived from. Similarly, $\Delta^2 f(x_i)$ is entered between and to the right of $\Delta f(x_{i+1})$ and $\Delta f(x_i)$ since $\Delta^2 f(x_i)$ is based upon these two first differences. Continuing in this fashion leads to the result shown in Table 5.2. Therefore, in using a difference table higher-order differences are derived by first developing the differences of lower order.

5 CALCULUS OF FINITE DIFFERENCES

TABLE 5.2
DIFFERENCE TABLE

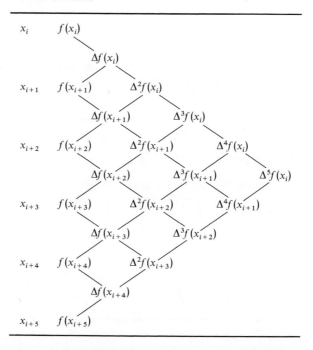

Example 5.34 Using the difference table, develop $\Delta^4 f(x)$ at $x = 1$, where
$$f(x) = (x + 1)^3, \qquad x = 0, 1, 2, \ldots$$

TABLE 5.3

x	$f(x)$	$\Delta f(x)$	$\Delta^2 f(x)$	$\Delta^3 f(x)$	$\Delta^4 f(x)$
1	8				
		$\dfrac{27 - 8}{2 - 1} = 19$			
2	27		$\dfrac{37 - 19}{3 - 1} = 9$		
		$\dfrac{64 - 27}{3 - 2} = 37$		$\dfrac{12 - 9}{4 - 1} = 1$	
3	64		$\dfrac{61 - 37}{4 - 2} = 12$		$\dfrac{1 - 1}{5 - 1} = 0$
		$\dfrac{125 - 64}{4 - 3} = 61$		$\dfrac{15 - 12}{5 - 2} = 1$	
4	125		$\dfrac{91 - 61}{5 - 3} = 15$		
		$\dfrac{216 - 125}{5 - 4} = 91$			
5	216				

5.6 DIFFERENCE EQUATIONS

From Eq. (5.67), to evaluate $\Delta^4 f(x)$ at $x = 1$, $f(x)$ for $x = 1, 2, 3, 4, 5$ must be defined (Table 5.3).

Example 5.35 Using a difference table evaluate $\Delta^3 f(x_i)$ at $i = 1$ where

i	0	1	2	3	4	5	6	7
x_i	-1	0	2	4	5	8	9	10
$f(x_i)$	0	1	3	8	7	2	0	-1

To evaluate the difference table, $f(x_i)$, $i = 1, 2, 3, 4$, must be determined (Table 5.4).

TABLE 5.4

i	x_i	$f(x_i)$	$\Delta f(x_i)$	$\Delta^2 f(x_i)$	$\Delta^3 f(x_i)$
1	0	1			
			$\dfrac{3-1}{2-0} = 1$		
2	2	3		$\dfrac{2.5-1}{4-0} = 0.375$	
			$\dfrac{8-3}{4-2} = 2.5$		$\dfrac{-1.167 - 0.375}{5-0} = -0.308$
3	4	8		$\dfrac{-1-2.5}{5-2} = -1.167$	
			$\dfrac{7-8}{5-4} = -1$		
4	5	7			

5.6 Difference Equations

Difference equations play an important role in operations research particularly in the theory of queues. The parallels already described between the calculus of finite differences and infinitesimal calculus would imply that the theory of difference equations should be similar in many respects to the theory of differential equations. A differential equation is an equation relating the derivatives of a continuous function $f(x)$. To illustrate

$$a_2 \frac{d^2}{dx^2} f(x) + a_1 \frac{d}{dx} f(x) + a_0 f(x) = z(x) \qquad (5.73)$$

is a *nonhomogeneous linear differential equation with constant coefficients*. On the other hand

$$a_2 \Delta^2 f(x) + a_1 \Delta f(x) + a_0 f(x) = z(x) \qquad (5.74)$$

is a *nonhomogeneous linear difference equation with constant coefficients*.

Definition 5.4 Let $f(x)$ be a function of the integer valued variable x. A difference equation is an equation relating $f(x)$ and any of its differences, $\Delta^i f(x)$, over a given range of values of x.

Definition 5.4 defines a difference equation only when x is an integer valued variable. This definition could be extended to include cases where x is not necessarily integer valued. However, consideration of such cases is beyond the scope of this text.

The solution of a difference equation is that function $f(x)$ which renders the difference equation valid for all values of x in the range specified for the equation. In seeking the solution of a difference equation the analyst must be certain that the solution he derives holds for all values of x in the range specified for the difference equation and not merely for some subset of values in this range.

Example 5.36 Show that

$$f(x) = cx$$

is a solution for the difference equation

$$\Delta f(x) = c, \quad x = 0, 1, 2, \ldots$$

Since the first difference of cx is c for all nonnegative integer values of x, the proposed solution satisfies the difference equation over the range of values of x for which it is defined, and, is therefore a solution for the difference equation given.

In attempting to solve a difference equation it is usually convenient to expand the differences $\Delta^i f(x)$. For example, expanding $\Delta^2 f(x)$ and $\Delta f(x)$ in Eq. (5.74) yields

$$\frac{a_2}{2} f(x+2) + (a_1 - a_2) f(x+1) + \left(a_0 - a_1 + \frac{a_2}{2}\right) f(x) = z(x)$$

It should be obvious that any difference equation can be expressed in terms of $f(x+i)$ through such an expansion, thus eliminating the difference operators from the equation. However, this expansion is frequently unnecessary since difference equations are usually expressed as relationships among the $f(x+i)$'s rather than as relationships among the $\Delta^i f(x)$'s.

Example 5.37 Solve the difference equation

$$2f(x+2) - f(x) = 2, \quad x = 0, 1, 2, \ldots$$

Expressing $f(x+2)$ as a function of $f(x)$ leads us to

$$f(x+2) = 1 + \frac{f(x)}{2}$$

5.6 DIFFERENCE EQUATIONS

Since $f(x+2)$ is a function of $f(x)$, we first attempt to find $f(x)$, for even x, in terms of $f(0)$ and then find $f(x)$, for odd x in terms of $f(1)$. Solving iteratively we have

$$f(2) = 1 + \frac{f(0)}{2}, \quad f(4) = 1 + \frac{f(2)}{2} = 1 + \frac{1}{2} + \frac{f(0)}{4}$$

$$f(6) = 1 + \frac{f(4)}{4} = 1 + \frac{1}{2} + \frac{1}{4} + \frac{f(0)}{8}$$

$$f(8) = 1 + \frac{f(6)}{2} = 1 + \frac{1}{2} + \frac{1}{4} + \frac{1}{8} + \frac{f(0)}{16}$$

We determine $f(x)$, for even x, by induction. Assume

$$f(x) = \sum_{i=0}^{x/2-1} \frac{1}{2^i} + \frac{f(0)}{2^{x/2}}$$

for even x. Now

$$f(x+2) = 1 + \frac{f(x)}{2} = 1 + \frac{1}{2} \sum_{i=0}^{x/2-1} \frac{1}{2^i} + \frac{f(0)}{2^{x/2+1}}$$

$$= \sum_{i=0}^{x/2} \frac{1}{2^i} + \frac{f(0)}{2^{x/2+1}}$$

and

$$f(x) = 2 - \left(\frac{1}{2}\right)^{x/2-1} + \left(\frac{1}{2}\right)^{x/2} f(0)$$

The same argument is used for odd x.

$$f(3) = 1 + \frac{f(1)}{2}, \quad f(5) = 1 + \frac{f(3)}{2} = 1 + \frac{1}{2} + \frac{f(1)}{4}$$

$$f(7) = 1 + \frac{f(5)}{2} = 1 + \frac{1}{2} + \frac{1}{4} + \frac{f(1)}{8}$$

$$f(9) = 1 + \frac{f(7)}{2} = 1 + \frac{1}{2} + \frac{1}{4} + \frac{1}{8} + \frac{f(1)}{16}$$

Induction leads to

$$f(x) = \sum_{i=0}^{(x-3)/2} \frac{1}{2^i} + \frac{f(1)}{2^{(x-1)/2}} = 2 - \left(\frac{1}{2}\right)^{(x-3)/2} + \left(\frac{1}{2}\right)^{(x-1)/2} f(1)$$

Therefore, the solution for the difference equation is given by

$$f(x) = \begin{cases} 2 - \left(\frac{1}{2}\right)^{(x-2)/2} + \left(\frac{1}{2}\right)^{x/2} f(0), & x = 2, 4, 6, \ldots \\ 2 - \left(\frac{1}{2}\right)^{(x-3)/2} + \left(\frac{1}{2}\right)^{(x-1)/2} f(1), & x = 3, 5, 7, \ldots \end{cases}$$

Example 5.37 illustrates two important points. As indicated above, the solution of a difference equation must satisfy the equation for all values of x for which it is defined. In Example 5.37, if $f(x)$, for even x, had been based upon $f(2)$ and $f(4)$ only, one might have been led to the solution

$$f(x) = 2 - \left(\frac{1}{2}\right)^{(x-2)/2} + \frac{f(0)}{x}$$

This solution is satisfactory for $x = 2, 4$, but does not satisfy the difference equation for even values of x greater than 4. Therefore in developing a solution iteratively, the analyst must make sure that the iterative process is sufficiently extensive to guarantee a solution. The second point illustrated in this example is that the solution given by $f(x)$ holds for all $f(0)$ and $f(1)$. To obtain a specific solution to the problem, $f(0)$ and $f(1)$ must be given in the form of initial conditions. That is, the numerical values of $f(0)$ and $f(1)$ must be specified.

In the remainder of the discussion of difference equations reference will be made to particular solutions, fundamental sets of solutions, and general solutions. A *particular solution* for a difference equation is any function, $f(x) \neq 0$, which reduces the difference equation to an identity. A solution satisfying the initial conditions for the problem is one such particular solution. However, any other function which satisfies the difference equation is also a particular solution whether it satisfies the initial conditions or not. Therefore

$$f(x) = (-\tfrac{1}{3})^x, \qquad x = 0, 1, 2, \ldots$$

is a particular solution for

$$3f(x+2) - 2f(x+1) - f(x) = 0, \qquad x = 0, 1, 2, \ldots$$

since it reduces this difference equation to an identity. Similarly

$$f(x) = 1, \qquad x = 0, 1, 2, \ldots$$

is also a particular solution. However, if $f(0) = 6$ and $f(1) = -\tfrac{2}{3}$ then neither of these particular solutions satisfies the difference equation and the imposed initial conditions. If we choose

$$f(x) = 5(-\tfrac{1}{3})^x + 1, \qquad x = 0, 1, 2, \ldots$$

then we have a paryicular solution which satisfies the initial conditions and is therefore a specific solution to the problem at hand.

Among the set of all particular solutions is a subset which is *linearly independent*. Let $f_i(x)$, $i = 1, 2, \ldots, n$, be linearly independent and such that any other particular solution, $f_{n+1}(x)$, can be expressed by

$$f_{n+1}(x) = c_1 f_1(x) + c_2 f_2(x) + \cdots + c_n f_n(x), \qquad x = 0, 1, 2, \ldots \quad (5.75)$$

where $c_1, c_2, \ldots, c_n$ are constants. Then, $f_i(x)$, $i = 1, 2, \ldots, n$, form a *fundamental set of solutions* for the difference equation. Therefore, the fundamental set of

5.6 DIFFERENCE EQUATIONS

solutions has the property that any particular solution belonging to the fundamental set cannot be expressed by a linear combination of the remaining particular solutions within the fundamental set. Further, every particular solution which does not belong to the fundamental set can be expressed by a linear combination of the particular solutions belonging to the fundamental set. Therefore, there is no particular solution which does not belong to the fundamental set which is linearly independent of the solutions belonging to the fundamental set.

The *general solution* to a difference equation is given by a linear combination of the particular solutions belonging to the fundamental set. The general solution has the property that any particular solution can be obtained by appropriate selection of the constant coefficients of the particular solutions included in the linear combination.

Example 5.38 Find the particular solution to the difference equation

$$f(x+1) - 4f(x) = 0, \quad x = 0, 1, 2, \ldots, \quad \text{when} \quad f(0) = 4$$

From the difference equation given,

$$f(x+1) = 4f(x)$$

For $x = 0, 1, 2$,

$$f(1) = 4f(0), \quad f(2) = 4f(1) = 4^2 f(0)$$
$$f(3) = 4f(2) = 4^3 f(0)$$

Assume

$$f(x) = 4^x f(0)$$

Then

$$f(x+1) = 4f(x) = 4^{x+1} f(0)$$

and $f(x) = 4^x f(0)$ is a particular solution. The particular solution when $f(0) = 4$ is given by

$$f(x) = 4^{x+1}$$

The balance of this discussion of difference equations will be devoted to linear difference equations with constant coefficients.

Definition 5.5 A difference equation defined over some set of values of x is said to be *linear* if it can be expressed in the form

$$a_n(x)f(x+n) + a_{n-1}(x)f(x+n-1) + \cdots + a_0(x)f(x) = z(x) \quad (5.76)$$

where $a_i(x)$, $i = 0, 1, 2, \ldots, n$, and $z(x)$ are functions of x but not of $f(x)$.

In Definition 5.5, if $a_n(x)$ and $a_0(x)$ are nonzero for all values of x for which Eq. 5.76 holds, then Eq. (5.76) is said to be a linear difference equation of order n.

Otherwise, the order of the equation is the difference between the largest and smallest arguments of $f(x)$. If $a_i(x)$ is a constant function for $i = 0, 1, 2, \ldots, n$, Eq. (5.76) is a *linear difference equation with constant coefficients*. If $z(x) = 0$ for all x for which Eq. (5.76) is defined, then Eq. (5.76) is said to be a *homogeneous linear difference equation*. Otherwise Eq. (5.76) is said to be a *nonhomogeneous linear difference equation*.

Example 5.39 Classify the following difference equations.

a. $f(x + 4) - 2f(x + 3) + 3f(x + 2) - f(x + 1) + f(x) = 0$
b. $f(x + 2) - f(x + 1) - 5f(x) = x^2$
c. $xf(x + 1) - f(x) = 0$
d. $xf(x + 1) - x^2 f(x) = 1 - x$
e. $f(x + 5) - 2f(x + 1) = 4$
f. $(x - 1)f(x + 2) - 4f(x + 1) = 0$

In each of the six cases the difference equations are defined for $x = 1, 2, 3, \ldots$. In part a the difference equation conforms to Definition 5.5, where

$$a_4(x) = 1, \quad a_3(x) = -2, \quad a_2(x) = 3, \quad a_1(x) = -1, \quad a_0(x) = 1, \quad z(x) = 0$$

Therefore the difference equation is linear. Since $a_i(x)$ is a constant function for $i = 0, 1, \ldots, 4$, it is a linear difference equation with constant coefficients. Since $z(x) = 0$ the difference equation is homogeneous and since $a_4(x) \neq 0$ and $a_0(x) \neq 0$ for all permissible values of x the difference equation is of order 4. In part b,

$$a_2(x) = 1, \quad a_1(x) = -1, \quad a_0(x) = -5, \quad z(x) = x^2$$

Hence the equation in b is a linear difference equation with constant coefficients. Since $z(x) \neq 0$ for all permissible values of x, this difference equation is nonhomogeneous. Finally, the equation is of order 2 since $a_2(x) \neq 0$ and $a_0(x) \neq 0$ for all x for which the equation is defined.

The difference equation in part c conforms to Definition 5.5 where

$$a_1(x) = x, \quad a_0(x) = -1, \quad z(x) = 0$$

Since $z(x) = 0$, $a_1(x) \neq 0$, and $a_0(x) \neq 0$ for all permissible values of x, this difference equation is homogeneous and of the first order. Therefore, the difference equation is first order, linear, and homogeneous, but does not have constant coefficients since $a_1(x)$ and $a_0(x)$ are not *both* constant functions. The difference equation given in part d is linear where

$$a_1(x) = x, \quad a_0(x) = -x^2, \quad z(x) = 1 - x$$

Since $z(x) \neq 0$ for $x > 1$, the difference equation is nonhomogeneous. Since $a_1(x) \neq 0$ and $a_0(x) \neq 0$ for all permissible values of x, the equation is of first order. Finally, since $a_1(x)$ and $a_0(x)$ are not constant functions, the difference equation does not have constant coefficients.

5.6 DIFFERENCE EQUATIONS

In part e the difference equation can be expressed in the form

$$\sum_{i=0}^{4} a_i(x) f(x+i) = 4, \qquad x = 2, 3, 4, \ldots$$

where

$$a_4(x) = 1, \quad a_3(x) = 0, \quad a_2(x) = 0, \quad a_1(x) = 0, \quad a_0(x) = -2$$

Since $z(x) \neq 0$ for all permissible values of x, the equation is nonhomogeneous and of order 4, since $a_4(x) \neq 0$ and $a_0(x) \neq 0$ for all permissible x. Since the $a_i(x)$ are constant functions for $i = 0, 1, \ldots, 4$ the difference equation is linear with constant coefficients.

By arguments similar to those just presented the difference equation given in part f can be classified as linear and homogeneous, and since

$$a_1(x) = x - 1, \qquad a_0(x) = -4$$

the equation does not have constant coefficients. The only problem remaining is that of defining the order of the equation. Since $a_1(x) = 0$ for $x = 1$ the order of this equation is given by $(x+2) - (x+1)$ and the equation is of order 1.

To summarize

a. Fourth-order, linear, homogeneous difference equation with constant coefficients;
b. Second-order, linear, nonhomogeneous difference equation with constant coefficients;
c. First-order, linear, homogeneous difference equation;
d. First-order, linear, nonhomogeneous difference equation;
e. Fourth-order, linear, nonhomogeneous difference equation with constant coefficients;
f. Linear, homogeneous difference equation with undefined order.

5.6.1 Homogeneous Linear Difference Equations with Constant Coefficients

In this section solutions of difference equations of the form

$$a_n f(x+n) + a_{n-1} f(x+n-1) + \cdots + a_0 f(x) = 0 \qquad (5.77)$$

will be developed where $a_0, a_1, \ldots, a_n$ are constants and $a_0 \neq 0$, $a_n \neq 0$. First let us treat the first order equation given by

$$a_1 f(x+1) + a_0 f(x) = 0 \qquad (5.78)$$

Solving for $f(x+1)$ yields

$$f(x+1) = c f(x) \qquad (5.79)$$

where $c = -a_0/a_1$. Since a_0 and a_1 are assumed to be nonzero, c is a nonzero finite constant. By induction it can be shown that

$$f(x) = c^x f(0) \tag{5.80}$$

is the required solution. If $f(0) = 1$ we have the particular solution

$$f(x) = c^x \tag{5.81}$$

Let us now consider the second order linear difference equation defined by

$$a_2 f(x + 2) + a_1 f(x + 1) + a_0 f(x) = 0 \tag{5.82}$$

Theorem 5.8 Let the difference equation given in Eq. (5.82) be defined for all integer values of x on the interval (b, c). Then there exists a number $k \neq 0$ such that

$$f(x) = k^x \tag{5.83}$$

is a particular solution of Eq. (5.82) where k satisfies the equation

$$a_2 k^2 + a_1 k + a_0 = 0 \tag{5.84}$$

Proof Substituting the proposed particular solution into Eq. (5.82) yields

$$a_2 k^{x+2} + a_1 k^{x+1} + a_0 k^x = 0$$

or, dividing by k^x

$$a_2 k^2 + a_1 k + a_0 = 0$$

Therefore, the value of k satisfying Eq. (5.82) must also satisfy Eq. (5.84) and is given by

$$k = \frac{-a_1 \pm \sqrt{a_1^2 - 4a_0 a_2}}{2a_2} \tag{5.85}$$

That is, in general there are two solutions for k,

$$k = \frac{-a_1 + \sqrt{a_1^2 - 4a_0 a_2}}{2a_2}, \quad \frac{-a_1 - \sqrt{a_1^2 - 4a_0 a_2}}{2a_2} \tag{5.86}$$

unless $a_1^2 - 4a_0 a_2 = 0$ in which case there is only one solution for k. When two solutions exist they form a fundamental set. ∎

Theorem 5.9 Particular Solutions Let $f_1(x)$ and $f_2(x)$ be two real solutions for the homogeneous linear difference equation given in Eq. (5.82). If, for all x

$$\begin{vmatrix} f_1(x) & f_2(x) \\ f_1(x+1) & f_2(x+1) \end{vmatrix} \neq 0 \tag{5.87}$$

5.6 DIFFERENCE EQUATIONS

then $f_1(x)$ and $f_2(x)$ are linearly independent and any other solution, $f_3(x)$, can be expressed by

$$f_3(x) = c_1 f_1(x) + c_2 f_2(x) \tag{5.88}$$

where c_1 and c_2 are suitably chosen constants.

Proof To prove this theorem we must establish the following

  a. Equation (5.87) implies $f_1(x)$ and $f_2(x)$ are linearly independent;
  b. If $f_1(x)$ and $f_2(x)$ are linearly independent then $c_1 f_1(x) + c_2 f_2(x)$ is a solution for Eq. (5.82).
  c. $f_3(x)$ is a solution for Eq. (5.82) only if it is linearly dependent upon $f_1(x)$ and $f_2(x)$.

Parts a and b are left as exercises for the reader. In part c we must show that a solution $f_3(x)$ is not linearly independent of $f_1(x)$ and $f_2(x)$. We will show that any other solution $f_3(x)$ is not linearly independent of $f_1(x)$ and $f_2(x)$ by contradiction. Assume $f_3(x)$ is a solution for the difference equation and is linearly independent of $f_1(x)$ and $f_2(x)$. That is

$$f_3(x) \neq c_1 f_1(x) + c_2 f_2(x)$$

for any choice of c_1 and c_2. Let

$$F = \begin{bmatrix} f_1(x) & f_2(x) & f_3(x) \\ f_1(x+1) & f_2(x+1) & f_3(x+1) \\ f_1(x+2) & f_2(x+2) & f_3(x+2) \end{bmatrix} \tag{5.89}$$

If $f_1(x), f_2(x)$, and $f_3(x)$ are linearly independent solutions, then

$$|F| \neq 0$$

For $a_0 \neq 0$ in Eq. (5.82)

$$a_0 |F| = \begin{vmatrix} a_0 f_1(x) & a_0 f_2(x) & a_0 f_3(x) \\ f_1(x+1) & f_2(x+1) & f_3(x+1) \\ f_1(x+2) & f_2(x+2) & f_3(x+2) \end{vmatrix} \tag{5.90}$$

Add a_1 times the second row of $a_0 |F|$ and a_2 times the third row of $a_0 |F|$ to the first row of $a_0 |F|$. Then

$$a_0 |F| = \begin{vmatrix} \sum_{i=0}^{2} a_i f_1(x+i) & \sum_{i=0}^{2} a_i f_2(x+i) & \sum_{i=0}^{2} a_i f_3(x+i) \\ f_1(x+1) & f_2(x+1) & f_3(x+1) \\ f_1(x+2) & f_2(x+2) & f_3(x+2) \end{vmatrix}$$

Since $f_1(x), f_2(x)$, and $f_3(x)$ are assumed to be solutions for the difference equation

$$\sum_{i=0}^{2} a_i f_j(x+i) = 0, \quad j = 1, 2, 3 \quad \text{and} \quad a_0 |F| = 0$$

However, since $a_0 \neq 0$, $|F| = 0$, and $f_3(x)$ cannot be linearly independent of $f_1(x)$ and $f_2(x)$. Therefore

$$f_3(x) = c_1 f_1(x) + c_2 f_2(x)$$

In Theorem 5.9, $f_1(x)$ and $f_2(x)$ are *particular* solutions of Eq. (5.82) while $f_3(x)$ is the *general* solution.

Example 5.40 Find the solution of

$$f(x+2) - 4f(x+1) + 3f(x) = 0, \quad x = 0, 1, 2, \ldots$$

subject to the initial conditions

$$f(0) = 1, \quad f(1) = 2$$

From Theorem 5.8, we know that

$$f(x) = k^x$$

is a particular solution where

$$k = \frac{4 \pm \sqrt{16 - 4(1)(3)}}{2} = 3, 1$$

Let the two solutions be defined by $f_1(x)$ and $f_2(x)$. That is

$$f_1(x) = 3^x, \quad f_2(x) = 1$$

Neither of these solutions satisfies the initial conditions given for the problem. Therefore, we attempt to find a third particular solution, $f_3(x)$, which satisfies these conditions. Since

$$\begin{vmatrix} f_1(x) & f_2(x) \\ f_1(x+1) & f_2(x+1) \end{vmatrix} = \begin{vmatrix} 3^x & 1 \\ 3^{x+1} & 1 \end{vmatrix} = 3^x(1-3) = -2(3)^x \neq 0$$

any solution, $f_3(x)$, can be expressed by

$$f_3(x) = c_1 3^x + c_2$$

from Theorem 5.9. From the initial conditions, c_1 and c_2 must satisfy the equations

$$c_1 + c_2 = 1, \quad 3c_1 + c_2 = 2$$

or

$$\begin{bmatrix} 1 & 1 \\ 3 & 1 \end{bmatrix} \begin{bmatrix} c_1 \\ c_2 \end{bmatrix} = \begin{bmatrix} 1 \\ 2 \end{bmatrix}$$

which leads to

$$\begin{bmatrix} c_1 \\ c_2 \end{bmatrix} = \begin{bmatrix} \frac{1}{2} \\ \frac{1}{2} \end{bmatrix}$$

5.6 DIFFERENCE EQUATIONS

Therefore

$$f_3(x) = \tfrac{1}{2}(3^x + 1)$$

is the desired solution. ◻

Example 5.41 Find the solution of

$$f(x+2) - a_1 f(x+1) + \tfrac{1}{4}(2a_1 a_0 - a_0^2) f(x) = 0, \qquad x = 0, 1, 2, \ldots$$

subject to the restrictions

$$f(1) = (a_1 - \tfrac{1}{2}a_0) f(0)$$

$$\sum_{x=0}^{\infty} f(x) = 1, \qquad a_0 > 2, \qquad 0 < (a_1 - \tfrac{1}{2}a_0) < 1$$

From Theorem 5.8,

$$f(x) = k^x$$

where

$$k = \frac{a_1 \pm \sqrt{a_1^2 - (2a_1 a_0 - a_0^2)}}{2} = \frac{a_1 \pm \sqrt{(a_1 - a_0)^2}}{2} = \frac{a_0}{2},\ a_1 - \frac{a_0}{2}$$

Let the two solutions be defined by

$$f_1(x) = (\tfrac{1}{2}a_0)^x, \qquad f_2(x) = (a_1 - \tfrac{1}{2}a_0)^x$$

Since $a_0 > 2$, $\tfrac{1}{2}a_0 > 1$

$$\sum_{x=0}^{\infty} (\tfrac{1}{2}a_0)^x = \infty$$

Therefore, $f_1(x)$ is not the required solution. For $f_2(x)$ we have

$$\sum_{x=0}^{\infty} f_2(x) = \sum_{x=0}^{\infty} (a_1 - \tfrac{1}{2}a_0)^x = \frac{1}{1 - (a_1 - \tfrac{1}{2}a_0)}$$

This summation equals unity only if $(a_1 - \tfrac{1}{2}a_0) = 0$, which violates the restriction that $0 < (a_1 - \tfrac{1}{2}a_0) < 1$. By Theorem 5.9, any other solution, $f_3(x)$, can be expressed by

$$f_3(x) = c_1 f_1(x) + c_2 f_2(x)$$

since

$$\begin{vmatrix} (\tfrac{1}{2}a_0)^x & (a_1 - \tfrac{1}{2}a_0)^x \\ (\tfrac{1}{2}a_0)^{x+1} & (a_1 - \tfrac{1}{2}a_0)^{x+1} \end{vmatrix} = (\tfrac{1}{2}a_0)^x (a_1 - \tfrac{1}{2}a_0)^x (a_1 - a_0) \neq 0$$

To find the values of c_1 and c_2 we solve the following two equations.

$$f_3(0) = c_1 + c_2$$
$$f_3(1) = c_1(\tfrac{1}{2}a_0) + c_2(a_1 - \tfrac{1}{2}a_0)$$

Since $f_3(1) = (a_1 - \frac{1}{2}a_0)f_3(0)$, we can solve for c_1 and c_2 in terms of $f_3(0)$.

$$\begin{bmatrix} c_1 \\ c_2 \end{bmatrix} = \frac{1}{a_1 - a_0} \begin{bmatrix} (a_1 - \frac{1}{2}a_0) & -1 \\ -\frac{1}{2}a_0 & 1 \end{bmatrix} \begin{bmatrix} f_3(0) \\ (a_1 - \frac{1}{2}a_0)f_3(0) \end{bmatrix} = \begin{bmatrix} 0 \\ f_3(0) \end{bmatrix}$$

and

$$f_3(x) = (a_1 - \tfrac{1}{2}a_0)^x f_3(0)$$

To find the expression for $f_3(0)$ we use the restriction

$$\sum_{x=0}^{\infty} f(x) = 1 = \sum_{x=0}^{\infty} (a_1 - \tfrac{1}{2}a_0)^x f_3(0) = \frac{f_3(0)}{1 - (a_1 - \tfrac{1}{2}a_0)}$$

Therefore

$$f_3(0) = 1 - (a_1 - \tfrac{1}{2}a_0)$$

and

$$f_3(x) = (1 - a_1 + \tfrac{1}{2}a_0)(a_1 - \tfrac{1}{2}a_0)^x \qquad \blacksquare$$

Theorem 5.10 Fundamental Sets of Solutions Let $f_1(x)$ and $f_2(x)$ be two particular solutions given by Theorem 5.8 for the homogeneous linear difference equation given in Eq. 5.82. If, for all x

$$\begin{vmatrix} f_1(x) & f_2(x) \\ f_1(x+1) & f_2(x+1) \end{vmatrix} = 0 \qquad (5.91)$$

then

1. $f_1(x) = f_2(x) = k^x$ \hfill (5.92)
2. The fundamental set of solutions is given by

$$f_1(x) = k^x \qquad (5.93)$$
$$f_2(x) = xk^x \qquad (5.94)$$

3. Any other solution, $f_3(x)$, can be expressed by

$$f_3(x) = c_1 f_1(x) + c_2 f_2(x) \qquad (5.95)$$

where $f_1(x)$ and $f_2(x)$ are defined in Eqs. (5.93) and (5.94).

Proof From Theorem 5.8, there are two solutions to (5.82) given by

$$f_1(x) = k_1^x, \qquad f_2(x) = k_2^x$$

$$\begin{vmatrix} f_1(x) & f_2(x) \\ f_1(x+1) & f_2(x+1) \end{vmatrix} = \begin{vmatrix} k_1^x & k_2^x \\ k_1^{x+1} & k_2^{x+1} \end{vmatrix} = k_1^x k_2^x (k_2 - k_1) = 0$$

Since $k_1 \neq 0$ and $k_2 \neq 0$, $k_1 = k_2$ and

$$f_1(x) = f_2(x)$$

5.6 DIFFERENCE EQUATIONS

To prove the second part of the theorem we note from Theorem 5.8 that there is one solution of the form

$$f_1(x) = k^x$$

To show that

$$f_2(x) = xk^x$$

is a solution for Eq. 5.82, note from Theorem 5.8 that

$$k = -\frac{a_1}{2a_2} \tag{5.96}$$

and

$$a_0 = \frac{a_1^2}{4a_2} \tag{5.97}$$

since there is only one solution for k. Therefore

$$f_2(x) = x\left(-\frac{a_1}{2a_2}\right)^x \tag{5.98}$$

Substituting $f_2(x)$ and the expression for a_0 in Eq. (5.97) into Eq. (5.82) yields

$$a_2(x+2)\left(-\frac{a_1}{2a_2}\right)^{x+2} + a_1(x+1)\left(-\frac{a_1}{2a_2}\right)^{x+1} + \frac{a_1^2}{4a_2}x\left(-\frac{a_1}{2a_2}\right)^x = 0 \tag{5.99}$$

Dividing by $(-a_1/2a_2)^x$ we have

$$a_2(x+2)\left(-\frac{a_1}{2a_2}\right)^2 - a_1(x+1)\left(\frac{a_1}{2a_2}\right) + a_0 x = 0$$

and

$$x\left(\frac{a_1^2}{4a_2} - \frac{a_1^2}{2a_2} + \frac{a_1^2}{4a_2}\right) + \left(\frac{a_1^2}{2a_2} - \frac{a_1^2}{2a_2}\right) = 0$$

The third part of the theorem may also be shown by direct substitution into Eq. (5.82), but is left as an exercise for the reader.

Example 5.42 Find the particular solution for

$$2f(x+2) + 4f(x+1) + 2f(x) = 0, \quad x = 0, 1, 2, \ldots$$

subject to the restrictions that

$$f(0) = 1, \quad f(1) = 2$$

From Theorem 5.8,

$$f(x) = k^x$$

yields at least one solution where

$$k = \frac{-4 \pm \sqrt{16 - 4(2)(2)}}{4} = -1$$

Therefore, Theorem 5.8 yields only one solution,

$$f_1(x) = (-1)^x$$

Since

$$f_1(0) = 1, \qquad f_1(1) = -1$$

$f_1(x)$ is not the required solution since $f_1(1) \neq 2$. From Theorem 5.10, a second solution is given by

$$f_2(x) = x(-1)^x$$

However, since

$$f_2(0) = 0$$

$f_2(x)$ is not the required solution. A third solution, $f_3(x)$ can be expressed by

$$f_3(x) = c_1(-1)^x + c_2 x(-1)^x$$

We determine c_1 and c_2 by solving the simultaneous equations resulting for $f_3(0)$ and $f_3(1)$.

$$1 = c_1, \qquad 2 = -c_1 - c_2$$

The values for c_1 and c_2 are given by

$$\begin{bmatrix} c_1 \\ c_2 \end{bmatrix} = \begin{bmatrix} 1 \\ -3 \end{bmatrix}$$

Therefore

$$f_3(x) = (1 - 3x)(-1)^x$$

The solution for a second-order, homogeneous, linear difference equation with constant coefficients can be extended to nth-order equations as follows.

Theorem 5.11 A particular solution of the nth-order difference equation defined by

$$a_n f(x + n) + a_{n-1} f(x + n - 1) + \cdots + a_0 f(x) = 0 \qquad (5.100)$$

where a_i is constant for $i = 0, 1, \ldots, n$, is given by

$$f(x) = k^x$$

The value of k is any one of the n roots of the equation.

$$a_n k^n + a_{n-1} k^{n-1} + \cdots + a_0 = 0 \qquad (5.101)$$

5.6 DIFFERENCE EQUATIONS

Example 5.43 Find the particular solutions of the difference equation defined by

$$f(x + 3) - 6f(x + 2) + 11f(x + 1) - 6f(x) = 0, \qquad x = 0, 1, 2, \ldots$$

From Theorem 5.11,

$$f(x) = k^x$$

is a particular solution where k satisfies the equation

$$k^3 - 6k^2 + 11k - 6 = 0$$

This equation may be factored as follows

$$k^3 - 6k^2 + 11k - 6 = (k - 1)(k^2 - 5k + 6) = (k - 1)(k - 2)(k - 3)$$

Therefore there are three distinct solutions.

$$f_1(x) = 1, \qquad f_2(x) = 2^x, \qquad f_3(x) = 3^x$$

Theorem 5.12 Fundamental Sets of Solution Associated with the nth-order difference equation

$$a_n f(x + n) + a_{n-1} f(x + n - 1) + \cdots + a_0 f(x) = 0$$

are n solutions, $f_i(x)$, $i = 1, 2, \ldots, n$. The set of n solutions satisfying

$$\begin{vmatrix} f_1(x) & f_2(x) & \cdots & f_n(x) \\ f_1(x + 1) & f_2(x + 1) & \cdots & f_n(x + 1) \\ \vdots & \vdots & & \vdots \\ f_1(x + n - 1) & f_2(x + n - 1) & \cdots & f_n(x + n - 1) \end{vmatrix} \neq 0 \qquad (5.102)$$

forms a fundamental set of solutions. Any other solution, $f_{n+1}(x)$, can be expressed by

1. $$f_{n+1}(x) = \sum_{i=1}^{n} c_i f_i(x) \qquad (5.103)$$

if the n solutions given in Theorem 5.11 are real and linearly independent.

2. $$f_{n+1}(x) = \sum_{i=1}^{m_1} c_i f_i(x) + \sum_{i=m_1+1}^{m_2} c_i x^{i-m_1} f_1(x)$$

$$+ \sum_{i=m_2+1}^{m_3} c_i x^{i-m_2} f_2(x) + \cdots$$

$$+ \sum_{i=m_{j-1}+1}^{m_j} c_i x^{i-m_{j-1}} f_{j-1}(x) + \sum_{i=m_j+1}^{n} c_i x^{i-m_j} f_j(x) \qquad (5.104)$$

if the n solutions given in Theorem 5.11 are such that

$$f_i(x) \neq f_k(x), \quad i \neq k \quad \text{and} \quad i, k = 1, 2, 3, \ldots, m_1$$
$$f_{m_1+1}(x) = f_{m_1+2}(x) = \cdots = f_{m_2}(x) = f_1(x)$$
$$f_{m_2+1}(x) = f_{m_2+2}(x) = \cdots = f_{m_3}(x) = f_2(x)$$
$$\vdots \qquad \vdots \qquad \vdots$$
$$f_{m_{j-1}+1}(x) = f_{m_{j-1}+2}(x) = \cdots = f_{m_j}(x) = f_{j-1}(x)$$
$$f_{m_j+1}(x) = f_{m_j+2}(x) = \cdots = f_n(x) = f_j(x), \quad \text{where } j \leq m_1$$

The solution $f_{n+1}(x)$ is a general solution for Eq. (5.100) where $c_1, c_2, \ldots, c_n$ are constants.

In part 2 of Theorem 5.12, the implication is that m_1 solutions given in Theorem 5.11 are distinct. In addition $(m_2 - m_1)$ solutions given in Theorem 5.11 are identical to each other and to one of the m_1 distinct solutions, $f_1(x)$, $(m_3 - m_2)$ are identical to each other and to one of the m_1 distinct solutions, $f_2(x), \ldots,$ and $(n - m_j)$ are identical to each other and to $f_j(x)$ which is one of the distinct solutions. To illustrate the result of Theorem 5.12, suppose that the set of solutions for a homogeneous, seventh-order, linear difference equation with constant coefficients is given by

$$f_i(x) = k_i^x, \quad i = 1, 2, \ldots, 7 \tag{5.105}$$

If

$$k_i \neq k_j, \quad i \neq j$$

then any solution $f_8(x)$ can be expressed by

$$f_8(x) = c_1 k_1^x + c_2 k_2^x + \cdots + c_7 k_7^x \tag{5.106}$$

On the other hand, if

$$k_1 = k_5 = k_6, \quad k_2 = k_7, \quad k_3 \neq k_4$$

and

$$k_1 \neq k_2, \quad k_1 \neq k_3, \quad k_1 \neq k_4, \quad k_2 \neq k_3, \quad k_2 \neq k_4$$

then any solution $f_8(x)$ can be expressed by

$$f_8(x) = c_1 k_1^x + c_2 k_2^x + c_3 k_3^x + c_4 k_4^x + (c_5 x + c_6 x^2) k_1^x + c_7 x k_2^x \tag{5.107}$$

Example 5.44 Find the general solution for

$$f(x + 8) - 12f(x + 6) + 30f(x + 4) - 28f(x + 2) + 9f(x) = 0$$

The general solution for this problem is given by either Eq. (5.103) or by Eq. (5.104). From Theorem 5.11, there are eight solutions of the form

$$f_i(x) = k_i^x, \quad i = 1, 2, \ldots, 8$$

where k_i must satisfy

$$k_i^8 - 12k_i^6 + 30k_i^4 - 28k_i^2 + 9 = 0$$

Factoring this expression yields

$$(k_i - 1)^3(k_i + 1)^3(k_i - 3)(k_i + 3) = 0$$

Let

$$k_1 = k_5 = k_6 = 1, \qquad k_2 = k_7 = k_8 = -1, \qquad k_3 = 3, \qquad k_4 = -3$$

Since these solutions are not linearly independent, the general solution is given by Eq. (5.104) and

$$f(x) = c_1 + c_2(-1)^x + c_3 3^x + c_4(-3)^x \\ + (c_5 x + c_6 x^2) + (c_7 x + c_8 x^2)(-1)^x$$

is the required solution.

5.6.2 Nonhomogeneous Linear Difference Equations with Constant Coefficients

In this section solutions for the equation

$$a_n f(x + n) + a_{n-1} f(x + n - 1) + \cdots + a_0 f(x) = z(x) \qquad (5.108)$$

are sought. In this equation $z(x)$ may or may not be a constant. A technique frequently employed in deriving particular solutions for equations of this type is the *method of undetermined coefficients*. This technique will be demonstrated where $z(x)$ is given by k^x, x^k, $x^k b^x$, or a linear combination of these terms.

The method of undetermined coefficients is basically a trial and error procedure. For the moment assume that $z(x)$ is given by either k^x, x^k, or $x^k b^x$ and let $g_i(x, c_1, c_2, \ldots, c_m)$ be the ith trial solution of Eq. (5.108) where $c_1, c_2, \ldots, c_m$ are undetermined constants. We first attempt to find values of $c_1, c_2, \ldots, c_m$ such that $g_1(x, c_1, c_2, \ldots, c_m)$ is a solution of Eq. (5.108). If this attempt fails we multiply $g_1(x, c_1, c_2, \ldots, c_m)$ by x to obtain a second trial solution. That is

$$g_2(x, c_1, c_2, \ldots, c_m) = x g_1(x, c_1, c_2, \ldots, c_m) \qquad (5.109)$$

If, again, values of $c_1, c_2, \ldots, c_m$ cannot be found such that $g_2(x, c_1, c_2, \ldots, c_m)$ satisfies Eq. (5.108), we multiply $g_2(x, c_1, c_2, \ldots, c_m)$ by x to obtain a third trial solution. Therefore, the ith trial solution is given by

$$g_i(x, c_1, c_2, \ldots, c_m) = x^{i-1} g_1(x, c_1, c_2, \ldots, c_m) \qquad (5.110)$$

The initial trial solutions suggested when $z(x)$ is given by each of the three functions mentioned above, are as follows

$$z(x) = k^x, \qquad g_1(x, c_1) = c_1 k^x \qquad (5.111)$$

$$z(x) = x^k, \qquad g_1(x, c_1, c_2, \ldots, c_k) = \sum_{i=1}^{k+1} c_i x^{i-1} \qquad (5.112)$$

$$z(x) = x^k b^x, \qquad g_1(x, c_1, c_2, \ldots, c_k) = b^x \sum_{i=1}^{k+1} c_i x^{i-1} \qquad (5.113)$$

Example 5.45 Find a particular solution for

$$2f(x+2) - 2f(x+1) + f(x) = 4^x$$

From Eq. (5.111), the initial trial solution is

$$g_1(x, c_1) = c_1 4^x$$

Letting

$$f(x) = c_1 4^x$$

we have

$$2c_1 4^{x+2} - 2c_1 4^{x+1} + c_1 4^x = 4^x \qquad \text{or} \qquad c_1[33 - 8] = 1$$

Therefore

$$f(x) = \frac{4^x}{25}$$

is the required particular solution.

If $z(x)$ is a linear combination of the functions $z_1(x)$ and $z_2(x)$ and $g(x, c_1, c_2, \ldots, c_m)$ is a solution corresponding to $z_1(x)$ and $h(x, b_1, b_2, \ldots, b_r)$ is a solution corresponding to $z_2(x)$, then the solution of the difference equation is given by $g(x, c_1, c_2, \ldots, c_m) + h(x, b_1, b_2, \ldots, b_r)$, where $c_1, c_2, \ldots, c_m$ and $b_1, b_2, \ldots, b_r$ are undetermined constants.

Example 5.46 Find a particular solution for the difference equation

$$f(x+2) - 3f(x+1) - f(x) = x + 3^x$$

Let

$$z_1(x) = x, \qquad z_2(x) = 3^x$$

The initial trial solutions corresponding to $z_1(x)$ and $z_2(x)$ are

$$g_1(x, c_1, c_2) = c_1 + c_2 x, \qquad h_1(x, b_1) = b_1 3^x$$

Therefore, the initial solution is given by $c_1 + c_2 x + b_1 3^x$. Substituting this solution in the difference equation yields us

$$[c_1 + c_2(x+2) + b_1 3^{x+2}] - 3[c_1 + c_2(x+1) + b_1 3^{x+1}]$$
$$- [c_1 + c_2 x + b_1 3^x] = x + 3^x$$

5.6 DIFFERENCE EQUATIONS

or
$$-(3c_1 + c_2) - 3c_2 x - b_1 3^x = x + 3^x$$

Therefore
$$3c_1 + c_2 = 0, \qquad -3c_2 = 1, \qquad -b_1 = 1$$

and
$$c_1 = \tfrac{1}{9}, \qquad c_2 = -\tfrac{1}{3}, \qquad b_1 = -1$$

The required solution is given by
$$f(x) = \tfrac{1}{9} - \tfrac{1}{3}x - 3^x$$

Thus far the discussion of nonhomogeneous linear difference equations with constant coefficients has been limited to the determination of particular solutions. The problem of finding a general solution for such equations is facilitated by the following theorem.

Theorem 5.13 General Solution Let $g_i(x)$, $i = 1, \ldots, n$, be a fundamental set of solutions for the homogeneous difference equation

$$a_n f(x + n) + a_{n-1} f(x + n - 1) + \cdots + a_0 f(x) = 0 \qquad (5.114)$$

and let $h(x)$ be a particular solution for

$$a_n f(x + n) + a_{n-1} f(x + n - 1) \cdots + a_0 f(x) = z(x) \qquad (5.115)$$

such that $h(x)$ is not a particular solution of Eq. (5.114). Then $\sum_{i=1}^{n} c_i g_i(x) + h(x)$ is a general solution for Eq. (5.115).

In applying Theorem 5.13, one first finds a fundamental set of solutions for Eq. (5.114). The next step is to find a particular solution for Eq. (5.115). If this solution is also a solution for Eq. (5.114), another particular solution for (5.115) must be found. The search for a particular solution for Eq. (5.115) continues until a solution is found which is not a solution for the homogeneous equation given in (5.114). At this point the general solution given in Theorem 5.13 may be defined.

Example 5.47 Find the general solution of the difference equation

$$4f(x + 2) - 7f(x + 1) + 3f(x) = 2^x$$

The particular solutions for

$$4f(x + 2) - 7f(x + 1) + 3f(x) = 0$$

which form a fundamental set are given by

$$g_i(x) = k^x, \qquad i = 1, 2$$

where k satisfies

$$4k^2 - 7k + 3 = 0 \qquad \text{or} \qquad k = 1, \tfrac{3}{4}$$

Therefore
$$g_1(x) = 1, \qquad g_2(x) = (\tfrac{3}{4})^x$$

Since $g_1(x)$ and $g_2(x)$ are not identical for all integer x, $g_1(x)$ and $g_2(x)$ form a fundamental set of solutions for the homogeneous equation. Let the first trial solution for the nonhomogeneous equation be

$$g_1(x, c_1) = c_1 2^x$$

Substituting this solution into the nonhomogeneous equation leads to

$$4c_1 2^{x+2} - 7c_1 2^{x+1} + 3c_1 2^x = 2^x$$

or

$$5c_1 = 1 \quad \text{and} \quad c_1 = \tfrac{1}{5}$$

Therefore

$$h(x) = \tfrac{1}{5} 2^x$$

Substituting this solution in the homogeneous equation we show that it is not a solution for the homogeneous equation.

$$\tfrac{4}{5} 2^{x+2} - \tfrac{7}{5} 2^{x+1} + \tfrac{3}{5} 2^x = \tfrac{1}{5} 2^x (16 - 14 + 3) = 2^x$$

From Theorem 5.13 the general solution is given by

$$f(x) = c_1 + c_2 (\tfrac{3}{4})^x + \tfrac{1}{5} 2^x$$

Example 5.48 Find a particular solution for the nonhomogeneous difference equation given in Example 5.47 such that

$$f(0) = 0, \qquad f(1) = 2$$

From the general solution found in Example 5.47,

$$f(0) = c_1 + c_2 + \tfrac{1}{5}, \qquad f(1) = c_1 + \tfrac{3}{4} c_2 + \tfrac{2}{5}$$

or

$$-\tfrac{1}{5} = c_1 + c_2, \qquad \tfrac{8}{5} = c_1 + \tfrac{3}{4} c_2$$

and

$$\begin{bmatrix} c_1 \\ c_2 \end{bmatrix} = \begin{bmatrix} \tfrac{35}{5} \\ -\tfrac{36}{5} \end{bmatrix}$$

Therefore

$$f(x) = \tfrac{35}{5} - \tfrac{36}{5} (\tfrac{3}{4})^x + \tfrac{1}{5} 2^x$$

is the required particular solution.

Example 5.49 Find the general solution of the following difference equation.

$$f(x+2) - 4f(x+1) + 3f(x) = 3^x$$

The solutions for the homogeneous equation are given by

$$f_i(x) = k^x, \quad i = 1, 2 \quad \text{where} \quad k = 3, 1$$

That is,

$$f_1(x) = 3^x, \quad f_2(x) = 1$$

The first trial solution for the nonhomogeneous equation is

$$g_1(x, c_1) = c_1 3^x$$

However, $g_1(x, c_1)$ is a particular solution of the homogeneous equation and is therefore discarded. The second trial solution is given by

$$g_2(x, c_1) = c_1 x 3^x$$

Substitution of $g_2(x, c_1)$ into the homogeneous equation will show that it is not a solution of the homogeneous equation. Solving for c_1

$$c_1(x+2)3^{x+2} - 4c_1(x+1)3^{x+1} + 3c_1 x 3^x = 3^x$$

or

$$c_1[x(9 - 12 + 3) + (18 - 12)] = 1 \quad \text{and} \quad c_1 = \tfrac{1}{6}$$

Therefore, the general solution is

$$f(x) = c_1 + c_2 3^x + \tfrac{1}{6} x 3^x$$

5.7 Applications

5.7.1 Numerical Methods: Interpolation and Approximation

Let $f(x)$ be a function tabulated only at $x_0, x_1, \ldots, x_n$. Suppose we wish to determine the value of the function at x where $x_i < x < x_{i+1}$, $i < n$. Since the value of the function is known only at the points $x_0, x_1, \ldots, x_n$, it will be assumed that its exact value is not known at x. Therefore, an attempt will be made to approximate $f(x)$. The simplest means of accomplishing this approximation would be linear interpolation. Suppose $f(x)$ is as shown in Fig. 5.6.
Through linear interpolation the points $x_i, f(x_i)$ and $x_{i+1}, f(x_{i+1})$ are connected by a straight line. The approximate value of $f(x)$, denoted by $\hat{f}(x)$, is given by the value of the linear function at x. Therefore $\hat{f}(x)$ is given by

$$\hat{f}(x) = a + bx \tag{5.116}$$

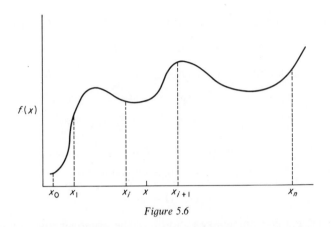

Figure 5.6

where a is the intercept and b is the slope. The slope of the straight line joining x_i, $f(x_i)$ and $x_{i+1}, f(x_{i+1})$ is given by

$$b = \frac{f(x_{i+1}) - f(x_i)}{x_{i+1} - x_i} \tag{5.117}$$

At $x = x_i$, $\hat{f}(x) = f(x_i)$. Therefore, a is determined as follows

$$f(x_i) = a + \frac{f(x_{i+1}) - f(x_i)}{x_{i+1} - x_i} x_i$$

and

$$a = f(x_i) - \frac{f(x_{i+1}) - f(x_i)}{x_{i+1} - x_i} x_i \tag{5.118}$$

Equation (5.116) now becomes

$$\hat{f}(x) = f(x_i) + \frac{(x - x_i)}{(x_{i+1} - x_i)} [f(x_{i+1}) - f(x_i)] \tag{5.119}$$

Noting that

$$\Delta f(x_i) = \frac{f(x_{i+1}) - f(x_i)}{x_{i+1} - x_i} \tag{5.120}$$

the approximation for $f(x)$ becomes

$$\hat{f}(x) = f(x_i) + (x - x_i) \Delta f(x_i) \tag{5.121}$$

It should be obvious that the degree to which $\hat{f}(x)$ accurately approximates $f(x)$ is largely dependent upon how well $f(x)$ can be approximated by a straight line on the interval (x_i, x_{i+1}). Therefore, when the interval (x_i, x_{i+1}) is large or when $f(x)$ is observed to fluctuate rapidly over the interval of tabulation, linear interpolation may lead to significant error. The reader will note that $\hat{f}(x)$ is based

5.7 APPLICATIONS

upon the values of x_i, x_{i+1} and $f(x_i), f(x_{i+1})$ only, and is therefore not influenced by remaining tabulated values of x and $f(x)$. Rapid gyration in $f(x)$ over the interval of tabulation would indicate that the influence of at least some of the remaining tabulated points should be reflected in the interpolation formula. This can be accomplished by approximating $f(x)$ by a polynomial relationship instead of the linear relationship given above. Newton's formula is among the most basic for polynomial approximation and is given without proof.

Let $f(x)$ be tabulated at $x_0, x_1, \ldots, x_n$ and let x be any point on the interval (x_0, x_n). Then

$$\hat{f}(x) = f(x_0) + \sum_{j=0}^{n-1} \left[\prod_{i=0}^{j} (x - x_i) \right] \Delta^{j+1} f(x_0) \qquad (5.122)$$

Therefore, as given in Eq. 5.122, Newton's formula gives an nth-degree polynomial approximation for $f(x)$.

Example 5.50 Given the following function (Table 5.5), find $\hat{f}(1)$ using linear, quadratic, and cubic interpolation.

TABLE 5.5

x	$f(x)$	x	$f(x)$
-3	0.001	0.5	0.691
-2	0.023	2	0.977
0	0.500	3	0.999

For linear interpolation, Newton's formula yields

$$\hat{f}(x) = f(x_0) + (x - x_0)\Delta f(x_0)$$

where $n = 1$. Therefore, x must lie on the interval (x_0, x_1). Among the tabulated values x lies on $(0.5, 2)$ since $x = 1$, and

$$x_0 = 0.5 \qquad x_1 = 2$$

Constructing a difference table we have

i	x_i	$f(x_i)$	$\Delta f(x_i)$
0	0.5	0.691	
			0.190
1	2	0.977	

and

$$\hat{f}(1) = 0.691 + (1 - 0.5)(0.19) = 0.786$$

For quadratic interpolation

$$\hat{f}(x) = f(x_0) + (x - x_0)\Delta f(x_0) + (x - x_0)(x - x_1)\Delta^2 f(x_0)$$

Again the interval (x_0, x_2) must include x. In this case $x_0 = 0$ or $x_0 = 0.5$. Let

$$x_0 = 0, \quad x_1 = 0.5, \quad x_2 = 2$$

Constructing the difference table yields us

i	x_i	$f(x_i)$	$\Delta f(x_i)$	$\Delta^2 f(x_i)$
0	0	0.500		
			0.382	
1	0.5	0.691		−0.096
			0.190	
2	2	0.977		

and

$$\hat{f}(1) = 0.500 + (1)(0.382) + (1)(0.5)(-0.096) = 0.834$$

Finally for cubic interpolation

$$\hat{f}(x) = f(x_0) + (x - x_0)\Delta f(x_0) + (x - x_0)(x - x_1)\Delta^2 f(x_0) \\ + (x - x_0)(x - x_1)(x - x_2)\Delta^3 f(x_0)$$

In this case we choose $x_0 = 0$, although $x_0 = -2$ or $x_0 = 0.5$ could also have been used. Again constructing a difference table, we have

i	x_i	$f(x_i)$	$\Delta f(x_i)$	$\Delta^2 f(x_i)$	$\Delta^3 f(x_i)$
0	0	0.500			
			0.382		
1	0.5	0.691		−0.096	
			0.190		0.009
2	2	0.977		−0.067	
			0.022		
3	3	0.999			

and

$$\hat{f}(1) = 0.500 + (1)(0.382) + (1)(0.5)(-0.096) + (1)(0.5)(-1)(0.009) = 0.829$$

The function used in this example is

$$f(x) = \int_{-\infty}^{x} \frac{1}{\sqrt{2\pi}} \exp\left[-\frac{y^2}{2}\right] dy$$

and $f(1) = 0.841$.

5.7.2 Numerical Methods: Integration

Suppose the integral $\int_a^b f(x)\, dx$ is to be evaluated, but cannot be evaluated analytically. In such a case the analyst must resort to numerical or approximate integration. The approach taken here to the development of methods for approximate integration involves first approximating $f(x)$ by a polynomial and then integrating the polynomial. Using Newton's formula $f(x)$ is approximated by $\hat{f}(x)$ where

$$\hat{f}(x) = f(x_0) + \sum_{j=0}^{n-1} \left[\prod_{i=0}^{j} (x - x_i) \right] \Delta^{j+1} f(x_0) \tag{5.123}$$

Then

$$\int_a^b f(x)\, dx \simeq \int_a^b \hat{f}(x)\, dx \tag{5.124}$$

where

$$\int_a^b \hat{f}(x)\, dx = (b-a)f(x_0) + \int_a^b \sum_{j=0}^{n-1} \prod_{i=0}^{j} [(x - x_i)] \Delta^{j+1} f(x_0)\, dx \tag{5.125}$$

Equation (5.125) leads to the well-known trapezoidal rule when $n = 1$. Let $x_0 = a$ and $x_n = b$. Then

$$\int_a^b \hat{f}(x)\, dx = \int_{x_0}^{x_n} \hat{f}(x)\, dx = \sum_{i=0}^{n-1} \int_{x_i}^{x_{i+1}} \hat{f}(x)\, dx \tag{5.126}$$

Therefore $f(x)$ is approximated by $\hat{f}(x)$ on each individual interval $[x_i, x_{i+1}]$. Evaluating the integral inside the summation yields

$$\int_{x_i}^{x_{i+1}} \hat{f}(x)\, dx = \int_{x_i}^{x_{i+1}} [f(x_i) + (x - x_i)\,\Delta f(x_i)]\, dx$$

$$= (x_{i+1} - x_i) f(x_i) + \left[\frac{x_{i+1}^2 - x_i^2}{2} - (x_{i+1} - x_i) x_i \right] \Delta f(x_i)$$

$$= (x_{i+1} - x_i) f(x_i) + \frac{(x_{i+1} - x_i)^2}{2} \Delta f(x_i)$$

$$= (x_{i+1} - x_i) f(x_i) + \frac{(x_{i+1} - x_i)}{2} [f(x_{i+1}) - f(x_i)]$$

$$= (x_{i+1} - x_i) \left[\frac{f(x_i) + f(x_{i+1})}{2} \right]$$

$$\tag{5.127}$$

The reader should note that $f(x_0)$ is replaced by $f(x_i)$ in Newton's formula since the interval containing x is $[x_i, x_{i+1}]$.

$$\int_{x_0}^{x_n} \hat{f}(x)\,dx = \sum_{i=0}^{n-1}(x_{i+1} - x_i)\left[\frac{f(x_i) + f(x_{i+1})}{2}\right]$$

$$= (x_1 - x_0)\left[\frac{f(x_0) + f(x_1)}{2}\right] + (x_2 - x_1)\left[\frac{f(x_1) + f(x_2)}{2}\right]$$

$$+ \cdots + (x_n - x_{n-1})\left[\frac{f(x_{n-1}) + f(x_n)}{2}\right]$$

$$= (x_1 - x_0)\frac{f(x_0)}{2} + \sum_{i=1}^{n-1}(x_{i+1} - x_{i-1})\frac{f(x_i)}{2}$$

$$+ (x_n - x_{n-1})\frac{f(x_n)}{2} \tag{5.128}$$

When $x_{i+1} - x_i = \Delta x$ for $i = 0, 1, 2, \ldots, n - 1$ we have the more familiar form of the trapezoidal rule given by

$$\int_{x_0}^{x_n} \hat{f}(x)\,dx = \frac{\Delta x}{2}[f(x_0) + 2f(x_1) + 2f(x_2) + \cdots + 2f(x_{n-1}) + f(x_n)] \tag{5.129}$$

Example 5.51 Using the trapezoidal rule, integrate the function $f(x)$ tabulated (Table 5.6) from 0.2 to 1.0.

TABLE 5.6

x	$f(x)$	x	$f(x)$
0.2	0.39	0.6	0.33
0.3	0.38	0.9	0.27
0.5	0.35	1.0	0.24

Since the tabulated increment for x is not constant the required integral is given by

$$\int_{x_0}^{x_n} \hat{f}(x)\,dx = (x_1 - x_0)\frac{f(x_0)}{2} + \sum_{i=1}^{4}(x_{i+1} - x_{i-1})\frac{f(x_i)}{2} + (x_5 - x_4)\frac{f(x_5)}{2}$$

where $x_0 = 0.2$ and $x_5 = 1.0$. Therefore

$$\int_{0.2}^{1.0} \hat{f}(x)\,dx = (0.3 - 0.2)\frac{0.39}{2} + (0.5 - 0.2)\frac{0.38}{2} + (0.6 - 0.3)\frac{0.35}{2}$$

$$+ (0.9 - 0.5)\frac{0.33}{2} + (1.0 - 0.6)\frac{0.27}{2} + (1.0 - 0.9)\frac{0.24}{2} = 0.261$$

5.7 APPLICATIONS

When the approximating function is a quadratic and the number of tabulated values of $f(x_i)$ is odd, Eq. (5.125) leads to Simpson's rule. In this case we express the integral to be evaluated as

$$\int_a^b \hat{f}(x)\,dx = \int_{x_0}^{x_2} \hat{f}(x)\,dx + \int_{x_2}^{x_4} \hat{f}(x)\,dx + \cdots + \int_{x_{n-2}}^{x_n} \hat{f}(x)\,dx \quad (5.130)$$

where $x_0 = a$ and $x_n = b$. The reason the limits of each integral in Eq. (5.130) are of the form x_i, x_{i+2} is that $\hat{f}(x)$ is a quadratic function and therefore must pass through three tabulated points, $f(x_i)$. Thus each interval of integration must contain three such points.

To simplify the derivation of Simpson's rule let $x_{i+1} - x_i = \Delta x$ for all i. Then

$$\int_{x_i}^{x_{i+2}} \hat{f}(x)\,dx = \int_{x_i}^{x_{i+2}} [f(x_i) + (x - x_i)\Delta f(x_i) + (x - x_i)(x - x_{i+1})\Delta^2 f(x_i)]\,dx$$

$$= 2(\Delta x) f(x_i) + 4(\Delta x)^2 \frac{\Delta f(x_i)}{2} + 4(\Delta x)^3 \frac{\Delta^2 f(x_i)}{6}$$

$$\int_{x_i}^{x_{i+2}} \hat{f}(x)\,dx = 2\Delta x \left[f(x_i) + f(x_{i+1}) - f(x_i) + \frac{f(x_{i+2}) - f(x_{i+1})}{6} - \frac{f(x_{i+1}) - f(x_i)}{6} \right]$$

$$= \frac{\Delta x}{3} [f(x_i) + 4f(x_{i+1}) + f(x_{i+2})] \quad (5.131)$$

where i is an even number.

$$\int_a^b \hat{f}(x)\,dx = \frac{\Delta x}{3} \{[f(x_0) + 4f(x_1) + f(x_2)] + [f(x_2) + 4f(x_3) + f(x_4)]$$

$$+ [f(x_4) + 4f(x_5) + f(x_6)] + \cdots$$

$$+ [f(x_{n-4}) + 4f(x_{n-3}) + f(x_{n-2})]$$

$$+ [f(x_{n-2}) + 4f(x_{n-1}) + f(x_n)]\}$$

$$= \frac{\Delta x}{3} [f(x_0) + 4f(x_1) + 2f(x_2) + 4f(x_3) + 2f(x_4) + \cdots$$

$$+ 2f(x_{n-2}) + 4f(x_{n-1}) + f(x_n)] \quad (5.132)$$

Example 5.52 By approximating $f(x)$ by a cubic equation on the intervals (x_i, x_{1+3}), $i = 0, 3, 6, 9, \ldots$, evaluate the following integral

$$\int_0^{1.2} f(x)\,dx$$

where

x	$f(x)$	x	$f(x)$
0.0	5.00	0.7	10.06
0.1	5.53	0.8	11.12
0.2	6.11	0.9	12.29
0.3	6.75	1.0	13.59
0.4	7.46	1.1	15.02
0.5	8.24	1.2	16.60
0.6	9.11		

Following the pattern used in developing the trapezoidal rule and Simpson's rule we evaluate $\int_{x_i}^{x_{i+3}} \hat{f}(x)\, dx$.

$$\int_{x_i}^{x_{i+3}} \hat{f}(x)\, dx = \int_{x_i}^{x_{i+3}} [f(x_i) + (x - x_i)\Delta f(x_i)$$
$$+ (x - x_i)(x - x_{i+1})\Delta^2 f(x_i)$$
$$+ (x - x_i)(x - x_{i+1})(x - x_{i+2})\Delta^3 f(x_i)]\, dx$$

Since $x_{i+1} - x_i = 0.1$ for all i, we have

$$\int_{x_i}^{x_{i+3}} \hat{f}(x)\, dx = 3\,\Delta x f(x_i) + 9(\Delta x)^2 \frac{\Delta f(x_i)}{2} + 9(\Delta x)^3 \frac{\Delta^2 f(x_i)}{2} + 9(\Delta x)^4 \frac{\Delta^3 f(x_i)}{4}$$

$$= 3\,\Delta x\{f(x_i) + \tfrac{3}{2}[f(x_{i+1}) - f(x_i)] + \tfrac{3}{4}[f(x_{i+2}) - f(x_{i+1})$$
$$- f(x_{i+1}) + f(x_i)] + \tfrac{3}{24}[f(x_{i+3}) - 3f(x_{i+2})$$
$$+ 3f(x_{i+1}) - f(x_i)]\}$$

$$= \frac{\Delta x}{8}[3f(x_i) + 9f(x_{i+1}) + 9f(x_{i+2}) + 3f(x_{i+3})] \qquad (5.133)$$

where $\Delta x = 0.1$. Therefore, if n is an integer multiple of 3, then

$$\int_{x_0}^{x_n} \hat{f}(x)\, dx = \int_{x_0}^{x_3} \hat{f}(x)\, dx + \int_{x_3}^{x_6} \hat{f}(x)\, dx + \cdots + \int_{x_{n-3}}^{x_n} \hat{f}(x)\, dx$$

$$= \frac{\Delta x}{8}[3f(x_0) + 9f(x_1) + 9f(x_2) + 6f(x_3) + 9f(x_4)$$
$$+ \cdots + 6f(x_{n-3}) + 9f(x_{n-2}) + 9f(x_{n-1}) + 3f(x_n)] \qquad (5.134)$$

For the problem of this example, $x_0 = 0.0$ and $x_n = 1.2$.

$$\int_0^{1.2} f(x)\, dx = (0.1/8)[3(5.00) + 9(5.53) + 9(6.11) + 6(6.75) + 9(7.46)$$
$$+ 9(8.24) + 6(9.11) + 9(10.06) + 9(11.12) + 6(12.29)$$
$$+ 9(13.59) + 9(15.02) + 3(16.60)]$$
$$= 11.598$$

5.7.3 Numerical Methods: Differentiation

The problem of numerical differentiation is resolved in the same fashion as that of numerical integration. If $f(x)$ is to be differentiated at a particular value of x we first approximate $f(x)$ using Newton's formula. The approximating function, $\hat{f}(x)$, is then differentiated with respect to x. To illustrate, let

$$\hat{f}(x) = f(x_0) + (x - x_0)\Delta f(x_0) + (x - x_0)(x - x_1)\Delta^2 f(x_0)$$

Taking the derivative of $\hat{f}(x)$ with respect to x

$$\frac{d}{dx}\hat{f}(x) = \Delta f(x_0) + (2x - x_0 - x_1)\Delta^2 f(x_0) \qquad (5.135)$$

Example 5.53 Approximate the following tabulated function (Table 5.7) by a third-degree polynomial and find the approximate derivative of the function at $x = 0$.

TABLE 5.7

x	$f(x)$	x	$f(x)$
-3.0	0.001	1.0	0.841
-1.5	0.067	1.5	0.933
-1.0	0.159	3.0	0.999

Using Newton's formula, we have

$$\hat{f}(x) = f(x_0) + (x - x_0)\Delta f(x_0) + (x - x_0)(x - x_1)\Delta^2 f(x_0)$$
$$+ (x - x_0)(x - x_1)(x - x_2)\Delta^3 f(x_0)$$

Taking the derivative of $\hat{f}(x)$

$$\frac{d}{dx}\hat{f}(x) = \Delta f(x_0) + (2x - x_0 - x_1)\Delta^2 f(x_0)$$
$$+ [3x^2 - 2x(x_0 + x_1 + x_2) + (x_0 x_1 + x_0 x_2 + x_1 x_2)]\Delta^3 f(x_0)$$

According to the approximation, $\hat{f}(x)$, x must lie on the interval (x_0, x_3). Letting $x_0 = -1.5$, we have the following difference table.

i	x_i	$f(x_i)$	$\Delta f(x_i)$	$\Delta^2 f(x_i)$	$\Delta^3 f(x_i)$
0	-1.5	0.067			
			0.184		
1	-1.0	0.159		0.063	
			0.341		-0.042
2	1.0	0.841		-0.063	
			0.184		
3	1.5	0.933			

Therefore

$$\frac{d}{dx}\hat{f}(0) = 0.184 + (1.5 + 1.0)(0.063)$$

$$+ [(-1.5)(-1.0) + (-1.5)(1.0) + (-1.0)(1.0)](-0.042)$$

$$= 0.384$$

5.7.4 Economic Systems: Inventory Models

Among the problems treated in operations research, inventory problems are among the most common. Consider the problem of designing a system through which an adequate and yet economic stock of products can be maintained. Assume that products are drawn from inventory uniformly at a rate of d per year. That is, the time interval between withdrawals from the inventory is $1/d$. To maintain a supply of the product, the inventory must be replenished periodically. If an adequate supply of product is to be maintained, each demand to the inventory must be met. That is, the inventory must not be empty when a demand occurs.

Specification of the inventory system consists of defining the points in time at which the inventory should be replenished and the number of units placed in inventory upon replenishment. One system for maintaining an adequate supply of product would be to place d units of product in inventory at the beginning of each year. Since d units are demanded per year this system would provide the supply of product necessary to meet each demand. However, this system would mean that $d/2$ units would be carried for one-half year. The problem here is that a cost is incurred when one unit is carried in inventory for any period of time and this cost is usually considered to be proportional to the period of time for which it is carried. The cost of carrying units in inventory is composed of expenses such as taxes, insurance, storage space, and so forth.

To reduce the inventory carrying cost the analyst might provide for replenishing the inventory by one unit every $1/d$ units of time. Such a system would result in a zero inventory carrying cost since the arrival of a unit in inventory coincides with a demand and therefore spends zero time in inventory. However, receipt of an order results in costs of transportation, handling, paperwork, etc. Therefore, the analyst must recognize a cost which is proportional to the frequency with which the inventory is replenished. Although this cost may be proportional to the number of units on the order replenishing the inventory, it will be considered independent of the size of the order here.

To specify the inventory system a mathematical model will be developed which represents the annual cost of maintaining any given system which meets the requirement of meeting all demand. By minimizing this model with respect to the decision variable involved, we can define that system which is economically

5.7 APPLICATIONS

optimal. Let Q be the number of units placed in inventory at each replenishment. Let C_0 be the cost of placing and receiving an order for replenishment regardless of the size of the order, and let C_I be the cost of carrying one unit of product for one year.

Assume that Q units are received at t_0 and that this receipt coincides with a demand. The next order should then be received at $t_0 + Q/d$. That is, the second order is received after the inventory reaches zero but before a demand occurs while the inventory is empty. A graph of the fluctuation of the inventory level over time is shown in Fig. 5.7.

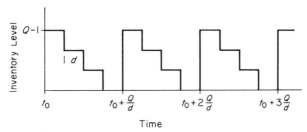

Figure 5.7 Fluctuation of Inventory Level with Time

The period between the receipt of two successive orders will be defined as a cycle and is therefore Q/d years in length. To determine the annual cost of any inventory system of the type discussed here we will determine the cost per cycle and multiply by the number of cycles per year. The cost of inventory replenishment per cycle is simply C_0. To determine the inventory carrying cost per cycle note that $Q - 1$ units are carried for $1/d$ years, $Q - 2$ units for $1/d$ years, $Q - 3$ units for $1/d$ years, etc. The cost of carrying $Q - i$ units for $1/d$ years is $C_I(Q - i)/d$. Therefore

$$\text{Carrying cost per cycle} = C_I \sum_{i=1}^{Q} \frac{Q - i}{d} = C_I \left[\frac{Q^2}{d} - \frac{Q(Q + 1)}{2d} \right] = C_I \frac{Q(Q - 1)}{2d} \tag{5.136}$$

since

$$\sum_{i=1}^{Q} i = \frac{Q(Q + 1)}{2}$$

Hence

$$\text{Total cost per cycle} = C_0 + C_I \frac{Q(Q - 1)}{2d} \tag{5.137}$$

The number of cycles per year is d/Q and the total annual cost of maintaining the inventory is given by $C_T(Q)$.

$$C_T(Q) = C_0 \frac{d}{Q} + C_I \frac{Q - 1}{2} \tag{5.138}$$

If $C_T(Q)$ has a local minimum at Q^* then
$$\Delta C_T(Q^* - 1) \leq 0 \leq \Delta C_T(Q^*) \tag{5.139}$$
where
$$\Delta C_T(Q) = \frac{C_I}{2} - \frac{C_0 d}{Q(Q+1)} \tag{5.140}$$

Example 5.54 For the inventory problem described above, find the optimal value of Q if
$$C_0 = \$10.00, \quad C_I = \$50.00, \quad d = 200$$
From Eq. (5.140),
$$\Delta C_T(Q) = 25 - \frac{2000}{Q(Q+1)}$$
The optimal value of Q, Q^*, must satisfy
$$\Delta C_T(Q^* - 1) \leq 0 \leq \Delta C_T(Q^*)$$
Considering the left-hand side inequality first
$$25 - \frac{2000}{(Q^* - 1)Q^*} \leq 0 \quad \text{or} \quad (Q^* - 1)Q^* \leq 80$$
Therefore,
$$Q^* \leq 9$$
For the inequality $\Delta C_T(Q^*) \geq 0$ we have
$$25 - \frac{2000}{Q^*(Q^* + 1)} \geq 0 \quad \text{or} \quad Q^*(Q^* + 1) \geq 80 \quad \text{and} \quad Q^* \geq 9$$
The only value of Q^* such that $9 \leq Q^* \leq 9$ is $Q^* = 9$. To show that this value of Q_0 minimizes $C_T(Q)$ note that
$$\Delta C_T(Q^* - 1) = 25 - \frac{2000}{72} = -2.78$$
and
$$\Delta C_T(Q^*) = 25 - \frac{2000}{90} = 2.78$$
Since $\Delta C_T(Q^* - 1) < 0 < \Delta C_T(Q^*)$, $Q^* = 9$ is the optimal order quantity.

The inventory system described here is a very simple one and does not include many characteristics often found in inventory problems. Demand was considered to be predictable, while in many problems demand is a random variable. In addition, stockouts were not permitted in the above system. Stockouts

5.7 APPLICATIONS 437

become particularly troublesome when demand is a random variable since stockouts cannot be avoided with certainty in such systems. Finally, problems of product deterioration during storage due to breakage, perishability, obsolescence, etc. are often encountered in inventory analysis.

5.7.5 Economic Systems: Quality Control Models

A quality control system is used to determine whether a unit or units of product meet quality standards. In the case which will be discussed here product is received in large lots and the acceptability of each lot is to be determined. One means of determining lot acceptability would be to inspect each unit in the lot, screening out unacceptable or defective units. However, if the number of units contained in each lot, the lot size, is large, inspecting each unit could be prohibitively expensive. For this reason the inspection is usually carried out by drawing, at random, a small number of units from the lot, inspecting each unit in this sample, and based upon the results of the inspection of the sample a conclusion is drawn regarding the acceptability of the lot as a whole.

Let L be the number of units of product contained in the lot, and let n be the number of units in the sample drawn from the lot. Therefore $n \leq L$. If every unit in the sample is defective there is certainly reason to suspect that the lot is of poor quality. On the other hand, if none of the units in the sample are defective, one might conclude that the lot is of acceptable quality. In either case the reliability of the conclusion would depend upon the size of the sample, n, since the larger the sample the more closely it represents the lot. For the purposes of this discussion the lot size, L, and the sample size, n, will be assumed to be fixed. The decision variable will be the acceptance number, c. The acceptance number defines the number of defective units which may be found in the sample without rejecting the lot. That is, if x is the number of defective units found in the sample, then the lot is accepted if $x \leq c$, and rejected if $x > c$.

In the discussion which follows we will develop a mathematical model which represents the cost of the quality control system per lot. This model will then be minimized with respect to c, the acceptance number, to determine the optimal or minimum cost quality control system. To develop such a model three costs must be recognized. First, there is the cost of the inspection of the sample. Let C_I be the total cost of inspecting a single unit of product. The total cost of inspection per lot is then given by

$$\text{inspection cost per lot} = C_I n \tag{5.141}$$

Since the inspection cost is not a function of c, this cost component could be ignored in this case. However, for the purpose of completeness it will be included in the model.

The second cost which the analyst must recognize is that associated with rejection of an entire lot. The magnitude of this cost depends upon the disposition of rejected lots. In some cases, the entire lot is inspected to screen out defective units upon rejection of a lot. In other cases, the rejected lot is returned to the supplier and an emergency order is placed to replace the rejected lot. Regardless of the situation encountered, a cost must be considered to arise whenever a lot is rejected. Otherwise the optimal quality control system would be to reject all lots without inspecting them. For the case considered here we will assume that it costs C_R for each unit contained in a rejected lot. Therefore when a lot is rejected a cost $C_R L$ is incurred.

As already indicated, the rejection or acceptance of a lot is dependent upon the number of defects found in the sample which in turn depends upon the number of defects in the lot. That is, as the number of defects in the lot increases we can expect the number of defects found in the sample to increase. We will assume that when a lot is received it contains a proportion of defective units q_1 or q_2. Lots of proportion defective q_1 are considered acceptable while a proportion defective of q_2 is considered unacceptable and should be rejected. Therefore, $0 < q_1 < q_2 < 1$ and a lot contains either $q_1 L$ or $q_2 L$ defective units. If one could predict with certainty the proportion defective for a given lot there would be no need for a quality control system. We will assume that the probability that a lot has a proportion defective q_1 is p. That is, the probability that a given lot is of acceptable quality is p and the probability that it is unacceptable is $1 - p$.

Let $p(x \,|\, q_i)$ be the probability that x defective units are found in a sample of size n selected from a lot of proportion defective q_i. In general, $p(x \,|\, q_i)$ can be represented by a Poisson distribution or

$$p(x \,|\, q_i) = \frac{(nq_i)^x}{x!} \exp[-nq_i] \tag{5.142}$$

Therefore,

probability of lot rejection = (prob. lot is q_1)(prob. $x > c$ if lot is q_1)
$$+ \text{(prob. lot is } q_2\text{)(prob. } x > c \text{ if lot is } q_2\text{)}$$

$$= p \sum_{x=c+1}^{n} \frac{(nq_1)^x}{x!} \exp[-nq_1]$$

$$+ (1 - p) \sum_{x=c+1}^{n} \frac{(nq_2)^x}{x!} \exp[-nq_2] \tag{5.143}$$

The expected cost of lot rejection is given by

expected cost of lot rejection

$$= C_R L \text{ (prob. of lot rejection)}$$

$$= C_R L \left[p \sum_{x=c+1}^{n} \frac{(nq_1)^x}{x!} \exp[-nq_1] + (1 - p) \sum_{x=c+1}^{n} \frac{(nq_2)^x}{x!} \exp[-nq_2] \right]$$

$$\tag{5.144}$$

5.7 APPLICATIONS

The third cost considered here is the cost of accepting a lot. A defect in a unit of product implies that there is something undesirable about the unit. We will assume that each defective unit of product in an accepted lot results in a cost C_A. That is, when a lot is accepted, every defective unit in the uninspected portion of the lot is accepted and every such unit results in a cost C_A. After the sample of size n is inspected, the number of units in the remaining lot is $L - n$. Therefore an accepted lot contains either $q_1(L - n)$ or $q_2(L - n)$ defective units. The expected cost of acceptance per lot is given by

expected cost of lot acceptance
$= C_A q_1(L - n)(\text{prob. lot is } q_1)(\text{prob. } x \leq c \text{ if lot is } q_1)$
$+ C_A q_2(L - n)(\text{prob. lot is } q_2)(\text{prob. } x \leq c \text{ if lot is } q_2)$

$$= C_A(L - n)\left[pq_1 \sum_{x=0}^{c} \frac{(nq_1)^x}{x!} \exp[-nq_1]\right.$$

$$\left. + (1 - p)q_2 \sum_{x=0}^{c} \frac{(nq_2)^x}{x!} \exp[-nq_2]\right] \quad (5.145)$$

Summing the three cost components gives us the expected total cost of the system per lot, $C_T(c)$.

$$C_T(c) = C_I n + C_R L \left[p \sum_{x=c+1}^{n} \frac{(nq_1)^x}{x!} \exp[-nq_1] \right.$$

$$+ (1 - p) \sum_{x=c+1}^{n} \frac{(nq_2)x}{x!} \exp[-nq_2] \Bigg]$$

$$+ C_A(L - n)\left[pq_1 \sum_{x=0}^{c} \frac{(nq_1)x}{x!} \exp[-nq_1]\right.$$

$$\left. + (1 - p)q_2 \sum_{x=0}^{c} \frac{(nq_2)^x}{x!} \exp[-nq_2]\right] \quad (5.146)$$

The optimal value of c, c^*, must satisfy the inequality

$$\Delta C_T(c^* - 1) \leq 0 \leq \Delta C_T(c^*) \quad (5.147)$$

where

$$\Delta C_T(c) = -C_R L \left[p \frac{(nq_1)^{c+1}}{(c + 1)!} \exp[-nq_1] + (1 - p) \frac{(nq_2)^{c+1}}{(c + 1)!} \exp[-nq_2] \right]$$

$$+ C_A(L - n)\left[pq_1 \frac{(nq_1)^{c+1}}{(c + 1)!} \exp[-nq_1]\right.$$

$$\left. + (1 - p)q_2 \frac{(nq_2)^{c+1}}{(c + 1)!} \exp[-nq_2]\right] \quad (5.148)$$

First, considering the condition $\Delta C_T(c^* - 1) \le 0$, we have

$$C_A(L-n)[pq_1(nq_1)^{c^*}\exp[-nq_1] + (1-p)q_2(nq_2)^{c^*}\exp[-nq_2]]$$
$$\le C_R L[p(nq_1)^{c^*}\exp[-nq_1] + (1-p)(nq_2)^{c^*}\exp[-nq_2]] \quad (5.149)$$

where $\Delta C_T(c^* - 1) \le 0$ is multiplied by $c^*!$.

$$(nq_1)^{c^*}\exp[-nq_1][C_A(L-n)pq_1 - C_R Lp] \le (nq_2)^{c^*}\exp[-nq_2][C_R L(1-p) - C_A(L-n)(1-p)q_2]$$

Dividing both sides by $(nq_2)^{c^*}\exp[-nq_2]$ and $[C_A(L-n)pq_1 - C_R Lp]$ leads to

$$\left(\frac{q_1}{q_2}\right)^{c^*}\exp[-n(q_1-q_2)] \ge \frac{(1-p)[C_R L - C_A(L-n)q_2]}{p[C_A(L-n)q_1 - C_R L]}$$

or

$$\left(\frac{q_1}{q_2}\right)^{c^*} \ge \frac{(1-p)}{p}\frac{C_R L - C_A(L-n)q_2}{C_A(L-n)q_1 - C_R L}\exp[n(q_1-q_2)]$$

and

$$c^* \le \frac{\ln\left(\frac{1-p}{p}\right) + \ln\left[\frac{C_R L - C_A(L-n)q_2}{C_A(L-n)q_1 - C_R L}\right] + n(q_1 - q_2)}{\ln\left(\frac{q_1}{q_2}\right)} \quad (5.150)$$

For the condition $\Delta C_T(c^*) \ge 0$,

$$C_A(L-n)[pq_1(nq_1)^{c^*+1}\exp[-nq_1] + (1-p)q_2(nq_2)^{c^*+1}\exp[-nq_2]]$$
$$\ge C_R L[p(nq_1)^{c^*+1}\exp[-nq_1] + (1-p)(nq_2)^{c^*+1}\exp[-nq_2]]$$

where $\Delta C_T(c^*) \ge 0$ is multiplied by $(c^* + 1)!$. Following an argument similar to the above we have

$$c^* \ge \frac{\ln\left(\frac{1-p}{p}\right) + \ln\left[\frac{C_R L - C_A(L-n)q_2}{C_A(L-n)q_1 - C_R L}\right] + n(q_1 - q_2)}{\ln\left(\frac{q_1}{q_2}\right)} - 1 \quad (5.151)$$

Let

$$\psi = \frac{\ln\left(\frac{1-p}{p}\right) + \ln\left[\frac{C_R L - C_A(L-n)q_2}{C_A(L-n)q_1 - C_R L}\right] + n(q_1 - q_2)}{\ln\left(\frac{q_1}{q_2}\right)} \quad (5.152)$$

Then the optimal value of c, c^*, must satisfy the inequality

$$\psi - 1 \le c^* \le \psi \quad (5.153)$$

5.7 APPLICATIONS

Example 5.55 For the quality control system discussed above let

$$p = 0.80, \quad q_1 = 0.01, \quad q_2 = 0.10, \quad L = 100{,}000$$
$$n = 100, \quad C_R = \$.01, \quad C_A = \$.50$$

Find the optimal value of c.
From Eq. (5.152),

$$\psi = \frac{\ln(0.25) + \ln\left[\dfrac{1000 - 4995}{499.5 - 1000}\right] + 100(0.01 - 0.10)}{\ln(0.10)} = 3.61$$

Therefore, $2.61 \le c^* \le 3.61$ and $c^* = 3$. Since $2.61 < c^* < 3.61$, c^* minimizes $C_T(c^*)$.

5.7.6 Queueing Models: Single Channel

Arrivals to a service channel occur in a Poisson fashion at an average rate of λ per unit time. That is, the probability of x arrivals in any time interval of length t is given by

$$p(x) = \frac{(\lambda t)^x}{x!} \exp[-\lambda t], \quad x = 0, 1, 2, \ldots$$

As shown in Fig. 5.8, when an arrival occurs it takes its place in the waiting line at the service channel unless there is no waiting line and the service channel is available to serve the entering arrival. When service is completed upon a unit, the unit is released from the service channel or facility, the unit at the head of the

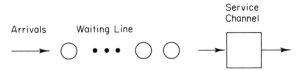

Figure 5.8 Single Channel Queueing System

waiting line is brought into the service channel, and service commences on that unit. If there is no waiting line when service is completed on a unit, the service channel becomes idle and remains idle until the next arrival occurs. The time to service a unit once it has entered the service channel is exponentially distributed with mean rate u per time unit. That is, the density function of time to complete service, t, is given by

$$f(t) = u \exp[-ux], \quad 0 < x < \infty$$

We will assume that $u > \lambda$ throughout this discussion.

Examples of the type of system considered here include the arrival of patrons at a theater ticket office, the arrival of airplanes at an airport, the arrival of patients at a doctor's office, and the arrival of cars at a toll booth. Of course the distribution of arrivals and service times will not always be Poisson and exponential, respectively.

Given the conditions specified above, the steady state probability that there are n units in the system, $P(n)$, at any time is given by

$$P(n) = \begin{cases} (1 - \lambda)P(0) + uP(1), & n = 0 \\ \lambda P(n - 1) + (1 - \lambda - u)P(n) + uP(n + 1), & n > 0 \end{cases} \quad (5.154)$$

The number of units in the system is comprised of the number in the waiting line plus the number in the service channel. Let us find an expression for $P(n)$ subject to the restrictions that

$$\sum_{n=0}^{\infty} P(n) = 1 \quad (5.155)$$

From Eq. (5.154) we have

$$P(1) = \frac{\lambda}{u} P(0) \quad (5.156)$$

and

$$uP(n + 1) - (\lambda + u)P(n) + \lambda P(n - 1) = 0, \quad n > 0 \quad (5.157)$$

From Theorem 5.8,

$$P(n) = k^n \quad (5.158)$$

is a solution for Eq. (5.157), where

$$k = \frac{(\lambda + u) \pm \sqrt{(\lambda + u)^2 - 4\lambda u}}{2u} = \frac{\lambda}{u}, 1$$

If $k = 1$,

$$\sum_{n=0}^{\infty} P(n) = \sum_{n=1}^{\infty} (1) + P(0) \neq 1$$

and this solution must be rejected. If $k = \lambda/u$ and $|\lambda/u| < 1$,

$$\sum_{n=0}^{\infty} P(n) = \sum_{n=0}^{\infty} \left(\frac{\lambda}{u}\right)^n = \frac{1}{1 - \lambda/u}$$

which satisfies Eq. (5.155) only if $\lambda/u = 0$. However, if $\lambda/u = 0$ either $\lambda = 0$ or $u = \infty$. In either case, the problem would be trivial. Therefore we seek another solution.

From Theorem 5.9, any solution to this difference equation can be expressed as

$$P_3(n) = c_1 P_1(n) + c_2 P_2(n) \quad (5.159)$$

5.7 APPLICATIONS

where

$$P_1(n) = 1, \quad P_2(n) = \left(\frac{\lambda}{u}\right)^n$$

if

$$\begin{vmatrix} P_1(n) & P_2(n) \\ P_1(n+1) & P_2(n+1) \end{vmatrix} \neq 0$$

But

$$\begin{vmatrix} 1 & \left(\frac{\lambda}{u}\right)^n \\ 1 & \left(\frac{\lambda}{u}\right)^{n+1} \end{vmatrix} = \left(\frac{\lambda}{u}\right)^n \left(\frac{\lambda}{u} - 1\right)$$

and is nonzero if $\lambda/u \neq 0$ and $\lambda/u \neq 1$. From Eqs. (5.156) and (5.159)

$$P_3(0) = c_1 + c_2 \tag{5.160}$$

$$\frac{\lambda}{u} P_3(0) = c_1 + c_2 \frac{\lambda}{u} \tag{5.161}$$

The values of c_1 and c_2 satisfying these equations are given by

$$\begin{bmatrix} c_1 \\ c_2 \end{bmatrix} = \frac{1}{\frac{\lambda}{u} - 1} \begin{bmatrix} \frac{\lambda}{u} - 1 \\ -1 & 1 \end{bmatrix} \begin{bmatrix} P_3(0) \\ \frac{\lambda}{u} P_3(0) \end{bmatrix} = \begin{bmatrix} 0 \\ P_3(0) \end{bmatrix}$$

Therefore,

$$P_3(n) = \left(\frac{\lambda}{u}\right)^n P_3(0) \tag{5.162}$$

To solve for $P_3(0)$ we use the restriction given in Eq. (5.155).

$$\sum_{n=0}^{\infty} P_3(n) = \sum_{n=0}^{\infty} \left(\frac{\lambda}{u}\right)^n P_3(0) = \frac{1}{1 - \lambda/u} P_3(0)$$

and

$$P_3(0) = 1 - \frac{\lambda}{u}$$

$P_3(n)$ is then given by

$$P_3(n) = \left(\frac{\lambda}{u}\right)^n (1 - \lambda/u) \tag{5.163}$$

Equation (5.163) is valid only if $0 < \lambda/u < 1$. It was assumed initially that $\lambda/u \leq 1$. If $\lambda \geq u$ the system will not reach a steady state condition and Eq. (5.153) is invalid. For example, if $\lambda > u$ the number in the system will tend to increase indefinitely. If $\lambda/u < 0$, either the arrival or service rate is negative which has no meaning from a practical point of view. Finally, if $\lambda/u = 0$, we have the trivial case already discussed.

5.7.7 Queueing Models: Single Channel with Impatient Arrivals

The queueing system considered here is the same as that just presented with the exception that an arrival may not join the waiting line if it is too long. Specifically, if the waiting line is equal to or exceeds $m > 1$ units, an arrival will join the waiting line with probability p. If the waiting line is of length less than m an arrival is certain to join the waiting line. The steady state probability that there are n units in the system is given by

$$P(n) = \begin{cases} (1 - \lambda)P(0) + uP(1), & n = 0 \\ \lambda P(n - 1) + (1 - \lambda - u)P(n) + uP(n + 1), & 0 < n \leq m + 1 \\ p\lambda P(n - 1) + [1 - p\lambda - u]P(n) + uP(n + 1), & n > m + 1 \end{cases}$$

(5.164)

subject to the restriction that

$$\sum_{n=0}^{\infty} P(n) = 1 \qquad (5.165)$$

From Eq. (5.164) we have

$$P(1) = \frac{\lambda}{u} P(0) \qquad (5.166)$$

$$\lambda P(n - 1) - (\lambda + u)P(n) + uP(n + 1) = 0, \qquad 0 < n \leq m + 1 \quad (5.167)$$

$$p\lambda P(n - 1) - [p\lambda + u]P(n) + uP(n + 1) = 0, \qquad n > m + 1 \quad (5.168)$$

From the previous model particular solutions for Eq. (5.167) are given by

$$P_1(n) = \left(\frac{\lambda}{u}\right)^n, \qquad P_2(n) = 1$$

Since $m > 1$, $P_2(n)$ cannot be the required solution since

$$\sum_{n=0}^{1} P_2(n) > 1$$

5.7 APPLICATIONS

Similarly, $P_1(n)$ is not the required solution since

$$\sum_{n=0}^{1} P_1(n) > 1$$

The third solution developed in the previous model is given by

$$P_3(n) = \left(\frac{\lambda}{u}\right)^n P(0) \quad \text{and} \quad \sum_{n=0}^{m+1} P_3(n) = \sum_{n=0}^{m+1} \left(\frac{\lambda}{u}\right)^n P(0) = \frac{1 - (\lambda/u)^{m+2}}{1 - \lambda/u} P(0)$$

which is valid for

$$P(0) < \frac{1 - \lambda/u}{1 - (\lambda/u)^{m+2}} \tag{5.169}$$

The explicit value of $P(0)$ will be specified after the solution for Eq. (5.168) is determined, using the restriction given in Eq. (5.165).

Particular solutions for Eq. (5.168) are given by

$$P(n) = k^n$$

where

$$k = \frac{(p\lambda + u) \pm \sqrt{(p\lambda + u)^2 - 4p\lambda u}}{2u} = \frac{p\lambda}{u}, 1 \tag{5.170}$$

Let

$$P_1(n) = \left(\frac{p\lambda}{u}\right)^n, \quad n > m + 1 \tag{5.171}$$

$$P_2(n) = 1, \quad n > m + 1 \tag{5.172}$$

$P_2(n)$ is not a solution for finite m since

$$\sum_{n=m+2}^{\infty} P_2(m) = \infty$$

If $P_1(n)$ is a solution it must satisfy Eqs. (5.167) and (5.168) for $n = m + 1$ and $n = m + 2$ respectively. That is, $P_1(n)$ must satisfy

$$\lambda P(m) - (\lambda + u)P(m + 1) + uP(m + 2) = 0 \tag{5.173}$$

and

$$\lambda p P(m + 1) - (p\lambda + m)P(m + 2) + uP(m + 3) = 0 \tag{5.174}$$

Letting

$$P(n) = \begin{cases} \left(\frac{\lambda}{u}\right)^n P(0), & 0 \leq n \leq m + 1 \\ \left(\frac{p\lambda}{u}\right)^n, & m + 2 \leq n \end{cases} \tag{5.175}$$

in Eq. (5.173) yields

$$\lambda\left(\frac{\lambda}{u}\right)^m P(0) - (\lambda + u)\left(\frac{\lambda}{u}\right)^{m+1} P(0) + u\left(\frac{p\lambda}{u}\right)^{m+2} = 0$$

which holds only if

$$P(0) = p^{m+2} \tag{5.176}$$

Substituting the results given in Eqs. (5.175) and (5.176) in Eq. (5.173) leads to

$$p^2\lambda\left(\frac{p\lambda}{u}\right)^{m+1} - (p\lambda + u)\left(\frac{p\lambda}{u}\right)^{m+2} + u\left(\frac{p\lambda}{u}\right)^{m+3} = 0$$

which holds only if $p = 0$ or $p = 1$. However, if $p = 0$, $P(0) = 0$, and

$$\sum_{n=0}^{\infty} P(n) = 0$$

If $p = 1$, $P(0) = 1$, and

$$\sum_{n=0}^{\infty} P(n) = \infty$$

Therefore, the solutions given by Eq. (5.175) do not satisfy the requirements of the problem.

From Theorem 5.9, if $P_3(n)$ is a solution of Eq. (5.168), then it can be represented by

$$P_3(n) = c_1 P_1(n) + c_2 P_2(n)$$

Therefore

$$P_3(m + 2) = c_1 \left(\frac{p\lambda}{u}\right)^{m+2} + c_2 \tag{5.177}$$

$$P_3(m + 3) = c_1 \left(\frac{p\lambda}{u}\right)^{m+3} + c_2 \tag{5.178}$$

From Eqs. (5.167) and (5.169),

$$P_3(m + 2) = \left[\frac{(\lambda + \mu)}{u}\left(\frac{\lambda}{u}\right)^{m+1} - \left(\frac{\lambda}{u}\right)^{m+1}\right] P(0)$$

$$= \left(\frac{\lambda}{u}\right)^{m+2} P(0) \tag{5.179}$$

From Eq. (5.168)

$$p\lambda\left(\frac{\lambda}{u}\right)^{m+1} P(0) - (p\lambda + u)\left(\frac{\lambda}{u}\right)^{m+2} P(0) + uP_3(m + 3) = 0$$

5.7 APPLICATIONS

and

$$P_3(m+3) = \left(\frac{\lambda}{u}\right)^{m+2}\left[1 - p\left(1 - \frac{\lambda}{\mu}\right)\right]P(0)$$

Solving for the values of the constants c_1 and c_2 yields

$$\begin{bmatrix} c_1 \\ c_2 \end{bmatrix} = \frac{1}{\left(\frac{p\lambda}{u}\right)^{m+2}\left(1 + \frac{p\lambda}{u}\right)} \begin{bmatrix} 1 & -1 \\ -\left(\frac{p\lambda}{u}\right)^{m+3} & \left(\frac{p\lambda}{u}\right)^{m+2} \end{bmatrix} \begin{bmatrix} \left(\frac{\lambda}{u}\right)^{m+2} P(0) \\ \left(\frac{\lambda}{u}\right)^{m+2}\left[1 - p\left(1 - \frac{\lambda}{u}\right)\right]P(0) \end{bmatrix}$$

$$= \begin{bmatrix} \dfrac{\left(1 - \dfrac{\lambda}{u}\right)P(0)}{p^{m+1}\left(1 - \dfrac{p\lambda}{u}\right)} \\ \\ \dfrac{(1-p)\left(\dfrac{\lambda}{u}\right)^{m+2} P(0)}{\left(1 - \dfrac{p\lambda}{u}\right)} \end{bmatrix} \tag{5.180}$$

Therefore,

$$P_3(n) = \frac{u - \lambda}{u - p\lambda} p^{n-m-1} \left(\frac{\lambda}{u}\right)^n P(0) + \frac{1-p}{u - p\lambda} u \left(\frac{\lambda}{u}\right)^{m+2} P(0)$$

or

$$P_3(n) = \frac{\left(\dfrac{\lambda}{u}\right)^n}{u - p\lambda} P(0)[(u - \lambda)p^{n-m-1} + (1-p)u]$$

The expression for $P(0)$ is determined using the restriction given by Eq. (5.165).

$$\sum_{n=0}^{\infty} P(n) = \sum_{n=0}^{m+1} \left(\frac{\lambda}{u}\right)^n P(0) + \sum_{n=m+2}^{\infty} \frac{\left(\dfrac{\lambda}{u}\right)^n}{u - p\lambda} [(u-\lambda)p^{n-m-1} + (1-p)u] P(0)$$

$$= \left[\frac{1 - \left(\dfrac{\lambda}{u}\right)^{m+2}}{1 - \dfrac{\lambda}{u}} + \frac{(u-\lambda)}{(u - p\lambda)p^{m+1}} \frac{\left(\dfrac{p\lambda}{u}\right)^{m+2}}{1 - \dfrac{p\lambda}{u}} + \frac{(1-p)u}{\mu - p\lambda} \frac{\left(\dfrac{\lambda}{u}\right)^{m+2}}{1 - \dfrac{\lambda}{u}} \right] P(0)$$

$$= \left[\frac{1}{1 - \dfrac{\lambda}{u}} + p\left(\frac{\lambda}{u}\right)^{m+2} \frac{\left(1 - \dfrac{\lambda}{u}\right) - \left(1 - \dfrac{p\lambda}{u}\right)}{\left(1 - \dfrac{p\lambda}{u}\right)^2} \right] P(0) \tag{5.181}$$

and

$$P(0) = \cfrac{1}{\cfrac{1}{(\lambda/u)} + p\left(\cfrac{\lambda}{u}\right)^{m+2} \cfrac{(1 - [\lambda/u]) - (1 - [p\lambda/u])}{(1 - [p\lambda/u])^2}} \qquad (5.182)$$

5.7.8 Engineering Applications

Example 5.56 Show that

$$\sum_{x=0}^{n} xa^x = \frac{a}{(a-1)^2}[(n+1)(a-1)a^n - a^{n+1} + 1]$$

using the relationship

$$\sum_{x=0}^{n} xa^x = a\frac{d}{da}\sum_{x=0}^{n} a^x$$

From Example 5.29, part b,

$$\sum_{x=0}^{n} a^x = \frac{a^{n+1} - 1}{a - 1}$$

Now

$$\frac{d}{da}\sum_{x=0}^{n} a^x = \frac{d}{da}\left[\frac{a^{n+1} - 1}{a - 1}\right] = \frac{(n+1)a^n}{(a-1)} - \frac{a^{n+1} - 1}{(a-1)^2}$$

$$= \frac{(n+1)(a^{n+1} - a^n) - a^{n+1} + 1}{(a-1)^2} = \frac{(n+1)(a-1)a^n - a^{n+1} + 1}{(a-1)^2}$$

Hence

$$a\frac{d}{da}\sum_{x=0}^{n} a^x = \frac{a}{(a-1)^2}[(n+1)(a-1)a^n - a^{n+1} + 1] = \sum_{x=0}^{n} xa^x \qquad \boxed{}$$

Example 5.57 In Fig. 5.9 the point 0 is maintained at a constant voltage $V(0)$ with respect to the ground. Find the voltage, $V(x)$, at the point x, $x = 0, 1, \ldots, n - 1$ $(V(n) = 0)$. Assume $c \geq 1$.

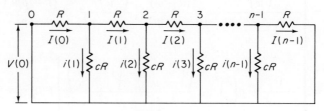

Figure 5.9

5.7 APPLICATIONS

Since the sum of the currents into the point x must equal the sum of the currents out of the point x, we have

$$I(x) = I(x + 1) + i(x + 1)$$

Expressing $I(x)$ in terms of voltage and resistance leads to

$$I(x) = \frac{V(x) - V(x + 1)}{R}, \qquad x = 0, 1, \ldots, n - 2$$

$$i(x) = \frac{V(x)}{cR}, \qquad x = 1, 2, \ldots, n - 1$$

Hence, we have

$$\frac{V(x) - V(x + 1)}{R} = \frac{V(x + 1) - V(x + 2)}{R} + \frac{V(x + 1)}{cR},$$

$$x = 0, 1, 2, \ldots, n - 2$$

Thus we have the difference equations

$$V(x + 2) - \left(\frac{2c + 1}{c}\right)V(x + 1) + V(x) = 0, \qquad x = 0, 1, 2, \ldots, n - 2$$

The fundamental set of solutions for this set of difference equations is given by

$$V(x) = k^x$$

where

$$k = \frac{\dfrac{2c + 1}{2} \pm \sqrt{\left(\dfrac{2c + 1}{2}\right)^2 - 4}}{2}$$

$$= \frac{2c + 1 + \sqrt{4c^2 + 4c - 15}}{4}, \frac{2c + 1 - \sqrt{4c^2 + 4c - 15}}{4}$$

and the general solution for $V(x)$ is

$$V(x) = \alpha \left[\frac{2c + 1 + \sqrt{4c^2 + 4c - 15}}{4}\right]^x + \beta \left[\frac{2c + 1 - \sqrt{4c^2 + 4c - 15}}{4}\right]^x$$

Since $V(0)$ is given and $V(n) = 0$, α and β satisfy the equations

$$V(0) = \alpha + \beta, \qquad 0 = \alpha A^n + \beta B^n$$

where

$$A = \frac{2c + 1 + \sqrt{4c^2 + 4c - 15}}{4}, \qquad B = \frac{2c + 1 - \sqrt{4c^2 + 4c - 15}}{4}$$

Thus

$$\begin{bmatrix} V(0) \\ 0 \end{bmatrix} = \begin{bmatrix} 1 & 1 \\ A^n & B^n \end{bmatrix} \begin{bmatrix} \alpha \\ \beta \end{bmatrix}$$

or

$$\begin{bmatrix} \alpha \\ \beta \end{bmatrix} = \begin{bmatrix} \dfrac{B^n}{B^n - A^n} & -\dfrac{1}{B^n - A^n} \\ -\dfrac{A^n}{B^n - A^n} & \dfrac{1}{B^n - A^n} \end{bmatrix} \begin{bmatrix} V(0) \\ 0 \end{bmatrix} = \begin{bmatrix} \dfrac{V(0)B^n}{B^n - A^n} \\ -\dfrac{V(0)A^n}{B^n - A^n} \end{bmatrix}$$

The solution for $V(x)$ is given by

$$V(x) = \frac{A^x B^n - A^n B^x}{B^n - A^n} V(0)$$

Example 5.58 A room is to be maintained at a constant temperature τ_0. However, perturbations continuously present inside and outside the room cause deviations from τ_0. Let

$T(t)$ = actual room temperature at time t;
$d(t)$ = actual deviation of $T(t)$ from τ_0, $T(t) - \tau_0$;
$e(t)$ = measured or observed deviation of $T(t)$ from τ_0, $d(t) + \epsilon(t)$, where $\epsilon(t)$ is normally distributed with mean zero and variance σ_1^2;
S = expected shift in the value of temperature between $t - 1$ and t;
s = actual shift in the value of temperature between $t - 1$ and t, $S + v(t)$, where $v(t)$ is normally distributed with mean zero and variance σ_2^2;
α = proportion of the observed deviation, $e(t)$, corrected at time t.

At $t = 1, 2, 3, \ldots$, the temperature of the room is measured and $e(t)$ is calculated. The temperature of the room is then altered by $-\alpha e(t)$. Find the mean of $d(t)$ for $t = 0, 1, 2, \ldots$, if

$$T(0) = \tau_0, \qquad e(0) = 0$$

and if $v(t)$ and $\epsilon(t)$ are independent random variables.

At any time $t = 1, 2, 3, \ldots$,

$$d(t + 1) = d(t) + s - \alpha e(t)$$

and

$$E[d(t + 1)] = E[d(t)] + E(s) - \alpha E[e(t)]$$
$$= E[d(t)] + S - \alpha E[d(t)] = (1 - \alpha)E[d(t)] + S$$

Hence, $E[d(t)]$ is the solution of the first order, linear, nonhomogeneous difference equation given by

$$E[d(t + 1)] - (1 - \alpha)E[d(t)] = S$$

5.7 APPLICATIONS

or

$$a_1 E[d(t+1)] - a_1(1-\alpha)E[d(t)] = 1$$

where

$$a_1 = \frac{1}{S}$$

Now, considering the homogeneous difference equation

$$a_1 E[d(t+1)] - a_1(1-\alpha)E[d(t)] = 0$$

we have the solution,

$$E[d_1(t)] = (1-\alpha)^t$$

and the general solution, $E[d(t)]$, of the nonhomogeneous difference equation is given by

$$E[d(t)] = c_1 + c_2(1-\alpha)^t$$

At $t = 0$ we have

$$E[d(0)] = 0$$

and at $t = 1$

$$E[d(1)] = E[d(0)] + S - \alpha E[d(0)] = S$$

Hence

$$0 = c_1 + c_2, \qquad S = c_1 + c_2(1-\alpha)$$

or

$$c_1 = \frac{S}{\alpha}, \qquad c_2 = -\frac{S}{\alpha}, \qquad \text{and} \qquad E[d(t)] = \frac{S}{\alpha}[1 - (1-\alpha)^t]$$

Now

$$\lim_{t \to \infty} E[d(t)] = \frac{S}{\alpha} \lim_{t \to \infty} [1 - (1-\alpha)^t]$$

However, the limit on the right converges if and only if $|1 - \alpha| < 1$. Hence, for $0 < \alpha < 2$,

$$\lim_{t \to \infty} E[d(t)] = \frac{S}{\alpha}$$

Problems

1. Let
$$f(x) = c$$
where x is a continuous variable, $f(x)$ is a continuous function of x, and c is a constant. Then
$$\frac{df(x)}{dx} = 0$$
Show that an identical relationship for the divided difference holds if x is treated as a discrete variable.

2. Let $f(x)$ be a function of the discrete variable x which may assume the values x_i, $i = 1, 2, \ldots$. If c is a constant show that
   a. $\Delta[cf(x_i)] = c\,\Delta f(x_i)$
   b. $\Delta\left[\dfrac{c}{f(x_i)}\right] = -\dfrac{c}{f(x_i)f(x_{i+1})}\,\Delta f(x_i)$

3. Find $\Delta f(x)$, where
$$f(x) = \tfrac{1}{2}x(x+1), \qquad x = 1, 2, \ldots$$

4. Find $\Delta f(x)$, where
$$f(x) = [\tfrac{1}{2}x(x-1)]^2, \qquad x = 1, 2, \ldots$$

5. Find $\Delta f(x)$, where
$$f(x) = (1/30)(x-1)x(2x-1)(3x^2 - 3x - 1)$$

6. Find $\Delta f(x)$ for each of the following relationships, where $x = 0, 1, 2, \ldots, n$. In each case
$$f(x) = \begin{cases} 0, & x = -1, -2, \ldots \\ 0, & x = n+1, n+2, \ldots \end{cases}$$
   a. $f(x) = 4x + 3$
   b. $f(x) = x^2 - 6$
   c. $f(x) = \dfrac{1}{x+1}$
   d. $f(x) = (x+1)(x+4)$
   e. $f(x) = x^2 - 5x + 10$
   f. $f(x) = \dfrac{x}{x+4}$
   g. $f(x) = \dfrac{(6x-1)(x+1)}{x^2+1}$

7. Find $\Delta f(x)$ for each of the following relationships where $x = 0, 0.5, 1, 1.5, \ldots, n$. In each case
$$f(x) = \begin{cases} 3, & x = -1, -2, \ldots \\ 0, & x = n+1, n+2, \ldots \end{cases}$$
   a. $f(x) = x^2 - 4x + 5$
   b. $f(x) = 2x^2 - 6x + 12$
   c. $f(x) = \dfrac{2x^2 - 6x + 12}{x+3}$

8. Show that

   a. $\Delta_x \left[\sin\left(\frac{x\theta}{2}\right) \sin\left(\frac{x\theta - \theta}{2}\right) \csc\left(\frac{\theta}{2}\right) \right] = \sin(x\theta)$

   b. $\Delta_x \left[\cos\left(\frac{x\theta}{2}\right) \sin\left(\frac{x\theta - \theta}{2}\right) \csc\left(\frac{\theta}{2}\right) \right] = \cos(x\theta)$

   where x is integer valued and where Δ_x denotes the difference with respect to x.

9. Show that

   a. $\Delta_x(a + bx)^{(n)} = bn(a + bx)^{(n-1)}$

   where $(a + bx)^{(n)} = [a + bx][a + b(x-1)] \cdots [a + b(x - n + 1)]$.

   b. $\Delta_x \left[\frac{1}{(a-1)x^{[a-1]}} \right] = -\frac{1}{x^{[a]}}$

10. Prove part c of Theorem 5.1.

11. Derive an expression for $\Delta C(x)$ similar to Eq. (5.14) where

    $$C(x) = \sum_{y=a}^{b(x)} f(x, y)$$

    and where x and y have unit increments, $a < b(x)$, and $b(x) < b(x + 1)$. Note that the lower limit of the summation is not a function of x.

12. Derive an expression for $\Delta C(x)$ similar to Eq. (5.13) where

    $$C(x) = \sum_{y=a(x)}^{b} f(x, y)$$

    where x and y have increments Δx and Δy, $a(x) < b$, and $a(x) < a(x + \Delta x)$. Note that the upper limit of the summation is not a function of x.

13. Show that if x_i is a unique maximum for $f(x)$ on (a', b') then

    $$\Delta f(x_{i-1}) > 0 > \Delta f(x_i)$$

    where x_{i-1}, x_i, and x_{i+1} lie on the interval (a', b').

14. Find the extreme points for $f(x)$ where

    $$f(x) = 3x^4 - 4x^3, \quad x = 0, \pm 1, \pm 2, \ldots$$

15. Evaluate the following sum

    $$\sum_{x=0}^{n} x^2 a^x$$

    where x assumes integer values only.

16. Evaluate the following where x assumes integer values only.

    a. $\sum_{x=a}^{b} x(x + 1)(x + 2)$ b. $\sum_{x=0}^{b} (x + 1)(x + 2)(x + 3)$

    c. $\sum_{x=1}^{b} \frac{1}{(2x - 1)(2x + 1)}$ d. $\sum_{x=1}^{b} \frac{1}{(x + 1)(x + 3)}$

17. Prove cases 2 and 4 of Theorem 5.5.

18. By induction show that
$$f(x) = c^x f(0)$$
is a solution of the difference equation
$$a_1 f(x+1) + a_0 f(x) = 0, \quad \text{where} \quad c = -\frac{a_0}{a_1} \quad \text{and} \quad a_0 \neq 0, \quad a_1 \neq 0$$

19. Classify the following difference equations with respect to linearity, homogeneiety, and order
   a. $f(x+1) - xf(x) = 0$
   b. $f(x+3) - 2f(x+1) = x - 4$
   c. $3f(x+5) + (x-1)f(x+2) + f(x) = 0$
   d. $2f(x+4) - 3f(x+1) - xf(x) = x^2 + 5$
   where $x = 1, 2, 3, \ldots$

20. Let $P(x)$ be the probability mass function of the random variable x, $x = 1, 2, \ldots, n$. If $P(x)$ satisfies the difference equation
$$P(x+1) - (1-p)P(x) = 0, \quad x = 1, 2, \ldots, n-1 \quad \text{and} \quad \sum_{x=1}^{n} P(x) = 1$$
define $P(x)$.

21. The probability mass function $P(x)$ satisfies the difference equation
$$(x+1)P(x+1) - \lambda P(x) = 0, \quad x = 0, 1, 2, \ldots$$
Solve for $P(x)$ recursively noting that
$$\sum_{i=0}^{\infty} \frac{a^i}{i!} = \exp[a]$$

22. Demand for a certain product is Poisson distributed with mean rate λ per unit time. Units are demanded one at a time. Whenever inventory position, inventory on hand plus units on order, falls to r units an order is immediately placed for $Q - r$ units, raising inventory position to Q units. Therefore inventory position, x, may vary between $r+1$ and Q units. If $P(x)$ is the probability that inventory position is x units at an arbitrary point in time, then $P(x)$ satisfies the relationships
$$P(x) = \lambda P(x+1) + (1-\lambda)P(x), \quad x = r+1, r+2, \ldots, Q-1,$$
$$P(Q) = \lambda P(r+1) + (1-\lambda)P(Q)$$
Define $P(x)$.

23. Let
$$\Delta^n f(x) = \frac{1}{\Delta x}[\Delta^{n-1} f(x + \Delta x) - \Delta^{n-1} f(x)]$$
Show that
$$\Delta^n f(x) = \frac{1}{(\Delta x)^n} \sum_{i=0}^{n} (-1)^i \frac{n!}{i!(n-i)!} f[x + (n-i)\Delta x]$$

24. Let $f_1(x)$ and $f_2(x)$ be two solutions to a homogeneous linear difference equation with constant coefficients. Show that $c_1 f_1(x) + c_2 f_2(x)$ is also a solution where c_1 and c_2 are constants.

25. Given the conditions of Theorem 5.9, show that $f_1(x)$ and $f_2(x)$ are linearly independent.

26. Prove part 3 of Theorem 5.10.

27. Find a fundamental set of solutions and a general solution for each of the following difference equations.
   a. $4f(x+2) - 5f(x+1) + f(x) = 0$
   b. $f(x+2) + 6f(x+1) + 5f(x) = 0$
   c. $9f(x+2) - 6f(x+1) + f(x) = 0$

PROBLEMS

28. Find the solutions for the difference equations given in Problem 27, subject to the restrictions
$$f(0) = 1, \quad f(1) = 10$$

29. Find a particular solution for each of the following difference equations.
    a. $f(x + 2) - 4f(x + 1) + 3f(x) = x$
    b. $5f(x + 2) - 7f(x + 1) - f(x) = x$
    c. $f(x + 2) + 9f(x + 1) + f(x) = 2^x$
    d. $f(x + 2) - 2f(x + 1) - 3f(x) = 3^x$
    e. $3f(x + 2) - f(x + 1) - f(x) = x + 5^x$

30. Find a general solution for each of the difference equations in Problem 29.

31. For each of the difference equations in Problem 29, find a solution such that
$$f(0) = 0, \quad f(1) = 1$$

32. Find a general solution for the difference equation
$$f(x + 3) - 3f(x + 2) - f(x + 1) + 3f(x) = 0$$

33. The probability generating function for the random variable x is defined by
$$A(s) = \sum_{x=0}^{n} P(x)s^x$$
where $P(x)$ is the probability mass function of x and
$$\sum_{x=0}^{n} P(x) = 1$$
If
$$P(x) = \frac{n!}{x!(n-x)!} p^x(1-p)^{n-x}, \quad x = 0, 1, 2, \ldots, n$$
find $A(s)$ if
$$\sum_{x=0}^{n} \frac{n!}{x!(n-x)!} a^x b^{n-x} = (a + b)^n$$

34. Let
$$f_1(x) = k_1^x, \quad f_2(x) = k_2^x$$
be two solutions for the difference equation
$$a_2 f(x + 2) + a_1 f(x + 1) + a_0 f(x) = 0$$
where $k_1 \neq k_2$. Show that
$$f_3(x) = xk^x$$
is not a solution for the difference equation.

35. Prove Theorem 5.11.

36. Find the general solution for
$$f(x + 3) - 6f(x + 2) + 9f(x + 1) - 4f(x) = 0$$

37. Let $(x_{i+1} - x_i) = \Delta x$ for all i. Then
$$\Delta^n f(x) = \frac{1}{n \Delta x} [\Delta^{n-1} f(x + \Delta x) - \Delta^{n-1} f(x)]$$

Let $\Delta f^n(x)$ be redefined as

$$\Delta^n f(x) = \frac{1}{\Delta x}[\Delta^{n-1} f(x + \Delta x) - \Delta^{n-1} f(x)]$$

In either case,

$$\frac{d}{dx} f(x) = \lim_{\Delta x \to 0} \Delta f(x)$$

Which definition of $\Delta^n f(x)$ leads to

$$\frac{d^n}{dx^n} f(x) = \lim_{\Delta x \to 0} \Delta^n f(x)?$$

38. If $\int_a^b f(x)\, dx$ exists, then

$$\int_a^b f(x)\, dx = \lim_{n \to \infty} \sum_{i=1}^n f(x_i)\, \Delta x$$

where

$$x_i = \left(i - \frac{1}{2}\right) \Delta x, \quad \Delta x = \frac{b-a}{n}$$

Show that

$$\int_a^b \exp[-x]\, dx = 1 - \exp[-b]$$

39. Let S_0 be the sum of money invested at time zero and let S_n be the value of that sum after n time periods. If i is the interest rate on the investment per time period, then

$$S_n = S_{n-1} + iS_{n-1}$$

Express S_n in terms of S_0.

References

Beckenbach, E. F. (ed.), *Modern Mathematics for the Engineer*. New York: McGraw-Hill, 1956.
Gelfond, A. O., *The Solution of Equations in Integers*. San Francisco: Freeman, 1961.
Giffin, W. C., *Introduction to Operations Engineering*. Homewood, Illinois: Irwin, 1971.
Goldberg, S., *Difference Equations*. New York: Wiley (Interscience) 1958.
Miller, K. S., *An Introduction to the Calculus of Finite Differences and Difference Equations*. New York: Holt, 1960.
Richardson, C. H., *An Introduction to the Calculus of Finite Differences*, Princeton, New Jersey: Van Nostrand, 1954.
Saaty, T. L., *Mathematical Methods of Operations Research*. New York: McGraw-Hill, 1959.
Wylie, C. R., *Advanced Engineering Mathematics*. New York: McGraw-Hill, 1960.

CHAPTER 6

COMPLEX VARIABLES AND TRANSFORM METHODS

6.1 Introduction

The reader familiar with probability theory may have encountered the problem of attempting to derive the distribution of the sum of n random variables. While direct derivation can be quite tedious, the use of an appropriate transform method often simplifies the problem considerably. Transforms can also be used to solve differential and integral equations with similar facility. The intent of this chapter is to present several transform methods which are useful in operations research. Since the transform methods discussed in later sections of this chapter are based upon complex variable theory, a fundamental knowledge of functions of a complex variable and the operations of differentiation and integration on such functions is essential. Hence the first half of this chapter will be devoted to an elementary treatment of complex variable theory.

6.2 Complex Variables

Consider the problem of solving the equation

$$(s-a)^2 = -b^2$$

for s. Taking the square root of both sides gives us

$$s = a + ib \tag{6.1}$$

where

$$i = \sqrt{-1} \tag{6.2}$$

Since $\sqrt{-1}$ is not a real number, s is not a real number. The number i is called an *imaginary number* and the product of i and any real number is also an imaginary number. Therefore, s is a number composed of a real part, a, and an imaginary part, ib, and is referred to as a *complex number*.

Every real number can be located on the real axis between $-\infty$ and ∞. Similarly we can define another axis, the imaginary axis, on which every imaginary number can be located. Let the real numbers be located on the x axis and the imaginary numbers on the y axis of a rectangular coordinate system. To avoid confusion, the y axis will be labeled iy. The two axes then intersect at $x = 0$, $iy = 0$. This coordinate system defines the *complex plane* and any point $s = x + iy$ can be located in this plane. We locate s by considering it in vector form. That is,

$$s = \begin{bmatrix} x \\ iy \end{bmatrix} \tag{6.3}$$

Therefore, Eq. (6.3) is equivalent to $s = x + iy$.

Example 6.1 Locate the following points in the complex plane.

a. $s = 1 + 3i$
b. $s = -2 - i$
c. $s = 2$
d. $s = i$

In parts a and b, s is a complex number since it has both real and imaginary components. In part c, the imaginary part of s can be represented by $0i$. Therefore s is real in this case and is located on the x or real axis. In part d, s is imaginary. That is, the real portion of s is zero, and s is located on the imaginary axis. These four points are shown graphically in Fig. 6.1.

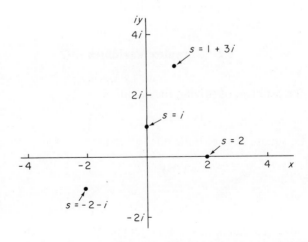

Figure 6.1 Representation of the Points Given in Example 6.1 in the Complex Plane

6.2 COMPLEX VARIABLES

The operations of addition, subtraction, multiplication, and division of complex numbers are carried out using the conventional rules of algebra. Therefore

$$(x + iy) \pm (v + iw) = (x \pm v) + i(y \pm w) \tag{6.4}$$

$$(x + iy)(v + iw) = (xv - yw) + i(xw + yv) \tag{6.5}$$

$$\frac{(x + iy)}{(v + iw)} = \frac{(xv + yw) + (yv - xw)i}{(v^2 + w^2)} \tag{6.6}$$

where $i^2 = -1$.

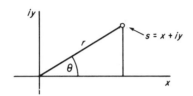

Figure 6.2 Graphical Representation of $s = x + iy$

Let the point $s = x + iy$ be represented as in Fig. 6.2, where r is the length of the line joining s and the origin and θ is the angle between this line and the real axis. The angle θ is often referred to as the *argument* of s, denoted arg(s). The distance of the point s from the origin, r, is defined by

$$r^2 = (x + iy)(x - iy) \tag{6.7}$$

where $x - iy$ is called the *complex conjugate* of s and is denoted by s^*. Then

$$r = \sqrt{ss^*} \tag{6.8}$$

or

$$r = \sqrt{x^2 + y^2} \tag{6.9}$$

Similarly, the length of the line joining $x + iy$ and iy is $\sqrt{x^2}$ or x while the length of the line joining $x + iy$ and x is given by $\sqrt{(iy)(-iy)}$ or y. Since

$$\sin(\theta) = \frac{y}{r}, \quad \cos(\theta) = \frac{x}{r}$$

we have

$$y = r \sin(\theta) \tag{6.10}$$

$$x = r \cos(\theta) \tag{6.11}$$

and

$$s = r[\cos(\theta) + i \sin(\theta)] \tag{6.12}$$

Equation (6.12) is a representation of s in *polar form* and will be useful throughout the discussion of complex variables.

In addition to polar form, s can be represented in *exponential form*. Expanding $\exp[i\theta]$ leads to

$$\exp[i\theta] = \sum_{k=0}^{\infty} \frac{(i\theta)^k}{k!}$$

$$= \sum_{k=0}^{\infty} \frac{(i\theta)^{2k}}{(2k)!} + \sum_{k=0}^{\infty} \frac{(i\theta)^{2k+1}}{(2k+1)!} \qquad (6.13)$$

But

$$\sum_{k=0}^{\infty} \frac{(i\theta)^{2k}}{(2k)!} = 1 - \frac{\theta^2}{2!} + \frac{\theta^4}{4!} - \frac{\theta^6}{6!} + \cdots = \cos(\theta)$$

and

$$\sum_{k=0}^{\infty} \frac{(i\theta)^{2k+1}}{(2k+1)!} = i\left[\theta - \frac{\theta^3}{3!} + \frac{\theta^5}{5!} - \frac{\theta^7}{7!} + \cdots\right] = i\sin(\theta)$$

Therefore

$$\exp[i\theta] = \cos(\theta) + i\sin(\theta) \qquad (6.14)$$

and

$$x + iy = r\exp[i\theta] \qquad (6.15)$$

We may also show that

$$\sin(\theta) = \frac{\exp[i\theta] - \exp[-i\theta]}{2i} \qquad (6.16)$$

$$\cos(\theta) = \frac{\exp[i\theta] + \exp[-i\theta]}{2} \qquad (6.17)$$

$$\sinh(\theta) = \frac{\exp[\theta] - \exp[-\theta]}{2} \qquad (6.18)$$

$$\cosh(\theta) = \frac{\exp[\theta] + \exp[-\theta]}{2} \qquad (6.19)$$

where $\sinh(\theta)$ and $\cosh(\theta)$ are the hyperbolic sine and cosine of θ.

Example 6.2 Prove Eqs. (6.16) and (6.19) where

$$\cosh(\theta) = \sum_{k=0}^{\infty} \frac{\theta^{2k}}{(2k)!}$$

To prove Eq. (6.16), we use Eq. (6.14).

$$\frac{\exp[i\theta] - \exp[-i\theta]}{2i} = \frac{1}{2i}[\cos(\theta) + i\sin(\theta) - \exp[-i\theta]]$$

Applying the result of Eq. (6.13), we have

$$\exp[-i\theta] = \sum_{k=0}^{\infty} \frac{(-i\theta)^{2k}}{(2k)!} + \sum_{k=0}^{\infty} \frac{(-i\theta)^{2k+1}}{(2k+1)!} = \cos(\theta) - i\sin(\theta)$$

Therefore

$$\frac{\exp[i\theta] - \exp[-i\theta]}{2i} = \frac{2i\sin(\theta)}{2i} = \sin(\theta)$$

To establish Eq. (6.19), we employ an approach similar to that used above. From Eq. (6.13),

$$\frac{\exp[\theta] + \exp[-\theta]}{2} = \frac{1}{2}\left[\sum_{k=0}^{\infty} \frac{\theta^k}{k!} + \sum_{k=0}^{\infty} \frac{(-\theta)^k}{k!}\right] = \frac{1}{2}\left[2\sum_{k=0}^{\infty} \frac{\theta^{2k}}{(2k)!}\right] = \cosh(\theta) \quad \blacksquare$$

6.3 Functions of a Complex Variable

In Chapter 3, functions of real variables and their properties were discussed at some length, and the ideas presented in that chapter can be extended to functions of a complex variable. Unless otherwise specified, $f(s)$ is to be taken as a single-valued function of the complex variable s. The limit of $f(s)$ as s approaches s_0 is expressed as $\lim_{s \to s_0} f(s)$. The reader should remember that s lies in the complex plane which will be denoted by S. Therefore, if we write

$$\lim_{s \to s_0} f(s) = L \tag{6.20}$$

we are implying that $f(s) \to L$ as $s \to s_0$ over all possible paths of approach of s to s_0 on S. On the other hand, if Eq. (6.20) is valid only along the path S' in S then Eq. (6.20) must be modified as follows.

$$\lim_{s \to s_0} f(s) = L, \quad s \in S' \tag{6.21}$$

If $\lim_{s \to s_0} f(s)$ exists and is equal to $f(s_0)$, then $f(s)$ is said to be continuous at s_0. Further if $f(s)$ is continuous at every point in the region or space S_0, then $f(s)$ is said to be continuous on S_0.

Definition 6.1 The derivative of $f(s)$ at s_0 is given by

$$f'(s_0) = \lim_{\Delta s \to 0} \frac{f(s_0 + \Delta s) - f(s_0)}{\Delta s} \tag{6.22}$$

if the limit exists and is independent of the path of approach of Δs to 0.

The implications of Definition 6.1 are twofold. Noting that $f(s)$ can be expressed as a function of x and iy, let

$$s_0 = x_0 + iy_0 \tag{6.23}$$

and
$$\Delta s = \Delta x + i\,\Delta y \qquad (6.24)$$

If $f(s_0)$ exists, then the limit in Eq. (6.22) must exist no matter how Δs approaches 0 in the S plane. Further, the limit in Eq. (6.22) for Δs on the paths S_0 and S_1 on S must be equal for all S_0 and S_1.

Example 6.3 Show that $df(s)/ds$ exists at s_0, where
$$f(s) = s^2$$

By Eq. (6.22),

$$f'(s_0) = \lim_{\Delta s \to 0} \frac{(s_0 + \Delta s)^2 - s_0^2}{\Delta s} = \lim_{\Delta s \to 0} \frac{s_0^2 + 2s_0\,\Delta s + (\Delta s)^2 - s_0^2}{\Delta s}$$

$$= \lim_{\Delta s \to 0} [2s_0 + \Delta s] = \lim_{\substack{\Delta x \to 0 \\ \Delta y \to 0}} [2(x_0 + iy_0) + (\Delta x + i\,\Delta y)] = 2s_0$$

since $(\Delta x + i\,\Delta y) \to 0$ as $\Delta x \to 0$ and $\Delta y \to 0$ regardless of the manner in which $\Delta s \to 0$. For example, if we let $\Delta x = 0$, then $\Delta s \to 0$ implies

$$\lim_{\Delta y \to 0} (0 + i\,\Delta y) = 0$$

Further, if we choose $y = 0$, then

$$\lim_{\Delta s \to 0} \Delta s = \lim_{\Delta x \to 0} [\Delta x + i(0)] = 0$$

Now let us consider a case where $f'(s_0)$ does not exist.

Example 6.4 Show that $f'(s^*)$ does not exist, where
$$f(s^*) = (s^*)^2$$

The reader will recall that s^* is the complex conjugate of s and is given by
$$s^* = x - iy$$
Then
$$f(s^*) = (x - iy)^2$$
Now

$$f'(s^*) = \lim_{\Delta s \to 0} \frac{f[(s + \Delta s)^*] - f(s^*)}{\Delta s}$$

$$= \lim_{\substack{\Delta x \to 0 \\ \Delta y \to 0}} \frac{(x - iy + \Delta x - i\,\Delta y)^2 - (x - iy)^2}{\Delta x + i\,\Delta y}$$

$$= \lim_{\substack{\Delta x \to 0 \\ \Delta y \to 0}} \frac{2(x - iy)(\Delta x - i\,\Delta y) + (\Delta x - i\,\Delta y)^2}{\Delta x + i\,\Delta y}$$

6.3 FUNCTIONS OF A COMPLEX VARIABLE

Let $\Delta x = 0$. Then

$$f'(s^*) = \lim_{\Delta y \to 0} \frac{-2(x - iy)i \Delta y + (i \Delta y)^2}{i \Delta y} = -2(x - iy)$$

Now if we let $\Delta y = 0$, we have

$$f'(x^*) = \lim_{\Delta x \to 0} \frac{2(x - iy) \Delta x + (\Delta x)^2}{\Delta x} = 2(x - iy)$$

Since $f'(s^*)$ takes on a different value for the two paths of approach selected, $f'(s^*)$ does not exist.

Definition 6.2 A function $f(s)$ is said to be analytic on the space S_0 if and only if $f'(s)$ exists at all points in S_0.

One may show that if $f'(s)$ exists on S_0, then $f''(s)$ also exists for all n on S_0. Thus, if $f(s)$ is analytic on S_0, then all derivatives of $f(s)$ exist on S_0. Analytic functions play an important role in complex variable theory, particularly with respect to contour integration which will be considered in later sections of this chapter. Using the definition of the derivative we can develop necessary conditions for $f(s)$ to be analytic on a space S_0. Before pursuing this development, it will be useful to introduce the notion of a conjugate function. Since s is composed of a real part, x, and an imaginary part iy, any function of s, $f(s)$ can be expressed by

$$f(x + iy) = g(x, y) + ih(x, y)$$

The functions $g(x, y)$ and $h(x, y)$ are called *conjugate functions*. To illustrate this concept consider the function given in Example 6.3. Setting $s = x + iy$ we have

$$f(s) = (x^2 - y^2) + 2ixy$$

Then

$$g(x, y) = x^2 - y^2$$

and

$$h(x, y) = 2xy$$

Theorem 6.1 Cauchy–Riemann Equations Let

$$f(s) = g(x, y) + ih(x, y) \tag{6.25}$$

If

$$\frac{\partial g(x, y)}{\partial x}, \quad \frac{\partial g(x, y)}{\partial y}, \quad \frac{\partial h(x, y)}{\partial x}, \quad \text{and} \quad \frac{\partial h(x, y)}{\partial y}$$

are continuous on S_0, then $f(s)$ is analytic on S_0, if and only if

$$\frac{\partial g(x,y)}{\partial x} = \frac{\partial h(x,y)}{\partial y}, \quad \frac{\partial g(x,y)}{\partial y} = -\frac{\partial h(x,y)}{\partial x} \tag{6.26}$$

for all s on S_0.

Proof Assume that $f(s)$ is analytic on S_0. Then $f'(s)$ must exist for all s on S_0. Hence

$$f'(s) = \lim_{\Delta s \to 0} \frac{f(s + \Delta s) - f(s)}{\Delta s}$$

$$= \lim_{\substack{\Delta x \to 0 \\ \Delta y \to 0}} \frac{\{g(x + \Delta x, y + \Delta y) - g(x,y) + i[h(x + \Delta x, y + \Delta y) - h(x,y)]\}}{\Delta x + i \Delta y}$$

Let $\Delta x = 0$. Then

$$f'(s) = \lim_{\Delta y \to 0} \frac{\{g(x, y + \Delta y) - g(x,y) + i[h(x, y + \Delta y) - h(x,y)]\}}{i \Delta y}$$

$$= \frac{1}{i}\frac{\partial g(x,y)}{\partial y} + \frac{\partial h(x,y)}{\partial y}$$

Now if we let $\Delta y = 0$ we have for $f'(s)$

$$f'(s) = \lim_{\Delta x \to 0} \frac{\{g(x + \Delta x, y) - g(x,y) + i[h(x + \Delta x, y) - h(x,y)]\}}{\Delta x}$$

$$= \frac{\partial g(x,y)}{\partial x} + i\frac{\partial h(x,y)}{\partial x}$$

However, by assumption $f'(s)$ exists. Thus $f'(s)$ must have the same value for the two paths of approach selected and

$$-i\frac{\partial g(x,y)}{\partial y} + \frac{\partial h(x,y)}{\partial y} = \frac{\partial g(x,y)}{\partial x} + i\frac{\partial h(x,y)}{\partial x} \tag{6.27}$$

The equality expressed in Eq. (6.27) implies that the respective real and imaginary expressions are also equal. Therefore,

$$\frac{\partial g(x,y)}{\partial x} = \frac{\partial h(x,y)}{\partial y} \quad \text{and} \quad \frac{\partial g(x,y)}{\partial y} = -\frac{\partial h(x,y)}{\partial x}$$

For a proof that $f(s)$ is analytic on S_0 if the equations in (6.26) hold, the reader should see LePage (1961).

Definition 6.3 A point s_0 at which $f(s)$ fails to be analytic is called a *singular point* or a *singularity* for $f(s)$.

6.3 FUNCTIONS OF A COMPLEX VARIABLE

Example 6.5 Determine whether the following functions are analytic in a closed circular region S_0 with center at zero and radius unity.

a. $f(s) = \dfrac{1}{s^2}$

b. $f(s) = \dfrac{s-1}{(s+1)}$

c. $f(s) = (s+2)^2$

To determine whether $f(s)$ in part a is analytic we apply the results of Theorem 6.1. Let

$$s = x + iy$$

Then

$$f(s) = \frac{1}{(x+iy)^2}$$

From Eq. (6.6),

$$\frac{1}{x+iy} = \frac{x-iy}{x^2+y^2}$$

and

$$f(s) = \left(\frac{x-iy}{x^2+y^2}\right)^2 = \frac{x^2-y^2}{(x^2+y^2)^2} - i\frac{2xy}{(x^2+y^2)^2} = g(x,y) + ih(x,y)$$

Taking partial derivatives of $g(x,y)$ and $h(x,y)$ with respect to x and y gives us

$$\frac{\partial g(x,y)}{\partial x} = \frac{2x(3y^2-x^2)}{(x^2+y^2)^3}, \qquad \frac{\partial g(x,y)}{\partial y} = \frac{2y(y^2-3x^2)}{(x^2+y^2)^3}$$

$$\frac{\partial h(x,y)}{\partial x} = \frac{2y(3x^2-y^2)}{(x^2+y^2)^3}, \qquad \frac{\partial h(x,y)}{\partial y} = \frac{2x(3y^2-x^2)}{(x^2+y^2)^3}$$

Since

$$\frac{\partial g(x,y)}{\partial x} = \frac{\partial h(x,y)}{\partial y} \quad \text{and} \quad \frac{\partial g(x,y)}{\partial y} = -\frac{\partial h(x,y)}{\partial x}$$

wherever these partial derivatives exist, $f(s)$ is analytic.

If we examine each partial derivative, we find that they fail to exist at $s = 0$. For example

$$\frac{\partial g(x,0)}{\partial x} = -\frac{2}{x^3}$$

At $x = 0$, $\partial g(x, 0)/\partial x = -\infty$ and therefore does not exist. A similar argument can be used to show that the remaining three partial derivatives do not exist at the origin. Since these partial derivatives do not exist at $s = 0$, $f'(s)$ does not exist at $s = 0$ and this point is a singular point for $f(s)$. Since $s = 0$ lies in the region S_0, $f(x)$ is not analytic in S_0.

For part b, let
$$f(s) = \frac{(x + iy - 1)}{(x + iy + 1)}$$

From Eq. (6.6)
$$f(s) = \frac{(x^2 - 1) + y^2}{(x + 1)^2 + y^2} + i\frac{2y}{(x + 1)^2 + y^2} = g(x, y) + ih(x, y)$$

Taking partial derivatives, we have
$$\frac{\partial g(x, y)}{\partial x} = \frac{2[(x + 1)^2 - y^2]}{[(x + 1)^2 + y^2]^2}, \qquad \frac{\partial g(x, y)}{\partial y} = \frac{4y(x + 1)}{[(x + 1)^2 + y^2]^2}$$

$$\frac{\partial h(x, y)}{\partial x} = -\frac{4y(x + 1)}{[(x + 1)^2 + y^2]^2}, \qquad \frac{\partial h(x, y)}{\partial y} = \frac{2[(x + 1)^2 - y^2]}{[(x + 1)^2 + y^2]^2}$$

Again, since
$$\frac{\partial g(x, y)}{\partial x} = \frac{\partial h(x, y)}{\partial y} \quad \text{and} \quad \frac{\partial g(x, y)}{\partial y} = -\frac{\partial h(x, y)}{\partial x}$$

$f(s)$ is analytic at all points in S_0 where the partial derivatives exist. Examining each of the partial derivatives indicates that each fails to exist at $s = -1$. For example
$$\frac{\partial g(x, 0)}{\partial x} = \frac{2}{(x + 1)^2}$$

For $x = -1$, $\partial g(x, 0)/\partial x = \infty$ and hence does not exist. Since S_0 was defined as closed and of radius unity, the singular point $s = -1$ lies in S_0 and $f(s)$ is not analytic in S_0.

For part c,
$$f(s) = (x + iy + 2)^2 = [(x + 2)^2 - y^2] + 2iy(x + 2) = g(x, y) + ih(x, y)$$

Taking the required partial derivatives gives us
$$\frac{\partial g(x, y)}{\partial x} = 2(x + 2), \qquad \frac{\partial g(x, y)}{\partial y} = -2y$$

$$\frac{\partial h(x, y)}{\partial x} = 2y, \qquad \frac{\partial h(x, y)}{\partial y} = 2(x + 2)$$

Therefore $f(s)$ is analytic wherever these partial derivatives exist on S_0. Since these partial derivatives exist for all finite x and y, $f(s)$ is analytic throughout S_0.

6.4 COMPLEX INTEGRATION

The reader will notice that it was not necessary to compute $f'(s)$ in the preceding example. However, according to Definition 6.2 if $f'(s)$ can be shown to exist throughout S_0, then $f(s)$ is analytic on S_0. Thus, one need not evaluate $f'(s)$ to show that $f(s)$ is analytic.

Definition 6.4 If s_0 is a singular point for $f(s)$ such that there exists a neighborhood of s_0 in which $f(s)$ is analytic, then s_0 is called an *isolated singularity*.

The singular points for $f(s)$ in parts a and b of Example 6.5 are isolated since, in each case, there were neighborhoods of the isolated points for which $f(s)$ was analytic. We will discuss singular points and their classification further. However, before pursuing this subject it will be useful to introduce some of the basic concepts of complex integration.

6.4 Complex Integration

To understand the meaning of the integral of a complex function it is necessary to introduce the *line integral*. Let $f(x, y)$ be an integrable function on the space $\mathscr{X}_2$ and let C be a continuous curve lying on the plane $\mathscr{X}_2$. Now suppose we wish to find the integral of $f(x, y)$ where the points (x, y) lie on the curve C. Such an integral is called a line integral. The integral of $f(x, y)$ on C is shown graphically in Fig. 6.3, where the shaded area, A, is the integral of $f(x, y)$ on C. If $s = (x, y)$ then the line integral of $f(s)$ on the curve C is denoted $\int_C f(s)\, ds$. We can define the line integral in a manner similar to that used in Chapter 2 to define the definite integral. Let $f(s)$ be defined on the curve C from the point a to b as shown

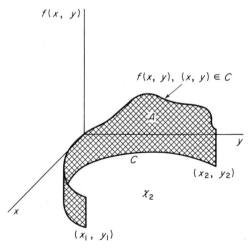

Figure 6.3 Line Integral of $f(x, y)$ on the Curve C

in Fig. 6.4 where a and b are values of s. Now we divide the curve into n arcs where the length of the ith arc is Δs_i and

$$\Delta s_i = |s_i - s_{i-1}| \tag{6.28}$$

Let

$$M_i = \max_{s_{i-1} \leq s \leq s_i} f(s) \tag{6.29}$$

$$m_i = \min_{s_{i-1} \leq s \leq s_i} f(s) \tag{6.30}$$

Then the upper and lower Riemann sums are given by

$$\bar{A}_{a,b,n}[f(s)] = \sum_{i=1}^{n} M_i \, \Delta s_i \tag{6.31}$$

$$\underline{A}_{a,b,n}[f(s)] = \sum_{i=1}^{n} m_i \, \Delta s_i \tag{6.32}$$

where $s_0 = a$ and $s_n = b$. Equations (6.31) and (6.32) lead directly to the definition of the *line integral* of $f(s)$.

Definition 6.5 The line integral of $f(s)$ on C exists and is given by

$$\int_C f(s) \, ds = \lim_{n \to \infty} \bar{A}_{a,b,n}[f(s)] = \lim_{n \to \infty} \underline{A}_{a,b,n}[f(s)] \tag{6.33}$$

if the limits exist.

We can define the projections of $\int_C f(s) \, ds$ on the x and y axes by defining the curve C by

$$y = g(x) \tag{6.34}$$

or

$$x = h(y) \tag{6.35}$$

Then the projection on the x axis is given by

$$\int_{a_x}^{b_x} f[x, g(x)] \, dx = \lim_{n \to \infty} \sum_{i=1}^{n} M_i \, \Delta x_i \tag{6.36}$$

and the projection on the y axis is given by

$$\int_{a_y}^{b_y} f[h(y), y] \, dy = \lim_{n \to \infty} \sum_{i=1}^{n} M_i \, \Delta y_i \tag{6.37}$$

The projections of $\int_C f(s) \, ds$ on the x and y axes are shown graphically in Fig. 6.4.

If we define the curve C by a relationship such as $y = g(x)$, we can reduce the line integral $\int_C f(s) \, ds$ to a form which is more easily dealt with from a computa-

6.4 COMPLEX INTEGRATION

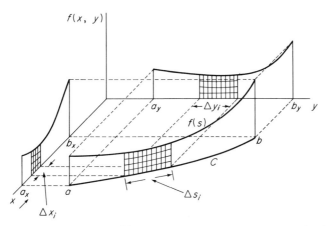

Figure 6.4 Line Integral of $f(s)$ on C and Its Projections on the x and y Axes

tional point of view. To simplify the line integral, we shall express $f(s)$ and ds in terms of x and y. Therefore

$$f(s) = f(x, y)$$

To define ds in terms of x and y consider the curve C shown in Fig. 6.5. Any arc on C can be approximated by a straight line joining the end points of the arc. If Δs is this approximation then

$$\Delta s = \sqrt{(\Delta x)^2 + (\Delta y)^2} \qquad (6.38)$$

or

$$\Delta s = \sqrt{1 + \left(\frac{\Delta y}{\Delta x}\right)^2}\, \Delta x$$

Then

$$ds = \sqrt{1 + [g'(x)]^2}\, dx \qquad (6.39)$$

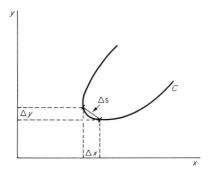

Figure 6.5 Approximation of an Arc by a Straight Line of Length Δs

and

$$\int_C f(s)\,ds = \int_{a_x}^{b_x} f[x, g(x)]\sqrt{1 + [g'(x)]^2}\,dx \qquad (6.40)$$

In a similar fashion, if $x = h(y)$

$$\int_C f(s)\,ds = \int_{a_y}^{b_y} f[h(y), y]\sqrt{1 + [h'(y)]^2}\,dy \qquad (6.41)$$

In Eqs. (6.40) and (6.41), $a_x < b_x$ and $a_y < b_y$.

Example 6.6 Find $\int_C f(s)\,ds$, where C is defined by

$$x^2 + y^2 = 25$$

from $x = 3$ to $x = 4$, where

$$f(s) = x + y - 2$$

Letting

$$y = \sqrt{25 - x^2}$$

we have

$$\frac{dy}{dx} = -\frac{x}{\sqrt{25 - x^2}}$$

and

$$f(s) = x + \sqrt{25 - x^2} - 2$$

Therefore

$$\begin{aligned}
\int_C f(s)\,ds &= \int_3^4 [x + \sqrt{25 - x^2} - 2]\sqrt{1 + \frac{x^2}{25 - x^2}}\,dx \\
&= 5\left[\int_3^4 \frac{x + \sqrt{25 - x^2} - 2}{\sqrt{25 - x^2}}\,dx\right] \\
&= 5\left[\int_3^4 \frac{x}{\sqrt{25 - x^2}}\,dx + \int_3^4 dx - 2\int_3^4 \frac{dx}{\sqrt{25 - x^2}}\right] \\
&= 5[1 + 1 - .58] = 7.10
\end{aligned}$$

We now solve the same problem by integrating over y instead of x. Letting

$$x = \sqrt{25 - y^2}$$

leads us to

$$\frac{dx}{dy} = -\frac{y}{\sqrt{25 - y^2}} \quad \text{and} \quad f(s) = y + \sqrt{25 - y^2} - 2$$

6.4 COMPLEX INTEGRATION

Therefore

$$\int_C f(s)\, ds = 5\int_3^4 \frac{y + \sqrt{25 - y^2} - 2}{\sqrt{25 - y^2}}\, dy = 7.10 \qquad \text{⑤}$$

Now let us turn our attention to line integration in the complex plane. From Eqs. (6.24) and (6.25)

$$f(s) = g(x, y) + ih(x, y), \qquad ds = dx + i\, dy$$

Then

$$\int_C f(s)\, ds = \int_C [g(x, y)\, dx - h(x, y)\, dy]$$

$$+ i \int_C [h(x, y)\, dx + g(x, y)\, dy] \qquad (6.42)$$

Example 6.7 Evaluate the integral $\int_C f(s)\, ds$ on the curves shown in Fig. 6.6 joining $s = 2 + i$ and $s = 7 + 6i$, where

$$f(s) = s^2$$

We can express $f(s)$ as

$$f(s) = (x + iy)^2 = (x^2 - y^2) + 2ixy$$

Then

$$\int_C f(s)\, ds = \int_C [(x^2 - y^2)\, dx - 2xy\, dy] + i \int_C [2xy\, dx + (x^2 - y^2)\, dy]$$

Over the curve C_1 we have the following relationships between x and y. For $x = 2$ on C_1, $1 \le y \le 6$ and for $y = 6$, $2 \le x \le 7$. On the vertical arc of C_1 $dx = 0$ while $dy = 0$ on the horizontal arc of C_1. Therefore,

$$\int_{C_1} f(x)\, dx = -\int_1^6 4y\, dy + i\int_1^6 (4 - y^2)\, dy + \int_2^7 (x^2 - 36)\, dx + i\int_2^7 12x\, dx$$

$$= -(70) + (-\tfrac{155}{3})i - (\tfrac{205}{3}) + (270)i = -\tfrac{415}{3} + \tfrac{655}{3}i \qquad |$$

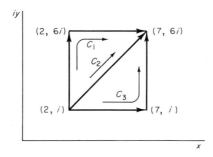

Figure 6.6 Curves C_1, C_2, and C_3 Connecting $s = 2 + i$ and $s = 7 + 6i$

Over the curve C_2, $x = y + 1$. Therefore, if we express x in terms of y,

$$dx = dy$$

and

$$\int_{C_2} f(s)\,ds = \int_1^6 [(2y+1) - 2(y^2+y)]\,dy + i\int_1^6 [2(y^2+y) + (2y+1)]\,dy$$
$$= \int_1^6 (1 - 2y^2)\,dy + i\int_1^6 (1 + 4y + 2y^2)\,dy = -\tfrac{415}{3} + \tfrac{655}{3}i$$

Over the curve C_3, $y = 1$ for $2 \le x \le 7$ and $dy = 0$. On the vertical arc of C_3, $x = 7$ for $1 \le y \le 6$ and $dx = 0$. Therefore

$$\int_{C_3} f(s)\,ds = \int_2^7 (x^2 - 1)\,dx + i\int_2^7 2x\,dx - \int_1^6 14y\,dy + i\int_1^6 (49 - y^2)\,dy$$
$$= (\tfrac{320}{3}) + (45)i - (245) + (\tfrac{420}{3})i = -\tfrac{415}{3} + \tfrac{655}{3}i \qquad \text{☒}$$

In Example 6.7 the integral of $f(s)$ was the same for all three curves connecting $s = 2 + i$ and $s = 7 + 6i$. In fact one can show that the integral of $f(s)$, in this case, is independent of the curve connecting $s = 2 + i$ and $s = 7 + 6i$. However, as demonstrated in the following example, this is not always the case.

Example 6.8 Let

$$f(s) = (x - iy)^2 = (x^2 - y^2) - 2ixy$$

Integrate $f(s)$ from $s = 0$ to $s = 1 + i$ on the curves $y = x$, C_1, and $y = \sqrt{x}$, C_2. The curves connecting the endpoints of integration are shown in Fig. 6.7.

From Eq. (6.42)

$$\int_C f(s)\,ds = \int_C [(x^2 - y^2)\,dx + 2xy\,dy] + i\int_C [-2xy\,dx + (x^2 - y^2)\,dy]$$

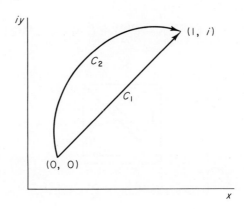

Figure 6.7 Curves C_1 and C_2 Connecting $s = 0$ and $s = 1 + i$

6.4 COMPLEX INTEGRATION

For C_1 we have $y = x$. Therefore, $dy = dx$ and

$$\int_{C_1} f(s)\,ds = \int_0^1 [0\,dx + 2x^2\,dx] + i\int_0^1 [-2x^2\,dx + 0\,dx] = \tfrac{2}{3}(1 - i)$$

For C_2 we have $y = \sqrt{x}$ and $dy = dx/2\sqrt{x}$. Therefore,

$$\int_{C_2} f(s)\,ds = \int_0^1 [(x^2 - x)\,dx + x\,dx] + i\int_0^1 [-x\,dx + (x^2 - x)\,dx]$$

$$= \int_0^1 x^2\,dx + i\int_0^1 (x^2 - 2x)\,dx = \tfrac{1}{3}(1 - 2i) \qquad \boxed{}$$

Many of the properties of ordinary integrals can be extended to line integrals in the complex plane. These properties are summarized as follows, where $f(s)$ and $g(s)$ are defined and integrable on the curve C.

$$\int_C [f(s) \pm g(s)]\,ds = \int_C f(s)\,ds \pm \int_C g(s)\,ds \tag{6.43}$$

$$\left| \int_C f(s)\,ds \right| \le \int_C |f(s)|\,|ds| \tag{6.44}$$

If s_1 and s_2 are the endpoints of C and $\int_{s_1}^{s_2} f(s)\,ds$ and $\int_{s_2}^{s_1} f(s)\,ds$ are line integrals on C, then

$$\int_{s_1}^{s_2} f(s)\,ds = -\int_{s_2}^{s_1} f(s)\,ds \tag{6.45}$$

If s_1, s_2, and s_3 are points on C such that $s_1 < s_2 < s_3$, then

$$\int_{s_1}^{s_3} f(s)\,ds = \int_{s_1}^{s_2} f(s)\,ds + \int_{s_2}^{s_3} f(s)\,ds \tag{6.46}$$

If, in Eq. (6.46), the curve connecting s_1 and s_2 is defined by C_1 and the curve connecting s_2 and s_3 is C_2 then $C = C_1 + C_2$ and

$$\int_C f(s)\,ds = \int_{C_1} f(s)\,ds + \int_{C_2} f(s)\,ds \tag{6.47}$$

In Example 6.7, we dealt with a function whose line integral between two points s_1 and s_2 was independent of the curve over which the integral was taken, at least for the three curves considered. However in Example 6.8, we found that the value of the line integral of the function considered was dependent upon the curve connecting the end points of integration. Conditions will be developed under which a line integral can be shown to be *independent of the path of integration*; but first we must discuss the problem of integrating a function over a closed curve.

In Fig. 6.8 the curves C_1, C_2, and C_4 are *closed* while C_3 is not closed. The curve C_1 is called a *simple closed curve* since it does not intersect itself. In discussing the integral of a function on a closed curve we will be concerned with simple

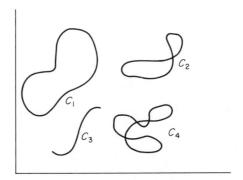

Figure 6.8 Curves Which Are Closed and Nonclosed

closed curves only. The integral of a function around a closed curve will be denoted by $\oint f(s)\,ds$, where $\oint$ is used to convey the idea that the path of integration is closed. Such integrals are often called *contour integrals*. We must now designate the *direction of integration around a closed path*. Let C be a simple closed curve enclosing the region R as shown in Fig. 6.9.

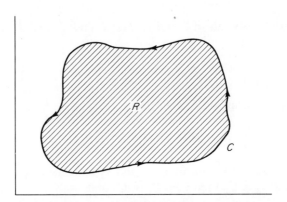

Figure 6.9 Positive Direction of Integration around C

Integration in a *counterclockwise direction* will be taken as *positive* while integration in a *clockwise direction* is *negative*. Of course, this does not imply that integration in a positive direction results in a positive value of the integral. One way of remembering that the positive direction of integration is counterclockwise is by noting that the region, R, enclosed by the curve C is always on the left as C is traversed in a positive direction.

Example 6.9 Find $\oint f(s)\,ds$ where

$$f(s) = x + iy$$

and the closed curve on which $f(s)$ is integrated is shown in Fig. 6.10.

6.4 COMPLEX INTEGRATION

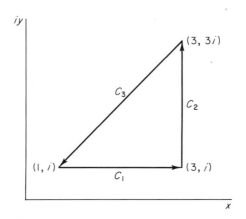

Figure 6.10 Curve of Integration for $f(s) = x + iy$

The closed curve of integration is given by $C = C_1 + C_2 + C_3$. Therefore,

$$\oint f(s)\, ds = \int_{C_1} f(s)\, ds + \int_{C_2} f(s)\, ds + \int_{C_3} f(s)\, ds$$

$$= \oint (x\, dx - y\, dy) + i \oint (x\, dx + y\, dy) \qquad (6.48)$$

On C_1, $dy = 0$ and $1 \le x \le 3$. On C_2, $dx = 0$ and $1 \le y \le 3$. On C_3, $x = y$ and $dx = dy$. Applying Eq. (6.42), $\oint f(s)\, ds$ becomes

$$\oint f(s)\, ds = \left[\int_1^3 x\, dx + i \int_1^3 x\, dx \right] + \left[\int_1^3 -y\, dy + i \int_1^3 y\, dy \right]$$

$$+ \left[\int_3^1 (y\, dy - y\, dy) + i \int_3^1 (y\, dy + y\, dy) \right]$$

Noting that

$$\int_3^1 y\, dy = -\int_1^3 y\, dy$$

we have

$$f(s)\, ds = [4(1 + i)] + [-4(1 - i)] - [0 + 8i] = 0$$

Example 6.10 Integrate

$$f(s) = (x - iy)^2$$

around the closed curve in Fig. (6.11), where $y = x^2$ on C_1 and $y = \sqrt{x}$ on C_2. For C_1, $dy = 2x\, dx$ and on C_2, $dy = dx/2\sqrt{x}$

$$f(s) = (x^2 - y^2) - 2ixy$$

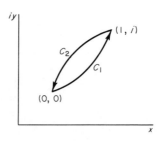

Figure 6.11

Therefore

$$\oint f(s)\,ds = \sum_{j=1}^{2} \int_{C_j} [(x^2 - y^2)\,dx + 2xy\,dy]$$
$$+ i\sum_{j=1}^{2} \int_{C_j} [-2xy\,dx + (x^2 - y^2)\,dy]$$

For C_1, we have

$$\int_{C_1} f(s)\,ds = \int_0^1 [(x^2 - x^4)\,dx + 4x^4\,dx] + i\int_0^1 [-2x^3\,dx + 2(x^3 - x^5)\,dx]$$
$$= \tfrac{14}{15} - \tfrac{1}{3}i$$

Noting the positive direction of integration on C_2, we have

$$\int_{C_2} f(s)\,ds = \int_1^0 [(x^2 - x)\,dx + x\,dx] + i\int_1^0 [-2x^{3/2}\,dx + \tfrac{1}{2}(x^{3/2} - x^{1/2})\,dx]$$
$$= -\tfrac{1}{3} + \tfrac{14}{15}i$$

and

$$\oint f(s)\,ds = \int_{C_1} f(s)\,ds + \int_{C_2} f(s)\,ds = \tfrac{3}{5}(1+i) \qquad \blacksquare$$

Theorem 6.2 Green's Theorem Let R be a closed region bounded by the simple closed curve C. If $g(x,y)$, $h(x,y)$, $\partial g(x,y)/\partial y$, and $\partial h(x,y)/\partial x$ are single-valued and continuous on R, then

$$\oint [g(x,y)\,dx + h(x,y)\,dy] = \iint_R \left[\frac{\partial h(x,y)}{\partial x} - \frac{\partial g(x,y)}{\partial y}\right]dx\,dy \qquad (6.49)$$

where $\iint_R$ denotes the integral over the region R.

Proof Assume that the closed curve C can be divided into two nonclosed curves C_1 and C_2 such that $C = C_1 + C_2$. Further, for $a_x \le x \le b_x$, let $y = v_1(x)$ on C_1 and $y = v_2(x)$ on C_2. This representation of C is shown graphically in Fig. 6.12.

6.4 COMPLEX INTEGRATION

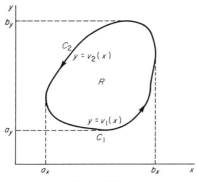

Figure 6.12

We will treat each part of the integral on the right-hand side of Eq. (6.49) separately.

$$\iint_R \frac{\partial g(x, y)}{\partial y} dx\, dy = \int_{a_x}^{b_x} \int_{v_1(x)}^{v_2(x)} \frac{\partial g(x, y)}{\partial y} dy\, dx$$

and

$$\int_{v_1(x)}^{v_2(x)} \frac{\partial g(x, y)}{\partial y} dy = g[x, v_2(x)] - g[x, v_1(x)]$$

But

$$\int_{a_x}^{b_x} g[x, v_2(x)]\, dx = -\int_{b_x}^{a_x} g[x, v_2(x)]\, dx$$

where $\int_{b_x}^{a_x} g[x, v_2(x)]\, dx$ is the line integral of $g(x, y)$ on the curve C_2 in the positive direction. Similarly $\int_{a_x}^{b_x} g[x, v_1(x)]\, dx$ is the line integral of $g(x, y)$ on C_1 in the positive direction. Therefore

$$\int_{a_x}^{b_x} \int_{v_1(x)}^{v_2(x)} \frac{\partial g(x, y)}{\partial y} dy\, dx = -\int_{C_1} g(x, y)\, dx - \int_{C_2} g(x, y)\, dx = -\oint g(x, y)\, dx$$

For $\iint_R \partial h(x, y)/\partial x\, dx\, dy$, we divide C into the curves C_1 and C_2 as shown in Fig. 6.13. The curve C_1, with endpoints a_y and b_y, will be described by $x = w_1(y)$, and C_2, with endpoints b_y and a_y, will be described by $x = w_2(y)$. Then

$$\iint_R \frac{\partial h(x, y)}{\partial x} dx\, dy = \int_{a_y}^{b_y} \int_{w_2(y)}^{w_1(y)} \frac{\partial h(x, y)}{\partial x} dx\, dy$$

$$= \int_{a_y}^{b_y} h[w_1(y), y]\, dy - \int_{a_y}^{b_y} h[w_2(y), y]\, dy$$

$$= \int_{a_y}^{b_y} h[w_1(y), y]\, dy + \int_{b_y}^{a_y} h[w_2(y), y]\, dy = \oint h(x, y)\, dy$$

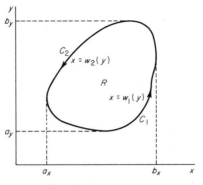

Figure 6.13

Therefore,

$$\iint_R \left[\frac{\partial h(x, y)}{\partial x} - \frac{\partial g(x, y)}{\partial y} \right] dx\, dy = \oint h(x, y)\, dy + \oint g(x, y)\, dx$$

$$= \oint [g(x, y)\, dx + h(x, y)\, dy] \qquad \blacksquare$$

Green's theorem provides the analyst with an alternate method for evaluating the line integral of a function on a simple closed curve. The application of Green's theorem is demonstrated in the following example.

Example 6.11 Solve the problem given in Example 6.10 using Green's theorem. From Example 6.10

$$\oint f(s)\, ds = \oint [(x^2 - y^2)\, dx + 2xy\, dy] + i \oint [-2xy\, dx + (x^2 - y^2)\, dy]$$

Applying Green's theorem to the first integral, let

$$g(x, y) = (x^2 - y^2), \qquad h(x, y) = 2xy$$

Then

$$\frac{\partial g(x, y)}{\partial y} = -2y, \qquad \frac{\partial h(x, y)}{\partial x} = 2y$$

and

$$\oint [g(x, y)\, dx + h(x, y)\, dy] = \int_0^1 \int_{x^2}^{\sqrt{x}} 4y\, dy\, dx = \int_0^1 2(x - x^4)\, dx = \tfrac{3}{5}$$

For the second integral, let

$$g(x, y) = -2xy, \qquad h(x, y) = (x^2 - y^2)$$

and
$$\frac{\partial g(x, y)}{\partial y} = -2x, \quad \frac{\partial h(x, y)}{\partial x} = 2x$$

Therefore,
$$i\oint [g(x, y)\, dx + h(x, y)\, dy] = i\int_0^1 \int_{x^2}^{\sqrt{x}} 4x\, dy\, dx = i\int_0^1 4x(\sqrt{x} - x^2)\, dx = \tfrac{3}{5}i$$

and
$$\oint f(s)\, ds = \tfrac{3}{5}(1 + i)$$

We now have the tools at hand to develop conditions under which the line integral between two given points, a and b, is independent of the path connecting the two points. That is, for a given region, R, containing the points a and b

$$\int_a^b f(s)\, ds = \int_C f(s)\, ds \tag{6.50}$$

for all curves C in R connecting a and b.

Theorem 6.3 Let C be any curve or path in a closed region R bounded by a simple closed curve. If C connects the points a and b in R, then the line integral

$$\int_a^b [g(x, y)\, dx + h(x, y)\, dy]$$

is independent of the path C connecting a and b if and only if in R

$$\frac{\partial g(x, y)}{\partial y} = \frac{\partial h(x, y)}{\partial x} \tag{6.51}$$

and $\partial g(x, y)/\partial y$ and $\partial h(x, y)/\partial x$ are continuous throughout R.

Proof Let C be any closed path bounding R', in R, passing through the points a and b. By Green's theorem

$$\oint [g(x, y)\, dx + h(x, y)\, dy] = \iint_{R'} \left[\frac{\partial h(x, y)}{\partial x} - \frac{\partial g(x, y)}{\partial y}\right] dx\, dy \tag{6.52}$$

Let the closed path of integration on the left-hand side of Eq. (6.52) be divided into two parts C_1 and C_2 such that

$$C = C_1 + C_2$$

where C_1 and C_2 are paths connecting a and b as shown in Fig. 6.14. If Eq. (6.51) is valid, then

$$\oint [g(x, y)\, dx + h(x, y)\, dy] = 0$$

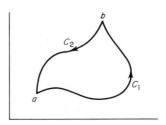

Figure 6.14

But

$$\oint [g(x, y)\, dx + h(x, y)\, dy] = \int_{C_1} [g(x, y)\, dx + h(x, y)\, dy]$$
$$+ \int_{C_2} [g(x, y)\, dx + h(x, y)\, dy]$$

Therefore

$$\int_{C_1} [g(x, y)\, dx + h(x, y)\, dy] = -\int_{C_2} [g(x, y)\, dx + h(x, y)\, dy]$$

and the integrals from a to b on C_1 and C_2 are equal. Since C is any curve in R passing through a and b, the line integral over any curves C_1 and C_2 yield the same result, and integration on R is independent of path.

To prove necessity, let

$$\int_{C_1} [g(x, y)\, dx + h(x, y)\, dy] = -\int_{C_2} [g(x, y)\, dx + h(x, y)\, dy]$$

where C_1 and C_2 are any paths in R connecting any two points in R, a and b. Then for the closed curve $C = C_1 + C_2$

$$\oint [g(x, y)\, dx + h(x, y)\, dy] = 0 \qquad (6.53)$$

Assume that $\partial g(x, y)/\partial y < \partial h(x, y)/\partial x$ for some subregion of R, R', bounded by the simple closed curve C'. By Green's theorem

$$\oint [g(x, y)\, dx + h(x, y)\, dy] = \iint_{R'} \left[\frac{\partial h(x, y)}{\partial x} - \frac{\partial g(x, y)}{\partial y} \right] dx\, dy > 0 \qquad (6.54)$$

where the line integral on the left is around C. But C must pass through two points in R, a and b. Since Eq. (6.53) is presumed to hold for all a and b in R and all paths C_1 and C_2 connecting a and b, Eq. (6.54) cannot be valid. A similar argument may be used when $\partial g(x, y)/\partial y > \partial h(x, y)/\partial x$. Therefore a contradiction of Eq. (6.53) results if

$$\frac{\partial g(x, y)}{\partial y} < \frac{\partial h(x, y)}{\partial x}$$

6.4 COMPLEX INTEGRATION

or if

$$\frac{\partial g(x, y)}{\partial y} > \frac{\partial h(x, y)}{\partial x}$$

and therefore

$$\frac{\partial g(x, y)}{\partial y} = \frac{\partial h(x, y)}{\partial x}$$

Let us now consider the application of Theorem 6.1 to functions of a complex variable. The reader will recall from Eq. (6.42) that for C belonging to the region S_0

$$\int_C f(s)\, ds = \int_C [g(x, y)\, dx - h(x, y)\, dy] + i \int_C [h(x, y)\, dx + g(x, y)\, dy]$$

Treating each integral on the right separately, the first integral is independent of the path C if and only if $\partial g(x, y)/\partial y$ and $\partial h(x, y)/\partial x$ are continuous on S_0 and

$$\frac{\partial g(x, y)}{\partial y} = -\frac{\partial h(x, y)}{\partial x} \tag{6.55}$$

Similarly, the second integral is independent of the path C if $\partial g(x, y)/\partial x$ and $\partial h(x, y)/\partial y$ are continuous on S_0 and

$$\frac{\partial g(x, y)}{\partial x} = \frac{\partial h(x, y)}{\partial y} \tag{6.56}$$

However, from Theorem 6.1 these are the conditions for $f(s)$ to be analytic on S_0. Thus we are led to the following theorem.

Theorem 6.4 Let S_0 be a region in the complex plane bounded by a simple closed curve and let s_1 and s_2 be any two points in S_0. The line integral $\int_{s_1}^{s_2} f(s)\, ds$ is independent of the path connecting s_1 and s_2 if the path belongs to S_0 and if $f(s)$ is analytic on S_0. Further

$$\oint f(s)\, ds = 0 \tag{6.57}$$

for every simple closed curve in S_0.

From Theorem 6.4, if C_1 and C_2 are any two simple closed curves on S_0 and if $f(s)$ is analytic there, then

$$\int_{C_1} f(s)\, ds = \int_{C_2} f(s)\, ds \tag{6.58}$$

since both of these contour integrals have value zero if S_0 is bounded by a simple closed curve. A region S_0 bounded by a simple closed curve is called a simply-connected region. Thus, if S_0 is *simply-connected*, any closed curve belonging to S_0

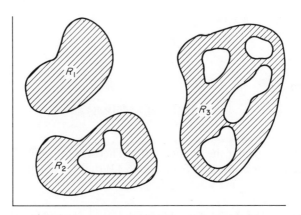

Figure 6.15 Simply- and Multiply-Connected Regions

can be shrunk to a point without ever leaving S_0. Any region which does not possess this property is said to be *multiply-connected*. In Fig. 6.15, the region R_1 is simply-connected, while R_2 and R_3 are multiply-connected. Suppose that R is a multiply-connected region bounded by two simple closed curves C_1 and C_2 as shown in Fig. 6.16. As already stated, the positive direction of integration around either C_1 or C_2 is the direction such that the region of interest is always on the left as the path of integration is traversed. Therefore, in Fig. 6.16 the positive direction of integration along C_1 is counterclockwise, while the positive direction of integration along C_2 is clockwise. We now show that Eq. (6.58) holds for multiply-connected as well as simply-connected regions.

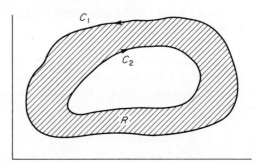

Figure 6.16 Multiply Connected Region R Bounded by the Simple Closed Curves C_1 and C_2

Theorem 6.5 Let $f(s)$ be analytic in the closed region S_0, bounded by the simple closed curves C_1 and C_2. Then

$$\int_{C_1} f(s)\, ds = \int_{C_2} f(s)\, ds \tag{6.59}$$

where C_1 and C_2 are traversed in a positive direction with respect to their interiors.

6.4 COMPLEX INTEGRATION

Proof Let S_0 be the multiply-connected region shown in Fig. 6.17. Let us connect the curves C_1 and C_2 by the curve C_3 as shown in Fig. 6.17. In Fig. 6.17 the positive directions along C_1, C_2, and C_3 are those with respect to S_0. (Note that

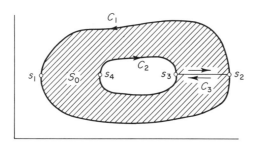

Figure 6.17 Multiply-Connected Region S_0

the positive direction along C_2 as defined in Eq. (6.59) is that with respect to its interior and not with respect to R). Therefore the curve connecting the points $s_1-s_2-s_3-s_4-s_3-s_2-s_1$ is a simple closed curve, and by Theorem 6.4,

$$\int_{s_2}^{s_1} f(s)\,ds + \int_{s_1}^{s_2} f(s)\,ds + \int_{s_2}^{s_3} f(s)\,ds + \int_{s_3}^{s_4} f(s)\,ds + \int_{s_4}^{s_3} f(s)\,ds + \int_{s_3}^{s_2} f(s)\,ds = 0$$

But

$$\int_{s_2}^{s_3} f(s)\,ds + \int_{s_3}^{s_2} f(s)\,ds = 0$$

since the line joining s_2 and s_3 is a region on which $f(s)$ is analytic, by assumption that $f(s)$ is analytic on S_0. Then

$$\int_{C_1} f(s)\,ds + \int_{C_2} f(s)\,ds = 0$$

or

$$\int_{C_1} f(s)\,ds = -\int_{C_2} f(s)\,ds \qquad (6.60)$$

where the direction of integration on C_1 and C_2 are positive with respect to S_0. If we reverse the direction of integration on C_2 we obtain the desired result. ∎

Two points are important in the proof of Theorem 6.5. First, a multiply-connected region can be converted to a simply-connected region. Second, in Theorem 6.4, we stated that $\oint f(s)\,ds = 0$ if $f(s)$ is analytic on the closed curve of integration and on the interior of the curve. In the proof of Theorem 6.5, we indicated that this result is preserved for a multiply-connected region by converting the multiply-connected region into a simply-connected region. The final result is that if we integrate in a positive direction around the boundary of a multiply-connected region then

$$\int_{C_1} f(s)\, ds + \int_{C_2} f(s)\, ds = 0 \tag{6.61}$$

Therefore, if $C_1, C_2, \ldots, C_n$ are nonoverlapping simple closed curves bounding the region S_0 on which $f(s)$ is analytic, as shown in Fig. 6.18, then

$$\int_{C_1} f(s)\, ds + \int_{C_2} f(s)\, ds + \cdots + \int_{C_n} f(s)\, ds = 0 \tag{6.62}$$

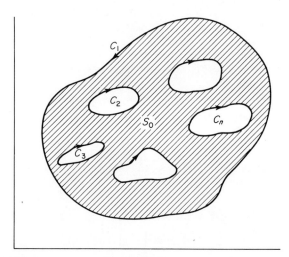

Figure 6.18 Multiply-Connected Region Bounded by n Nonoverlapping Simple Closed Curves

where integration around each closed curve is in a positive direction with respect to S_0, or

$$\int_{C_1} f(s)\, ds = -\int_{C_2} f(s)\, ds - \int_{C_3} f(s)\, ds - \cdots - \int_{C_n} f(s)\, ds \tag{6.63}$$

However, $\oint_{C_i} f(s)\, ds$ is not necessarily zero, since $f(s)$ may not be analytic on the interior of C_i. In particular, if $f(s)$ is analytic on C_i and its interior as well as on S_0, then

$$\int_{C_i} f(s)\, ds = 0 \tag{6.64}$$

for all i.

Example 6.12 Let S_0 be a region bounded by circles of radius 1 and 4 both with centers at the origin. Show that

$$\int_{C_1} f(s)\, ds = \int_{C_2} f(s)\, ds$$

6.4 COMPLEX INTEGRATION

where C_1 is the curve defined by the circle of radius 1, C_2 is the curve defined by the circle of radius 4, both integrals are in the positive direction with respect to their interiors, and where

$$f(s) = \frac{1}{s}$$

From Eqs. (6.12) and (6.13),

$$s = r \exp[i\theta] \quad \text{and} \quad ds = ir \exp[i\theta] \, d\theta$$

Therefore, for C_1 ($r = 1$),

$$\int_{C_1} f(s) \, ds = \int_0^{2\pi} \frac{i \exp[i\theta]}{\exp[i\theta]} \, d\theta = i \int_0^{2\pi} d\theta = 2\pi i$$

For C_2 ($r = 4$),

$$\int_{C_2} f(s) \, ds = \int_0^{2\pi} \frac{i 4 \exp[i\theta]}{4 \exp[i\theta]} \, d\theta = i \int_0^{2\pi} d\theta = 2\pi i \qquad \square$$

In Example 6.12, $f(s)$ was analytic throughout the interior of C_1 and C_2 except at $s = 0$. Therefore, $s = 0$ is a singular point for $f(s)$. Suppose that the points $s_1, s_2, \ldots, s_n$ are singular points for $f(s)$ on S_0, which do not lie on the boundary of S_0. Let C be a simple closed curve bounding S_0 and let $C_1, C_2, \ldots, C_n$ be closed curves with interiors including the singular points $s_1, s_2, \ldots, s_n$. We will further assume that $C, C_1, C_2, \ldots, C_n$ are nonoverlapping. Then

$$\oint_C f(s) \, ds = \sum_{i=1}^n \oint_{C_i} f(s) \, ds \qquad (6.65)$$

where the direction of integration on $C, C_1, C_2, \ldots, C_n$ is positive with respect to their interiors. The reader will note that the integral $\oint_{C_i} f(s) \, ds$ is not necessarily zero since $f(s)$ is not analytic throughout the interior of C_i, $i = 1, 2, \ldots, n$.

Example 6.13 Let

$$f(s) = \frac{1}{(s-1)(s-2)}$$

on S_0 where S_0 is bounded by a circle, C, with radius 5 and center at $(0, 0)$. Find

$$\oint_C f(s) \, ds$$

Let

$$f(s) = \frac{1}{[(x-1) + iy][(x-2) + iy]}$$

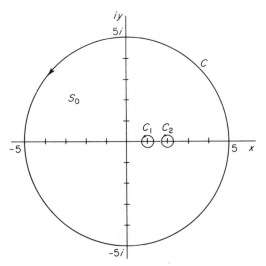

Figure 6.19 Region S_0 with Singular Points for $f(s)$ at $s = (1, 0)$ and $s = (2, 0)$

Then $f(s) \to \infty$ at $s = (1, 0)$ and $s = (2, 0)$, and $f(s)$ is not analytic throughout S_0. Let us construct circular contours, C_1 and C_2, about each of these points, each having radius $r < 0.5$ as shown in Fig. 6.19. Then

$$\oint_C f(s)\,ds = \oint_{C_1} f(s)\,ds + \oint_{C_2} f(s)\,ds$$

where the direction of integration on each circle is positive with respect to its interior. On C_j

$$(x - j) + iy = r[\cos(\theta) + i\sin(\theta)]$$

where

$$\sin(\theta) = \frac{y}{r}, \qquad \cos(\theta) = \frac{x - j}{r}, \qquad \text{and} \qquad (x - j) + iy = r\exp[i\theta]$$

Therefore we have

$$f(s) = \frac{1}{r\exp[i\theta](r\exp[i\theta] - 1)}$$

on C_1, and

$$f(s) = \frac{1}{(r\exp[i\theta] + 1)r\exp[i\theta]}$$

on C_2. Finally,

$$\oint_C f(s)\,ds = i\int_0^{2\pi} \frac{d\theta}{(r\exp[i\theta] - 1)} + i\int_0^{2\pi} \frac{d\theta}{(r\exp[i\theta] + 1)}$$

6.4 COMPLEX INTEGRATION

where

$$ds = ir \exp[i\theta] \, d\theta$$

Therefore,

$$\oint_C f(s) \, ds = i\left[-\theta + \frac{1}{i}\ln(-1 + r\exp[i\theta])\right]\Big|_0^{2\pi}$$

$$+ i\left[\theta - \frac{1}{i}\ln(1 + r\exp[i\theta])\right]\Big|_0^{2\pi}$$

$$= \ln\left[\frac{r\exp[2\pi i] - 1}{r\exp[2\pi i] + 1}\right] - \ln\left[\frac{r-1}{r+1}\right]$$

$$= \ln\left[\frac{r^2 \exp[2\pi i] + r\exp[2\pi i] - r - 1}{r^2 \exp[2\pi i] - r\exp[2\pi i] + r - 1}\right]$$

Since

$$\exp[2\pi i] = \cos(2\pi) + i\sin(2\pi) = 1$$

then

$$\oint_C f(s) \, ds = \ln\left[\frac{r^2 - 1}{r^2 - 1}\right] = 0$$

The above example illustrates the fact that the integral of a function around a simple closed curve may be zero, even though the function is not analytic throughout the region enclosed by the curve. Of course this is not always the case as shown in the following example.

Example 6.14 Find

$$\oint_C \frac{s}{(s^2 + 4)} \, ds$$

where C is a circle of radius 3 with center at the origin.

Clearly, this function fails to be analytic at $s = (0, 2i)$, and $s = (0, -2i)$. As in Example 6.13, we again enclose each singular point on the interior of C by a circle of radius r, as shown in Fig. 6.20, such that the three resulting curves do not overlap. Expressing $f(s)$ as

$$f(s) = \frac{s}{(s - 2i)(s + 2i)}$$

on C_1 let

$$s - 2i = r\exp[i\theta]$$

and on C_2 let

$$s + 2i = r\exp[i\theta]$$

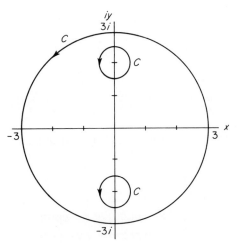

Figure 6.20

In both cases

$$ds = ir \exp[i\theta]\, d\theta$$

Then

$$\oint_C \frac{ds}{(s^2+4)} = \int_0^{2\pi} \frac{ir\exp[i\theta](r\exp[i\theta]-2i)}{r\exp[i\theta](r\exp[i\theta]-4i)}\, d\theta$$

$$+ \int_0^{2\pi} \frac{ir\exp[i\theta](r\exp[i\theta]+2i)}{(r\exp[i\theta]+4i)(r\exp[i\theta])}\, d\theta$$

$$= i\int_0^{2\pi} \frac{r\exp[i\theta]}{(r\exp[i\theta]-4i)}\, d\theta + 2\int_0^{2\pi} \frac{d\theta}{(r\exp[i\theta]-4i)}$$

$$+ i\int_0^{2\pi} \frac{r\exp[i\theta]}{(r\exp[i\theta]+4i)}\, d\theta - 2\int_0^{2\pi} \frac{d\theta}{(r\exp[i\theta]+4i)}$$

$$= i\int_0^{2\pi} \frac{d\theta}{\left(1 - \dfrac{4i}{r}\exp[-i\theta]\right)} + 2\int_0^{2\pi} \frac{d\theta}{(r\exp[i\theta]-4i)}$$

$$+ i\int_0^{2\pi} \frac{d\theta}{\left(1 + \dfrac{4i}{r}\exp[-i\theta]\right)} - 2\int_0^{2\pi} \frac{d\theta}{(r\exp[i\theta]+4i)}$$

$$= i(2\pi) + 2\left(-\frac{\pi}{2i}\right) + i(2\pi) - 2\left(\frac{\pi}{2i}\right) = 2\pi i$$

6.4 COMPLEX INTEGRATION

6.4.1 The Method of Residues

As the reader has undoubtedly recognized, line integration in the complex plane can be quite tedious. In this section a method of complex integration will be introduced which can greatly simplify this problem. This method of integration is called the *method of residues*. To fully understand the method of residues we must discuss Cauchy's integral theorem and the Laurent series.

Theorem 6.6 Cauchy's Integral Theorem Let S_0 be a closed region bounded by the simple closed curve C and let $f(s)$ be analytic on S_0. If s_0 is any interior point for S_0, then

$$f(s_0) = \frac{1}{2\pi i} \oint_C \frac{f(s)}{s - s_0} ds \qquad (6.66)$$

where integration on C is in the positive direction with respect to S_0.

Proof Let us construct a circular curve, C_1, about s_0 with radius r such that C and C_1 do not overlap. Then

$$\oint_C \frac{f(s)}{s - s_0} ds = \oint_{C_1} \frac{f(s)}{s - s_0} ds$$

We may construct the circular curve C_1 such that it does not overlap C since s_0 is assumed to be an interior point for S_0. Letting

$$s - s_0 = r \exp[i\theta]$$

we have

$$\oint_{C_1} \frac{f(s)}{s - s_0} ds = \int_0^{2\pi} \frac{f(s_0 + r \exp[i\theta])}{r \exp[i\theta]} ir \exp[i\theta] d\theta = i \int_0^{2\pi} f(s_0 + r \exp[i\theta]) d\theta$$

However, this equation is valid for every circle about s_0 provided it does not intersect C. Therefore

$$\int_0^{2\pi} f(s_0 + r \exp[i\theta]) d\theta = \lim_{r \to 0} \int_0^{2\pi} f(s_0 + r \exp[i\theta]) d\theta$$

$$= \int_0^{2\pi} f(s_0) d\theta = 2\pi f(s_0)$$

and

$$\oint_C \frac{f(s)}{s - s_0} ds = 2\pi i f(s_0) \quad \text{or} \quad f(s_0) = \frac{1}{2\pi i} \oint_C \frac{f(s)}{s - s_0} ds \qquad \blacksquare$$

Theorem 6.7 Laurent Series Let S_0 be a closed region bounded by two concentric circles C_1 and C_2 with radii r_1 and r_2 ($r_1 > r_2$) and both having centers s_0. If $f(s)$ is analytic on S_0, then for any s in S_0

$$f(s) = \sum_{j=-\infty}^{\infty} a_j(s - s_0)^j \qquad (6.67)$$

where

$$a_j = \frac{1}{2\pi i} \oint_C \frac{f(s)}{(s - s_0)^{j+1}} ds, \quad j = 0, \pm 1, \pm 2, \ldots \qquad (6.68)$$

for any circle C with center s_0 and radius r, where $r_2 < r < r_1$.

Now suppose that s_0 is an isolated singular point for $f(s)$ on S_0. Since s_0 is an isolated singularity, we can construct concentric circles, C_1 and C_2, about s_0 such that $f(s)$ is analytic on the region S_0' bounded by the concentric circles as shown in Fig. 6.21.

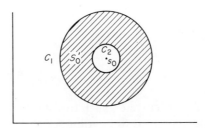

Figure 6.21 Region, S_0', on Which $f(s)$ Is Analytic

By Theorem 6.7, $f(s)$ can be expressed by a Laurent series for s on the interior of S_0' and

$$f(s) = \sum_{j=1}^{\infty} \frac{a_{-j}}{(s - s_0)^j} + \sum_{j=0}^{\infty} a_j(s - s_0)^j$$

If we choose C_2 with radius zero, then $f(s)$ is analytic throughout S_0 except at s_0.

Definition 6.6 In Eq. (6.67)

$$\sum_{j=1}^{\infty} \frac{a_{-j}}{(s - s_0)^j}$$

is called the *principal part* of the Laurent series and

$$\sum_{j=0}^{\infty} a_j(s - s_0)^j$$

is called the *analytic part* of the Laurent series.

6.4 COMPLEX INTEGRATION

Definition 6.7 Let $f(s)$ be analytic throughout the closed region S_0 except at the interior point s_0. The isolated singularity at s_0 is called a *pole of order n* if

$$\sum_{j=1}^{\infty} \frac{a_{-j}}{(s-s_0)^j} = \sum_{j=1}^{n} \frac{a_{-j}}{(s-s_0)^j} \qquad (6.69)$$

The implication of Definition 6.7 is that if the principal part of the Laurent expansion about s_0 is such that all terms from the $(n+1)$st on are zero, the isolated singularity at s_0 is called a pole of order n. The following definition presents an alternative method of identifying a pole and its order.

Definition 6.8 Let $f(s)$ be analytic throughout the closed region S_0 except at the interior point s_0. The isolated singularity at s_0 is a pole of order n if there exists an integer $n > 0$ such that

$$\lim_{s \to s_0} (s-s_0)^n f(s) = L \qquad (6.70)$$

where L is finite and $L \neq 0$.

To demonstrate the equivalence of Definitions 6.7 and 6.8, note that

$$(s-s_0)^n f(s) = \sum_{j=1}^{\infty} \frac{a_{-j}}{(s-s_0)^{j-n}} + \sum_{j=0}^{\infty} a_j (s-s_0)^{j+n}$$

But by Definition 6.7,

$$(s-s_0)^n \sum_{j=n+1}^{\infty} \frac{a_{-j}}{(s-s_0)^j} = 0$$

Therefore,

$$(s-s_0)^n f(s) = \sum_{j=1}^{n} \frac{a_{-j}}{(s-s_0)^{j-n}} + \sum_{j=0}^{\infty} a_j (s-s_0)^{j+n}$$

Taking the limit as $s \to s_0$, we have

$$\lim_{s \to s_0} (s-s_0)^n f(s) = \sum_{j=1}^{n} \lim_{s \to s_0} \frac{a_{-j}}{(s-s_0)^{j-n}}$$

Since $a_{-n} \neq 0$ by Definition 6.7, we have

$$\lim_{s \to s_0} (s-s_0)^n f(s) = \sum_{j=1}^{n-1} \lim_{s \to s_0} \frac{a_{-j}}{(s-s_0)^{j-n}} + a_{-n} = a_{-n} \qquad (6.71)$$

where

$$\lim_{s \to s_0} \frac{a_{-j}}{(s-s_0)^{j-n}} = 0, \qquad j < n \qquad (6.72)$$

The coefficient a_{-1} is called the *residue* at the singular point s_0 and will play an important role in the discussion which follows on complex integration. From Eq. (6.68)

$$a_{-1} = \frac{1}{2\pi i} \oint_C f(s)\, ds \qquad (6.73)$$

However Eq. (6.73) requires evaluation of $\oint_C f(s)\, ds$. Given an alternative method of evaluating a_{-1}, we could evaluate $\oint_C f(s)\, ds$ by

$$\oint_C f(s)\, ds = 2\pi i a_{-1} \qquad (6.74)$$

which is exactly what we are seeking. That is, if s_0 is an isolated singularity we would like to find a simple method through which a_{-1} can be evaluated so that we can evaluate the integral in Eq. (6.74) without going through the formal process of integration.

Theorem 6.8 Let S_0 be a closed region bounded by the circle C with center s_0 and let $f(s)$ be analytic throughout S_0 except at s_0. If s_0 is a pole of order n, then

$$a_{-1} = \lim_{s \to s_0} \frac{1}{(n-1)!} \frac{d^{n-1}}{ds^{n-1}} [(s - s_0)^n f(s)] \qquad (6.75)$$

Proof If we multiply $f(s)$ by $(s - s_0)^n$ we obtain

$$(s - s_0)^n f(s) = \sum_{j=1}^{n} a_{-j}(s - s_0)^{n-j} + \sum_{j=0}^{\infty} a_j (s - s_0)^{n+j}$$

Taking the $(n - 1)$st partial derivative of $(s - s_0)^n f(s)$ with respect to s yields

$$\frac{d^{n-1}}{ds^{n-1}} [(s - s_0)^n f(s)] = (n - 1)! a_{-1} + \sum_{j=2}^{n} a_{-j} \frac{d^{n-1}}{ds^{n-1}} (s - s_0)^{n-j}$$

$$+ \sum_{j=0}^{\infty} a_j \frac{(n+j)!}{(j+1)!} (s - s_0)^{j+1}$$

$$= (n - 1)! a_{-1} + \sum_{j=0}^{\infty} a_j \frac{(n+j)!}{(j+1)!} (s - s_0)^{j+1} \qquad (6.76)$$

since

$$\frac{d^{n-1}}{ds^{n-1}} (s - s_0)^{n-j} = 0$$

for $1 < j \leq n$. Taking the limit of both sides of Eq. (6.76) we have

$$a_{-1} = \frac{1}{(n-1)!} \lim_{s \to s_0} \frac{d^{n-1}}{ds^{n-1}} [(s - s_0)^n f(s)]$$

6.4 COMPLEX INTEGRATION

Example 6.15 Identify the poles of each of the following functions and calculate the residue at each pole.

a. $f(s) = \dfrac{s}{(s-1)(s-2)}$

b. $f(s) = \dfrac{1}{(s+1)(s+2)^2}$

c. $f(s) = \dfrac{s^2}{(s^2+1)}$

d. $f(s) = \dfrac{s}{(s^2+4)(s-1)^4}$

In part a, $f(s)$ has poles of order 1 at $s = 1$ and $s = 2$ since

$$\lim_{s \to 1} [(s-1)f(s)] = \lim_{s \to 1} \left(\frac{s}{s-2}\right) = -1$$

and

$$\lim_{s \to 2} [(s-2)f(s)] = \lim_{s \to 2} \left(\frac{s}{s-1}\right) = 2$$

At the pole $s = 1$, Eq. 6.75 yields the residue

$$a_{-1} = \frac{1}{0!} \lim_{s \to 1} [(s-1)f(s)] = -1$$

At $s = 2$ the residue is given by

$$a_{-1} = \frac{1}{0!} \lim_{s \to 2} [(s-2)f(s)] = 2$$

The function in part b has a pole of order 1 at $s = -1$, and a pole of order 2 at $s = -2$ since

$$\lim_{s \to -1} [(s+1)f(s)] = \lim_{s \to -1} \frac{1}{(s+2)^2} = 1$$

and

$$\lim_{s \to -2} [(s+2)^2 f(s)] = \lim_{s \to -2} \frac{1}{s+1} = -1$$

The residue at $s = -1$ is given by

$$a_{-1} = \frac{1}{0!} \lim_{s \to -1} [(s+1)f(s)] = 1$$

since $s = -1$ is a pole of order 1. The pole at $s = -2$ is of order 2 and the residue at this point is given by

$$a_{-1} = \frac{1}{1!} \lim_{s \to -2} \frac{d}{ds} [(s+2)^2 f(s)]$$

$$= \lim_{s \to -2} \frac{d}{ds} \left[\frac{1}{s+1}\right] = \lim_{s \to -2} \left[-\frac{1}{(s+1)^2}\right] = -1$$

The function in part c has poles of order 1 at $s = -i$ and $s = i$.

$$\lim_{s \to -i} [(s+i)f(s)] = \lim_{s \to -i} \left[\frac{s^2(s+i)}{s^2+1}\right] = \lim_{s \to -i} \left(\frac{s^2}{s-i}\right) = \frac{1}{2i}$$

At $s = i$,

$$\lim_{s \to i} [(s-i)f(s)] = \lim_{s \to i} \left[\frac{s^2(s-i)}{s^2+1}\right] = \lim_{s \to i} \left(\frac{s^2}{s+i}\right) = -\frac{1}{2i}$$

Since each of these poles are of order 1, then

$$a_{-1} = \frac{1}{2i}$$

at $s = -i$ and

$$a_{-1} = -\frac{1}{2i}$$

at $s = i$.

In part d, $f(s)$ has poles of order 1 at $s = -2i$ and $s = 2i$ and a pole of order 4 at $s = 1$. The residue at $s = -2i$ is given by

$$a_{-1} = \lim_{s \to -2i} [(s+2i)f(s)] = \lim_{s \to -2i} \left[\frac{s}{(s-2i)(s-1)^4}\right] = \frac{1}{2(1+2i)^4}$$

At $s = 2i$,

$$a_{-1} = \lim_{s \to 2i} [(s-2i)f(s)] = \lim_{s \to 2i} \left[\frac{s}{(s+2i)(s-1)^4}\right] = \frac{1}{2(2i-1)^4}$$

At $s = 1$, the pole is of order 4 and the residue at this point is given by

$$a_{-1} = \frac{1}{3!} \lim_{s \to 1} \frac{d^3}{ds^3} [(s-1)^4 f(s)]$$

$$a_{-1} = \frac{1}{6} \lim_{s \to 1} \frac{d^3}{ds^3} \left[\frac{s}{s^2+4}\right] = \frac{1}{6} \lim_{s \to 1} \left[-\frac{6(s^4 - 24s^2 + 16)}{(s^2+4)^4}\right] = \frac{7}{625}$$

We now have the foundation necessary to develop a simple method through which contour integrals may be resolved.

6.4 COMPLEX INTEGRATION

Theorem 6.9 Residue Theorem Let S_0 be a closed region bounded by the simple closed curve C. If $f(s)$ is analytic on S_0 except at the singular points $s_0, s_1, s_2, \ldots, s_m$, lying on the interior of C, then

$$\oint_C f(s)\, ds = 2\pi i \sum_{j=0}^{m} a_{-1,j} \qquad (6.77)$$

where $a_{-1,j}$ is the residue associated with s_j, $j = 0, 1, 2, \ldots, m$, and where integration around C is in the positive direction.

Proof Let the singular point s_j be inclosed by a circle C_j, $j = 0, 1, 2, \ldots, m$, such that the circles $C_0, C_1, C_2, \ldots, C_m$ do not overlap each other or C. Then

$$\oint_C f(s)\, ds = \sum_{j=0}^{m} \oint_{C_j} f(s)\, ds$$

from Eq. (6.65), where integration on each curve is positive with respect to its interior. From Eq. (6.74)

$$\oint_{C_j} f(s)\, ds = 2\pi i a_{-1,j}$$

Therefore,

$$\oint_C f(s)\, ds = 2\pi i \sum_{j=0}^{m} a_{-1,j} \qquad \blacksquare$$

Example 6.16 Evaluate $\oint_C f(s)\, ds$ for each of the functions $f(s)$ in Example 6.15 where C is a circle of radius 3 with center at $s = 0$.

In Example 6.15, the following functions were given.

a. $f(s) = \dfrac{s}{(s-1)(s-2)}$

b. $f(s) = \dfrac{1}{(s+1)(s+2)^2}$

c. $f(s) = \dfrac{s^2}{(s^2+1)}$

d. $f(s) = \dfrac{s}{(s^2+4)(s-1)^4}$

For each function,

$$\oint_C f(s)\, ds = 2\pi i \text{ (sum of the residues within } C\text{)}$$

For part a, poles of order 1 exist at $s = 1$ and $s = 2$. Since both of these poles lie within C,

$$\oint_C f(s)\, ds = 2\pi i \text{ (residue at } s = 1 + \text{residue at } s = 2\text{)}$$

From Example 6.15, the residue at $s = 1$ is -1 and the residue at $s = 2$ is 2. Therefore

$$\oint_C \frac{s}{(s-1)(s-2)} ds = 2\pi i(-1+2) = 2\pi i$$

For part b, a first-order pole exists at $s = -1$ while a second-order pole exists at $s = -2$. Again, both of these poles lie within C and

$$\oint_C f(s)\, ds = 2\pi i \text{ (residue at } s = -1 + \text{ residue at } s = -2)$$

From Example 6.15, the residue at $s = -1$ is 1 while that at $s = -2$ is -1 and

$$\oint_C \frac{1}{(s+1)(s+2)^2} ds = 2\pi i(1-1) = 0$$

In part c, two first-order poles exist and lie at $s = -i$ and $s = i$, both lying within C. The residue at $s = -i$ is $1/2i$ and the residue at $s = i$ is $-1/2i$. Therefore

$$\oint_C \frac{s^2}{(s^2+1)} ds = 2\pi i\left(\frac{1}{2i} - \frac{1}{2i}\right) = 0$$

In part d, first-order poles are located at $s = -2i$ and $s = 2i$ while a fourth-order pole is located at $s = 1$. Since each pole is located within the region bounded by C, from the results of Example 6.15 we have

$$\oint_C \frac{s}{(s^2+16)(s-1)^4} ds = 2\pi i[\text{sum of the residues at } s = \pm 2i, \; s = 1]$$

$$= 2\pi i\left[\frac{1}{2(1+2i)^4} + \frac{1}{2(2i-1)^4} + \frac{7}{625}\right] = 0 \quad \text{◺}$$

So far we have used the residue theorem to evaluate the integral of a function around a closed curve. However, under appropriate conditions we may apply the residue theorem to integrals over nonclosed curves. Of particular interest are integrals of the form

$$\int_{C_1} f(s)\, ds = \int_{a-i\infty}^{a+i\infty} f(s)\, ds \qquad (6.78)$$

and

$$\int_{C_2} f(s)\, ds = \int_{-\infty+ia}^{\infty+ia} f(s)\, ds \qquad (6.79)$$

In Eq. (6.78), the integral is taken over a vertical line in the complex plane, while the integral in Eq. (6.79) is taken over a horizontal line in the complex plane. The

6.4 COMPLEX INTEGRATION

reader will recall that the integrals in Eqs. (6.78) and (6.79) are improper integrals and may be written as

$$\int_{C_1} f(s)\, ds = \lim_{r \to \infty} \int_{a-ir}^{a+ir} f(s)\, ds \tag{6.80}$$

$$\int_{C_2} f(s)\, ds = \lim_{r \to \infty} \int_{-r+ia}^{r+ia} f(s)\, ds \tag{6.81}$$

The paths of integration in Eqs. (6.80) and (6.81) are shown in Fig. 6.22. Let us now examine the conditions under which these integrals can be evaluated using the residue theorem.

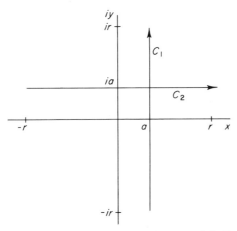

Figure 6.22 Paths of Integration, C_1 and C_2, for the Integrals in Eqs. (6.80) and (6.81) Where $r \to \infty$

Theorem 6.10 Let $f(s)$ be a function which is analytic to the left of the line, C_1, connecting $a - i\infty$ and $a + i\infty$ in the complex plane except at a finite number of poles, with residues $a_{-1,j}$, $j = 1, 2, \ldots, n$, none of which lie on the line C_1. Then

$$\int_{a-i\infty}^{a+i\infty} f(s)\, ds = 2\pi i \sum_{j=1}^{n} a_{-1,j} \tag{6.82}$$

if

$$\lim_{r \to \infty} [r \exp[i\theta] f(r \exp[i\theta])] = 0 \tag{6.83}$$

uniformly for $\theta_0 \leq \theta \leq 2\pi - \theta_0$ where

$$s = r \exp[i\theta]$$

and θ_0 is the angle between $a + ir$ and the x axis for all r.

Proof From Eq. (6.80), we have

$$\int_{a-i\infty}^{a+i\infty} f(s)\, ds = \lim_{r \to \infty} \int_{a-ir}^{a+ir} f(s)\, ds$$

Initially we will work with the integral over the limits $a - ir$ to $a + ir$. Let C_1 be the line connecting the points $a - ir$ and $a + ir$ and let us connect these points by a circular arc, C_2, of radius R and with center at $s = 0$ as shown in Fig. (6.23). We

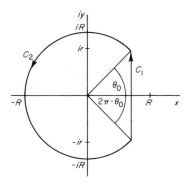

Figure 6.23 Closed Curve $C_1 + C_2$ Passing Through $a - ir$ and $a + ir$

therefore have a closed curve passing through $a - ir$ and $a + ir$. If we choose r large enough to enclose all of the poles to the left of C_1

$$\int_{C_1} f(s)\, ds + \int_{C_2} f(s)\, ds = 2\pi i \sum_{j=1}^{n} a_{-1,j} \tag{6.84}$$

by the residue theorem. Rearranging the terms in Eq. (6.84) and taking the absolute value of both sides of the resulting equation yields

$$\left| \int_{C_1} f(s)\, ds - 2\pi i \sum_{j=1}^{n} a_{-1,j} \right| = \left| -\int_{C_2} f(s)\, ds \right| \tag{6.85}$$

Considering the integral on the right-hand side, we have

$$\left| -\int_{C_2} f(s)\, ds \right| \leq \int_{C_2} |f(s)|\, |ds|$$

from Eq. (6.44). Let

$$s = R \exp[i\theta]$$

That is, s is a point on a circle of radius R with center at the origin. Then

$$\int_{C_2} |f(s)|\, |ds| = \int_{\theta_0}^{2\pi - \theta_0} |i|\, |R \exp[i\theta] f(R \exp[i\theta])|\, |d\theta|$$

Since $\lim_{r \to \infty} [r \exp[i\theta] f(r \exp[i\theta])] = 0$ uniformly for $\theta_0 \leq \theta \leq 2\pi - \theta_0$, there exists R_0 corresponding to an arbitrarily small $\epsilon > 0$ such that

$$|r \exp[i\theta] f(r \exp[i\theta])| < \epsilon, \qquad r > R_0$$

6.4 COMPLEX INTEGRATION

Since $R > r$, $r > R_0$ implies

$$|R \exp[i\theta] f(R \exp[i\theta])| < \epsilon$$

Choosing $r > R_0$, gives us

$$\int_{C_2} |f(s)| \, |ds| \leq \int_{\theta_0}^{2\pi - \theta_0} \epsilon \, d\theta = 2\pi\epsilon$$

where

$$|i| = \sqrt{(i)(-i)} = 1$$

As $r \to \infty$, $R \to \infty$, and $\epsilon \to 0$. Therefore,

$$\lim_{r \to \infty} \left| \int_{C_1} f(s) \, ds - 2\pi i \sum_{j=1}^{n} a_{-1,j} \right| = \lim_{r \to \infty} \left| -\int_{C_2} f(s) \, ds \right| = 0$$

and

$$\int_{a-i\infty}^{a+i\infty} f(s) \, ds = 2\pi i \sum_{j=1}^{n} a_{-1,j}$$

Example 6.17 Evaluate

a. $\displaystyle\int_{5-i\infty}^{5+i\infty} \frac{1}{s^2 - 4} \, ds$

b. $\displaystyle\int_{3-i\infty}^{3+i\infty} \frac{\exp[-st]}{s - 2} \, ds$

c. $\displaystyle\int_{b-i\infty}^{b+i\infty} \frac{a \exp[-st]}{s(s^2 + a^2)} \, ds, \quad b > 0$

Theorem 6.10 applies to part a since

$$\lim_{r \to \infty} [r \exp[i\theta] f(r \exp[i\theta])] = \lim_{r \to \infty} \left[\frac{r \exp[i\theta]}{r^2 \exp[2i\theta] - 4} \right]$$

$$= \lim_{r \to \infty} \left[\frac{1}{r \exp[i\theta] - \frac{4}{r} \exp[-i\theta]} \right] = 0$$

The function $1/(s^2 - 4)$ has simple poles at $s = -2$ and $s = 2$, both of which lie to the left of the line joining $5 - i\infty$ and $5 + i\infty$. The residue at $s = -2$, $a_{-1, 1}$ is given by

$$a_{-1,1} = \lim_{s \to -2} \left[\frac{s + 2}{s^2 - 4} \right] = -\frac{1}{4}$$

At $s = 2$, the residue $a_{-1,2}$ is given by

$$a_{-1,2} = \lim_{s \to 2} \left[\frac{s-2}{s^2 - 4} \right] = \frac{1}{4}$$

Therefore,

$$\int_{5-i\infty}^{5+i\infty} f(s)\, ds = 2\pi i [-\tfrac{1}{4} + \tfrac{1}{4}] = 0$$

In part b,

$$\lim_{r \to \infty} [r \exp[i\theta] f(r \exp[i\theta])] = \lim_{r \to \infty} \left[\frac{r \exp[i\theta] \exp[-tr \exp[i\theta]]}{r \exp[i\theta] - 2} \right]$$

$$= \lim_{r \to \infty} \left[\frac{\exp[i\theta] \exp[-tr \exp[i\theta]]}{\exp[i\theta] - \dfrac{2}{r}} \right] = 0$$

and the only pole for this function is at $s = 2$ and is of order 1. Therefore

$$a_{-1} = \exp[-2t] \quad \text{and} \quad \int_{3-i\infty}^{3+i\infty} \frac{\exp[-st]}{s-2} = 2\pi i \exp[-2t]$$

For the function in part c,

$$\lim_{r \to \infty} [r \exp[i\theta] f(r \exp[i\theta])] = \lim_{r \to \infty} \left[\frac{ar \exp[i\theta] \exp[-tr \exp[i\theta]]}{r \exp[i\theta](r^2 \exp[2i\theta] + a^2)} \right] = 0$$

The poles for this function are $s = 0$, $s = -ia$, and $s = ia$; all are simple poles and lie to the left of the line joining $b - i\infty$ and $b + i\infty$ for $b > 0$. The residues at $s = 0$, $a_{-1,1}$, $s = -ia$, $a_{-1,2}$, and $s = ia$, $a_{-1,3}$, are given by

$$a_{-1,1} = \lim_{s \to 0} \left[\frac{a \exp[-st]}{(s^2 + a^2)} \right] = \frac{1}{a}$$

$$a_{-1,2} = \lim_{s \to -ia} \left[\frac{a(s + ia) \exp[-st]}{s(s^2 + a^2)} \right] = -\frac{1}{2a} \exp[iat]$$

$$a_{-1,3} = \lim_{s \to ia} \left[\frac{a(s - ia) \exp[-st]}{s(s^2 + a^2)} \right] = -\frac{1}{2a} \exp[-iat]$$

Therefore

$$\int_{b-i\infty}^{b+i\infty} \frac{a \exp[-st]}{s(s^2 + a^2)}\, ds = \frac{2\pi i}{a} \left[1 - \frac{\exp[iat] + \exp[-iat]}{2} \right] = \frac{2\pi i}{a} [1 - \cos(at)] \quad \boxed{}$$

Theorem 6.11 Let $f(s)$ be a function which is analytic above the line, C_1, connecting $-\infty + ia$ and $\infty + ia$ in the complex plane except at a finite number of

6.4 COMPLEX INTEGRATION

poles with residues $a_{-1,j}, j = 1, 2, \ldots, n$, none of which lie on the line C_1. Then

$$\int_{-\infty+ia}^{\infty+ia} f(s)\, ds = 2\pi i \sum_{j=1}^{n} a_{-1,j} \tag{6.86}$$

if

$$\lim_{r\to\infty} [r \exp[i\theta] f(r \exp[i\theta])] = 0$$

uniformly for $\theta_0 \leq \theta \leq \pi - \theta_0$, where

$$s = r \exp[i\theta]$$

and θ_0 is the angle between $r + ia$ and the x axis for all r.

Proof The proof of this theorem is much the same as that given for Theorem 6.10. We will only sketch the proof here and will leave a formal proof as an exercise for the reader. As in the proof of Theorem 6.10 we will start by expressing the integral in (6.86) as

$$\int_{-\infty+ia}^{\infty+ia} f(s)\, ds = \lim_{r\to\infty} \int_{-r+ia}^{r+ia} f(s)\, ds \tag{6.87}$$

Let C_1 be the path connecting $-r + ia$ and $r + ia$. We form a closed curve passing through $-r + ia$ and $r + ia$ by defining the circular arc, C_2, as shown in Fig. 6.24. That is, C_2 is a circular arc of radius R with center at $s = 0$. Then

$$\int_{C_1} f(s)\, ds + \int_{C_2} f(s)\, ds = 2\pi i \sum_{j=1}^{n} a_{-1,j}$$

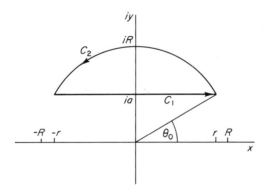

Figure 6.24 Closed Curve $C_1 + C_2$ Passing Through $r - ia$ and $r + ia$

The remainder of the proof lies in showing that

$$\lim_{r\to\infty} \left| \int_{C_2} f(s)\, ds \right| = 0 \qquad \blacksquare$$

The reader should note that the conditions given in Theorems 6.10 and 6.11 are sufficient conditions only. In later sections of this chapter we will find it necessary to deal with integrals of the forms given in Eqs. (6.78) and (6.79) where the conditions given in Theorems 6.10 and 6.11 are not valid. These integrals take the form

$$\int_{C_1} \exp[bs] f(s) \, ds = \int_{a-i\infty}^{a+i\infty} \exp[bs] f(s) \, ds \tag{6.88}$$

and

$$\int_{C_2} \exp[bs] f(s) \, ds = \int_{-\infty+ia}^{\infty+ia} \exp[bs] f(s) \, ds \tag{6.89}$$

For example, consider the function $\exp[bs]/(s-c)$, where $b > 0$. According to Theorem 6.10, the integral $\int_{a-i\infty}^{a+i\infty} \exp[bs] \, ds/(s-c)$ can be resolved using the residue theorem if $\lim_{r \to \infty} [r \exp[i\theta] f(r \exp[i\theta])] = 0$ uniformly for $\theta_0 \leq \theta \leq 2\pi - \theta_0$. But

$$\lim_{r \to \infty} [r \exp[i\theta] f(r \exp[i\theta])] = \lim_{r \to \infty} \left[\frac{r \exp[i\theta] \exp[br \exp[i\theta]]}{r \exp[i\theta] - c} \right] = \infty$$

and Theorem 6.10 does not apply to the evaluation of the above integral. However, in the following two theorems, we shall develop conditions under which the residue theorem can be used for functions similar to that just described.

Theorem 6.12 Let $\exp[st] f(s)$ be a function which is analytic to the left of the line, C_1, connecting $a - i\infty$ and $a + i\infty$ in the complex plane except at a finite number of poles with residues $a_{-1,j}, j = 1, 2, \ldots, n$, none of which lie on the line C_1. Then

$$\int_{a-i\infty}^{a+i\infty} \exp[st] f(s) \, ds = 2\pi i \sum_{j=1}^{n} a_{-1,j} \tag{6.90}$$

if $t > 0$ and

$$\lim_{r \to \infty} [f(r \exp[i\theta])] = 0 \tag{6.91}$$

uniformly for $\theta_0 \leq \theta \leq 2\pi - \theta_0$, where

$$s = r \exp[i\theta]$$

and θ_0 is the angle between $a + ir$ and the x axis for all r.

Theorem 6.13 Let $\exp[st] f(s)$ be a function which is analytic above the line, C_1, connecting $-\infty + ia$ and $\infty + ia$ in the complex plane except at a finite number of poles with residues $a_{-1,j}, j = 1, 2, \ldots, n$, none of which lie on the line C_1. Then

$$\int_{-\infty+ia}^{\infty+ia} \exp[st] f(s) \, ds = 2\pi i \sum_{j=1}^{n} a_{-1,j} \tag{6.92}$$

if $t > 0$ and

$$\lim_{r \to \infty} [f(r \exp[i\theta])] = 0$$

uniformly for $\theta_0 \leq \theta \leq \pi - \theta_0$, where

$$s = r \exp[i\theta]$$

and θ_0 is the angle between $a + ir$ and the x axis for all r.

Example 6.18 Evaluate the following

a. $\int_{a-i\infty}^{a+i\infty} \frac{\exp[st]}{(s-b)^n} ds, \qquad a > b, \quad t > 0$

b. $\int_{a-i\infty}^{a+i\infty} \frac{s \exp[st]}{(s^2 - b^2)} ds, \qquad a > b, \quad t > 0$

c. $\int_{a-i\infty}^{a+i\infty} \frac{(s-b)\exp[st]}{(s-b)^2 - c^2} ds, \qquad a > b + c, \quad b > 0, \quad c > 0, \quad t > 0$

In part a, let

$$f(s) = \frac{1}{(s-b)^n}$$

Then

$$\lim_{r \to \infty} [f(r \exp[i\theta])] = \lim_{r \to \infty} \left[\frac{1}{(r \exp[i\theta] - b)^n} \right] = 0$$

The function $\exp[st]/(s-b)^n$ has one pole at $s = b$ of order n. Since $s = b$ lies to the left of the line connecting $a - i\infty$ and $a + i\infty$, Theorem 6.12 may be used to evaluate the integral in part a. The residue at $s = b$ is

$$a_{-1} = \frac{1}{(n-1)!} \lim_{s \to b} \frac{d^{n-1}}{ds^{n-1}} [(s-b)^n \exp[st] f(s)] = \frac{t^{n-1}}{(n-1)!} \exp[bt]$$

and

$$\int_{a-i\infty}^{a+i\infty} \frac{\exp[st]}{(s-b)^n} ds = 2\pi i \frac{t^{n-1}}{(n-1)!} \exp[bt]$$

For part b, let

$$f(s) = \frac{s}{(s^2 - b^2)}$$

Then

$$\lim_{r \to \infty} [f(r \exp[i\theta])] = \lim_{r \to \infty} \left[\frac{r \exp[i\theta]}{r^2 \exp[2i\theta] - b^2} \right] = 0$$

The poles of $s \exp[st]/(s^2 - b^2)$ lie at $s = -b$ and $s = b$ and are of order 1. Since all of the poles lie to the left of the line joining $a - i\infty$ and $a + i\infty$, Theorem 6.12 may be used to evaluate the required integral. The residues at the poles given above are

$$a_{-1,1} = \lim_{s \to -b} [(s + b) \exp[st] f(s)] = \lim_{s \to -b} \left[\frac{s \exp[st]}{(s - b)}\right] = \frac{1}{2} \exp[-bt]$$

at $s = -b$, and

$$a_{-1,2} = \lim_{s \to b} [(s - b) \exp[st] f(s)] = \lim_{s \to b} \left[\frac{s \exp[st]}{(s + b)}\right] = \frac{1}{2} \exp[bt]$$

at $s = b$. Therefore,

$$\int_{a-i\infty}^{a+i\infty} \frac{s \exp[st]}{(s^2 - b^2)} ds = 2\pi i \left(\frac{\exp[-bt] + \exp[bt]}{2}\right) = 2\pi i \cosh(bt)$$

For part c, let

$$f(s) = \frac{s - b}{(s - b)^2 - c^2}$$

Then

$$\lim_{r \to \infty} [f(r \exp[i\theta])] = \lim_{r \to \infty} \left[\frac{r \exp[i\theta] - b}{(r \exp[i\theta] - b)^2 - c^2}\right] = 0$$

This function has one pole at $s = b + c$ and one at $s = b - c$, each of the first order. The residue at $s = b + c$ is

$$a_{-1,1} = \lim_{s \to b+c} [(s - b - c) \exp[st] f(s)] = \lim_{s \to b+c} \left[\frac{(s - b) \exp[st]}{(s - b) + c}\right]$$

$$= \frac{1}{2} \exp[(b + c)t]$$

The residue at $s = b - c$ is given by

$$a_{-1,2} = \lim_{s \to b-c} [(s - b + c) \exp[st] f(s)] = \lim_{s \to b-c} \left[\frac{(s - b) \exp[st]}{(s - b) - c}\right]$$

$$= \frac{1}{2} \exp[(b - c)t]$$

and

$$\int_{a-i\infty}^{a+i\infty} \frac{(s - b) \exp[st]}{(s - b)^2 - c^2} ds = 2\pi i \left[\frac{\exp[(b + c)t] + \exp[(b - c)t]}{2}\right]$$

$$= 2\pi i \exp[bt] \cosh(ct)$$

6.4 COMPLEX INTEGRATION

In the proof of Theorem 6.10, we showed that under certain restricted conditions

$$\lim_{r \to \infty} \int_C f(s) \, ds = 0 \tag{6.93}$$

where C is a semicircular path with center at the origin. It will be useful to generalize this result beyond the context of Theorem 6.10. Specifically we would like to know the conditions under which Eq. (6.93) is valid.

Theorem 6.14 Let C_1 and C_2 be semicircular paths each with radius r and center at the origin, as shown in Fig. 6.25. If $f(s)$ is analytic on C_1 and C_2 and

$$\lim_{r \to \infty} [rf(r \exp[i\theta])] = 0 \tag{6.94}$$

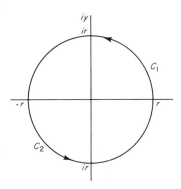

Figure 6.25 Semicircular Curves C_1 and C_2 Referred to in Theorem 6.14

uniformly for $-\pi/2 \le \theta \le \pi/2 \, (C_1)$ and $\pi/2 \le \theta \le 3\pi/2 \, (C_2)$, then

$$\lim_{r \to \infty} \int_{C_1} f(s) \, ds = 0 \tag{6.95}$$

$$\lim_{r \to \infty} \int_{C_2} f(s) \, ds = 0 \tag{6.96}$$

Further if

$$\lim_{r \to \infty} [f(r \exp[i\theta])] = 0 \tag{6.97}$$

uniformly for $-\pi/2 \le \theta \le \pi/2 \, (C_1)$ and $\pi/2 \le \theta \le 3\pi/2 \, (C_2)$, then

$$\lim_{r \to \infty} \int_{C_1} f(s) \exp[-as] \, ds = 0 \tag{6.98}$$

$$\lim_{r \to \infty} \int_{C_2} f(s) \exp[as] \, ds = 0 \tag{6.99}$$

where a is real and positive.

6 COMPLEX VARIABLES AND TRANSFORM METHODS

It should be noted that the results of Theorem 6.14 can be extended to arcs other than semicircles. The application of Theorem 6.14 is illustrated in the following examples.

Example 6.19 Show that

$$\int_{-\infty}^{\infty} \frac{\sin(s)\cos(s)}{s} \, ds = \frac{\pi}{2}$$

We will use Theorem 6.14 to evaluate this integral. Let

$$z = is$$

Then

$$\int_{-\infty}^{\infty} \frac{\sin(s)\cos(s)}{s} \, ds = \int_{-i\infty}^{i\infty} \frac{\sin(-iz)\cos(-iz)}{z} \, dz$$

$$= \lim_{r \to \infty} \int_{-ir}^{ir} \frac{\sin(-iz)\cos(-iz)}{z} \, dz$$

Let C be the path connecting $-ir$ and ir. Let C_1 and C_2 be the semicircles shown in Fig. 6.26, having radius r and center $z = 0$. Expressing $\sin(-iz)$ and $\cos(-iz)$ in exponential form, we have

$$\sin(-iz) = \frac{\exp[z] - \exp[-z]}{2i} \quad \text{and} \quad \cos(-iz) = \frac{\exp[z] + \exp[-z]}{2}$$

with the result that

$$\int_{-ir}^{ir} \frac{\sin(-iz)\cos(-iz)}{z} \, dz = \int_{-ir}^{ir} \frac{\exp[2z] - \exp[-2z]}{4iz} \, dz$$

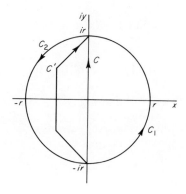

Figure 6.26 Closed Curve $C_1 + C_2$ about $z = 0$

6.4 COMPLEX INTEGRATION

The function $(\exp[2z] - \exp[-2z])/4iz$, has no poles on the finite s plane and is analytic there. Therefore the integral from $-ir$ to ir is independent of path. That is,

$$\int_C \frac{\exp[2z] - \exp[-2z]}{4iz} dz = \int_{C'} \frac{\exp[2z] - \exp[-2z]}{4iz} dz$$

where C' is any path connecting $-ir$ and ir. Therefore

$$\int_C \frac{\exp[2z] - \exp[-2z]}{4iz} dz = \int_{C'} \frac{\exp[2z]}{4iz} dz - \int_{C'} \frac{\exp[-2z]}{4iz} dz \quad (6.100)$$

The functions $\exp[2z]/4iz$ and $\exp[-2z]/4iz$ have poles at $z = 0$. Let C' be the path from $-ir$ to ir shown in Fig. 6.26. Then

$$\oint_{C'+C_2} \frac{\exp[2z]}{4iz} dz = 2\pi i \left[\text{sum of residues of } \frac{\exp[2z]}{4iz} \text{ in the interior of } C' + C_2 \right]$$

and

$$\int_{C_1} \frac{\exp[-2z]}{4iz} dz - \int_{C'} \frac{\exp[-2z]}{4iz} dz = 2\pi i \left[\text{sum of residues of } \frac{\exp[-2z]}{4iz} \right.$$
$$\left. \text{in the interior of } C_1 + C' \right]$$

From Theorem 6.14, if

$$\lim_{r \to \infty} [f(r \exp[i\theta])] = 0$$

uniformly for C_1 and C_2, then

$$\lim_{r \to \infty} \int_{C_1} f(z) \exp[-az] dz = 0 \quad \text{and} \quad \lim_{r \to \infty} \int_{C_2} f(z) \exp[az] dz = 0$$

But if

$$f(z) = \frac{1}{4iz}$$

then

$$\lim_{r \to \infty} [f(r \exp[i\theta])] = \lim_{r \to \infty} \left[\frac{1}{4ir \exp[i\theta]} \right] = 0$$

Therefore,

$$\lim_{r \to \infty} \int_{C_1} \frac{\exp[-2z]}{4iz} dz = 0 \quad \text{and} \quad \lim_{r \to \infty} \int_{C_2} \frac{\exp[2z]}{4iz} dz = 0$$

Hence

$$\lim_{r \to \infty} \int_{C'} \frac{\exp[2z]}{4iz} dz = 2\pi i \left[\text{sum of residues of } \frac{\exp[2z]}{4iz} \text{ in the interior of } C' + C_2 \right]$$

But this function has no poles on the interior of $C + C_2$ and therefore

$$\lim_{r \to \infty} \int_{C'} \frac{\exp[2z]}{4iz} dz = 0$$

The reader will note that

$$\oint_{C'+C_2} \frac{\exp[2z]}{4iz} dz = 0$$

since this function is analytic on $C' + C_2$. Consider the remaining function where the direction of integration on C' is positive with respect to the interior of $C_1 + C'$.

$$\lim_{r \to \infty} \int_C \frac{\exp[-2z]}{4iz} dz = 2\pi i [\text{sum of residues on the interior of } C_1 + C']$$

For this function, a pole of order 1 exists at $z = 0$. Therefore

$$a_{-1} = \lim_{z \to 0} [zf(z)] = \lim_{z \to 0} \left[\frac{\exp[-2z]}{4i} \right] = \frac{1}{4i}$$

and

$$\lim_{r \to \infty} \int_{C'} \frac{\exp[-2z]}{4iz} dz = \frac{\pi}{2}$$

Integration on $C_1 + C'$ was in the positive direction. Therefore,

$$\lim_{r \to \infty} \int_{C'} \frac{\exp[-2z]}{4iz} dz = \frac{\pi}{2} = \lim_{r \to \infty} \int_{ir}^{-ir} \frac{\exp[-2z]}{2iz} dz$$

Thus the direction of integration on C' is opposite to that originally intended on C. Thus, taking the direction on C' from $-ir$ to ir,

$$\lim_{r \to \infty} \int_{C'} \frac{\exp[-2z]}{4iz} dz = -\frac{\pi}{2}$$

By Eq. (6.100),

$$\lim_{r \to \infty} \int_C \frac{\exp[2z] - \exp[-2z]}{4iz} dz = \frac{\pi}{2}$$

and

$$\int_{-\infty}^{\infty} \frac{\sin(s)\cos(s)}{s} ds = \lim_{r \to \infty} \int_{-ir}^{ir} \frac{\exp[2z] - \exp[-2z]}{4iz} = \frac{\pi}{2}$$

6.4 COMPLEX INTEGRATION

The reader should notice that in applying Theorem 6.14 in Example 6.19, care was taken to integrate $\exp[2z]/4iz$ around $C' + C_2$ and $\exp[-2z]/4iz$ around $C_1 + C'$ since the conclusions of Theorem 6.14 apply only for such paths. That is, we could show that

$$\lim_{r \to \infty} \int_{C_2} \frac{\exp[2z]}{4iz} dz = 0 \tag{6.101}$$

and

$$\lim_{r \to \infty} \int_{C_1} \frac{\exp[-2z]}{4iz} dz = 0 \tag{6.102}$$

However, Theorem 6.14 would not allow us to draw these conclusions if we were to interchange the paths C_1 and C_2 in Eqs. (6.101) and (6.102). A further point of interest with respect to Example 6.19 is that we were not able to use Theorem 6.12 to resolve the integrals $\int_{-i\infty}^{i\infty} \exp[2z]\, dz/4iz$ and $\int_{-i\infty}^{i\infty} \exp[-2z]\, dz/4iz$, since the pole of each function lies on the path connecting $-i\infty$ and $i\infty$. These points are further illustrated in Example 6.20.

Example 6.20 Evaluate

$$\int_{-\infty}^{\infty} \frac{\sin(s)}{s} ds$$

The reader will notice that this integral corresponds to that in Eq. (6.86) where $a = 0$. However Theorem 6.11 does not apply since $\sin(s)/s$ has a pole at $s = 0$ which lies on the line connecting $-\infty$ and ∞. From Eq. (6.16)

$$\sin(s) = \frac{\exp[is] - \exp[-is]}{2i}$$

Let

$$z = is$$

Then

$$\int_{-\infty}^{\infty} \frac{\sin(s)}{s} ds = \int_{-i\infty}^{i\infty} \frac{\exp[z] - \exp[-z]}{2iz} dz = \int_{-i\infty}^{i\infty} \frac{\sinh(z)}{iz} dz$$

Since $\sinh(z)/z$ is analytic, the value of the integral is independent of the path taken. Let us construct a circular closed path with center at $z = 0$ and radius r as shown in Fig. 6.27. The closed curve consists of the semicircles C_1 and C_2. The curve C connects $-i\infty$ and $i\infty$ by a straight line. The reader should note that although $\sinh(z)/z$ is analytic throughout S_0 in Fig. 6.27, $\exp[z]/z$ and $\exp[-z]/z$ are not. Since $\sinh(z)/z$ is analytic on S_0

$$\oint_{C_1+C_2} \frac{\sinh(z)}{z} dz = 0$$

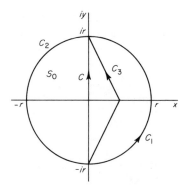

Figure 6.27 Closed Curve $C_1 + C_2$ about $z = 0$

Now let us construct the path C_3 shown in Fig. 6.27 connecting $-ir$ and ir. Integrating in the positive direction, we have

$$\int_{C_1} \frac{\sinh(z)}{z} dz - \int_{C_3} \frac{\sinh(z)}{z} dz = 0 \quad \text{and} \quad \int_{C_2} \frac{\sinh(z)}{z} dz + \int_{C_3} \frac{\sinh(z)}{z} dz = 0$$

But

$$\int_{C_3} \frac{\sinh(z)}{z} dz = \int_{C_3} \frac{\exp[z]}{2z} dz - \int_{C_3} \frac{\exp[-z]}{2z} dz$$

Let us attempt to resolve the integrals $\int_{C_3} \exp[z] \, dz/z$ and $\int_{C_3} \exp[-z] \, dz/z$. To accomplish this we refer to Theorem 6.14. We know that around any closed curve including C_3 the integral of either function is $2\pi i$ times the residues at the poles of the function in the interior of the closed curve. Thus, we will form such a closed curve by closing C_3 by a semicircle. If we can show, by Theorem 6.14, that the integral around the semicircle approaches zero as $r \to \infty$, then the integral along the path C_3 is simply $2\pi i$ times the sum of the residues at the poles of the function in the interior of the closed curve. To apply Theorem 6.14 we must show that

$$\lim_{r \to \infty} [f(r \exp[i\theta])] = 0$$

uniformly, where $f(z) = 1/z$ for both of the integrals. Therefore

$$\lim_{r \to \infty} [f(r \exp[i\theta])] = \lim_{r \to \infty} \left[\frac{1}{r \exp[i\theta]} \right] = 0$$

Since the power of e in the function $\exp[z]/z$ is positive we must close C_3 by C_2 in order to show that the integral of $\exp[z]/z$ along a semicircular path approaches zero as $r \to \infty$ by Eq. (6.99). Therefore,

$$\lim_{r \to \infty} \int_{C_2} \frac{\exp[-z]}{z} dz = 0$$

6.4 COMPLEX INTEGRATION

The only pole on the interior of $C_2 + C_3$ is at $z = 0$ and has residue given by

$$a_{-1} = \lim_{z \to 0} [\exp[z]] = 1$$

The power of e in the function $\exp[-z]/z$ is negative and by Eq. (6.98) we must close C_3 by the semicircle C_1 to evaluate the behavior of the function on a semicircle as $r \to \infty$.

$$\lim_{r \to \infty} \int_{C_1} \frac{\exp[-z]}{z} dz = 0$$

However, $\exp[-z]/z$ is analytic on the interior of $C_1 + C_3$ and therefore has no poles on the interior. But in such a region

$$\lim_{r \to \infty} \int_{C_3} \frac{\exp[-z]}{z} dz = \lim_{r \to \infty} \int_{C'} \frac{\exp[-z]}{z} dz$$

by Theorem 6.5, where C' is any path connecting the endpoints of C_3. However, C_1 is such a path and

$$\lim_{r \to \infty} \int_{C_3} \frac{\exp[-z]}{z} dz = \lim_{r \to \infty} \int_{C_1} \frac{\exp[-z]}{z} dz = 0$$

Therefore,

$$\lim_{r \to \infty} \int_{C_3} \frac{\sinh(z)}{z} dz = \lim_{r \to \infty} \int_{C_3} \frac{\exp[z]}{2z} dz = \pi i$$

and

$$\lim_{r \to \infty} \int_{C_3} \frac{\sinh(z)}{iz} dz = \pi$$

But this integral is independent of path and

$$\int_{-\infty}^{\infty} \frac{\sin(s)}{s} ds = \int_{-i\infty}^{i\infty} \frac{\sinh(z)}{iz} dz = \int_{C_3} \frac{\sinh(z)}{iz} dz = \pi \qquad \boxed{}$$

To summarize the approach taken to evaluate the integrals in Examples 6.19 and 6.20, we first broke the respective analytic functions into parts such that each part contained a pole. The next step was to distort the path of integration from the original path such that the poles of the component functions did not lie on the path of integration. This distortion of the path of integration was permissible since the integral of the sum of the component functions was independent of the path connecting the end points of integration. We next closed the distorted path of integration by two semicircular arcs to the left and right of the distorted path of integration. At this point we could apply the residue theorem to the integration of each function around the closed paths. Finally, we were able to show that the integral along each semicircular arc approached zero as the radius of the arc approached infinity. The integral of the function was then given by the sum of the integrals of the components along the distorted path of integration. The reader should note that care was taken in Examples 6.19 and 6.20 in integrating the components of the analytic functions over the semicircular arcs closing the distorted path connecting the end points of integration. Specifically these paths of integration were selected in conformity with Theorem 6.14 so that we would be able to show that the integrals around both semicircles go to zero as the radius of the semicircles approached infinity.

6.5 The Fourier Transforms

Let $f(t)$ be a function of the real variable t. Then

$$F_t(u) = \int_{-\infty}^{\infty} f(t) \exp[-iut]\, dt \qquad (6.103)$$

is the *Fourier transform* or the *Fourier integral* of $f(t)$. The functions $F_t(u)$ and $f(t)$ form a *transform pair*. Our primary concern in this chapter will be with the applications of the Fourier transform in probability theory. However before discussing this application of the Fourier transform, it will be useful to investigate some of the properties of the integral in Eq. (6.103). In the following theorem we define a sufficient condition for the existence of the Fourier transform of $f(t)$.

Theorem 6.15 If $f(t)$ is integrable on $(-\infty, \infty)$, and

$$\int_{-\infty}^{\infty} |f(t)|\, dt < \infty \qquad (6.104)$$

then the Fourier transform of $f(t)$ exists.

The Fourier transform of $f(t)$ is quite useful in obtaining the moments of $f(t)$. Let the nth *moment* of $f(t)$ about zero be m_n. Then

$$m_n = \int_{-\infty}^{\infty} t^n f(t)\, dt \qquad (6.105)$$

In the following theorem we show how these moments can be obtained from the Fourier transform of $f(t)$.

Theorem 6.16 If the Fourier transform of $f(t)$ exists, then the nth moment of $f(t)$ about zero is given by

$$m_n = \frac{1}{(-i)^n} \frac{d^n}{du^n} F_t(u) \bigg|_{u=0} \qquad (6.106)$$

6.5 THE FOURIER TRANSFORMS

Proof The nth derivative of $F_t(u)$ with respect to u is given by

$$\frac{d^n}{du^n} F_t(u) = \int_{-\infty}^{\infty} (-it)^n f(t) \exp[-iut] \, dt$$

Therefore,

$$\frac{d^n}{du^n} F_t(u) \bigg|_{u=0} = (-i)^n \int_{-\infty}^{\infty} t^n f(t) \, dt$$

Dividing by $(-i)^n$ leads to the desired result. ∎

Example 6.21 Let

$$f(t) = \begin{cases} \exp[-at], & t \geq 0 \\ 0, & t < 0 \end{cases}$$

where $a > 0$. Find the Fourier transform of $f(t)$ and the first and second moments of $f(t)$ about zero.

Since

$$\int_{-\infty}^{\infty} |f(t)| \, dt = \int_{-\infty}^{0} (0) \, dt + \int_{0}^{\infty} \exp[-at] \, dt = \frac{1}{a}$$

the Fourier transform of $f(t)$ exists and is given by

$$F_t(u) = \int_{-\infty}^{\infty} f(t) \exp[-iut] \, dt = \int_{-\infty}^{0} (0) \exp[-iut] \, dt + \int_{0}^{\infty} \exp[-t(a+iu)] \, dt$$

$$= \frac{1}{a+iu}$$

The first moment about zero is given by

$$m_1 = \frac{1}{-i} \frac{d}{du} [(a+iu)^{-1}] \bigg|_{u=0} = -\frac{1}{i}[-i(a+iu)^{-2}]\bigg|_{u=0} = \frac{1}{a^2}$$

For m_2, we have

$$m_2 = \frac{1}{(-i)^2} \frac{d^2}{du^2} [(a+iu)^{-1}] \bigg|_{u=0} = -[-2(a+iu)^{-3}]\bigg|_{u=0} = \frac{2}{a^3} \quad \blacksquare$$

In probability theory we are often interested in the density function of the sum of n random variables. For example, let T_1 and T_2 be the profit derived from two operations with associated density functions $f(t_1)$ and $h(t_2)$. Suppose that we wish to find the density function, $g(t)$, of the total profit derived from the two operations, $T = T_1 + T_2$. Then

$$g(t) = \int_{-\infty}^{\infty} f(t_1) h(t - t_1) \, dt_1 \quad (6.107)$$

Of course problems of this nature arise in areas other than probability theory.

We could derive the function $g(t)$ from the relationship given in Eq. (6.107), or, under appropriate conditions, we can use Fourier transforms to obtain $g(t)$. Definition of $g(t)$ through Fourier transforms is a two-step process. First we must find the Fourier transform of $g(t)$ and then we must invert this transform to find the function, $g(t)$, associated with it. Let us now examine the problem of identifying the Fourier transform of $g(t)$.

Theorem 6.17 Convolution Theorem Let $f(t_1)$ and $h(t_2)$ be bounded and continuous at all but a finite number of points on every closed interval $[a, b]$.
Assume that $\int_{-\infty}^{\infty} |f(t_1)|\, dt_1$ and $\int_{-\infty}^{\infty} |h(t_2)|\, dt_2$ exist. If

$$g(t) = \int_{-\infty}^{\infty} f(t_1) h(t - t_1)\, dt_1$$

and if $F_{t_1}(u)$ and $F_{t_2}(u)$ are the Fourier transforms of $f(t_1)$ and $h(t_2)$, then the Fourier transform of $g(t)$, $F_t(u)$, is given by

$$F_t(u) = F_{t_1}(u) F_{t_2}(u) \tag{6.108}$$

Proof We will prove this theorem for the case where $f(t_1)$ and $h(t_2)$ are continuous. The Fourier transform of $g(t)$ is given by

$$F_t(u) = \int_{-\infty}^{\infty} \int_{-\infty}^{\infty} f(t_1) h(t - t_1) \exp[-iut]\, dt_1\, dt \tag{6.109}$$

if the integral exists. But since $f(t_1)$ and $h(t_2)$ are bounded, $f(t_1)h(t - t_1)$ is bounded and continuous and $\int_{-\infty}^{\infty} f(t_1) h(t - t_1)\, dt_1$ exists, is continuous, and is bounded. Therefore, $F_t(u)$ exists. Let

$$t = t_1 + t_2$$

Then

$$F_t(u) = \int_{-\infty}^{\infty} \int_{-\infty}^{\infty} f(t_1) h(t_2) \exp[-iu(t_1 + t_2)]\, dt_1\, dt_2$$

$$= \int_{-\infty}^{\infty} f(t_1) \exp[-iut_1]\, dt_1 \int_{-\infty}^{\infty} h(t_2) \exp[-iut_2]\, dt_2 = F_{t_1}(u) F_{t_2}(u) \quad \blacksquare$$

The results of the convolution theorem can be extended to an arbitrary number of functions. Let $f_1(t_1), f_2(t_2), \ldots, f_n(t_n)$ have the Fourier transforms $F_{t_1}(u)$, $F_{t_2}(u), \ldots, F_{t_n}(u)$, and let

$$g(t) = \int_{-\infty}^{\infty} \int_{-\infty}^{\infty} \cdots \int_{-\infty}^{\infty} \int_{-\infty}^{\infty} g(t - u_{n-1}) g(u_{n-1} - u_{n-2})$$

$$\cdots g(u_2 - u_1) g(u_1)\, du_1\, du_2 \cdots du_{n-1} \tag{6.110}$$

6.5 THE FOURIER TRANSFORMS

where

$$u_k = \sum_{j=1}^{k} t_j, \quad k < n, \quad t = \sum_{j=1}^{n} t_j$$

Then the Fourier transform of $g(t)$, $F_t(u)$, is given by

$$F_t(u) = \prod_{j=1}^{n} F_{t_j}(u) \quad (6.111)$$

Having obtained the Fourier transform of $g(t)$, we still have the problem of identifying $g(t)$ itself. We accomplish this by inverting $F_t(u)$.

Theorem 6.18 Let $F_t(u)$ be the Fourier transform of $f(t)$. If

$$\int_{-\infty}^{\infty} |F_t(u)| \, du < \infty \quad (6.112)$$

then

$$f(t) = \frac{1}{2\pi} \int_{-\infty}^{\infty} F_t(u) \exp[iut] \, du \quad (6.113)$$

for all values of t for which $f(t)$ is continuous.

Proof Let

$$G(t + \Delta t) = \int_{-\infty}^{t+\Delta t} f(y) \, dy \quad (6.114)$$

Then

$$G(t + \Delta t) - G(t - \Delta t) = \int_{t-\Delta t}^{t+\Delta t} f(y) \, dy$$

Since $G(t)$ is the antiderivative of $f(t)$, we have

$$f(t) = \lim_{\Delta t \to 0} \frac{G(t + \Delta t) - G(t)}{\Delta t} = \lim_{\Delta t \to 0} \frac{G(t) - G(t - \Delta t)}{\Delta t}$$

or

$$f(t) = \lim_{\Delta t \to 0} \frac{G(t + \Delta t) - G(t - \Delta t)}{2 \Delta t}$$

Suppose that $G(t + \Delta t) - G(t - \Delta t)$ is defined by

$$G(t + \Delta t) - G(t - \Delta t) = \lim_{T \to \infty} \frac{1}{\pi} \int_{-T}^{T} \frac{\sin(u \, \Delta t)}{u} F_t(u) \exp[iut] \, du \quad (6.115)$$

Dividing both sides by $2\,\Delta t$ and taking the limit as $\Delta t \to 0$ yields

$$f(t) = \lim_{T \to \infty} \lim_{\Delta t \to 0} \frac{1}{2\pi} \int_{-T}^{T} \frac{\sin(u\,\Delta t)}{u\,\Delta t} F_t(u) \exp[iut]\,du$$

$$= \lim_{T \to \infty} \frac{1}{2\pi} \int_{-T}^{T} F_t(u) \exp[iut]\,du$$

$$= \frac{1}{2\pi} \int_{-\infty}^{\infty} F_t(u) \exp[iut]\,du \qquad (6.116)$$

since

$$\lim_{\Delta t \to 0} \frac{\sin(u\,\Delta t)}{u\,\Delta t} = \lim_{\Delta t \to 0} \frac{1}{u\,\Delta t} \left[u\,\Delta t - \frac{(u\,\Delta t)^3}{3!} + \frac{(u\,\Delta t)^5}{5!} - \cdots \right] = 1$$

Therefore, the proof of this theorem lies in establishing the validity of Eq. (6.115). Let

$$H(T, t, \Delta t) = \frac{1}{\pi} \int_{-T}^{T} \frac{\sin(u\,\Delta t)}{u} F_t(u) \exp[iut]\,du$$

Since

$$F_t(u) = \int_{-\infty}^{\infty} f(x) \exp[-iux]\,dx$$

we have

$$H(T, t, \Delta t) = \frac{1}{\pi} \int_{-T}^{T} \frac{\sin(u\,\Delta t)}{u} \int_{-\infty}^{\infty} f(x) \exp[iu(t-x)]\,dx\,du$$

Reversing the order of integration and letting

$$\exp[iu(t-x)] = \cos[u(t-x)] + i\sin[u(t-x)]$$

yields

$$H(T, t, \Delta t) = \frac{1}{\pi} \int_{-\infty}^{\infty} f(x) \int_{-T}^{T} \frac{\sin(u\,\Delta t)}{u} \{\cos[u(t-x)] + i\sin[u(t-x)]\}\,du\,dx$$

Since $\sin(u\,\Delta t)/u$ is an even function and $\sin[u(t-x)]$ is an odd function, their product is an odd function and

$$i\int_{-T}^{T} \frac{\sin(u\,\Delta t)}{u} \sin u(t-x) = 0$$

Hence

$$H(T, t, \Delta t) = \frac{1}{\pi} \int_{-\infty}^{\infty} f(x) \int_{-T}^{T} \frac{\sin(u\,\Delta t)}{u} \cos[u(t-x)]\,du\,dx$$

6.5 THE FOURIER TRANSFORMS

Let

$$\sin(u\,\Delta t)\cos[u(t-x)] = \tfrac{1}{2}\{\sin[u(t-x+\Delta t)] - \sin[u(t-x-\Delta t)]\}$$

Since $\sin(au)/u$ is even

$$\int_{-T}^{T} \frac{\sin(au)}{u}\,du = 2\int_{0}^{T} \frac{\sin(au)}{u}\,du$$

and

$$H(T, t, \Delta t) = \frac{1}{\pi}\int_{-\infty}^{\infty} f(x)\int_{0}^{T} \frac{\sin[u(t-x+\Delta t)] - \sin[u(t-x-\Delta t)]}{u}\,du\,dx$$

Since $\sin(au)/u$ and $\sin(bu)/u$ are even functions where

$$a = t - x + \Delta t, \qquad b = t - x - \Delta t$$

we have the result

$$\lim_{T\to\infty} \int_{0}^{T} \frac{\sin(au)}{u}\,du = \frac{1}{2}\lim_{T\to\infty}\int_{-T}^{T} \frac{\sin(au)}{u}\,du$$

A similar result holds for $\sin(bu)/u$. Let

$$z = au$$

Then

$$\lim_{T\to\infty} \int_{0}^{T} \frac{\sin(au)}{u}\,du = \frac{1}{2}\lim_{T\to\infty}\int_{-aT}^{aT} \frac{\sin(z)}{z}\,dz$$

By Example 6.20,

$$\lim_{T\to\infty}\int_{-aT}^{aT} \frac{\sin(z)}{z}\,dz = \begin{cases} -\pi/2, & a < 0 \\ 0, & a = 0 \\ \pi/2, & a > 0 \end{cases}$$

This result is obvious for $a > 0$. For $a = 0$, the limits of integration are equal and the integral is therefore zero. For $a < 0$, the limits of the integral are aT to $-aT$ and the resulting integral is $-\pi/2$ by Example 6.20.

A similar result holds for the integral of $\sin(bu)/u$. Therefore,

$$\lim_{T\to\infty}\int_{0}^{T} \frac{\sin[u(t-x+\Delta t)] - \sin[u(t-x-\Delta t)]}{u}\,du$$

$$= \begin{cases} 0, & x < t - \Delta t \\ \pi/2, & x = t - \Delta t \\ \pi, & t - \Delta t < x < t + \Delta t \\ \pi/2, & x = t + \Delta t \\ 0, & x > t + \Delta t \end{cases}$$

Then

$$\lim_{T \to \infty} H(T, t, \Delta t) = \int_{t-\Delta t}^{t+\Delta t} f(x)\, dx = G(t + \Delta t) - G(t - \Delta t)$$

and Eq. (6.115) is verified and

$$f(t) = \frac{1}{2\pi} \int_{-\infty}^{\infty} F_t(u) \exp[iut]\, du$$

by Eq. (6.116).

Example 6.22 Let

$$f(t) = \frac{\lambda^n}{2\pi(\lambda - it)^n}$$

where n is a positive integer greater than unity and $\lambda > 0$. Show that

$$F_t(u) = \begin{cases} \dfrac{\lambda}{(n-1)!} (\lambda u)^{n-1} \exp[-\lambda u], & u > 0 \\ 0, & u \le 0 \end{cases}$$

and that inversion of $F_t(u)$ yields $f(t)$.

The Fourier transform of $f(t)$ is given by

$$F_t(u) = \frac{1}{2\pi} \int_{-\infty}^{\infty} \frac{\lambda^n \exp[-iut]}{(\lambda - it)^n}\, dt$$

Let

$$z = (\lambda - it)u$$

Then

$$F_t(u) = -\frac{iu^{n-1}}{2\pi} \int_{u\lambda - i\infty}^{u\lambda + i\infty} \frac{\lambda^n \exp[-\lambda u + z]}{z^n}\, dz$$

$$= -\frac{i\lambda(u\lambda)^{n-1}}{2\pi} \exp[-\lambda u] \int_{u\lambda - i\infty}^{u\lambda + i\infty} \frac{\exp[z]}{z^n}\, dz \qquad (6.117)$$

Since $\exp[z]/z^n$ has a pole of order n at $z = 0$, and since this pole lies to the left of the line joining $u\lambda - i\infty$ and $u\lambda + i\infty$ for $u > 0$, we may apply the results of Theorem 6.12. The residue at $z = 0$ is given by

$$a_{-1} = \frac{1}{(n-1)!} \lim_{z \to 0} \frac{d^{n-1}}{dz^{n-1}} \exp[z] = \frac{1}{(n-1)!}$$

Therefore

$$F_t(u) = 2\pi i \left[-\frac{i\lambda(u\lambda)^{n-1}}{2\pi(n-1)!} \exp[-\lambda u] \right] = \frac{\lambda}{(n-1)!} (\lambda u)^{n-1} \exp[-\lambda u], \qquad u > 0$$

6.5 THE FOURIER TRANSFORMS

For $u < 0$, the pole at $z = 0$ lies to the right of the straight line connecting $u\lambda - i\infty$ and $u\lambda + i\infty$. That is,

$$F_t(u) = -\frac{iu^{n-1}}{2\pi}\int_{u\lambda+i\infty}^{u\lambda-i\infty}\frac{\lambda^n \exp[-\lambda u + z]}{z^n}dz = \frac{iu^{n-1}}{2\pi}\int_{u\lambda-i\infty}^{u\lambda+i\infty}\frac{\lambda^n \exp[-\lambda u + z]}{z^n}dz$$

To find $F_t(u)$ for $u < 0$, consider the straight line, C, connecting $u\lambda - ir$ and $u\lambda + ir$ shown in Fig. 6.28. Now construct a circular arc, C_1, with center at $z = 0$ and radius R connecting $u\lambda - ir$ and $u\lambda + ir$. Then

$$\oint_{C+C_1}\frac{\exp[z]}{z^n}dz = \int_C \frac{\exp[z]}{z^n}dz + \int_{C_1}\frac{\exp[z]}{z^n}dz = 0$$

since $\exp[z]/z^n$ is analytic on the interior of $C + C_1$.

By Theorem 6.14,

$$\int_{C_1}\frac{\exp[z]}{z^n}dz = 0$$

since

$$\lim_{R\to\infty}[f(R\exp[i\theta])] = \lim_{R\to\infty}\left[\frac{1}{R^n \exp[ni\theta]}\right] = 0$$

Therefore,

$$\int_C \frac{\exp[z]}{z^n}dz = 0 \quad\text{and}\quad F_t(u) = 0, \quad u < 0$$

For $u = 0$,

$$F_t(u) = 0$$

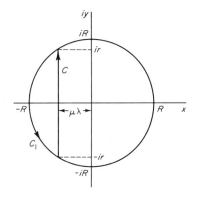

Figure 6.28 Closed Curve $C + C_1$ for $u < 0$

by Eq. (6.117). Therefore,

$$F_t(u) = \begin{cases} \dfrac{\lambda}{(n-1)!}(\lambda u)^{n-1}\exp[-\lambda u], & u > 0 \\ 0, & u \le 0 \end{cases}$$

To find $f(t)$ we use the inversion formula.

$$f(t) = \frac{1}{2\pi}\int_{-\infty}^{\infty} F_t(u)\exp[iut]\,du = \frac{1}{2\pi}\int_0^{\infty} \frac{\lambda}{(n-1)!}(\lambda u)^{n-1}\exp[u(it-\lambda)]\,du$$

Since

$$\int_0^{\infty} x^n \exp[-ax]\,dx = \frac{n!}{a^{n+1}}$$

we have

$$f(t) = \frac{\lambda^n}{2\pi(n-1)!}\int_0^{\infty} u^{n-1}\exp[-u(\lambda-it)]\,du = \frac{\lambda^n}{2\pi(\lambda-it)^n}$$

Example 6.23 Let

$$g(t_1) = \frac{1}{\sqrt{2\pi}}\exp\left[-\frac{(t_1-m_1)^2}{2}\right], \qquad -\infty < t_1 < \infty$$

$$h(t_2) = \frac{1}{\sqrt{2\pi}}\exp\left[-\frac{(t_2-m_2)^2}{2}\right], \qquad -\infty < t_2 < \infty$$

Show that

$$f(t) = \frac{1}{2\sqrt{\pi}}\exp\left[-\frac{(t-m_1-m_2)^2}{4}\right], \qquad -\infty < t < \infty$$

where

$$f(t) = \int_{-\infty}^{\infty} g(t_1)h(t-t_1)\,dt_1 \quad \text{and} \quad \int_{-\infty}^{\infty} g(t_1)\,dt_1 = \int_{-\infty}^{\infty} h(t_2)\,dt_2 = 1$$

From the convolution theorem,

$$F_t(u) = F_{t_1}(u)F_{t_2}(u)$$

$F_{t_j}(u)$ is given by

$$F_{t_j}(u) = \int_{-\infty}^{\infty} \frac{1}{\sqrt{2\pi}}\exp\left[-\frac{(t_j-m_j)^2}{2}\right]\exp[-iut_j]\,dt_j, \qquad j=1,2$$

Let

$$z = t_j - m_j + iu$$

6.5 THE FOURIER TRANSFORMS

Then

$$F_{t_j}(u) = \int_{-\infty + iu}^{\infty + iu} \frac{1}{\sqrt{2\pi}} \exp\left[-\frac{z^2}{2} - ium_j - \frac{u^2}{2}\right] dz$$

Letting $z = x + iu$ we have

$$F_{t_j}(u) = \exp\left[-ium_j - \frac{u^2}{2}\right] \int_{-\infty}^{\infty} \frac{1}{\sqrt{2\pi}} \exp\left[-\frac{(x+iu)^2}{2}\right] dx = \exp\left[-ium_j - \frac{u^2}{2}\right]$$

Therefore

$$F_t(u) = \exp[-iu(m_1 + m_2) - u^2], \qquad -\infty < u < \infty$$

To determine $f(t)$ we must find the inverse of $F_t(u)$. From Eq. (6.113),

$$f(t) = \frac{1}{2\pi} \int_{-\infty}^{\infty} F_t(u) \exp[iut] \, du = \frac{1}{2\pi} \int_{-\infty}^{\infty} \exp[-iu(m_1 + m_2 - t) - u^2] \, du$$

Let

$$z = u + \frac{i(m_1 + m_2 - t)}{2}$$

Then

$$f(t) = \frac{1}{2\pi} \int_{-\infty + ia}^{\infty + ia} \exp\left[-z^2 - \frac{(m_1 + m_2 - t)^2}{4}\right] dz$$

where

$$a = \frac{m_1 + m_2 - t}{2}$$

Then

$$f(t) = \frac{1}{2\pi} \exp\left[-\frac{(t - m_1 - m_2)^2}{4}\right] \int_{-\infty + ia}^{\infty + ia} \exp[-z^2] \, dz$$

Let

$$z = \frac{x}{\sqrt{2}} + ia$$

Then

$$f(t) = \frac{1}{2\pi} \exp\left[-\frac{(t - m_1 - m_2)^2}{4}\right] \int_{-\infty}^{\infty} \frac{1}{\sqrt{2}} \exp\left[-\frac{(x + \sqrt{2}\,ia)^2}{2}\right] dx$$

$$= \frac{1}{2\sqrt{\pi}} \exp\left[-\frac{(t - m_1 - m_2)}{4}\right]$$

From the definition of the Fourier transform and its inverse, one can show that $f(t)$ and its Fourier transform form a *unique pair*. That is, if $g(t_1)$ and $h(t_2)$ are continuous functions with Fourier transforms $F_{t_1}(u)$ and $F_{t_2}(u)$ respectively, then if

$$g(t_1) \neq h(t_2)$$

for some $t_1 = t_2$,

$$F_{t_1}(u) \neq F_{t_2}(u)$$

Similarly if

$$g(t_1) = h(t_2)$$

for $t_1 = t_2$, then

$$F_{t_1}(u) = F_{t_2}(u)$$

Now suppose that $g(t_1) = h(t_2)$ for $t_1 = t_2$ except at a finite number of points of discontinuity of $g(t_1)$ and $h(t_2)$. For example, let $g(t_1)$ and $h(t_2)$ be discontinuous at $t_1 = t_2 = t$ and continuous elsewhere. Then

$$F_{t_1}(u) = \lim_{a \to t-} \int_{-\infty}^{a} g(t_1) \exp[-iut_1] \, dt_1 + \lim_{a \to t+} \int_{a}^{\infty} g(t_1) \exp[-iut_1] \, dt_1$$

and

$$F_{t_2}(u) = \lim_{a \to t-} \int_{-\infty}^{a} h(t_2) \exp[-iut_2] \, dt_2 + \lim_{a \to t+} \int_{a}^{\infty} h(t_2) \exp[-iut_2] \, dt_2$$

Since $g(t_1) = h(t_2)$ for $t_1 = t_2 \neq t$

$$F_{t_1}(u) = F_{t_2}(u)$$

Therefore we may state that if the Fourier transforms of two functions are equal, then the two functions are equal except possibly at a *finite number of points of discontinuity* of the two functions.

6.6 The Laplace Transform

As we have already seen the Fourier transform of the function $f(t)$ exists if $f(t)$ is absolutely integrable on $(-\infty, \infty)$. However, transforms of functions failing to satisfy this condition are often of interest. One need only consider polynomial functions. For example, let

$$f(t) = a + bt^2$$

6.6 THE LAPLACE TRANSFORM

It is immediately obvious that $\int_{-\infty}^{\infty} |f(t)|\, dt$ does not exist. Thus we seek a transform similar in some respects to the Fourier transform, and yet one which is guaranteed to exist under conditions less restrictive than those given for the Fourier transform.

The *two-sided Laplace transform* defined by

$$\mathscr{L}[f(t)] = \int_{-\infty}^{\infty} f(t) \exp[-st]\, dt \qquad (6.118)$$

possesses these properties where

$$s = \sigma + i\omega$$

Thus the two-sided Laplace transform of $f(t)$ is simply the Fourier transform $f(t) \exp[-\sigma t]$. Then the two-sided Laplace transform of $f(t)$ exists when $\int_{-\infty}^{\infty} |f(t)| \exp[-\sigma t]\, dt$ exists. It is immediately obvious that the two-sided Laplace transform of $f(t)$ may exist even when its Fourier transform does not exist. For example, consider the function given by

$$f(t) = \begin{cases} 0, & t < 0 \\ a, & t \geq 0 \end{cases}$$

The Fourier transform of $f(t)$ is given by

$$F_t(u) = \int_0^{\infty} a \exp[-iut]\, dt = -\frac{a}{iu} \exp[-iut]\Big|_0^{\infty} = -\frac{a}{iu}[\cos(ut) - i\sin(ut)]\Big|_0^{\infty}$$

Since this integral does not converge, $F_t(u)$ does not exist. Now consider the two-sided Laplace transform of this function.

$$\mathscr{L}[f(t)] = \int_{-\infty}^{\infty} a \exp[-st]\, dt = -\frac{a}{s} \exp[-st]\Big|_0^{\infty}$$

$$= -\frac{a}{\sigma + i\omega} \exp[-\sigma t][\cos(\omega t) - i\sin(\omega t)]\Big|_0^{\infty}$$

If $\sigma < 0$, $\mathscr{L}[f(t)]$ fails to exist. However, for $\sigma > 0$

$$\mathscr{L}[f(t)] = \frac{a}{\sigma + i\omega} = \frac{a}{s}$$

Example 6.24 Find the two-sided Laplace transform of

$$f(t) = \begin{cases} a, & t < 0 \\ \exp[at], & t \geq 0 \end{cases}$$

and the values of σ for which the Laplace transform exists where $a < 0$.

The two-sided Laplace transform of $f(t)$ is given by

$$\mathscr{L}[f(t)] = \int_{-\infty}^{0} a \exp[-st]\, dt + \int_{0}^{\infty} \exp[at]\exp[-st]\, dt$$

$$= -\frac{a}{s}\exp[-st]\Big|_{-\infty}^{0} + \frac{1}{a-s}\exp[(a-s)t]\Big|_{0}^{\infty}$$

$$= -\frac{a}{\sigma + i\omega}\exp[-\sigma t][\cos(\omega t) - i\sin(\omega t)]\Big|_{-\infty}^{0}$$

$$+ \frac{1}{(a-\sigma-i\omega)}\exp[(a-\sigma)t][\cos(\omega t) - i\sin(\omega t)]\Big|_{0}^{\infty}$$

The first integral converges for $\sigma < 0$ and the second converges for $\sigma > a$. Therefore both integrals exist for $a < \sigma < 0$, and

$$\mathscr{L}[f(t)] = \frac{a}{s} - \frac{1}{a-s} = \frac{a(a-s)-s}{s(a-s)}$$

Of somewhat more interest than the two-sided Laplace transform is the *one-sided Laplace transform* or simply the *Laplace transform* defined by

$$\mathscr{L}[f(t)] = \int_{0}^{\infty} f(t)\exp[-st]\, dt \tag{6.119}$$

If this integral exists for some value of s then the Laplace transform of $f(t)$ is said to exist.

Theorem 6.19 Let $f(t)$ be continuous at all but a finite number of points on every closed interval $[0, a]$. If σ and b_0 are real numbers and

$$\lim_{t \to \infty}[f(t)\exp[-\sigma t]] = 0 \tag{6.120}$$

for $\sigma > b_0$, then

$$\mathscr{L}[f(t)] = \int_{0}^{\infty} f(t)\exp[-st]\, dt$$

exists for $\sigma > b_0$.

Example 6.25 Let the Laplace transform of $f(t)$ be given by

$$\mathscr{L}[f(t)] = G(s)$$

Show that

$$\mathscr{L}[\exp[at]f(t)] = G(s-a)$$

6.6 THE LAPLACE TRANSFORM

From the definition of the one-sided Laplace transform

$$\mathscr{L}[\exp[at]f(t)] = \int_0^\infty \exp[at]f(t)\exp[-st]\,dt = \int_0^\infty f(t)\exp[-(s-a)t]\,dt$$

Since

$$\int_0^\infty f(t)\exp[-st]\,dt = G(s)$$

we have

$$\int_0^\infty f(t)\exp[-(s-a)t]\,dt = G(s-a)$$

if the integral exists. However since $\mathscr{L}[f(t)]$ exists, there exists a value of s for which $\int_0^\infty f(t)\exp[-st]\,dt$ exists. Let $s = b_0 + i\omega$ be this value. Then $\int_0^\infty f(t)\exp[-(s-a)t]\,dt$ exists at least for $s - a = b_0 + i\omega$ and therefore $\mathscr{L}[\exp[at]f(t)]$ exists.

Example 6.26 Show that the Laplace transform of the nth derivative of $f(t)$ is given by

$$\mathscr{L}[f^n(t)] = s^n\mathscr{L}[f(t)] - \sum_{j=1}^n s^{n-j}f^{j-1}(0) \tag{6.121}$$

if $\lim_{t\to\infty}[f^j(t)\exp[-\sigma t]] = 0, j = 1, 2, \ldots, n$ for some σ.

Let us consider the Laplace transform of $f'(t)$

$$\mathscr{L}[f'(t)] = \int_0^\infty f'(t)\exp[-st]\,dt$$

We resolve the integral on the right by parts. Let

$$u = e^{-st}, \qquad dv = f'(t)\,dt$$

Then

$$\int_0^\infty f'(t)\exp[-st]\,dt = f(t)\exp[-st]\Big|_0^\infty + s\int_0^\infty f(t)\exp[-st]\,dt$$

If

$$\lim_{t\to\infty}[f(t)\exp[-\sigma t]] = 0$$

for some σ, then

$$\int_0^\infty f'(t)\exp[-st]\,dt = s\mathscr{L}[f(t)] - f(0)$$

The Laplace transform of $f''(t)$ is given by

$$\mathscr{L}[f''(t)] = \int_0^\infty f''(t)\exp[-st]\,dt$$

Again integrating by parts we have

$$\mathscr{L}[f''(t)] = f'(t)\exp[-st]\Big|_0^\infty + s\int_0^\infty f'(t)\exp[-st]\,dt$$
$$= -f'(0) + s\mathscr{L}[f'(t)] = s^2\mathscr{L}[f(t)] - sf(0) - f'(0)$$

Assume that

$$\mathscr{L}[f^k(t)] = s^k\mathscr{L}[f(t)] - \sum_{j=1}^{k} s^{k-j}f^{j-1}(0)$$

Then

$$\mathscr{L}[f^{k+1}(t)] = \int_0^\infty f^{k+1}(t)\exp[-st]\,dt = f^k(t)\exp[-st]\Big|_0^\infty + s\int_0^\infty f^k(t)\exp[-st]\,dt$$
$$= -f^k(0) + s\mathscr{L}[f^k(t)]$$
$$= s^{k+1}\mathscr{L}[f(t)] - \sum_{j=1}^{k+1} s^{k+1-j}f^{j-1}(0)$$

and Eq. (6.121) is verified by induction. ◰

Example 6.27 Show that

$$\mathscr{L}\left[\int_0^t f(x)\,dx\right] = \frac{1}{s}\mathscr{L}[f(t)] \qquad (6.122)$$

if $\int_0^\infty f(x)\,dx$ is bounded.

The Laplace transform of $\int_0^t f(x)\,dx$ is

$$\mathscr{L}\left[\int_0^t f(x)\,dx\right] = \int_0^\infty \int_0^t f(x)\exp[-st]\,dx\,dt$$

Integrating by parts let

$$u = \int_0^t f(x)\,dx, \qquad du = f(t)\,dt, \qquad dv = \exp[-st]\,dt, \qquad v = \frac{\exp[-st]}{-s}$$

Then

$$\mathscr{L}\left[\int_0^t f(x)\,dx\right] = \frac{\exp[-st]}{-s}\int_0^t f(x)\,dx\Big|_0^\infty + \frac{1}{s}\int_0^\infty f(t)\exp[-st]\,dt = \frac{1}{s}\mathscr{L}[f(t)]$$

where

$$\lim_{t\to\infty}\left[\exp[-st]\int_0^t f(x)\,dx\right] = 0$$

since $\int_0^\infty f(x)\,dx$ is bounded, and

$$\lim_{t\to 0}\left[\exp[-st]\int_0^t f(x)\,dx\right] = 0 \qquad ◰$$

6.6 THE LAPLACE TRANSFORM

Example 6.28 Find the Laplace transforms of the following functions.

a. $t^n \exp[ct]$

b. $\dfrac{\exp[c_1 t] - \exp[c_2 t]}{c_1 - c_2}$

For part a,

$$\mathscr{L}[t^n \exp[ct]] = \int_0^\infty t^n \exp[ct] \exp[-st]\, dt$$

$$= \int_0^\infty t^n \exp[-(s-c)t]\, dt = \frac{\Gamma(n+1)}{(s-c)^{n+1}}$$

For part b,

$$\mathscr{L}\left[\frac{\exp[c_1 t] - \exp[c_2 t]}{c_1 - c_2}\right] = \int_0^\infty \frac{\exp[c_1 t] - \exp[c_2 t]}{c_1 - c_2} \exp[-st]\, dt$$

$$= \frac{1}{c_1 - c_2}\left[\frac{1}{(s - c_1)} - \frac{1}{(s - c_2)}\right] = \frac{1}{(s - c_1)(s - c_2)}$$

As will be demonstrated in later applications, the Laplace transform is usually useful only if its inverse is known. Presently we will develop the *inversion formula* for Laplace transforms. However, the process of finding the inverse of a Laplace transform can be greatly simplified with a table of Laplace transforms such as that given in Table 3 of the Appendix. More extensive tables can be found in most texts on the Laplace transform.

Example 6.29 Let

$$g(t) = t^2, \qquad h(t) = t$$

Using Table 3, Appendix, find $f(t)$ where

$$f(t) = \int_0^t g(x) h(t - x)\, dx$$

From Table 3, Appendix,

$$\mathscr{L}[g(t)] = \frac{\Gamma(3)}{s^3} = \frac{2}{s^3} \qquad \text{and} \qquad \mathscr{L}[h(t)] = \frac{\Gamma(2)}{s^2} = \frac{1}{s^2}$$

Also from Table 3, Appendix,

$$\mathscr{L}[f(t)] = \mathscr{L}[g(t)]\mathscr{L}[h(t)] = \frac{2}{s^5}$$

To find the function $f(t)$ from its Laplace transform, we again refer to Table 3, Appendix, from which

$$L(s) = \frac{c}{s^n} \Rightarrow f(t) = \frac{ct^{n-1}}{\Gamma(n)}$$

Therefore,

$$L(s) = \frac{2}{s^5} \Rightarrow f(t) = \frac{2t^4}{24} \quad \text{or} \quad f(t) = \frac{t^4}{12}$$

One of the most frequent applications of the Laplace transform lies in the solution of differential equations. In the following example this important application is demonstrated.

Example 6.30 Solve the following differential equations using Laplace transforms.

a. $2f(t) + f'(t) - f''(t) = 0$, $f(0) = 1$, $f'(0) = 2$
b. $f(t) + f''(t) - 2 = t^2$, $f(0) = 0$, $f'(0) = 0$
c. $f'(t) + f''(t) + \sin(t) = \cos(t)$, $f(0) = 0$, $f'(0) = 1$

From Table 3, Appendix, we have

$$\mathscr{L}[f(t)] = L(s), \qquad \mathscr{L}[f'(t)] = sL(s) - f(0)$$
$$\mathscr{L}[f''(t)] = s^2 L(s) - sf(0) - f'(0)$$

For part a, we have

$$\mathscr{L}[2f(t) + f'(t) - f''(t)] = 2L(s) + sL(s) - f(0) - s^2 L(s) + sf(0) + f'(0) = 0$$

since

$$\mathscr{L}[g(t) \pm h(t)] = \mathscr{L}[g(t)] \pm \mathscr{L}[h(t)] \quad \text{and} \quad \mathscr{L}(0) = 0$$

Solving for $L(s)$ leads us to

$$L(s)[2 + s - s^2] = f(0)(1 - s) - f'(0)$$

But

$$f(0) = 1, \qquad f'(0) = 2$$

Therefore

$$L(s) = -\frac{s+1}{(s+1)(2-s)} = \frac{1}{(s-2)}$$

From Table 3, Appendix, the function $f(t)$ with Laplace transform $1/(s-2)$ is given by

$$f(t) = \exp[2t]$$

6.6 THE LAPLACE TRANSFORM

For part b, we have

$$\mathscr{L}[f(t) + f''(t) - 2] = L(s)(s^2 + 1) - sf(0) - f'(0) - \frac{2}{s} = \mathscr{L}(t^2) = \frac{2}{s^3}$$

Since

$$f(0) = f'(0) = 0$$

$L(s)$ is defined by

$$L(s) = \frac{2(s^2 + 1)}{s^3(s^2 + 1)} = \frac{2}{s^3}$$

From Table 3, Appendix,

$$f(t) = t^2$$

For part c,

$$\mathscr{L}[f'(t) - f''(t) + \sin(t)] = L(s)(s + s^2) - f(0)(1 + s) - f'(0) + \frac{1}{s^2 + 1}$$

$$= \mathscr{L}[\cos(t)] = \frac{s}{s^2 + 1}$$

Since

$$f(0) = 0, \qquad f'(0) = 1$$

we have

$$L(s)s(1 + s) - 1 + \frac{1}{s^2 + 1} = \frac{s}{s^2 + 1} \qquad \text{or} \qquad L(s) = \frac{1}{s^2 + 1}$$

The function $f(t)$ with Laplace transform $1/(s^2 + 1)$ is

$$f(t) = \sin(t) \qquad \qquad ▣$$

When the inverse Laplace transform is not available from tables of Laplace transforms, we can use an inversion formula similar to that given for the Fourier transform. Although other methods of inversion are available, they will not be discussed here. For a discussion of these methods the reader should see Savant (1962) or Hall et al. (1959).

Theorem 6.20 Let $L(s)$ be the Laplace transform of $f(t)$ where $f(t)$ satisfies the conditions given in Theorem 6.19. Then

$$f(t) = \frac{1}{2\pi i}\int_{a-i\infty}^{a+i\infty} L(s)\exp[st]\,ds, \qquad t > 0 \qquad (6.123)$$

where a is a constant chosen such that the singularities of $L(s)$ lie to the left of the straight line joining $a - i\infty$ and $a + i\infty$.

Proof To prove this theorem we note that the Laplace transform of $f(t)$ is also the Fourier transform of $f(t)\exp[-\sigma t]$. Then

$$F_t(u) = \int_0^\infty f(t)\exp[-\sigma t]\exp[-iut]\,dt$$

where $f(t) = 0$ for $t < 0$. By Eq. 6.113,

$$f(t)\exp[-at] = \frac{1}{2\pi}\int_{-\infty}^\infty F_t(u)\exp[iut]\,du$$

or

$$f(t) = \frac{1}{2\pi}\int_{-\infty}^\infty F_t(u)\exp[(a+iu)t]\,du$$

Since

$$\mathscr{L}[f(t)] = F_t(u) = L(s)$$

we have

$$f(t) = \frac{1}{2\pi}\int_{-\infty}^\infty L(s)\exp[(a+iu)t]\,du$$

Let

$$s = a + iu$$

Then

$$ds = i\,du$$

since a is a constant, and

$$f(t) = \frac{1}{2\pi i}\int_{a-i\infty}^{a+i\infty} L(s)\exp[st]\,ds$$

Example 6.31 Let

$$L(s) = \frac{1}{s-c}$$

Show that

$$f(t) = \exp[ct]$$

From Theorem 6.20,

$$f(t) = \frac{1}{2\pi i}\int_{a-i\infty}^{a+i\infty} \frac{\exp[st]}{s-c}\,ds$$

where a is chosen such that the poles of $L(s)$ lie to the left of the line joining $a - i\infty$ and $a + i\infty$. Since $L(s)$ has only one pole, at $s = c$, we must choose $a > c$.

6.7 APPLICATIONS

Applying the results of Theorem 6.12

$$\int_{a-i\infty}^{a+i\infty} \frac{\exp[st]}{s-c} ds = 2\pi i \left[\text{sum of residues of } \frac{\exp[st]}{s-c} \right] = 2\pi i \exp[ct]$$

Therefore,

$$f(t) = \exp[ct]$$

Example 6.32 Let

$$L(s) = \frac{s}{s^2 + c^2}$$

Show that

$$f(t) = \cos(ct)$$

From Theorem 6.20,

$$f(t) = \frac{1}{2\pi i} \int_{a-i\infty}^{a+i\infty} \frac{s \exp[st]}{s^2 + c^2} ds$$

The poles of $L(s)$ lie at $s = -ic$ and $s = ic$. Therefore $a > 0$. Applying the results of Theorem 6.12,

$$\int_{a-i\infty}^{a+i\infty} \frac{s \exp[st]}{s^2 + c^2} ds = 2\pi i \left[\text{sum of residues of } \frac{s \exp[st]}{s^2 + c^2} \right]$$

The residue at $s = -ic$ is given by

$$a_{-1,1} = \lim_{s \to -ic} \left[\frac{s \exp[st]}{s - ic} \right] = \frac{1}{2} \exp[-ict]$$

At $s = ic$ we have

$$a_{-1,2} = \lim_{s \to ic} \left[\frac{s \exp[st]}{s + ic} \right] = \frac{1}{2} \exp[ict]$$

Therefore

$$f(t) = \frac{\exp[ict] + \exp[-ict]}{2} = \cos(ct)$$

6.7 Applications

6.7.1 The Characteristic Function

Let X be a continuous random variable with density function given by $f(x)$, where $\int_{-\infty}^{\infty} f(x) dx = 1$. The *characteristic function* of X is given by $\phi_X(u)$ where

$$\phi_X(u) = \int_{-\infty}^{\infty} \exp[iux] f(x) dx \qquad (6.124)$$

If X is a discrete, integer-valued random variable with probability mass function given by $p(x)$, where $\sum_{x=-\infty}^{\infty} p(x) = 1$, then its characteristic function is given by

$$\phi_X(u) = \sum_{x=-\infty}^{\infty} \exp[iux] p(x) \tag{6.125}$$

The characteristic function plays an important role in probability theory. If the characteristic function of a random variable is known, the moments of the random variable can be derived in a relatively simple manner. Of perhaps more importance is the application of the characteristic function in finding the density function or probability mass function of the sum of random variables.

The reader will recall that the Fourier transform and its inverse formed a unique pair. In a similar manner the characteristic function and its corresponding density or probability mass function form a unique pair. Therefore, if the characteristic function of a given random variable is known, we should be able to derive the corresponding density or probability mass function from the characteristic function. If $\phi_X(u)$ is the characteristic function of the continuous random variable X, then the density function of X is given by

$$f(x) = \frac{1}{2\pi} \int_{-\infty}^{\infty} \exp[-iux] \phi_X(u)\, du \tag{6.126}$$

for every X for which $f(x)$ is continuous. If x is a discrete random variable, then its probability mass function is given by

$$p(x) = \lim_{t \to \infty} \frac{1}{2t} \int_{-t}^{t} \exp[-iux] \phi_X(u)\, du \tag{6.127}$$

A comparison of the transform pair given in Eqs. (6.124) and (6.126) with the Fourier transform pair as given in Eqs. (6.103) and (6.113) illustrates the close relationship between these two transformations.

Example 6.33 Let X be a normally distributed random variable with mean m, variance σ^2, and density function given by

$$f(x) = \frac{1}{\sigma\sqrt{2\pi}} \exp\left[\frac{-(x-m)^2}{2\sigma^2}\right], \qquad -\infty < x < \infty$$

Show that the characteristic function of this random variable is given by

$$\phi_X(u) = \exp\left[-ium - \frac{u^2 \sigma^2}{2}\right]$$

and that

$$f(x) = \frac{1}{2\pi} \int_{-\infty}^{\infty} \exp[-iux] \phi_X(u)\, du$$

6.7 APPLICATIONS

To determine the characteristic function of X we apply Eq. [6.124], since X is a continuous random variable.

$$\phi_X(u) = \int_{-\infty}^{\infty} \exp[iux] \frac{1}{\sigma\sqrt{2\pi}} \exp\left[-\frac{(x-m)^2}{2\sigma^2}\right] dx$$

$$= \frac{1}{\sigma\sqrt{2\pi}} \int_{-\infty}^{\infty} \exp\left[-\frac{(x-m)^2}{2\sigma^2} + iux\right] dx$$

The term in the exponent may be expressed as

$$-\frac{(x-m)^2}{2\sigma^2} + iux = -\frac{[x-(m+i\sigma^2 u)]^2}{2\sigma^2} + imu - \frac{\sigma^2 u^2}{2}$$

and

$$\phi_X(u) = \frac{1}{\sigma\sqrt{2\pi}} \exp\left[imu - \frac{\sigma^2 u^2}{2}\right] \int_{-\infty}^{\infty} \exp\left[-\frac{[x-(m+i\sigma^2 u)]^2}{2\sigma^2}\right] dx$$

From Example 6.23, the integral on the right-hand side is equal to $\sigma\sqrt{2\pi}$. Therefore,

$$\phi_X(u) = \exp\left[imu - \frac{\sigma^2 u^2}{2}\right]$$

To show that the original density function, $f(x)$, can be obtained from this characteristic function, we apply Eq. (6.126).

$$f(x) = \frac{1}{2\pi} \int_{-\infty}^{\infty} \exp[-iux] \exp\left[imu - \frac{\sigma^2 u^2}{2}\right] du$$

$$= \frac{1}{2\pi} \int_{-\infty}^{\infty} \exp\left[-iu(x-m) - \frac{\sigma^2 u^2}{2}\right] du$$

Let

$$y = \sigma u$$

Then

$$f(x) = \frac{1}{2\pi\sigma} \int_{-\infty}^{\infty} \exp\left[-\frac{iy(x-m)}{\sigma} - \frac{y^2}{2}\right] dy$$

The term in the exponent may be expressed by

$$-\frac{iy(x-m)}{\sigma} - \frac{y^2}{2} = -\frac{1}{2}\left[y - \frac{i(m-x)}{\sigma}\right]^2 - \frac{(x-m)^2}{2\sigma^2}$$

and

$$f(x) = \frac{1}{2\pi\sigma} \exp\left[-\frac{(x-m)^2}{2\sigma^2}\right] \int_{-\infty}^{\infty} \exp\left[-\frac{1}{2}\left[y - \frac{i(m-x)}{\sigma}\right]^2\right] dy$$

From Example 6.23,

$$\frac{1}{\sqrt{2\pi}}\int_{-\infty}^{\infty} \exp\left[-\frac{1}{2}\left[y - \frac{i(m-x)}{\sigma}\right]^2\right] dy = 1$$

and

$$f(x) = \frac{1}{\sigma\sqrt{2\pi}} \exp\left[-\frac{(x-m)^2}{2\sigma^2}\right]$$

Example 6.34 Let X be a Poisson random variable with probability mass function given by

$$p(x) = \begin{cases} \dfrac{\lambda^x}{x!}\exp[-\lambda], & x = 0, 1, 2, \ldots \\ 0, & \text{otherwise} \end{cases}$$

Show that the characteristic function of X is given by

$$\phi_X(u) = \exp[\lambda(\exp[iu] - 1)]$$

and that this characteristic function may be inverted using Eq. (6.127) to yield the expression for $p(x)$.

From Eq. (6.125)

$$\phi_X(u) = \sum_{x=0}^{\infty} \exp[iux]\frac{\lambda^x}{x!}\exp[-\lambda]$$

$$= \exp[-\lambda]\sum_{x=0}^{\infty} \frac{(\lambda\exp[iu])^x}{x!} = \exp[\lambda(\exp[iu] - 1)]$$

Using Eq. (6.127), we have

$$p(x) = \lim_{t\to\infty} \frac{1}{2t}\int_{-t}^{t} \exp[-iux]\exp[\lambda(\exp[iu] - 1)]\, du$$

Letting

$$\exp[\lambda(\exp[iu] - 1)] = \sum_{n=0}^{\infty} \frac{[\lambda(\exp[iu] - 1)]^n}{n!}$$

gives us

$$p(x) = \sum_{n=0}^{\infty} \frac{\lambda^n}{n!} \lim_{t\to\infty} \frac{1}{2t}\int_{-t}^{t} \exp[-iux](\exp[iu] - 1)^n\, du$$

The term $(\exp[iu] - 1)^n$ may be expressed through the binomial expansion as

$$(\exp[iu] - 1)^n = \sum_{j=0}^{n} \binom{n}{j} \exp[iju](-1)^{n-j}$$

6.7 APPLICATIONS

which leads to

$$p(x) = \sum_{n=0}^{\infty} \frac{\lambda^n}{n!} \sum_{j=0}^{n} \binom{n}{j}(-1)^{n-j} \lim_{t \to \infty} \frac{1}{2t} \int_{-t}^{t} \exp[iu(j-x)] \, du \quad (6.128)$$

Expressing $\exp[iu(j-x)]$ as $\cos[u(j-x)] + i\sin[u(j-x)]$, we obtain for the integral on the right

$$\int_{-t}^{t} \exp[iu(j-x)] \, du = \int_{-t}^{t} \{\cos[(j-x)u] + i\sin[(j-x)u]\} \, du$$

$$= 2 \int_{0}^{t} \cos[(j-x)u] \, du$$

since $\cos[(j-x)u]$ is an even function and $\sin[u(j-x)]$ is an odd function. Integrating over u, we have

$$\int_{-t}^{t} \exp[iu(j-x)] \, du = \frac{2\sin[(j-x)t]}{(j-x)}$$

and

$$\lim_{t \to \infty} \frac{1}{2t} \int_{-t}^{t} \exp[iu(j-x)] \, du = \lim_{t \to \infty} \frac{\sin[(j-x)t]}{t(j-x)}$$

The limit as $t \to \infty$ is resolved by recalling the sine expansion.

$$\sin(\theta) = \theta - \frac{\theta^3}{3!} + \frac{\theta^5}{5!} - \frac{\theta^7}{7!} + \cdots$$

Then

$$\frac{\sin[(j-x)t]}{t(j-x)} = 1 - \frac{[(j-x)t]^2}{3!} + \frac{[(j-x)t]^4}{5!} - \frac{[(j-x)t]^6}{7!} + \cdots$$

For $j = x$,

$$\lim_{t \to \infty} \frac{\sin[(j-x)t]}{t(j-x)} = 1$$

and for $j \neq x$,

$$\lim_{t \to \infty} \frac{\sin[(j-x)t]}{t(j-x)} = 0$$

Therefore

$$p(x) = \sum_{n=x}^{\infty} \frac{\lambda^n}{n!} \binom{n}{x}(-1)^{n-x}$$

since for every value of $j \neq x$ the corresponding term in the summation goes to zero. Multiplying and dividing by λ^x leads to

$$p(x) = \frac{\lambda^x}{x!} \sum_{n=x}^{\infty} \frac{(-\lambda)^{n-x}}{(n-x)!}$$

Let

$$m = n - x$$

Then

$$p(x) = \frac{\lambda^x}{x!} \sum_{m=0}^{\infty} \frac{(-\lambda)^m}{m!} = \frac{\lambda^x}{x!} \exp[-\lambda], \qquad x = 0, 1, 2, \ldots$$

If x is not a nonnegative integer, each term in the summation of Eq. (6.128) goes to zero since $\lim_{t \to \infty} \{\sin[(j-x)t]/t(j-x)\}$ approaches zero for $j \neq x$, and since j can assume only nonnegative integer values. Therefore

$$p(x) = \begin{cases} \dfrac{\lambda^x}{x!} \exp[-\lambda], & x = 0, 1, 2, \ldots \\ 0, & \text{otherwise} \end{cases}$$

Example 6.35 Let

$$\phi_X(u) = \frac{\exp[iub] - \exp[iua]}{iu(b-a)}, \qquad a < b$$

be the characteristic function of the random variable X. Show that X has a uniform distribution with density function given by

$$f(x) = \begin{cases} \dfrac{1}{b-a}, & a < x < b \\ 0, & \text{otherwise} \end{cases}$$

From Eq. (6.126),

$$f(x) = \frac{1}{2\pi} \int_{-\infty}^{\infty} \exp[-iux] \frac{\exp[iub] - \exp[iua]}{iu(b-a)} du$$

$$= \frac{1}{2\pi i(b-a)} \left[\int_{-\infty}^{\infty} \frac{\exp[-iu(x-b)]}{u} du - \int_{-\infty}^{\infty} \frac{\exp[-iu(x-a)]}{u} du \right]$$

Letting

$$\exp[-iu(x-b)] = \cos[u(b-x)] + i \sin[u(b-x)]$$

the first integral on the right-hand side becomes

$$\int_{-\infty}^{\infty} \frac{\exp[-iu(x-b)]}{u} du = \int_{-\infty}^{\infty} \frac{i \sin[u(b-x)]}{u} du$$

6.7 APPLICATIONS

since $\cos[u(b - x)]/u$ is an odd function for $x \neq b$ and $\sin[u(b - x)]/u$ is an even function for $x \neq b$. Treating the second integral on the right-hand side in a similar fashion, we have

$$f(x) = \frac{1}{2\pi(b-a)} \int_{-\infty}^{\infty} \frac{\sin[u(b-x)] - \sin[u(a-x)]}{u} du, \quad x \neq a, \; x \neq b$$

Let
$$y = u(b - x) \quad \text{and} \quad z = u(a - x)$$

For $x < a$,

$$f(x) = \frac{1}{2\pi(b-a)} \left[\int_{-\infty}^{\infty} \frac{\sin(y)}{y} dy - \int_{-\infty}^{\infty} \frac{\sin(z)}{z} dz \right] = \frac{1}{2\pi(b-a)} [\pi - \pi] = 0$$

by Example 6.20. For $a < x < b$,

$$f(x) = \frac{1}{2\pi(b-a)} \left[\int_{-\infty}^{\infty} \frac{\sin(y)}{y} dy - \int_{\infty}^{-\infty} \frac{\sin(z)}{z} dz \right]$$

$$f(x) = \frac{1}{2\pi(b-a)} \left[\int_{-\infty}^{\infty} \frac{\sin(y)}{y} dy + \int_{-\infty}^{\infty} \frac{\sin(z)}{z} dz \right] = \frac{1}{b-a}$$

For $x > b$,

$$f(x) = \frac{1}{2\pi(b-a)} \left[\int_{\infty}^{-\infty} \frac{\sin(y)}{y} dy - \int_{\infty}^{-\infty} \frac{\sin(z)}{z} dz \right]$$

$$= \frac{1}{2\pi(b-a)} \left[-\int_{-\infty}^{\infty} \frac{\sin(y)}{y} dy + \int_{-\infty}^{\infty} \frac{\sin(z)}{z} dz \right] = 0$$

At $x = a$ and $x = b$, the function $f(x)$ is discontinuous and Eq. (6.126) does not apply. At such points it is customary to define $f(a)$ and $f(b)$ as either $f(a + \epsilon)$ or $f(a - \epsilon)$ and $f(b + \epsilon)$ or $f(b - \epsilon)$. Therefore,

$$\dot{f}(x) = \begin{cases} \dfrac{1}{b-a}, & a < x < b \\ 0, & \text{otherwise} \end{cases}$$

Two properties of characteristic functions are noteworthy. First the characteristic function of every density function and probability mass function is unique. Second, the characteristic function always exists. Therefore, associated with every density function and probability mass function is one and only one characteristic function.

The importance of the characteristic function in determining the moments of a random variable has already been mentioned. The kth moment of the random variable X is referred to as the expected value of X^k and is denoted $E(X^k)$ as pointed out in Chapter 1. Hence

$$E(X^k) = \int_{-\infty}^{\infty} x^k f(x) \, dx \tag{6.129}$$

if X is a continuous random variable, and

$$E(X^k) = \sum_{x=-\infty}^{\infty} x^k p(x) \qquad (6.130)$$

if X is a discrete random variable. In terms of the characteristic function, $E(X^k)$ can be expressed by

$$E(X^k) = \frac{1}{i^k} \frac{d^k}{du^k} \phi_X(u) \bigg|_{u=0} \qquad (6.131)$$

The first two moments of a random variable are used to determine the mean and variance of the random variable. Specifically, the mean, m, of X is given by $E(X)$, while the variance of X is defined by $E(X-m)^2$. In terms of the moments about zero, the variance, σ^2, of a random variable X can be expressed by

$$\sigma^2 = E(X^2) - [E(X)]^2 \qquad (6.132)$$

Example 6.36 Let X be a Poisson distributed random variable with the characteristic function given in Example 6.34. Find the mean, m, and variance, σ^2, of X.

The mean of X is defined by $E(X)$. From Eq. (6.131)

$$E(X) = \frac{1}{i} \frac{d}{du} \phi_X(u) \bigg|_{u=0} = \frac{1}{i} \frac{d}{du} [\exp[\lambda(\exp[iu] - 1)]] \bigg|_{u=0}$$

$$= \frac{1}{i} [i\lambda \exp[iu] \exp[\lambda(\exp[iu] - 1)]] \bigg|_{u=0} = \lambda$$

From Eqs. (6.131) and (6.132), the variance of X is given by

$$\sigma^2 = E(X^2) - m^2 = \frac{1}{i^2} \frac{d^2}{du^2} \phi_X(u) \bigg|_{u=0} - \lambda^2$$

$$= \frac{1}{i^2} [(i\lambda \exp[iu])^2 \exp[\lambda(\exp[iu] - 1)]$$

$$+ i^2\lambda \exp[iu] \exp[\lambda(\exp[iu] - 1)]] \bigg|_{u=0} - \lambda^2$$

$$= \lambda^2 + \lambda - \lambda^2 = \lambda \qquad \blacksquare$$

In Theorem 6.17, we showed that if $t = t_1 + t_2$, then

$$F_t(u) = F_{t_1}(u) F_{t_2}(u) \qquad (6.133)$$

Similarly, if X_1 and X_2 are random variables with characteristic functions $\phi_{X_1}(u)$ and $\phi_{X_2}(u)$, respectively, and if $Y = X_1 + X_2$, then

$$\phi_Y(u) = \phi_{X_1}(u) \phi_{X_2}(u) \qquad (6.134)$$

Normally one is interested in identifying the density or probability mass function of y rather than the characteristic function of Y. Therefore, it becomes necessary

6.7 APPLICATIONS

to find the inverse of $\phi_Y(u)$ either mathematically using Eq. (6.126) or (6.127), or through the use of a table of characteristic functions.

Example 6.37 Let X_j be a Poisson random variable with parameter $\lambda_j, j = 1, 2, \ldots, n$. Show that

$$y = \sum_{j=1}^{n} x_j$$

is also Poisson distributed with parameter γ where

$$\gamma = \sum_{j=1}^{n} \lambda_j$$

If Y is a Poisson random variable with parameter γ, then the characteristic function of Y is given by

$$\phi_Y(u) = \exp[\gamma(\exp[iu] - 1)]$$

from Example 6.34 since the characteristic function of a random variable is unique. The characteristic function of X_j is given by

$$\phi_{X_j}(u) = \exp[\lambda_j(\exp[iu] - 1)]$$

From Eq. (6.134), the characteristic function of Y is given by

$$\phi_Y(u) = \prod_{j=1}^{n} \exp[\lambda_j(\exp[iu] - 1)]$$

$$= \exp\left[\sum_{j=1}^{n} \lambda_j(\exp[iu] - 1)\right] = \exp[\gamma(\exp[iu] - 1)] \quad \text{▨}$$

Example 6.38 Let X_j be an exponential random variable with density function given by

$$f(x_j) = \begin{cases} \lambda \exp[-\lambda x_j], & 0 < x_j < \infty, \quad j = 1, 2, \ldots, n \\ 0, & \text{otherwise} \end{cases}$$

where $\lambda > 0$. Show that

$$Y = \sum_{j=1}^{n} X_j$$

is a gamma random variable with density function given by

$$f(y) = \begin{cases} \dfrac{\lambda^n}{(n-1)!} y^{n-1} \exp[-\lambda y], & 0 < y < \infty \\ 0, & \text{otherwise} \end{cases}$$

To find the density function of Y, we will first determine the characteristic of Y and then invert this characteristic function using Eq. (6.126) to obtain the density function of Y. From Eq. (6.134),

$$\phi_Y(u) = \prod_{j=1}^{n} \phi_{X_j}(u)$$

The characteristic function of X_j is given by

$$\phi_{X_j}(u) = \int_0^\infty \exp[iux_j]\lambda \exp[-\lambda x_j]\, dx_j = \left(1 - \frac{iu}{\lambda}\right)^{-1}$$

Therefore,

$$\phi_Y(u) = \prod_{j=1}^{n} \left(1 - \frac{iu}{\lambda}\right)^{-1} = \left(1 - \frac{iu}{\lambda}\right)^{-n}$$

From Eq. (6.126)

$$f(y) = \frac{1}{2\pi} \int_{-\infty}^{\infty} \exp[-iuy]\left(1 - \frac{iu}{\lambda}\right)^{-n} du$$

Let

$$z = (\lambda - iu)y$$

Then

$$f(y) = -\frac{i\lambda^n y^{n-1}}{2\pi} \exp[-\lambda y] \int_{\lambda y - i\infty}^{\lambda y + i\infty} \frac{\exp[z]}{z^n} dz$$

From Example 6.22, $\exp[z]/z^n$ has an nth order pole at $z = 0$, which lies to the left of the line joining $\lambda y - i\infty$ and $\lambda y + i\infty$ for $\lambda y > 0$. The residue at $z = 0$ is

$$a_{-1} = \frac{1}{(n-1)!} \lim_{z \to \infty} \frac{d^{n-1}}{dz^{n-1}} (\exp[z]) = \frac{1}{(n-1)!}$$

and

$$f(y) = 2\pi i \left[-\frac{i\lambda^n y^{n-1}}{2\pi(n-1)!} \exp[-\lambda y] \right] = \frac{\lambda^n}{(n-1)!} y^{n-1} \exp[-\lambda y], \quad 0 < y < \infty$$

From Example 6.22, $f(y) = 0$ for $y < 0$. ▨

6.7.2 The Z Transform

With the exception of the characteristic function, the transform methods discussed thus far apply to the analysis of continuous systems. Even when dealing with discrete random variables, the characteristic function is not always the most convenient transform to use. A transform which may be applied to discrete

6.7 APPLICATIONS

systems is the *z transform*. Let $f(x)$ be a function of the integer valued variable x. The z transform of $f(x)$, $\psi_x(z)$, is defined by

$$\psi_x(z) = \sum_{x=-\infty}^{\infty} z^x f(x) \qquad (6.135)$$

if the sum converges.

In the remainder of this treatment of the z transform we shall be concerned with functions $f(x)$, where x is nonnegative. In this case $\psi_x(z)$ is given by

$$\psi_x(z) = \sum_{x=0}^{\infty} z^x f(x) \qquad (6.136)$$

The transform defined in Eq. (6.136) is sometimes called the *generating function* of $f(x)$. If the sum in Eq. (6.136) converges for all z on some interval including $(-z_0, z_0)$, then the z transform of $f(x)$ is said to exist and, like the transforms discussed previously, it is unique.

Theorem 6.21 If $\sum_{x=0}^{\infty} f(x)$ converges, where x is integer valued, then the z transform of $f(x)$ exists.

Proof The z transform of $f(x)$ is given by

$$\psi_x(z) = \sum_{x=0}^{\infty} z^x f(x)$$

For all z on the interval $(-1, 1)$

$$z^x f(x) \leq f(x)$$

Therefore

$$\psi_x(z) \leq \sum_{x=0}^{\infty} f(x)$$

and $\psi_x(z)$ exists for all z on the interval $(-1, 1)$. ∎

The result given in Theorem 6.21 is particularly useful in dealing with probability mass functions. If $p(x)$ is a probability mass function where $x = 0, 1, 2, \ldots$, then the z transform of $p(x)$ always exists since

$$\sum_{x=0}^{\infty} p(x) = 1$$

Example 6.39 Find the z transforms of the following functions, where $x = 0, 1, 2, \ldots$ in all cases.

a. $f(x) = 1$
b. $f(x) = r^x$
c. $f(x) = \dfrac{\lambda^x}{x!}$

For part a,

$$\psi_x(z) = \sum_{x=0}^{\infty} z^x = \frac{1}{1-z}$$

which exists for all z on the interval $(-1, 1)$. The reader will notice that $\psi_x(z)$ exists in this case even though $\sum_{x=0}^{\infty} f(x)$ does not converge. The z transform of $f(x)$ in part b is given by

$$\psi_x(z) = \sum_{x=0}^{\infty} z^x r^x = \frac{1}{1 - rz}$$

and $\psi_x(z)$ exists for all z on the interval $(-1/r, 1/r)$.

For part c,

$$\psi(z) = \sum_{x=0}^{\infty} z^x \frac{\lambda^x}{x!} = \sum_{x=0}^{\infty} \frac{(z\lambda)^x}{x!} = \exp[z]$$

Therefore, $\psi_x(z)$ exists for all z on every finite interval. ∎

Theorem 6.22 If $\psi_x(z)$ is the z transform of $f(x)$, where $x = 0, 1, 2, \ldots$, then

$$f(x) = \frac{1}{x!} \frac{d^x}{dz^x} \psi_x(z) \bigg|_{z=0} \tag{6.137}$$

Proof From Eq. (6.136)

$$\frac{1}{x!} \frac{d^x}{dz^x} \psi_x(z) \bigg|_{z=0} = \frac{1}{x!} \frac{d^x}{dz^x} \sum_{y=0}^{\infty} z^y f(y) \bigg|_{z=0} = \frac{1}{x!} \sum_{y=0}^{\infty} \frac{d^x}{dz^x} z^y f(y) \bigg|_{z=0}$$

For $y < x$,

$$\frac{d^x}{dz^x} z^y = 0$$

and

$$\frac{1}{x!} \frac{d^x}{dz^x} \psi_x(z) \bigg|_{z=0} = \frac{1}{x!} \sum_{y=x}^{\infty} \frac{d^x}{dz^x} z^y f(y) \bigg|_{z=0}$$

$$= \frac{1}{x!} \sum_{y=x}^{\infty} y(y-1) \cdots (y - x + 1) z^{y-x} f(y) \bigg|_{z=0}$$

For $y > x$,

$$z^{y-x} \bigg|_{z=0} = 0$$

For $y = x$,

$$z^{y-x} \bigg|_{z=0} = 1$$

6.7 APPLICATIONS

Therefore

$$\frac{1}{x!}\frac{d^x}{dz^x}\psi_x(z)\bigg|_{z=0} = f(x)$$

Example 6.40 Find the functions $f(x)$, $x = 0, 1, 2, \ldots$, corresponding to the following z transforms.

a. $\psi_x(z) = \dfrac{1}{1 - rz}$

b. $\psi_x(z) = \dfrac{z}{(1 - z)^2}$

c. $\psi_x(z) = \left(\dfrac{p}{1 - qz}\right)^r$, $\quad q = 1 - p$

Applying Eq. (6.137) to the function in part a, we find

$$f(x) = \frac{1}{x!}\frac{d^x}{dz^x}\left[\frac{1}{1 - rz}\right]\bigg|_{z=0}$$

$$= \frac{1}{x!}(-1)(-2)(-3)\cdots(-x)(-r)^x[1 - rz]^{-x}\bigg|_{z=0} = r^x$$

For part b,

$$f(x) = \frac{1}{x!}\frac{d^x}{dz^x}\left[\frac{z}{(1-z)^2}\right]\bigg|_{z=0}$$

When $x = 0$

$$f(0) = 0$$

For $x = 1$,

$$f(1) = \frac{d}{dz}\left[\frac{z}{(1-z)^2}\right]\bigg|_{z=0} = \left[\frac{1}{(1-z)^2} + 2\frac{z}{(1-z)^3}\right]\bigg|_{z=0} = 1$$

For $x = 2$,

$$f(2) = \frac{1}{2}\frac{d^2}{dz^2}\left[\frac{z}{(1-z)^2}\right]\bigg|_{z=0} = \frac{1}{2}\left[\frac{4}{(1-z)^3} + 6\frac{z}{(1-z)^4}\right]\bigg|_{z=0} = 2$$

For $x = 3$,

$$f(3) = \frac{1}{3!}\frac{d^3}{dz^3}\left[\frac{z}{(1-z)^2}\right]\bigg|_{z=0} = \frac{1}{6}\left[\frac{18}{(1-z)^4} + 24\frac{z}{(1-z)^5}\right]\bigg|_{z=0} = 3$$

We will attempt to determine $(d^x/dz^x)[z/(1 - z)^2]$ by induction. Assume that

$$\frac{d^x}{dz^x}\left[\frac{z}{(1-z)^2}\right] = \frac{xx!}{(1-z)^{x+1}} + \frac{(x+1)!z}{(1-z)^{x+2}}$$

Then

$$\frac{d^{x+1}}{dz^{x+1}}\left[\frac{z}{(1-z)^2}\right] = \frac{x(x+1)! + (x+1)!}{(1-z)^{x+2}} + \frac{(x+2)!z}{(1-z)^{x+3}}$$

$$= \frac{(x+1)(x+1)!}{(1-z)^{x+2}} + \frac{(x+2)!z}{(1-z)^{x+3}}$$

and the proof by induction is complete. Therefore

$$f(x) = \frac{1}{x!}\left[\frac{xx!}{(1-z)^{x+1}} + \frac{(x+1)!z}{(1-z)^{x+2}}\right]\bigg|_{z=0} = x$$

For part c, $d^x\psi_x(z)/dz^x$ is given by

$$\frac{d^x}{dz^x}\psi_x(z) = p^r q^x(r)(r+1)\cdots(r+x-1)(1-qz)^{-(r+x)}$$

Then

$$f(x) = \frac{1}{x!}\frac{d^x}{dz^x}\left[\frac{p}{1-qz}\right]^r\bigg|_{z=0} = \frac{(r+x-1)!}{x!(r-1)!}p^r q^x$$

The reader may recognize that $f(x)$ is the probability mass function of the negative binomial random variable. ▣

Just as the characteristic function is useful in finding the distribution of the sum of random variables, so also is the z transform when the random variables comprising the sum are discrete. In general, if $f(x_1)$ and $g(x_1)$ are functions of the integer-valued variables x_1 and x_2, and if

$$h(y) = \sum_{x=0}^{y} f(x)g(y-x) \qquad (6.138)$$

then

$$\psi_y(z) = \psi_{x_1}(z)\psi_{x_2}(z) \qquad (6.139)$$

if the z transforms $\psi_{x_1}(z)$ and $\psi_{x_2}(z)$ exist. Of course, if X_1 and X_2 are random variables, the existence of their z transforms is no problem. Thus it is always possible to find the z transform of the sum of discrete random variables through Eq. (6.139).

Example 6.41 Let the random variables X_i, $i = 1, 2, \ldots, n$, have probability mass functions given by

$$p(x_i) = \begin{cases} p^{x_i}(1-p)^{1-x_i}, & x_i = 0, 1, \quad i = 1, 2, \ldots, n \\ 0, & \text{otherwise} \end{cases}$$

Using z transforms, find $h(y)$ the probability mass function of Y, $h(y)$ where

$$Y = \sum_{i=1}^{n} X_i$$

6.7 APPLICATIONS

The z transform of X_i is given by

$$\psi_{X_i}(z) = \sum_{x_i=0}^{1} z^{x_i} p(x_i)$$

$$= \sum_{x_i=0}^{1} z^{x_i} p^{x_i} (1-p)^{1-x_i} = 1 - p(1-z)$$

Therefore, from Eq. (6.139)

$$\psi_Y(z) = [1 - p(1-z)]^n$$

Now

$$h(y) = \frac{1}{y!} \frac{d^y}{dz^y} [1 - p(1-z)]^n \bigg|_{z=0}$$

$$= \begin{cases} \dfrac{1}{y!} n(n-1) \cdots (n-y+1) p^y [1-p(1-z)]^{n-y} \bigg|_{z=0}, & y = 0, 1, \ldots, n \\ 0, & y > n \end{cases}$$

Hence

$$h(y) = \begin{cases} \dfrac{n!}{y!(n-y)!} p^y (1-p)^{n-y}, & y = 0, 1, 2, \ldots, n \\ 0, & y > n \end{cases}$$

and Y is binomially distributed with parameters n and p. ∎

Example 6.42 Let

$$p(x_i) = \frac{\lambda_i^{x_i}}{x_i!} \exp[-\lambda_i], \quad x_i = 0, 1, 2, \ldots, \quad i = 1, 2, \quad \text{and} \quad Y = X_1 + X_2$$

Using Eqs. (6.138) and (6.139), show that

$$h(y) = \frac{(\lambda_1 + \lambda_2)^y}{y!} \exp[-(\lambda_1 + \lambda_2)], \quad y = 0, 1, 2, \ldots$$

From Eq. (6.138)

$$h(y) = \sum_{x=0}^{y} \frac{\lambda_1^x}{x!} \exp[-\lambda_1] \frac{\lambda_2^{y-x}}{(y-x)!} \exp[-\lambda_2]$$

$$= \exp[-(\lambda_1 + \lambda_2)] \sum_{x=0}^{y} \frac{\lambda_1^x \lambda_2^{y-x}}{x!(y-x)!}$$

Multiply the right-hand side of this equation by $y!(\lambda_1 + \lambda_2)^y / y!(\lambda_1 + \lambda_2)^y$.

$$h(y) = \frac{(\lambda_1 + \lambda_2)^y}{y!} \exp[-(\lambda_1 + \lambda_2)] \sum_{x=0}^{y} \frac{y!}{x!(y-x)!} \left(\frac{\lambda_1}{\lambda_1 + \lambda_2}\right)^x \left(\frac{\lambda_2}{\lambda_1 + \lambda_2}\right)^{y-x}$$

Since the term to the right of the summation is the expression for the probability mass function of the binomial random variable with parameters y and $\lambda_1/(\lambda_1 + \lambda_2)$, and since the sum of the probability mass function over all permissible values of x is unity, we have

$$h(y) = \frac{(\lambda_1 + \lambda_2)^y}{y!} \exp[-(\lambda_1 + \lambda_2)], \qquad y = 0, 1, 2, \ldots$$

To demonstrate the same result through the use of z transforms, we first find $\psi_{X_i}(z)$, $i = 1, 2$.

$$\psi_{X_i}(z) = \sum_{x=0}^{\infty} z^{x_i} \frac{\lambda_i^{x_i}}{x_i!} \exp[-\lambda_i] = \exp[\lambda_i(z-1)], \qquad i = 1, 2$$

Then, from Eq. (6.139)

$$\psi_Y(z) = \exp[\lambda_1(z-1)]\exp[\lambda_2(z-1)] = \exp[(\lambda_1 + \lambda_2)(z-1)]$$

From Eq. (6.137)

$$h(y) = \frac{1}{y!} \frac{d^y}{dz^y} [\exp[(\lambda_1 + \lambda_2)(z-1)]]\bigg|_{z=0}$$

$$= \frac{(\lambda_1 + \lambda_2)^y}{y!} \exp[(\lambda_1 + \lambda_2)(z-1)]\bigg|_{z=0}$$

$$= \frac{(\lambda_1 + \lambda_2)^y}{y!} \exp[-(\lambda_1 + \lambda_2)], \qquad y = 0, 1, 2, \ldots \qquad \blacksquare$$

6.7.3 Transient State Analysis of Queueing Systems

In Chapter 5 we discussed the steady state solutions for certain basic queueing systems. We will now turn our attention to the transient state analysis of a simple queueing model. The system presented is elementary. However, the method of analysis may be extended to more complex systems, although the tedium involved in analyzing such systems increases significantly.

Consider a single channel queueing system where a waiting line is not allowed. That is, if an arrival occurs when there is one unit in the service channel, the arriving unit is refused entry to the system. Otherwise, the arrival enters the system, taking its place in the service channel. Assume that the probability of an arrival in a small interval of time, Δt, is approximately $\lambda \Delta t$, the approximation approaching the true probability of an arrival as Δt approaches zero. We will also assume that the probability of a service completion in Δt is approximately $u \Delta t$, given that a unit was being serviced when the interval of time Δt began. The parameters λ and u represent the arrival rate and service rate respectively. Finally, we will assume that two events cannot occur in Δt. That is, Δt is chosen small enough that the probability of two or more arrivals, two or more services, or an

6.7 APPLICATIONS

arrival and a service is zero. Let $P_n(t)$ be the probability that there are n units in the system at time t, $n = 0, 1, 2$. The probability that there are zero units in the system at time $t + \Delta t$ is given by

$$P_0(t + \Delta t) = P_0(t) \cdot P(0 \text{ arrivals in } \Delta t) + P_1(t) \cdot P(\text{service in } \Delta t) \quad (6.140)$$

since no more than one event can occur in Δt. Then

$$P_0(t + \Delta t) = (1 - \lambda \Delta t)P_0(t) + u \Delta t P_1(t) \quad (6.141)$$

Subtracting $P_0(t)$ from both sides of Eq. (6.141) and dividing by Δt leads to

$$\frac{P_0(t + \Delta t) - P_0(t)}{\Delta t} = -\lambda P_0(t) + u P_1(t) \quad (6.142)$$

Since

$$\lim_{\Delta t \to 0} \frac{P_0(t + \Delta t) - P_0(t)}{\Delta t} = \frac{d}{dt} P_0(t)$$

we have from Eq. (6.142)

$$\frac{d}{dt} P_0(t) = -\lambda P_0(t) + u P_1(t) \quad (6.143)$$

and

$$P_1(t) = \frac{\lambda}{u} P_0(t) + \frac{1}{u} P_0'(t) \quad (6.144)$$

For any time t,

$$\sum_{j=0}^{1} P_j(t) = 1$$

Therefore

$$\frac{\lambda + u}{u} P_0(t) + \frac{1}{u} P_0'(t) = 1 \quad (6.145)$$

If $L(s)$ is the Laplace transform of $P_0(t)$, then

$$\mathscr{L}[P_0'(t)] = sL(s) - P_0(0)$$

If we assume that the system is empty at $t = 0$, then

$$P_0(0) = 1 \quad \text{and} \quad \frac{\lambda + u}{u} L(s) + \frac{1}{u}[sL(s) - 1] = \frac{1}{s}$$

or

$$L(s) = \frac{u}{s(\lambda + u + s)} + \frac{1}{\lambda + u + s} = \frac{u + s}{s(\lambda + u + s)} \quad (6.146)$$

From Table 3, Appendix,

$$P_0(t) = \frac{u + \lambda \exp[-(\lambda + u)t]}{\lambda + u} \qquad (6.147)$$

For $P_1(t)$ we have

$$P_1(t) = 1 - P_0(t) = \frac{\lambda}{\lambda + u}[1 - \exp[-(\lambda + u)t]] \qquad (6.148)$$

If we let $t \to \infty$ in Eqs. (6.147) and (6.148) we obtain the corresponding steady state probabilities, P_0 and P_1.

$$P_0 = \lim_{t \to \infty} P_0(t) = \frac{u}{\lambda + u} \qquad (6.149)$$

$$P_1 = \lim_{t \to \infty} P_1(t) = \frac{\lambda}{\lambda + u} \qquad (6.150)$$

The variation of $P_0(t)$ and $P_1(t)$ with t is shown in Fig. 6.29 for various values of λ and u.

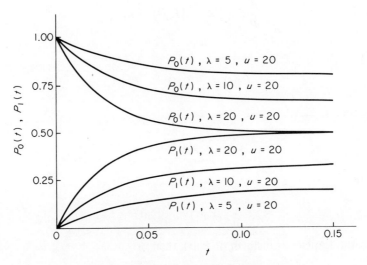

Figure 6.29 Variation of $P_0(t)$ and $P_1(t)$ with t

6.7.4 Replacement Analysis and Renewal Processes

Consider a piece of equipment which is used continuously until it fails. Upon failure it is immediately replaced by an identical piece of equipment. Let $T_1, T_2, \ldots, T_n$ be the times at which the first n pieces of equipment fail. If the time intervals between successive failure, $T_1, (T_2 - T_1), \ldots, (T_n - T_{n-1})$, are identically and

6.7 APPLICATIONS

independently distributed random variables, then the failure process is called a renewal process.

We will examine two aspects of the renewal process described. First, we will attempt to determine the expected number of failures by some time t. Then we will determine the density function of the life remaining in a piece of equipment, T, given that it was observed to be operating at some time τ. The expected number of failures in the interval of time t is often referred to as the mean value function and will be denoted by $m(t)$. The density function of excess life will be defined by $g(t \mid \tau)$. To determine $m(t)$ and $g(t \mid \tau)$ we will use the following equation,

$$h(t) = H(t) + \int_0^t h(t-y) f(y)\, dy \tag{6.151}$$

where $H(t)$ and $f(y)$ are known functions. To determine $h(t)$ we use Laplace transforms. Let $L(s)$ be the Laplace transform of $h(t)$. Then, from Table 3, Appendix,

$$L(s) = \mathscr{L}[H(t)] + L(s)\mathscr{L}[f(t)] \tag{6.152}$$

and

$$L(s) = \frac{\mathscr{L}[H(t)]}{1 - \mathscr{L}[f(t)]} \tag{6.153}$$

Let $f(t)$ be the density function of time until failure for any piece of equipment. The mean value function can then be expressed by

$$m(t) = \int_0^t f(y)\, dy + \int_0^t m(t-y) f(y)\, dy \tag{6.154}$$

As our first example, assume that time until failure is exponentially distributed with parameter λ. Then

$$f(t) = \lambda \exp[-\lambda t], \quad 0 < t < \infty$$

The mean value function is then expressed by

$$m(t) = \int_0^t \lambda \exp[-\lambda y]\, dy + \int_0^t m(t-y) \lambda \exp[-\lambda y]\, dy$$

$$= 1 - \exp[-\lambda t] + \int_0^t m(t-y) \lambda \exp[-\lambda y]\, dy \tag{6.155}$$

Taking the Laplace transform of Eq. (6.155) we have

$$L(s) = \frac{1}{s} - \frac{1}{(s+\lambda)} + L(s) \frac{\lambda}{(s+\lambda)} \tag{6.156}$$

where $L(s)$ is the Laplace transform of $m(t)$. Solving for $L(s)$ leads to

$$L(s) = \frac{\lambda}{s^2} \tag{6.157}$$

and from Table 3, Appendix,

$$m(t) = \lambda t$$

Now let us turn our attention to the case where the time between failures has the gamma distribution specified by

$$f(t) = \lambda^2 t \exp[-\lambda t], \quad 0 < t < \infty \tag{6.158}$$

The mean value function is given by

$$m(t) = \int_0^t \lambda^2 y \exp[-\lambda y] \, dy + \int_0^t m(t-y) \lambda^2 y \exp[-\lambda y] \, dy$$

$$= 1 - (1 + \lambda t) \exp[-\lambda t] + \int_0^t m(t-y) \lambda^2 y \exp[-\lambda y] \, dy \tag{6.159}$$

The Laplace transform of Eq. (6.159) is

$$L(s) = \frac{1}{s} - \frac{1}{(s+\lambda)} - \frac{\lambda}{(s+\lambda)^2} + L(s) \frac{\lambda^2}{(s+\lambda)^2}$$

and

$$L(s) = \frac{\lambda^2}{s^2(s+2\lambda)}$$

From Table 3, Appendix,

$$\frac{1}{s^2} = \mathscr{L}(t) \quad \text{and} \quad \frac{1}{(s+2\lambda)} = \mathscr{L}[\exp[-2\lambda t]]$$

Since

$$\mathscr{L}[f(t)]\mathscr{L}[g(t)] = \mathscr{L}\left[\int_0^t f(x)g(t-x) \, dx\right]$$

we have

$$\mathscr{L}[t]\mathscr{L}[\exp[-2\lambda t]] = \mathscr{L}\left[\int_0^t (t-x)\exp[-2\lambda x] \, dx\right]$$

Therefore

$$m(t) = \lambda^2 \int_0^t (t-x)\exp[-2\lambda x] \, dx$$

or

$$m(t) = \frac{\lambda t}{2} - \frac{\exp[-2\lambda t]}{4} - \frac{1}{4} \tag{6.160}$$

The renewal equation can also be used to determine the distribution of excess life remaining in an item. As already indicated, $g(t \mid \tau)$ is the density function of

6.7 APPLICATIONS

time until failure, T, given that the item was operating at τ. Let $G(t\,|\,\tau)$ be the complementary distribution function of t. Then

$$G(t\,|\,\tau) = \int_{t}^{\infty} g(x\,|\,\tau)\,dx \qquad (6.161)$$

and let $f(y)$ be the density function of time until failure. $G(t\,|\,\tau)$ can be defined by the renewal equation

$$G(t\,|\,\tau) = \int_{t+\tau}^{\infty} f(y)\,dy + \int_{0}^{\tau} G(t\,|\,\tau - y)f(y)\,dy \qquad (6.162)$$

Equation (6.162) is the renewal equation through which the density function, $g(t\,|\,\tau)$, will be determined. Let

$$f(y) = \lambda \exp[-\lambda y], \qquad 0 < y < \infty$$

Then

$$G(t\,|\,\tau) = \int_{t+\tau}^{\infty} \lambda \exp[-\lambda y]\,dy + \int_{0}^{\tau} G(t\,|\,\tau - y)\lambda \exp[-\lambda y]\,dy$$

$$= \exp[-\lambda(t + \tau)] + \int_{0}^{\tau} G(t\,|\,\tau - y)\lambda \exp[-\lambda y]\,dy \qquad (6.163)$$

We take the Laplace transform of Eq. (6.163) where

$$L(s) = \mathscr{L}[G(t\,|\,\tau)] = \int_{0}^{\infty} G(t\,|\,\tau)\exp[-s\tau]\,d\tau$$

Then

$$L(s) = \exp[-\lambda t]\mathscr{L}[\exp[-\lambda\tau]] + L(s)\frac{\lambda}{(s+\lambda)} = \frac{\exp[-\lambda t]}{(s+\lambda)} + L(s)\frac{\lambda}{s+\lambda}$$

and

$$L(s) = \frac{\exp[-\lambda t]}{s}$$

Therefore

$$G(t\,|\,\tau) = \exp[-\lambda t]$$

The density function, $g(t\,|\,\tau)$, can be expressed as the first derivative of the distribution function which is given by $1 - G(t\,|\,\tau)$. Therefore

$$g(t\,|\,\tau) = \frac{d}{dt}[1 - G(t\,|\,\tau)] = \lambda \exp[-\lambda t], \qquad 0 < t < \infty \qquad (6.164)$$

The reader will notice that the density functions of excess life and time until failure are identical. This property of the exponential random variable is the basis for characterizing the exponential random variable as a variable without memory as already indicated in Chapter 1. That is, the density function of life remaining in equipment subject to exponential failure is the same no matter how long the equipment has been in operation. The exponential random variable is the only continuous random variable possessing this property. ∎

A Dynamic Pricing Problem Let $p(t)$ be the sale price for a given product at time t. Suppose that the demand for the product at t is linearly dependent upon the price at t and the rate of change in the price at t, $p'(t)$. If $d(t)$ is the demand at t, then

$$d(t) = a_0 + a_1 p(t) + a_2 p'(t) \tag{6.165}$$

We are assuming that a consumer's decision to buy or not to buy the product is dependent upon the current price of the product and the rate at which the price is rising or falling. Let us also assume that the volume of product supplied by the seller at t is similarly dependent upon $p(t)$ and $p'(t)$. Therefore, if $s(t)$ is the supply of product at t, then

$$s(t) = b_0 + b_1 p(t) + b_2 p'(t) \tag{6.166}$$

If the seller is to meet the demand at t, then $s(t) = d(t)$, and

$$(a_0 - b_0) + (a_1 - b_1)p(t) + (a_2 - b_2)p'(t) = 0 \tag{6.167}$$

We wish to solve this differential equation for the pricing policy $p(t)$. Let $L(s)$ be the Laplace transform of $p(t)$. Then

$$\mathscr{L}[p'(t)] = sL(s) - p(0)$$

where $p(0)$ is the initial price for which the product is offered. Taking the Laplace transform of Eq. (6.167) yields

$$\frac{a_0 - b_0}{s} + (a_1 - b_1)L(s) + (a_2 - b_2)[sL(s) - p(0)] = 0 \tag{6.168}$$

and

$$L(s) = \frac{sp(0)(a_2 - b_2) - (a_0 - b_0)}{s[s(a_2 - b_2) + (a_1 - b_1)]} \tag{6.169}$$

or

$$L(s) = \frac{sp(0) - \dfrac{(a_0 - b_0)}{(a_2 - b_2)}}{s\left[s + \dfrac{(a_1 - b_1)}{(a_2 - b_2)}\right]} = p(0) \frac{s - \dfrac{(a_0 - b_0)}{p(0)(a_2 - b_2)}}{s\left[s + \dfrac{(a_1 - b_1)}{(a_2 - b_2)}\right]} \tag{6.170}$$

From Table 3, Appendix,

$$p(t) = \frac{a_0 - b_0}{b_1 - a_1} - \left[\frac{a_0 - b_0}{b_1 - a_1} - p(0)\right] \exp\left[-\left(\frac{a_1 - b_1}{a_2 - b_2}\right)t\right] \tag{6.171}$$

6.7 APPLICATIONS

If $(a_1 - b_1)/(a_2 - b_2) > 0$, $p(t)$ attains a steady state value of $(a_0 - b_0)/(b_1 - a_1)$. Otherwise, the transient state persists.

6.7.5 Engineering Applications

Example 6.43 The voltage, $v(t)$, applied to a series circuit composed of an inductance l, a resistance R, and a capacitance C is given by

$$v(t) = V_0 \cos(\omega t + \alpha)$$

The resulting current, $j(t)$, is given by

$$j(t) = I_0 \cos(\omega t + \beta)$$

where ω is the radian frequency and α and β are phase angles. Define current and voltage transforms as

$$V = \frac{V_0}{\sqrt{2}} \exp[i\alpha] \quad \text{and} \quad J = \frac{J_0}{\sqrt{2}} \exp[i\beta]$$

where $i = \sqrt{-1}$. If $v(t)$ and $j(t)$ are related by

$$v(t) = Rj(t) + l\frac{dj(t)}{dt} + \frac{1}{C}\int_0^t j(z)\, dz$$

show that V and J are related by the equation

$$V = \left(R + i\omega l + \frac{1}{i\omega C}\right) J$$

From Eq. (6.14)

$$\exp[i\theta] = \cos(\theta) + i \sin(\theta)$$

Let $\mathrm{Re}(\exp[i\theta])$ represent the real part of $\exp[i\theta]$. That is,

$$\mathrm{Re}(\exp[i\theta]) = \cos(\theta)$$

Then

$$v(t) = V_0 \, \mathrm{Re}[\exp[i(\omega t + \alpha)]] = \mathrm{Re}(\sqrt{2}\, V \exp[i\omega t])$$

and

$$j(t) = \mathrm{Re}(\sqrt{2}\, J \exp[i\omega t])$$

We then have

$$\mathrm{Re}(\sqrt{2}\,V\exp[i\omega t]) = R\,\mathrm{Re}(\sqrt{2}\,J\exp[i\omega t]) + l\frac{d}{dt}\mathrm{Re}(\sqrt{2}\,J\exp[i\omega t])$$
$$+ \frac{1}{C}\int_0^t \mathrm{Re}(\sqrt{2}\,J\exp[i\omega t])\,dt$$
$$= \sqrt{2}\,JR\,\mathrm{Re}(\exp[i\omega t]) + \sqrt{2}\,Jl\frac{d}{dt}\mathrm{Re}(\exp[i\omega t])$$
$$+ \frac{\sqrt{2}\,J}{C}\int_0^t \mathrm{Re}(\exp[i\omega t])\,dt$$
$$= \sqrt{2}\,JR\,\mathrm{Re}(\exp[i\omega t]) + \sqrt{2}\,Jl\,\mathrm{Re}\left(\frac{d}{dt}\exp[i\omega t]\right)$$
$$+ \frac{\sqrt{2}\,J}{C}\mathrm{Re}\left(\int_0^t \exp[i\omega t]\,dt\right)$$

Finally,

$$\sqrt{2}\,V\,\mathrm{Re}(\exp[i\omega t]) = \sqrt{2}\,J\left[R\,\mathrm{Re}(\exp[i\omega t]) + l\,\mathrm{Re}(i\omega \exp[i\omega t])\right.$$
$$\left. + \frac{1}{C}\mathrm{Re}\left(\frac{1}{i\omega}\exp[i\omega t]\right)\right]$$
$$= \mathrm{Re}\left[\sqrt{2}\,J\left(R + i\omega l + \frac{1}{i\omega C}\right)\exp[i\omega t]\right]$$

Since the real components of both sides of this equation are equal, the complex components of both sides must also be equal. Thus we may write

$$\sqrt{2}\,V\exp[i\omega t] = \sqrt{2}\,J\left(R + i\omega l + \frac{1}{i\omega C}\right)\exp[i\omega t]$$

or

$$V = \left(R + i\omega l + \frac{1}{i\omega C}\right)J \qquad \blacksquare$$

Example 6.44 Let

$$f(s) = g(x, y) + ih(x, y)$$

where $f(s)$ is analytic. In fluid mechanics $g(x, y)$ is referred to as the velocity potential and $h(x, y)$ as the stream function. If

$$g(x, y) = x^2 - y^2 + 2x + y$$

find $h(x, y)$.

6.7 APPLICATIONS

Since $f(s)$ is analytic, from the Cauchy–Riemann equations we have

$$\frac{\partial g(x, y)}{\partial x} = \frac{\partial h(x, y)}{\partial y} \quad \text{and} \quad \frac{\partial g(x, y)}{\partial y} = -\frac{\partial h(x, y)}{\partial x}$$

Now

$$\frac{\partial g(x, y)}{\partial x} = 2x + 2, \qquad \frac{\partial g(x, y)}{\partial y} = -2y + 1$$

Hence

$$\frac{\partial h(x, y)}{\partial y} = 2x + 2 \quad \text{and} \quad \frac{\partial h(x, y)}{\partial x} = 2y - 1$$

Integrating $\partial h(x, y)/\partial y$ and $\partial h(x, y)/\partial x$ leads us to

$$h(x, y) = 2xy + 2y + C_1(x), \qquad h(x, y) = 2xy - x + C_2(y)$$

where $C_1(x)$ and $C_2(y)$ constants with respect to y and x respectively. Hence

$$C_1(x) = -x + C, \qquad C_2(y) = 2y + C$$

where C is a constant. Thus

$$h(x, y) = 2xy - x + 2y + C$$

Example 6.45 A building of the type shown in Fig. 6.30 is to be constructed. The roof of the structure satisfies the equation $f(x, y)$, where

$$f(x, y) = 4(x + y) - (x^2 + y^2)$$

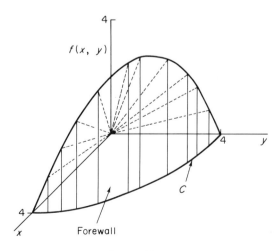

Figure 6.30

The location of the forewall of the building is given by
$$x^2 + y^2 = 16, \quad 0 \leq x \leq 4$$
Find the area of the forewall.

To find the area required we must integrate the function $f(x, y)$ between $(4, 0)$ and $(0, 4)$ on the curve $x^2 + y^2 = 16$, C. Noting that
$$y = \sqrt{16 - x^2}$$
we have by Example 6.6

$$\int_C f(s)\, ds = \int_4^0 [4(x + \sqrt{16 - x^2}) - 16] \sqrt{1 + \frac{x^2}{16 - x^2}}\, dx$$

$$= 16 \int_4^0 \frac{x + \sqrt{16 - x^2}}{\sqrt{16 - x^2}}\, dx - 64 \int_4^0 \frac{dx}{\sqrt{16 - x^2}}$$

$$= 100.48 + 16 \int_4^0 \frac{x}{\sqrt{16 - x^2}}\, dx + 16 \int_4^0 dx$$

Hence

$$\int_C f(s)\, ds = 100.48 - 16\sqrt{16 - x^2}\,\Big|_4^0 - 64 = 100.48$$

Example 6.46 The voltage across elements having resistance R and inductance l at time t is given by
$$V = RI(t) + l \frac{dI(t)}{dt}, \quad \text{where} \quad I(0) = I_0$$

Find $I(t)$.

Taking the Laplace transform of each component of the expression for $V(t)$ yields
$$\mathscr{L}(V) = \frac{V}{s}, \quad \mathscr{L}[I(t)] = L(s), \quad \mathscr{L}\left[\frac{dI(t)}{dt}\right] = sL(s) - I(0)$$

Then
$$\frac{V}{s} = RL(s) + lsL(s) - lI_0 \quad \text{and} \quad L(s) = \frac{V}{s(R + ls)} + \frac{lI_0}{R + ls}$$

or
$$I(t) = \frac{V}{R} - \frac{V}{R} \exp\left[-\frac{R}{l} t\right] + I_0 \exp\left[-\frac{R}{l} t\right]$$

$$= \frac{V}{R}\left(1 - \exp\left[-\frac{R}{l} t\right]\right) + I_0 \exp\left[-\frac{R}{l} t\right]$$

6.7 APPLICATIONS

Example 6.47 An electromotive force V is applied at $t = 0$ to a circuit consisting of an inductance l, resistance R, and capacitance C connected in series. Find the current at t, $I(t)$, if

$$V = RI(t) + l\frac{dI(t)}{dt} + \frac{1}{C}\int_0^t I(t)\, dt$$

where $I(0) = 0$.

Taking the Laplace transform of both sides of the differential equation yields

$$\frac{V}{s} = RL(s) + l\,[sL(s) - I(0)] + \frac{L(s)}{Cs}, \qquad \text{where} \quad L(s) = \mathscr{L}[I(t)]$$

Then

$$L(s) = \frac{V}{l\left(s^2 + \dfrac{R}{l}s + \dfrac{1}{lC}\right)}$$

since $I(0) = 0$. Noting that

$$\left(s^2 + \frac{R}{l}s + \frac{1}{lC}\right) = \left[s + \frac{R}{2l} - \sqrt{\frac{R^2}{4l^2} - \frac{1}{lC}}\right]\left[s + \frac{R}{2l} + \sqrt{\frac{R^2}{4l^2} - \frac{1}{lC}}\right]$$

Hence $I(t)$ is given by

$$I(t) = \frac{V}{l}\frac{\exp[C_1 t] - \exp[C_2 t]}{C_1 - C_2}$$

where

$$C_1 = -\frac{R}{2l} + \sqrt{\frac{R^2}{4l^2} - \frac{1}{lC}} \quad \text{and} \quad C_2 = -\frac{R}{2l} - \sqrt{\frac{R^2}{4l^2} - \frac{1}{lC}} \quad \blacksquare$$

Example 6.48 A load of mass m is attached to a spring and brought to rest. A force equal to $A \sin(\omega t)$ is then applied to the load. The displacement, $x(t)$, of the load at time t may be related to the mass of the load and the applied force by the differential equation

$$\frac{m}{g}\frac{d^2 x(t)}{dt^2} + kx(t) = A\sin(\omega t)$$

where g is the acceleration due to gravity, k is a positive constant related to the spring, and $A > 0$. Find the displacement, $x(t)$, at time t if $x(t) = 0$ and $dx(t)/dt = 0$ at $t = 0$.

Taking the Laplace transform of the differential equation yields

$$\frac{m}{g}[s^2 L(s) - sx(0) - x'(0)] + kL(s) = \frac{A\omega}{s^2 + \omega^2}$$

or

$$L(s)\left[\frac{m}{g}s^2 + k\right] = \frac{A\omega}{s^2 + \omega^2} \quad \text{and} \quad L(s) = \frac{gA\omega/m}{(s^2 + \omega^2)\left(s^2 + \dfrac{kg}{m}\right)}$$

Hence

$$L(s) = \frac{gA\omega/m}{\left(\omega^2 - \dfrac{kg}{m}\right)}\left[\frac{1}{\left(s^2 + \dfrac{kg}{m}\right)} - \frac{1}{(s^2 + \omega^2)}\right]$$

and

$$x(t) = \frac{gA\omega}{m\left(\omega^2 - \dfrac{kg}{m}\right)}\left[\sqrt{\frac{m}{kg}}\sin\left(\sqrt{\frac{kg}{m}}\,t\right) - \frac{\sin(\omega t)}{\omega}\right]$$

$$= \sqrt{\frac{g}{mk}}\,A\omega\,\frac{\sin\left(\sqrt{\dfrac{kg}{m}}\,t\right)}{\left(\omega^2 - \dfrac{kg}{m}\right)} - \frac{Ag}{m}\,\frac{\sin(\omega t)}{\left(\omega^2 - \dfrac{kg}{m}\right)}$$

Example 6.49 Let $f(t)$ be a periodic function with period 2π. Examples of such functions are shown in Fig. 6.31. Functions of this type can be represented by the *Fourier Series* given by

$$f(t) = a_0 + \sum_{n=1}^{\infty}[a_n \cos(nt) + b_n \sin(nt)] \tag{6.172}$$

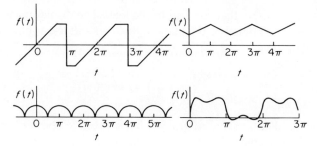

Figure 6.31 Periodic Functions with Period 2π

6.7 APPLICATIONS

Show that a_0 and a_n, b_n, $n = 1, 2, 3, \ldots$, are given by

$$a_0 = \frac{1}{2\pi} \int_0^{2\pi} f(t) \, dt \tag{6.173}$$

$$a_n = \frac{1}{\pi} \int_0^{2\pi} f(t) \cos(nt) \, dt \tag{6.174}$$

$$b_n = \frac{1}{\pi} \int_0^{2\pi} f(t) \sin(nt) \, dt \tag{6.175}$$

Integrating $f(t)$ yields

$$\int_0^{2\pi} f(t) \, dt = \int_0^{2\pi} a_0 \, dt + \int_0^{2\pi} \sum_{n=1}^{\infty} [a_n \cos(nt) + b_n \sin(nt)] \, dt$$

$$= 2\pi a_0 + \sum_{n=1}^{\infty} \left[a_n \int_0^{2\pi} \cos(nt) \, dt + b_n \int_0^{2\pi} \sin(nt) \, dt \right]$$

But

$$\int_0^{2\pi} \cos(nt) \, dt = \frac{1}{n} \int_0^{2n\pi} \cos(y) \, dy = \frac{1}{n} [\sin(2n\pi) - \sin(0)] = 0$$

and

$$\int_0^{2\pi} \sin(nt) \, dt = \frac{1}{n} \int_0^{2n\pi} \sin(y) \, dy$$

$$= \frac{1}{n} [-\cos(2n\pi) + \cos(0)] = \frac{1}{n} [-1 + 1] = 0$$

Hence

$$a_0 = \frac{1}{2\pi} \int_0^{2\pi} f(t) \, dt$$

To determine a_n we evaluate $\int_0^{2\pi} f(t) \cos(nt) \, dt$.

$$\int_0^{2\pi} f(t) \cos(nt) \, dt = \int_0^{2\pi} \cos(nt) \left\{ a_0 + \sum_{j=1}^{\infty} [a_j \cos(jt) + b_j \sin(jt)] \right\} dt$$

Now

$$\int_0^{2\pi} \cos(nt) \, dt = 0$$

$$\int_0^{2\pi} \cos(nt)\cos(jt) \, dt = \left. \frac{\sin[(n-j)t]}{2(n-j)} \right|_0^{2\pi} + \left. \frac{\sin[(n+j)t]}{2(n+j)} \right|_0^{2\pi} = 0, \quad j \neq n$$

$$\int_0^{2\pi} \cos(nt)\sin(jt) \, dt = -\left. \frac{\cos[(j-n)t]}{2(j-n)} \right|_0^{2\pi} - \left. \frac{\cos[(n+j)t]}{2(n+j)} \right|_0^{2\pi} = 0, \quad j \neq n$$

$$\int_0^{2\pi} \cos(nt)\sin(nt) \, dt = \frac{1}{n} \int_0^{2n\pi} \cos(y)\sin(y) \, dy = \left. \frac{1}{2n} \sin^2(y) \right|_0^{2\pi} = 0$$

and
$$\int_0^{2\pi} \cos^2(nt)\,dt = \frac{1}{n}\int_0^{2n\pi} \cos^2(y)\,dy$$
$$= \frac{1}{n}\left[\frac{1}{2}y + \frac{1}{4}\sin(2y)\right]\Big|_0^{2n\pi} = \pi$$

Thus
$$\int_0^{2\pi} f(t)\cos(nt)\,dt = a_n\pi \quad \text{or} \quad a_n = \frac{1}{\pi}\int_0^{2\pi} f(t)\cos(nt)\,dt$$

To determine b_n we evaluate $\int_0^{2\pi} f(t)\sin(nt)\,dt$.

$$\int_0^{2\pi} f(t)\sin(nt)\,dt = \int_0^{2\pi} \sin(nt)\left\{a_0 + \sum_{j=1}^{\infty}[a_j\cos(jt) + b_j\sin(jt)]\right\}dt$$

But
$$\int_0^{2\pi} \sin(nt) = 0$$
$$\int_0^{2\pi} \sin(nt)\sin(jt)\,dt = \frac{\sin[(n-j)t]}{2(n-j)}\Big|_0^{2\pi} - \frac{\sin[(n+j)t]}{2(n+j)}\Big|_0^{2\pi} = 0, \quad n \neq j$$

and
$$\int_0^{2\pi} \cos(nt)\sin(jt)\,dt = \int_0^{2\pi} \cos(nt)\sin(nt)\,dt = 0$$

from above. Finally,
$$\int_0^{2\pi} \sin^2(nt)\,dt = \frac{1}{n}\int_0^{2n\pi} \sin^2(y)\,dy$$
$$= \frac{1}{n}\left[\frac{1}{2}y - \frac{1}{4}\sin(2y)\right]\Big|_0^{2n\pi} = \pi$$

and
$$b_n = \frac{1}{\pi}\int_0^{2\pi} f(t)\sin(nt)\,dt$$

Example 6.50 *(Continuation of Example 6.49)* Let
$$f(t) = \begin{cases} 1, & 0 < t < \pi \\ -1, & \pi < t < 2\pi \end{cases}$$
Determine the Fourier series expression for $f(t)$.

From Eq. (6.173) we have

$$a_0 = \frac{1}{2\pi} \int_0^{2\pi} f(t)\, dt = \frac{1}{2\pi}\left[\int_0^{\pi} dt - \int_{\pi}^{2\pi} dt\right] = 0$$

For a_n,

$$a_n = \frac{1}{\pi} \int_0^{2\pi} f(t) \cos(nt)\, dt = \frac{1}{\pi}\left[\int_0^{\pi} \cos(nt)\, dt - \int_{\pi}^{2\pi} \cos(nt)\, dt\right]$$

$$= \frac{1}{n\pi}\left[\sin(t)\Big|_0^{n\pi} - \sin(t)\Big|_{n\pi}^{2n\pi}\right] = 0$$

Finally, b_n is given by

$$b_n = \frac{1}{\pi} \int_0^{2\pi} f(t) \sin(nt)\, dt = \frac{1}{\pi}\left[\int_0^{\pi} \sin(nt)\, dt - \int_{\pi}^{2\pi} \sin(nt)\, dt\right]$$

$$= \frac{1}{n\pi}\left[-\cos(t)\Big|_0^{n\pi} + \cos(t)\Big|_{n\pi}^{2n\pi}\right] = \frac{1}{n\pi}[-2\cos(n\pi) + 1 + \cos(2n\pi)]$$

For odd n,

$$\cos(n\pi) = -1$$

and for even n,

$$\cos(n\pi) = 1$$

Since

$$\cos(2n\pi) = 1$$

for all n, we have

$$b_n = \begin{cases} 4/n\pi, & \text{odd } n \\ 0, & \text{even } n \end{cases}$$

Therefore,

$$f(t) = \frac{4}{\pi} \sum_{n=1}^{\infty} \frac{\sin[(2n-1)t]}{2n-1}$$

Problems

1. Show that

$$\cos(\theta) = \frac{\exp[i\theta] + \exp[-i\theta]}{2}$$

2. Evaluate $f'(s)$ for the following functions.
   a. $f(s) = s^2 + 4s + 1$
   b. $f(s) = \exp[s]$

3. Let S_0 be the s plane. Determine whether the following functions are analytic on S_0.

   a. $f(s) = s^2 - 4$ b. $f(s) = \dfrac{s+4}{s-3}$

4. Identify the poles of each of the following functions and specify the order of each.

   a. $\dfrac{1}{s-5}$ b. $\dfrac{1}{(s^2-9)}$ c. $\dfrac{1}{s^2(s-8)^3}$

   d. $\dfrac{s}{(s^2+4)^2}$ e. $\dfrac{s}{(s^2+4)^2-64}$ f. $\dfrac{(s+1)}{(s^2-1)}$

5. Determine whether Green's theorem can be applied to the function in Example 6.12.
6. Integrate the following functions on the simple closed curve C, where the path C is a circle of radius 5 and center at $s = 0$.

   a. $f(s) = \dfrac{1}{(s^2-4)}$ b. $f(s) = \dfrac{1}{(s^2-36)^2}$ c. $f(s) = \dfrac{1}{(s^2+9)^2(s-6)}$

7. Show that

$$\int_0^\infty \frac{\cos(mx)}{1+x^2}\,dx = \frac{\pi}{2}\exp[-m]$$

where $m > 0$.

8. Evaluate

$$\int_0^\infty \frac{\tan(x)}{x}\,dx$$

9. Evaluate

$$\int_0^\infty \frac{a}{a^2+x^2}\,dx$$

10. Show that

$$F_t(u) = \int_{-\infty}^\infty f(t)\cos(ut)\,dt$$

if $f(t)$ is an even function.

11. Find the Fourier transform of the nth derivative of $f(t)$, given that the nth derivative and its Fourier transform exist.
12. Find the Fourier transforms of the following functions

   a. $f(t) = \begin{cases} 0, & t < 0 \\ t, & t > 0 \end{cases}$

   b. $f(t) = \dfrac{\exp[it]}{1+t^2}$, $-\infty < t < \infty$

   c. $f(t) = \begin{cases} \int_0^t \dfrac{(t-y)^{n-1}}{(n-1)!}\exp[-y]\,dy, & t > 0 \\ 0, & \text{otherwise} \end{cases}$

PROBLEMS

13. Show that
$$F_r(u) = -i \int_{-\infty}^{\infty} f(t) \sin(ut) \, dt$$
if $f(t)$ is an odd function.

14. Find the function $f(t)$ such that the Fourier transform of $f(t)$ is given by
$$F_t(u) = \pi \exp[-|u-1|]$$

15. Derive the Laplace transforms of the following functions
    a. $\exp[c_1 t] \sinh(c_2 t)$ b. $\frac{1}{2}t^2 \cos(ct)$

16. Solve the following differential equations:
    a. $2f^3(t) + f^2(t) + 2f'(t) = 0$, $f^k(0) = 1$, $k = 0, 1, 2$
    b. $f^4(t) - 16f(t) = \exp[t]$, $f^k(0) = 0$, $k = 0, 1, 2, 3$

17. Solve the following differential equations:
    a. $f^2(t) = 5$, $f^k(0) = 4$, $k = 0, 1$
    b. $f^2(t) - f(t) = 0$, $f'(0) = 0$, $f(0) = 4$
    c. $f^2(t) + 8f'(t) = \exp[4t]$, $f'(0) = 1$, $f(0) = 30$

18. Let $\mathscr{L}[f(t)]$ and $\mathscr{L}[g(t)]$ be the Laplace transforms of $f(t)$ and $g(t)$. Show that
$$\mathscr{L}[f(t) \pm g(t)] = \mathscr{L}[f(t)] \pm \mathscr{L}[g(t)]$$

19. Let
$$\mathscr{L}[f(t)] = G(s)$$
Show that
$$\mathscr{L}[f(at)] = \frac{1}{a} G\left(\frac{s}{a}\right)$$

20. Using Eq. (6.123), find the functions $f(t)$ for which the following are Laplace transforms.
    a. $\dfrac{c}{s^n}$, $n > 0$ b. $\dfrac{1}{(s-c)^n}$, $n > 0$
    c. $\dfrac{2cs}{(s^2 + c^2)^2}$

21. Find the characteristic functions corresponding to the following density functions.
    a. $f(x) = \lambda^2 x \exp[-\lambda x]$, $0 < x < \infty$
    b. $f(x) = \frac{1}{2} \exp[-|x|]$, $-\infty < x < \infty$
    c. $f(x) = \dfrac{1}{\pi} \dfrac{\beta}{\beta^2 + (x - \alpha)^2}$, $-\infty < x < \infty$, $\beta > 0$

22. Find the characteristic functions corresponding to the following probability mass functions.
    a. $p(x) = \dfrac{1}{b - a + 1}$, $x = a, a+1, \ldots, b$, $a < b$
    b. $p(x) = p(1-p)^{x-1}$, $x = 1, 2, \ldots$, $0 < p < 1$
    c. $p(x) = \dfrac{n!}{x!(n-x)!} p^x (1-p)^{n-x}$, $x = 0, 1, \ldots, n$, $0 < p < 1$

23. Given the following characteristic functions, find the mean variance of the corresponding random variables.

   a. $[(1-p) + p\exp[iu]]^n$ (binomial)

   b. $\dfrac{p\exp[iu]}{[1-(1-p)\exp[iu]]}$ (geometric)

   c. $\left[\dfrac{p}{1-(1-p)\exp[iu]}\right]^n$ (negative binomial)

   d. $\dfrac{\exp[iu\lambda]}{1+u^2}$ (Laplace)

24. Show that
$$E(X^k) = \frac{1}{i^k}\frac{d^k}{du^k}\phi_x(u)\bigg|_{u=0}$$

25. Given the following characteristic functions, find the mean of each random variable.

   a. $\phi_X(u) = \dfrac{p_1 p_2 \exp[2iu]}{1 - (2 - p_1 - p_2)\exp[iu] + (1-p_1)(1-p_2)\exp[2iu]}$

   b. $\phi_X(u) = \dfrac{\exp[iun\lambda]}{(1+u^2)^n}$

26. Let X_j be a binomial random variable with parameters $0 < p < 1$ and n (positive integer) and characteristic function
$$\phi_{X_j}(u) = [(1-p) + p\exp[iu]]^n, \quad j = 1, 2, \ldots, m$$
Show that the random variable y, defined by
$$Y = \sum_{j=1}^m X_j$$
is binomially distributed with parameters p and mn.

27. Show that the characteristic function given by
$$\phi_X(u) = \frac{\exp[iu\lambda]}{1+u^2}$$
is the characteristic function of the Laplace random variable with density function
$$f(x) = \tfrac{1}{2}\exp[-|x-\lambda|], \quad -\infty < x < \infty$$

28. Show that
$$\phi_X(u) = \left(1 - \frac{iu}{\lambda}\right)^{-1}, \quad \lambda > 0$$
is the characteristic function of the exponential random variable with density function
$$f(x) = \lambda\exp[-\lambda x], \quad 0 < x < \infty$$

29. Show that
$$\phi_X(u) = [(1-p) + p\exp[iu]]^n, \quad 0 < p < 1, \quad n = \text{positive integer}$$

PROBLEMS

is the characteristic function of the binomial random variable with probability mass function given by

$$p(x) = \frac{n!}{x!(n-x)!} p^x (1-p)^{n-x}, \qquad x = 0, 1, 2, \ldots, n$$

30. Let X_1 and X_2 be exponential random variables with density functions given by

$$f(x_j) = \lambda_j \exp[-\lambda_j x_j], \qquad \lambda_j > 0, \quad j = 1, 2$$

Using characteristic functions find the density function of $X_1 + X_2$.

31. Let X_k be a Poisson random variable with characteristic function given by

$$\phi_{X_k}(u) = \exp[\lambda_k [\exp[iu] - 1]]$$

where $\lambda_k = \exp[-k]$. Let

$$Y = \sum_{k=1}^{\infty} k X_k$$

If

$$\phi_{kX_k}(u) = E[\exp[iukx_k]] \qquad \text{and} \qquad \gamma = \sum_{k=1}^{\infty} \lambda_k$$

show that

$$\phi_Y(u) = \exp\left[\gamma \left[\frac{(1 - \exp[-1])\exp[iu]}{1 - \exp[iu - 1]} - 1\right]\right]$$

32. Find the z transforms of the following probability mass functions.

a. $p(x) = \begin{cases} \dfrac{n!}{x!(n-x)!} p^x (1-p)^{n-x}, & x = 0, 1, 2, \ldots, n \\ 0, & \text{otherwise} \end{cases}$

where $0 < p < 1$ and n is a positive integer.

b. $p(x) = \begin{cases} \dfrac{2x}{n(n+1)} & x = 0, 1, 2, \ldots, n \\ 0, & \text{otherwise} \end{cases}$

33. Let $p(x)$ be the probability mass function of the random variable X, where X may assume all integer values. If m and σ^2 are the mean and variance of X respectively, show that

$$m = \frac{d}{dz} \psi_X(z) \bigg|_{z=1}$$

$$\sigma^2 = \frac{d^2}{dz^2} \psi_X(z) \bigg|_{z=1} + \frac{d}{dz} \psi_X(z) \bigg|_{z=1} - \left[\frac{d}{dz} \psi_X(z) \bigg|_{z=1}\right]^2$$

34. Find the functions $f(x)$ corresponding to the following z transforms

a. $\psi_x(z) = \dfrac{p}{1 - z(1-p)}, \quad 0 < p < 1$ b. $\psi_x(z) = \exp[z]$ c. $\psi_x(z) = \dfrac{z^2}{1-z}$

35. Let X be a discrete random variable with time dependent probability mass function $P_X(t)$ satisfying the difference equations

$$P_0(t + \Delta t) = (1 - \lambda \Delta t) P_0(t)$$
$$P_x(t + \Delta t) = (1 - \lambda \Delta t) P_x(t) + \lambda \Delta t P_{x-1}(t), \qquad x = 1, 2, \ldots$$

where

$$P_0(0) = 1, \qquad P_x(0) = 0, \qquad x = 1, 2, \ldots$$

Show that

$$P_x(t) = \frac{(\lambda t)^x}{x!} \exp[-\lambda t], \qquad x = 0, 1, 2, \ldots$$

by converting the above difference equations to differential equations.

36. A certain piece of equipment is replaced as it fails. The density function of time until failure is given by

$$f(t) = \lambda^2 t \exp[-\lambda t], \qquad 0 < t < \infty$$

Find the density function of life remaining in the equipment given that it is operating at τ, $g(t \mid \tau)$.

References

Allen, R. G. D., *Mathematical Analysis for Economists*. London: Macmillan, 1962.
Brown, G. B., *Smoothing, Forecasting and Prediction of Discrete Time Series*. Englewood Cliffs, New Jersey: Prentice-Hall, 1963.
Elmaghraby, S. E., *The Design of Production Systems*. New York: Reinhold, 1966.
Feller, W., *An Introduction to Probability Theory and Its Applications*. New York: Wiley (Interscience), 1957.
Hall, D. L., Maple, C. G., and Vinograde, B., *Introduction to the Laplace Transform*. New York: Appleton, 1959.
Howard, R. A., *Dynamic Programming and Markov Processes*. New York: Wiley (Interscience), 1960.
LePage, W. R., *Complex Variables and the Laplace Transform for Engineers*. New York: McGraw-Hill, 1961.
Lukacs, E., *Characteristic Functions*. London: Griffin, 1960.
Morris, W. T., *Analysis for Materials Handling Management*. Homewood, Illinois: Irwin, 1962.
Morse, P. M., *Queues, Inventories, and Maintenance*. New York: Wiley (Interscience), 1962.
Papoulis, A., *The Fourier Integral*. New York: McGraw-Hill, 1962.
Parzen, E., *Stochastic Processes*. San Francisco: Holden-Day, 1965.
Parzen, E., *Modern Probability Theory and Its Applications*. New York: Wiley (Interscience), 1960.
Savant, C. J., *Fundamentals of the Laplace Transformation*. New York: McGraw-Hill, 1962.
Weintraub, S., *Price Theory*. New York: Pitman, 1949.
Wylie, C. R., *Advanced Engineering Mathematics*. New York: McGraw-Hill, 1960.

APPENDIX

TABLES

TABLE 1
Cumulative Distribution Function, $F(z)$, of the Standard Normal Random Variable, Z

Z	$F(z)$	Z	$F(z)$	Z	$F(z)$
−4.000	0.0000	−3.500	0.0002	−3.000	0.0014
−3.990	0.0000	−3.490	0.0002	−2.990	0.0014
−3.980	0.0000	−3.480	0.0003	−2.980	0.0014
−3.970	0.0000	−3.470	0.0003	−2.970	0.0015
−3.960	0.0000	−3.460	0.0003	−2.960	0.0015
−3.950	0.0000	−3.450	0.0003	−2.950	0.0016
−3.940	0.0000	−3.440	0.0003	−2.940	0.0016
−3.930	0.0000	−3.430	0.0003	−2.930	0.0017
−3.920	0.0000	−3.420	0.0003	−2.920	0.0018
−3.910	0.0001	−3.410	0.0003	−2.910	0.0018
−3.900	0.0001	−3.400	0.0003	−2.900	0.0019
−3.890	0.0001	−3.390	0.0004	−2.890	0.0019
−3.880	0.0001	−3.380	0.0004	−2.880	0.0020
−3.870	0.0001	−3.370	0.0004	−2.870	0.0021
−3.860	0.0001	−3.360	0.0004	−2.860	0.0021
−3.850	0.0001	−3.350	0.0004	−2.850	0.0022
−3.840	0.0001	−3.340	0.0004	−2.840	0.0023
−3.830	0.0001	−3.330	0.0004	−2.830	0.0023
−3.820	0.0001	−3.320	0.0005	−2.820	0.0024
−3.810	0.0001	−3.310	0.0005	−2.810	0.0025
−3.800	0.0001	−3.300	0.0005	−2.800	0.0026
−3.790	0.0001	−3.290	0.0005	−2.790	0.0026
−3.780	0.0001	−3.280	0.0005	−2.780	0.0027
−3.770	0.0001	−3.270	0.0005	−2.770	0.0028
−3.760	0.0001	−3.260	0.0006	−2.760	0.0029
−3.750	0.0001	−3.250	0.0006	−2.750	0.0030
−3.740	0.0001	−3.240	0.0006	−2.740	0.0031
−3.730	0.0001	−3.230	0.0006	−2.730	0.0032
−3.720	0.0001	−3.220	0.0006	−2.720	0.0033
−3.710	0.0001	−3.210	0.0007	−2.710	0.0034
−3.700	0.0001	−3.200	0.0007	−2.700	0.0035
−3.690	0.0001	−3.190	0.0007	−2.690	0.0036
−3.680	0.0001	−3.180	0.0007	−2.680	0.0037

TABLE 1 (*Continued*)

Z	F(z)	Z	F(z)	Z	F(z)
−3.670	0.0001	−3.170	0.0008	−2.670	0.0038
−3.660	0.0001	−3.160	0.0008	−2.660	0.0039
−3.650	0.0001	−3.150	0.0008	−2.650	0.0040
−3.640	0.0001	−3.140	0.0009	−2.640	0.0041
−3.630	0.0001	−3.130	0.0009	−2.630	0.0043
−3.620	0.0002	−3.120	0.0009	−2.620	0.0044
−3.610	0.0002	−3.110	0.0009	−2.610	0.0045
−3.600	0.0002	−3.100	0.0010	−2.600	0.0047
−3.590	0.0002	−3.090	0.0010	−2.590	0.0048
−3.580	0.0002	−3.080	0.0010	−2.580	0.0049
−3.570	0.0002	−3.070	0.0011	−2.570	0.0051
−3.560	0.0002	−3.060	0.0011	−2.560	0.0052
−3.550	0.0002	−3.050	0.0012	−2.550	0.0054
−3.540	0.0002	−3.040	0.0012	−2.540	0.0055
−3.530	0.0002	−3.030	0.0012	−2.530	0.0057
−3.520	0.0002	−3.020	0.0013	−2.520	0.0059
−3.510	0.0002	−3.010	0.0013	−2.510	0.0060
−2.500	0.0062	−2.000	0.0227	−1.500	0.0668
−2.490	0.0064	−1.990	0.0233	−1.490	0.0681
−2.480	0.0066	−1.980	0.0239	−1.480	0.0695
−2.470	0.0067	−1.970	0.0244	−1.470	0.0708
−2.460	0.0069	−1.960	0.0250	−1.460	0.0722
−2.450	0.0071	−1.950	0.0256	−1.450	0.0735
−2.440	0.0073	−1.940	0.0262	−1.440	0.0750
−2.430	0.0075	−1.930	0.0268	−1.430	0.0764
−2.420	0.0078	−1.920	0.0274	−1.420	0.0778
−2.410	0.0080	−1.910	0.0281	−1.410	0.0793
−2.400	0.0082	−1.900	0.0287	−1.400	0.0808
−2.390	0.0084	−1.890	0.0294	−1.390	0.0823
−2.380	0.0086	−1.880	0.0301	−1.380	0.0838
−2.370	0.0089	−1.870	0.0307	−1.370	0.0854
−2.360	0.0091	−1.860	0.0314	−1.360	0.0869
−2.350	0.0094	−1.850	0.0322	−1.350	0.0885
−2.340	0.0096	−1.840	0.0329	−1.340	0.0901
−2.330	0.0099	−1.830	0.0336	−1.330	0.0918
−2.320	0.0102	−1.820	0.0344	−1.320	0.0934
−2.310	0.0104	−1.810	0.0352	−1.310	0.0951
−2.300	0.0107	−1.800	0.0359	−1.300	0.0968
−2.290	0.0110	−1.790	0.0367	−1.290	0.0985
−2.280	0.0113	−1.780	0.0375	−1.280	0.1003
−2.270	0.0116	−1.770	0.0384	−1.270	0.1021
−2.260	0.0119	−1.760	0.0392	−1.260	0.1038
−2.250	0.0122	−1.750	0.0401	−1.250	0.1057
−2.240	0.0125	−1.740	0.0409	−1.240	0.1075
−2.230	0.0129	−1.730	0.0418	−1.230	0.1094
−2.220	0.0132	−1.720	0.0427	−1.220	0.1112
−2.210	0.0135	−1.710	0.0436	−1.210	0.1132
−2.200	0.0139	−1.700	0.0446	−1.200	0.1151

Z	F(z)	Z	F(z)	Z	F(z)
−2.190	0.0143	−1.690	0.0455	−1.190	0.1170
−2.180	0.0146	−1.680	0.0465	−1.180	0.1190
−2.170	0.0150	−1.670	0.0475	−1.170	0.1210
−2.160	0.0154	−1.660	0.0485	−1.160	0.1230
−2.150	0.0158	−1.650	0.0495	−1.150	0.1251
−2.140	0.0162	−1.640	0.0505	−1.140	0.1272
−2.130	0.0166	−1.630	0.0516	−1.130	0.1293
−2.120	0.0170	−1.620	0.0526	−1.120	0.1314
−2.110	0.0174	−1.610	0.0537	−1.110	0.1335
−2.100	0.0179	−1.600	0.0548	−1.100	0.1357
−2.090	0.0183	−1.590	0.0559	−1.090	0.1379
−2.080	0.0188	−1.580	0.0571	−1.080	0.1401
−2.070	0.0192	−1.570	0.0582	−1.070	0.1423
−2.060	0.0197	−1.560	0.0594	−1.060	0.1446
−2.050	0.0202	−1.550	0.0606	−1.050	0.1469
−2.040	0.0207	−1.540	0.0618	−1.040	0.1492
−2.030	0.0212	−1.530	0.0630	−1.030	0.1515
−2.020	0.0217	−1.520	0.0643	−1.020	0.1539
−2.010	0.0222	−1.510	0.0655	−1.010	0.1563
−1.000	0.1587	−0.500	0.3086	0.000	0.5000
−0.990	0.1611	−0.490	0.3121	0.010	0.5040
−0.980	0.1636	−0.480	0.3157	0.020	0.5080
−0.970	0.1660	−0.470	0.3192	0.030	0.5120
−0.960	0.1685	−0.460	0.3228	0.040	0.5160
−0.950	0.1711	−0.450	0.3264	0.050	0.5200
−0.940	0.1736	−0.440	0.3300	0.060	0.5240
−0.930	0.1762	−0.430	0.3336	0.070	0.5279
−0.920	0.1788	−0.420	0.3373	0.080	0.5319
−0.910	0.1814	−0.410	0.3409	0.090	0.5359
−0.900	0.1841	−0.400	0.3446	0.100	0.5399
−0.890	0.1867	−0.390	0.3483	0.110	0.5438
−0.880	0.1894	−0.380	0.3520	0.120	0.5478
−0.870	0.1922	−0.370	0.3557	0.130	0.5517
−0.860	0.1949	−0.360	0.3595	0.140	0.5557
−0.850	0.1977	−0.350	0.3632	0.150	0.5596
−0.840	0.2005	−0.340	0.3670	0.160	0.5636
−0.830	0.2033	−0.330	0.3707	0.170	0.5675
−0.820	0.2061	−0.320	0.3745	0.180	0.5714
−0.810	0.2090	−0.310	0.3783	0.190	0.5754
−0.800	0.2119	−0.300	0.3821	0.200	0.5793
−0.790	0.2148	−0.290	0.3859	0.210	0.5832
−0.780	0.2177	−0.280	0.3898	0.220	0.5871
−0.770	0.2207	−0.270	0.3936	0.230	0.5910
−0.760	0.2236	−0.260	0.3975	0.240	0.5949
−0.750	0.2266	−0.250	0.4013	0.250	0.5987
−0.740	0.2297	−0.240	0.4052	0.260	0.6026
−0.730	0.2327	−0.230	0.4091	0.270	0.6064
−0.720	0.2358	−0.220	0.4130	0.280	0.6103

TABLE 1 (*Continued*)

Z	F(z)	Z	F(z)	Z	F(z)
−0.710	0.2389	−0.210	0.4169	0.290	0.6141
−0.700	0.2420	−0.200	0.4208	0.300	0.6179
−0.690	0.2451	−0.190	0.4247	0.310	0.6217
−0.680	0.2483	−0.180	0.4286	0.320	0.6255
−0.670	0.2515	−0.170	0.4325	0.330	0.6293
−0.660	0.2547	−0.160	0.4365	0.340	0.6331
−0.650	0.2579	−0.150	0.4404	0.350	0.6368
−0.640	0.2611	−0.140	0.4444	0.360	0.6406
−0.630	0.2644	−0.130	0.4483	0.370	0.6443
−0.620	0.2677	−0.120	0.4523	0.380	0.6480
−0.610	0.2710	−0.110	0.4562	0.390	0.6517
−0.600	0.2743	−0.100	0.4602	0.400	0.6554
−0.590	0.2776	−0.090	0.4642	0.410	0.6591
−0.580	0.2810	−0.080	0.4681	0.420	0.6628
−0.570	0.2844	−0.070	0.4721	0.430	0.6664
−0.560	0.2878	−0.060	0.4761	0.440	0.6700
−0.550	0.2912	−0.050	0.4801	0.450	0.6737
−0.540	0.2946	−0.040	0.4841	0.460	0.6773
−0.530	0.2981	−0.030	0.4881	0.470	0.6808
−0.520	0.3016	−0.020	0.4920	0.480	0.6844
−0.510	0.3051	−0.010	0.4960	0.490	0.6879
0.500	0.6915	1.000	0.8414	1.500	0.9332
0.510	0.6950	1.010	0.8438	1.510	0.9345
0.520	0.6985	1.020	0.8462	1.520	0.9357
0.530	0.7020	1.030	0.8485	1.530	0.9370
0.540	0.7054	1.040	0.8509	1.540	0.9382
0.550	0.7089	1.050	0.8532	1.550	0.9394
0.560	0.7123	1.060	0.8554	1.560	0.9406
0.570	0.7157	1.070	0.8577	1.570	0.9418
0.580	0.7191	1.080	0.8599	1.580	0.9429
0.590	0.7224	1.090	0.8622	1.590	0.9441
0.600	0.7258	1.100	0.8643	1.600	0.9452
0.610	0.7291	1.110	0.8665	1.610	0.9463
0.620	0.7324	1.120	0.8687	1.620	0.9474
0.630	0.7357	1.130	0.8708	1.630	0.9484
0.640	0.7389	1.140	0.8729	1.640	0.9495
0.650	0.7422	1.150	0.8749	1.650	0.9505
0.660	0.7454	1.160	0.8770	1.660	0.9515
0.670	0.7486	1.170	0.8790	1.670	0.9525
0.680	0.7518	1.180	0.8810	1.680	0.9535
0.690	0.7549	1.190	0.8830	1.690	0.9545
0.700	0.7581	1.200	0.8849	1.700	0.9554
0.710	0.7612	1.210	0.8869	1.710	0.9564
0.720	0.7643	1.220	0.8888	1.720	0.9573
0.730	0.7673	1.230	0.8907	1.730	0.9582
0.740	0.7704	1.240	0.8925	1.740	0.9591
0.750	0.7734	1.250	0.8944	1.750	0.9599
0.760	0.7764	1.260	0.8962	1.760	0.9608

Z	F(z)	Z	F(z)	Z	F(z)
0.770	0.7794	1.270	0.8980	1.770	0.9616
0.780	0.7823	1.280	0.8997	1.780	0.9625
0.790	0.7853	1.290	0.9015	1.790	0.9633
0.800	0.7882	1.300	0.9032	1.800	0.9641
0.810	0.7911	1.310	0.9049	1.810	0.9648
0.820	0.7939	1.320	0.9066	1.820	0.9656
0.830	0.7968	1.330	0.9082	1.830	0.9664
0.840	0.7996	1.340	0.9099	1.840	0.9671
0.850	0.8024	1.350	0.9115	1.850	0.9678
0.860	0.8051	1.360	0.9131	1.860	0.9686
0.870	0.8079	1.370	0.9147	1.870	0.9693
0.880	0.8106	1.380	0.9162	1.880	0.9699
0.890	0.8133	1.390	0.9177	1.890	0.9706
0.900	0.8160	1.400	0.9192	1.900	0.9713
0.910	0.8186	1.410	0.9207	1.910	0.9719
0.920	0.8212	1.420	0.9222	1.920	0.9726
0.930	0.8238	1.430	0.9236	1.930	0.9732
0.940	0.8264	1.440	0.9251	1.940	0.9738
0.950	0.8290	1.450	0.9265	1.950	0.9744
0.960	0.8315	1.460	0.9279	1.960	0.9750
0.970	0.8340	1.470	0.9292	1.970	0.9756
0.980	0.8365	1.480	0.9306	1.980	0.9762
0.990	0.8389	1.490	0.9319	1.990	0.9767
2.000	0.9773	2.500	0.9938	3.000	0.9986
2.010	0.9778	2.510	0.9940	3.010	0.9987
2.020	0.9783	2.520	0.9941	3.020	0.9987
2.030	0.9788	2.530	0.9943	3.030	0.9988
2.040	0.9793	2.540	0.9945	3.040	0.9988
2.050	0.9798	2.550	0.9946	3.050	0.9988
2.060	0.9803	2.560	0.9948	3.060	0.9989
2.070	0.9808	2.570	0.9949	3.070	0.9989
2.080	0.9812	2.580	0.9951	3.080	0.9990
2.090	0.9817	2.590	0.9952	3.090	0.9990
2.100	0.9821	2.600	0.9953	3.100	0.9990
2.110	0.9826	2.610	0.9955	3.110	0.9991
2.120	0.9830	2.620	0.9956	3.120	0.9991
2.130	0.9834	2.630	0.9957	3.130	0.9991
2.140	0.9838	2.640	0.9959	3.140	0.9991
2.150	0.9842	2.650	0.9960	3.150	0.9992
2.160	0.9846	2.660	0.9961	3.160	0.9992
2.170	0.9850	2.670	0.9962	3.170	0.9992
2.180	0.9854	2.680	0.9963	3.180	0.9993
2.190	0.9857	2.690	0.9964	3.190	0.9993
2.200	0.9861	2.700	0.9965	3.200	0.9993
2.210	0.9865	2.710	0.9966	3.210	0.9993
2.220	0.9868	2.720	0.9967	3.220	0.9994
2.230	0.9871	2.730	0.9968	3.230	0.9994
2.240	0.9875	2.740	0.9969	3.240	0.9994

TABLE 1 (*Continued*)

Z	F(z)	Z	F(z)	Z	F(z)
2.250	0.9878	2.750	0.9970	3.250	0.9994
2.260	0.9881	2.760	0.9971	3.260	0.9994
2.270	0.9884	2.770	0.9972	3.270	0.9995
2.280	0.9887	2.780	0.9973	3.280	0.9995
2.290	0.9890	2.790	0.9974	3.290	0.9995
2.300	0.9893	2.800	0.9974	3.300	0.9995
2.310	0.9896	2.810	0.9975	3.310	0.9995
2.320	0.9898	2.820	0.9976	3.320	0.9995
2.330	0.9901	2.830	0.9977	3.330	0.9996
2.340	0.9904	2.840	0.9977	3.340	0.9996
2.350	0.9906	2.850	0.9978	3.350	0.9996
2.360	0.9909	2.860	0.9979	3.360	0.9996
2.370	0.9911	2.870	0.9979	3.370	0.9996
2.380	0.9914	2.880	0.9980	3.380	0.9996
2.390	0.9916	2.890	0.9981	3.390	0.9996
2.400	0.9918	2.900	0.9981	3.400	0.9997
2.410	0.9920	2.910	0.9982	3.410	0.9997
2.420	0.9922	2.920	0.9982	3.420	0.9997
2.430	0.9925	2.930	0.9983	3.430	0.9997
2.440	0.9927	2.940	0.9984	3.440	0.9997
2.450	0.9929	2.950	0.9984	3.450	0.9997
2.460	0.9931	2.960	0.9985	3.460	0.9997
2.470	0.9932	2.970	0.9985	3.470	0.9997
2.480	0.9934	2.980	0.9986	3.480	0.9997
2.490	0.9936	2.990	0.9986	3.490	0.9998
3.500	0.9998	3.670	0.9999	3.840	0.9999
3.510	0.9998	3.680	0.9999	3.850	0.9999
3.520	0.9998	3.690	0.9999	3.860	0.9999
3.530	0.9998	3.700	0.9999	3.870	0.9999
3.540	0.9998	3.710	0.9999	3.880	0.9999
3.550	0.9998	3.720	0.9999	3.890	0.9999
3.560	0.9998	3.730	0.9999	3.900	0.9999
3.570	0.9998	3.740	0.9999	3.910	0.9999
3.580	0.9998	3.750	0.9999	3.920	1.0000
3.590	0.9998	3.760	0.9999	3.930	1.0000
3.600	0.9998	3.770	0.9999	3.940	1.0000
3.610	0.9998	3.780	0.9999	3.950	1.0000
3.620	0.9998	3.790	0.9999	3.960	1.0000
3.630	0.9999	3.800	0.9999	3.970	1.0000
3.640	0.9999	3.810	0.9999	3.980	1.0000
3.650	0.9999	3.820	0.9999	3.990	1.0000
3.660	0.9999	3.830	0.9999	4.000	1.0000

TABLE 2
Antidifference Table (In All Cases x May Assume Positive Integer Values)

$f(x)$	$Sf(x)$
a	ax
ax	$(a/2)x(x-1)$
x^2	$(\frac{1}{6})x(x-1)(2x-1)$
x^3	$(\frac{1}{4})[x(x-1)]^2$
x^4	$(\frac{1}{30})x(x-1)(2x-1)[3x(x-1)-1]$
a^x	$\dfrac{a^x - 1}{a - 1}$
$x(x+1)$	$(\frac{1}{3})(x-1)x(x+1)$
$2x+1$	x^2
$x(x+1)(x+2)$	$(\frac{1}{4})(x-1)x(x+1)(x+2)$
$x(x+1)(x+2)(x+3)$	$(\frac{1}{5})(x-1)x(x+1)(x+2)(x+3)$
$2^x\left[\dfrac{x(x+1)}{2}+1\right]$	$2^{x-1}(x^2-3x+6)$
$(2x-1)(2x+1)(2x+3)$	$(x-1)(x+1)(2x^2-3)$
$x(x+2)(x+1)^2$	$(\frac{1}{10})(x-1)x(x+1)(x+2)(2x+1)$
$x(3x+1)$	$(x-1)x^2$
$\dfrac{1}{x(x+1)}$	$\dfrac{x-1}{x}$
$\dfrac{1}{(2x-1)(2x+1)}$	$\dfrac{x-1}{2x-1}$
$\dfrac{1}{(3x-2)(3x+1)}$	$\dfrac{x-1}{3x-2}$
$\dfrac{1}{x(x+2)}$	$\dfrac{(x-1)(3x+2)}{4x(x+1)}$
$\dfrac{1}{(2x-1)(2x+1)(2x+3)}$	$\dfrac{(x^2-1)}{3(2x-1)(2x+1)}$
$\dfrac{1}{(3x-2)(3x+1)(3x+4)}$	$\dfrac{(x-1)(3x+2)}{8(3x-2)(3x+1)}$
$\dfrac{x}{(x+1)(x+2)}2^{x-1}$	$\dfrac{2^{x-1}}{(x+1)}$
$\dfrac{x+3}{x(x+1)(x+2)}$	$-\dfrac{2x+3}{2x(x+1)}$
$\dfrac{x+2}{x(x+1)2^x}$	$\dfrac{1}{x2^{x-1}}$
$\dfrac{x^2}{(x+1)(x+2)}4^x$	$\dfrac{(x-2)4^x}{3(x+1)}$
$\dfrac{2x+3}{x(x+1)3^x}$	$-\dfrac{1}{x3^{x-1}}$

TABLE 2 (*Continued*)

$f(x)$	$Sf(x)$
$\dfrac{1}{(a+bx)[a+b(x+1)]}$	$\dfrac{x-1}{(a+b)(a+bx)}$
$\dfrac{(b-1)b^{x-1}}{(a-b)(1+b^{x-1})(1+b^x)}$	$\dfrac{b^{x-1}}{(a-b)(1+b^{x-1})}$
$xx!$	$x!$
$x(x!)^2(x+2)$	$(x!)^2$
$\dfrac{x}{(x+1)!}$	$-\dfrac{1}{x!}$
$\dfrac{(4x+1)(x+1)!}{(2x+2)!}$	$-\dfrac{x!}{(2x)!}$
$\dfrac{(x+1)(2x)!}{x!}$	$\dfrac{(2x)!}{x!}$
$\dfrac{n!}{x!(n-x)!}\dfrac{(n-2x-1)}{(x+1)}$	$\dfrac{n!}{x!(n-x)!},\quad x \le n$
$\dfrac{(a-1)a^{x-1}}{(1-a^x)(1-a^{x-1})}$	$\dfrac{a^x}{1-a^x}$
$\dfrac{(4x+5)(x+1)!}{(2x+3)!}$	$-\dfrac{x!}{(2x+1)!}$
xa^x	$\dfrac{a}{(a-1)^2}[xa^{x-1}(a-1)-(a^x-1)]$
$ax^{(a-1)}$	$x^{(a)}$
$x^{(a)}$	$\dfrac{x^{(a+1)}}{a+1}$
$*bn(a+bx)^{(n-1)}$	$(a+bx)^{(n)}$
$*(a+bx)^{(n)}$	$\dfrac{(a+bx)^{(n+1)}}{b(n+1)}$
$\dfrac{1}{x^{[a]}}$	$-\dfrac{1}{(a-1)x^{[a-1]}},\quad a \ne 1$
$\dfrac{a}{x^{[a+1]}}$	$\dfrac{1}{x^{[a]}}$
$\sin(x\theta)$	$\sin\left(\dfrac{x\theta}{2}\right)\sin\left(\dfrac{x\theta-\theta}{2}\right)\operatorname{cosec}\left(\dfrac{\theta}{2}\right)$
$\cos(x\theta)$	$\cos\left(\dfrac{x\theta}{2}\right)\sin\left(\dfrac{x\theta-\theta}{3}\right)\operatorname{cosec}\left(\dfrac{\theta}{2}\right)$

* $(a+bx)^{(n)} = [a+bx][a+b(x-1)][a+b(x-2)] \cdots [a+b(x-n+1)]$.

TABLES

TABLE 3
TABLE OF LAPLACE TRANSFORMS WHERE $L(s) = \mathscr{L}[f(t)]$

$f(t)$	$L(s)$	$f(t)$	$L(s)$
$cf(t)$	$cL(s)$	c	$\dfrac{c}{s}$
$c_1 g(t) + c_2 h(t)$	$c_1 \mathscr{L}[g(t)] + c_2 \mathscr{L}[h(t)]$	$\dfrac{c}{\Gamma(n)} t^{n-1}$	$\dfrac{c}{s^n}, \quad n > 0$
$f'(t)$	$sL(s) - f(0)$	e^{ct}	$\dfrac{1}{s-c}$
$f^n(t)$	$s^n L(s) - \sum_{j=1}^{n} s^{n-j} f^{j-1}(0)$	$\dfrac{e^{ct} t^{n-1}}{\Gamma(n)}$	$\dfrac{1}{(s-c)^n}, \quad n > 0$
$\exp[at] f(t)$	$L(s-a)$	$\dfrac{\exp[c_1 t] - \exp[c_2 t]}{c_1 - c_2}$	$\dfrac{1}{(s-c_1)(s-c_2)}$
$\int_0^t f(x)\,dx$	$\dfrac{1}{s} L(s)$	$\dfrac{(a+c_1)\exp[c_1 t] - (a+c_2)\exp[c_2 t]}{c_1 - c_2}$	$\dfrac{s+a}{(s-c_1)(s-c_2)}$
$\int_0^t f(x) g(t-x)\,dx$	$\mathscr{L}[f(t)]\mathscr{L}[g(t)]$	$\sin(ct)$	$\dfrac{c}{s^2 + c^2}$
$-t f(t)$	$L'(s)$	$\cos(ct)$	$\dfrac{s}{s^2 + c^2}$
$(-t)^n f(t)$	$L^n(s)$	$\exp[c_1 t]\sin(c_2 t)$	$\dfrac{c_2}{(s-c_1)^2 + c_2^2}$
$\dfrac{1}{t} f(t)$	$\int_s^{\infty} L(x)\,dx$	$\exp[c_1 t]\cos(c_2 t)$	$\dfrac{s - c_1}{(s-c_1)^2 + c_2^2}$
$f(t-c), \quad t > c$	$\exp[-cs] L(s)$	$\sinh(ct)$	$\dfrac{c}{s^2 - c^2}$
$t \cos(ct)$	$\dfrac{s^2 - c^2}{(s^2 + c^2)^2}$	$\cosh(ct)$	$\dfrac{s}{s^2 - c^2}$
$t \sinh(ct)$	$\dfrac{2cs}{(s^2 - c^2)^2}$	$\exp[c_1 t]\sinh(c_2 t)$	$\dfrac{c_2}{(s-c_1)^2 - c_2^2}$
$t \cosh(ct)$	$\dfrac{s^2 + c^2}{(s^2 - c^2)^2}$	$\exp[c_1 t]\cosh(c_2 t)$	$\dfrac{s - c_1}{(s-c_1)^2 - c_2^2}$
		$t \sin(ct)$	$\dfrac{2cs}{(s^2 + c^2)^2}$

INDEX

A

Adjoint,
 definition of, 138
 properties, 139–140
Antidifference, *see* Summation of series

B

Bayes' theorem, 17

C

Characteristic equation, 161
Characteristic function,
 convolution, 538–539
 definition of, 531–532
 inversion, 532
 moments, 537–538
Characteristic root, *see* Eigenvalue
Circuits, 104, 200–202, 352–353, 448–450, 553–554, 556–557
Cofactor method, *see* Determinant
Complex variables,
 analytic functions, 463–464
 Cauchy's integral theorem, 489
 complex conjugate, 459
 complex number, 457–458
 complex plane, 458
 conjugate functions, 463
 derivative, 461–462
 imaginary number, 457–458
 independence of path of integration, 481–482
 integration, 467–512
 isolated singularity, 467, 491
 Laurent series, 489, 490
 limit, 461
 line integral, *see* Integration
 method of residues, 489–512
 multiply-connected region, 482, 483
 path of integration, *see* Integration
 pole, 491, 492
 residue, 492
 residue theorem, 495
 simply-connected region, 481–482, 483
 singularity, 464
Conditional,
 density function, 44–45
 distribution function, 44
 expectation, *see* Expectation
 probability mass function, 45
Confidence intervals, 97–98
Continuity, 234–242
Convex,
 combination, 155
 set, 155–157, 197–199

D

Derivative
 chain rule, 248–250
 complex variables, *see* Complex variables
 definition of, 242
 directional, 253–254
 higher-order, 251
 Laplace transform, 525–526
 left-hand, 242
 mean-value theorem, 289
 mean-value theorem, several variables, 303
 partial, 252–254
 product, 244–246
 ratio, 244–246
 right-hand, 242
 Rolle's theorem, 288
 sum, 244–246
 Taylor's mean-value theorem, 296–298
 Taylor's theorem, several variables, 305–306
 total differential, 318

Derived distribution,
  several random variables, 56–74
  single random variable, 49–56
Determinant,
  definition of, 127
  cofactor method, 132–136
  evaluation, 127–129, 132–136
  properties, 128–134
Difference equation,
  constant coefficients, 410, 411–425
  definition of, 406
  fundamental solutions 408–409, 416–417, 419–420
  general solution, 409, 414, 423
  homogeneous, 410, 411–421
  linear, 409, 411–425
  nonhomogeneous, 405, 421–425
  order, 409–410
  particular solution, 412–414, 418
  undetermined coefficients, 421–422
Discrete functions,
  convex, 382
  extreme point, 383–385, 388–391
  point of inflection, 385
Divided difference,
  definition of, 365
  difference table, 403–404
  higher-order differences, 400–403
  product, 367–368
  ratio, 367–368
  simple difference, 365
  sum, 367–368, 371–381

E

Eigenvalue,
  definition of, 161
  multiplicity, 163–166
  quadratic form, 178
Eigenvector,
  definition of, 161
  orthogonal, 162–167
Entropy, 108–109
Equality constraints, *see* Extreme point
Error,
  Type I, 93–97
  Type II, 93–97
Estimation, 92–93
Event,
  dependent, 5
  general discussion of, 4–5
  independent, 5
  mutually exclusive, 5
  probability of, 8–18
Expectation,
  conditional, 106–108
  covariance, 75–76
  definition of, 75
  joint, 75–77
  mean, 75, 83–86, 90–93, 97–98
  variance, 75, 83–86, 90–93, 106–108
Experimental design, 186–189
Extreme point
  definition of, 195–199, 284
  equality constraints, 316–336
  global, 285–286
  Hessian, 311
  inequailty constraints, 336–341
  interior, 282, 302
  Kuhn–Tucker conditions, 337
  Lagrange multipliers, 324–326
  local, 285–286, 302
  maximum, 282, 302
  minimum, 282, 302
  necessary conditions, 291–292, 308–309, 318–321, 324–326, 337
  point of inflection, 292, *see also* Discrete functions
  saddle point, 310–311
  stationary point, 292
  sufficient conditions, 293–295, 299–300, 311–312, 327–331

F

Facility location, 355
Finite difference, *see* Divided difference
Fourier series, 558–561
Fourier transform,
  convolution, 513–515
  definition of, 512
  inversion, 515–518
  moments, 512–513
  transform pair, 512, 522
Function,
  analytic, *see* Complex variables
  conjugate, *see* Complex variables
  convexity, 282–283
  definition of, 220–222
  monotonic, 284–285

G

Gradient method, 346–350
Greatest lower bound, 220

INDEX 579

H

Half space, 157–161
Hessian, *see* Extreme point
Hyperplane, 157, 196–197
Hypothesis testing, 93–97, 341–344

I

Inequality constraints, *see* Extreme point
Information theory, 108–109
Integration,
  antiderivative, 267
  by parts, 270–271
  closed curve, 476–478
  complex, *see* Complex variables
  continuity, 267–270
  definition of, 258–259, 263
  direction, 474
  even function, 264–266
  improper, 275
  line integral, 467–470, 473, 479–481
  odd function, 264–266
  path of integration, 473–474
  properties, 263–264
  Riemann sum, 258
Interval, 216–217
Inventory, 1, 28–29, 98–99, 281, 344–346, 434–437
Inverse,
  definition of, 137
  properties, 140-141
Inversion, 126–129

J

Joint density function, 41–42
Joint distribution, 40–48
Joint distribution function, 41–42
Joint probability mass function, 41–42

K

Kuhn–Tucker conditions, *see* Extreme point

L

Lagrange multipliers, *see* Extreme point
Laplace transform,
  definition of, 523–524
  derivative, *see* Derivative
  differential equations, 528–529
  inversion, 527, 529–530
  one-sided, 524
  two-sided, 523
Latent root, *see* Eigenvalue
Leading principal minor,
  definition of, 174
  quadratic forms, 174–178
Least upper bound, 220
Limit,
  complex variables, *see* Complex variables
  left-hand, 223
  L'Hospital's rule, 246–247
  product, 231–234
  ratio, 231–234
  right-hand, 225
  several variables, 229–231
  single variable, 222–229
  sum, 231–234
  two-sided, 225
Linear programing, 193–200

M

Maintenance, 29
Marginal,
  density function, 42–44
  distribution function, 42
  probability mass function, 42
Markov chains, 179–186
Matrix,
  addition, 115–116
  adjoint, *see* Adjoint
  congruent, 166
  definition of, 114
  determinant, *see* Determinant
  diagonal, 166–167
  equality, 115
  inverse, *see* Inverse
  multiplication, 116–120
  order, 115
  orthogonal, 163
  periodic, 125–126
  quadratic form, *see* Quadratic form
  rank, 136–137
  scalar, 125
  similar, 166
  singular, 128
  skew–symmetric, 123–124
  square, 115
  symmetric, 123–124, 161–178
  transpose, 121–123
  vector, *see* Vector

Maximum, *see* Extreme point
Minimum, *see* Extreme point
Minor, 132–136
Moments, 83–86, 90–98
Moment generating function, 81–89

N

Neighborhood, 217
Numerical methods,
 approximation, 425–428
 differentiation, 433–434
 integration, 429–432
 interpolation, 425–428
 Newton's formula, 429

O

Optimum, *see* Extreme point
Outcome, 2, 4, 5, 7–8

P

Permutation, 126–127
Point of inflection *see* Extreme point
Probability,
 basic, 6–18
 conditional, 9, 10
 definition of, 7
 joint, 9, 10–11, 14–15
 properties, 8
Probability distributions,
 applications, 28–29
 Bernoulli, 28, 29, 30, 32, 544–545
 beta, 29, 31, 37
 binomial, 28, 30, 32, 39–40, 78–79, 82–83, 88–89, 544–545
 χ^2, 29, 31, 35, 55, 89
 cumulative distribution function, *see* Distribution function
 density function, 21–28
 derived, *see* Derived distributions
 distribution function, 18–28
 exponential, 29, 31, 36, 38, 513, 539–540
 F, 29, 35–36
 gamma, 29, 31, 36–37, 39, 59–60, 78–79, 82–83, 518–520, 539–540
 geometric, 28, 30, 32
 hypergeometric, 28, 30, 33–34, 39–40
 joint normal, 43–44, 46–48
 negative binomial, 28, 30, 32–33, 543–544
 noncentral χ^2, 63–68
 normal, 29, 31, 34–35, 68–74, 82–83, 87–88, 520–521, 532–534
 Poisson, 28, 30, 34, 39–40, 61–62, 78–79, 82–83, 534–536, 538–539, 545–546
 Probability density function, *see* Density function
 probability mass function, 21–28, 29–30, 32–34
 rectangular, 29, 30, 34
 t, 29, 31, 35, 57–59
 uniform, 29, 31, 37–38
 Weibull, 29, 31, 37, 38, 54
Production systems, 190–193, 344–346

Q

Quadratic form,
 definition of, 167–168
 indefinite, 176, 178
 negative definite, 176, 177–178
 negative semidefinite, 176, 178
 positive definite, 176–178
 positive semidefinite, 176, 178
 sum of squares, 168–174
Quality control, 1, 28–29, 46–48, 93–97, 99–101, 437–441
Queueing, 1, 28–29, 441–448, 546–548

R

Random experiment, 2
Random variable,
 continuous, 21–22
 discrete, 21–22, 29–30, 32–34
 general discussion, 2–6
 independent, 42–44
 sum, 87–89
Reliability, 1, 29, 99, 281

S

Saddle point, *see* Extreme point
Sample space, 2–4
Search technique, 282
Service systems, *see* Queueing
Sets,
 definition of, 207
 operations, 208–209
 Venn diagram, 212–214
Simulation, 28–29, 49

Space,
  boundary point, 218
  bounded, 220
  open, 218
  closed, 218
  vector, *see* Vector space
Stationary point, *see* Extreme point
Summation, *see* Divided difference
Summation of series,
  antidifference, 392
  evaluation, 395–396
  summation by parts, 396–397

T

Tables,
  antidifference, 573–574
  cumulative normal distribution, 567–572
  Laplace transforms, 575
Total differential, *see* Derivative
Traffic flow, 1, 105–106, 353–354
Transforms,
  characteristic function, *see* Characteristic function
  Fourier, *see* Fourier transform
  joint moment generating function, 86–87
  Laplace, *see* Laplace trransform
  marginal moment generation function, 86–87
  moment generating function, *see* Moment generating function
  Z transform, *see* Z transform

V

Vector,
  characteristic, *see* Eigenvector
  column, 121
  eigenvector, *see* Eigenvector
  invariant, *see* Eigenvector
  latent, *see* Eigenvector
  length, 144–145, 217
  linear combination, 143
  linearly independent, 150–153
  orthogonal, 146–147, 150–151
  row, 121
  space, *see* Vector space
Vector space,
  basis, 152–153
  definition of, 147
  linear transformation, 153–155
  span, 148–150, 152–153
  subspace, 148

Z

Z transform,
  convolution, 544
  definition of, 541
  inversion, 542–543